Reinforcement Learning

Zhiqing Xiao

Reinforcement Learning

Theory and Python Implementation

Zhiqing Xiao
Beijing, China

Jointly published with China Machine Press.
The print edition is not for sale in China (Mainland). Customers from China (Mainland) please order the print book from: China Machine Press.
ISBN of the Co-Publisher's edition: 978-7-111-63177-4

ISBN 978-981-19-4932-6   ISBN 978-981-19-4933-3   (eBook)
https://doi.org/10.1007/978-981-19-4933-3

This Springer imprint is published by the registered company Springer Nature Singapore Pte Ltd.
The registered company address is: 152 Beach Road, #21-01/04 Gateway East, Singapore 189721, Singapore

If disposing of this product, please recycle the paper.

Preface

Reinforcement Learning (RL) is a type of Artificial Intelligence (AI) that changes our lives: RL players have defeated human in many games such as the game of Go and StarCraft; RL controllers are driving varied robots and unmanned vehicles; RL traders are making tons of money in financial markets, and the large language model with RL such as ChatGPT have been used in many business applications. Since the same RL algorithm with the same parameter setting can solve very different tasks, RL is also regarded as an important way to general AI. Here I sincerely invite you to learn RL to surf in these AI waves.

Synopsis

This book is a tutorial on RL, with explanation of theory and Python implementation. It consists of the following three parts.

- Chapter 1: Introduce the background of RL from scratch, and introduce the environment library Gym.
- Chapters 2–14: Introduce the mainstream RL theory and algorithms. Based on the most influential RL model–discounted return discrete-time Markov decision process, we derive the fundamental theory mathematically. Upon the theory we introduce algorithms, including both classical RL algorithms and deep RL algorithms, and then implement those algorithms in Python.
- Chapters 15–16: Introduce other RL models and extensions of RL models, including average-reward, continuous-time, non-homogenous, semi-Markov, partial observability, preference-based RL, and imitation learning, to have a complete understanding of the landscape of RL and its extension.

Features

This book comprehensively introduces the mainstream RL theory.

- This book introduces the trunk of the modern RL theory in a systematically way. All major results are accompanied with proofs. We introduce the algorithms based on the theory, which covers all mainstream RL algorithms, including key technology of ChatGPT such as Proximal Policy Optimization (PPO) and Reinforcement Learning with Human Feedbacks (RLHF).
- This book uses a consistent set of mathematical notations, which are compatible with mainstream RL tutorials.

All chapters are accompanied with Python codes.

- Easy to understand: All codes are implemented in a consistent and concise way, which directly maps to the explanation of algorithms.
- Easy to check: All codes and running results are shown in GitHub. We can either browse them in the web browser, or download locally to run them. Every algorithm is implemented in a self-contained standalone file, which can be browsed and executed individually.
- Diverse environments: We not only consider the built-in tasks in the library Gym, but also consider the third-party extension of the Gym. We even create environments for our own tasks.
- Highly compatible: All codes can be run in any one of all three major operating systems (Windows, macOS, and Linux). The methods to setup the environments are provided. Deep RL algorithms are implemented based on both TensorFlow 2 and PyTorch, so that readers can choose any one among the two or have a one-to-one comparison.
- Based on latest versions of software: All codes are based on the latest version of Python and its extension packages. The codes in GitHub will be updated according to the software update.
- Little hardware requirement: All codes can be run in a PC without GPU.

Errata, Codes, and Exercise Answers

Errata, codes (and their running results), the reference answers, explanation of Gym source codes, bibliography, and some other resources for exercises can be found in: `https://github.com/zhiqingxiao/rl-book`. The author will update them irregularly.

If you have any suggestions, comments, or questions after you have Googled and checked the GitHub repo, you can open an issue or a discussion on the GitHub repo. The email address of the author is: `xzq.xiaozhiqing@gmail.com`.

Zhiqing Xiao 肖智清
(He/Him/His)

Acknowledgements

I sincerely thank everyone who contributes to this book. In the first place, please allow me to thank Wei Zhu, Celine Chang, Ellen Seo, Sudha Ramachandran, Veena Perumal and other editors with Springer Nature, and Jingya Gao, Zhang Zhou, Yuli Wei, and other editors with China Machine Press for their critical contributions to successfully publish the book, who also helped make this publishing happen. Besides, Lakshmidevi Srinivasan, Pavitra Arulmurugan, and Arul Vani with Straive also worked on the publishing process. The following people help proofreading an earlier version of this book (Xiao, 2019): Zhengyan Tong, Yongjin Zhao, Yongjie Huang, Wei Li, Yunlong Ma, Junfeng Huang, Yuezhu Li, Ke Li, Tao Long, Qinghu Cheng, and Changhao Wang. Additionally, I want to thank my parents for their most selfless help, and thank my bosses and colleagues for their generous supports.

Thank you for choosing this book. Happy reading!

Contents

Notations

General rules

- Upper-case letters are random events or random numbers, while lower-case letters are deterministic events or deterministic variables.
- The serif typeface, such as X, denotes numerical values. The sans typeface, such as X, denotes events in general, which can be either numerical or not numerical.
- Bold letters denote vectors (such as $\mathbf{w}$) or matrices (such as $\mathbf{F}$), where matrices are always upper-case, even they are deterministic matrices.
- Calligraph letters, such as $\mathcal{X}$, denote sets.
- Fraktur letters, such as $\mathfrak{f}$, denote mappings.

In the sequel are notations throughout the book. We also occasionally follow other notations defined locally.

English Letters

A, a	advantage
A, a	action
$\mathcal{A}$	action space
B, b	baseline in policy gradient; numerical belief in partially observable tasks; (lower case only) bonus; behavior policy in off-policy learning
B, b	belief in partially observable tasks
$\mathfrak{B}_\pi, \mathfrak{b}_\pi$	Bellman expectation operator of policy π (upper case only used in distributional RL)
$\mathfrak{B}_*, \mathfrak{b}_*$	Bellman optimal operator (upper case only used in distributional RL)
$\mathcal{B}$	a batch of transition generated by experience replay; belief space in partially observable tasks
$\mathcal{B}^+$	belief space with terminal belief in partially observable tasks
c	counting; coefficients in linear programming

Cov	covariance
d, d_∞	metrics
d_f	f-divergence
d_{KL}	KL divergence
d_{JS}	JS divergence
d_{TV}	total variation
D_t	indicator of episode end
$\mathcal{D}$	set of experience
e	the constant e $\approx$ 2.72
e	eligibility trace
E	expectation
f	a mapping
F	Fisher Information Matrix (FIM)
G, g	return
g	gradient vector
h	action preference
H	entropy
I	identity matrix
k	index of iteration
ℓ	loss
$\mathbb{N}$	set of natural numbers
o	observation probability in partially observable tasks; infinitesimal in asymptotic notations
$O, \tilde{O}$	infinite in asymptotic notations
O, o	observation
$\mathcal{O}$	observation space
O	zero matrix
p	probability, dynamics
P	transition matrix
Pr	probability
Q, q	action value
Q_π, q_π	action value of policy π (upper case only used in distributional RL)
Q_*, q_*	optimal action values (upper case only used in distributional RL)
q	vector representation of action values
R, r	reward
$\mathcal{R}$	reward space
$\mathbb{R}$	set of real numbers
S, s	state
$\mathcal{S}$	state space
$\mathcal{S}^+$	state space with terminal state
T	steps in an episode
T, t	trajectory
$\mathcal{T}$	time index set
u	belief update operator in partially observable tasks
U, u	TD target; (lower case only) upper bound

V, v	state value
V_π, v_π	state value of the policy π (upper case only used in distributional RL)
V_*, v_*	optimal state values (upper case only used in distributional RL)
$\mathbf{v}$	vector representation of state values
Var	variance
$\mathbf{w}$	parameters of value function estimate
X, x	an event
$\mathcal{X}$	event space
$\mathbf{z}$	parameters for eligibility trace

Greek Letters

α	learning rate
β	reinforce strength in eligibility trace; distortion function in distributional RL
γ	discount factor
Δ, δ	TD error
ε	parameters for exploration
η	state visitation frequency
$\boldsymbol{\eta}$	vector representation of state visitation frequency
λ	decay strength of eligibility trace
$\boldsymbol{\theta}$	parameters for policy function estimates
ϑ	threshold of numerical iteration
π	the constant $\pi \approx 3.14$
Π, π	policy
π_*	optimal policy
π_E	expert policy in imitation learning
ρ	state–action visitation frequency; important sampling ratio in off-policy learning
$\boldsymbol{\rho}$	vector representation of state–action visitation frequency
ϕ	quantile in distributional RL
$\mathcal{T}, \tau$	sojourn time of SMDP
Ψ	Generalized Advantage Estimate (GAE)
Ω, ω	cumulative probability in distributional RL; (lower case only) conditional probability for partially observable tasks

Other Operators

$\mathbf{0}$	zero vector
$\mathbf{1}$	a vector all of whose entries are one
$\overset{\text{a.e.}}{=}$	equal almost everywhere

$\overset{d}{=}$	share the same distribution
$\overset{def}{=}$	define
$\leftarrow$	assign
$<, \leq, \geq, >$	compare numbers; element-wise comparison
$\prec, \preccurlyeq, \succcurlyeq, \succ$	partial order of policy
$\ll$	absolute continuous
$\varnothing$	empty set
∇	gradient
$\sim$	obey a distribution; utility equivalence in distributional RL
$\|\ \|$	absolute value of a real number; element-wise absolute values of a vector or a matrix; the number of elements in a set
$\|\|\ \|\|$	norm

Chapter 1
Introduction of Reinforcement Learning (RL)

This chapter covers

- definition of Reinforcement Learning (RL)
- key elements of RL
- applications of RL
- taxonomy of RL
- performance metrics of RL algorithms
- usage of Gym, a library for RL environments

1.1 What is RL?

Reinforcement Learning (**RL**) is a type of machine learning task where decision-makers try to maximize long-term rewards or minimize long-term costs. In an RL task, decision-makers observe the environments, and act according to the observations. After the actions, the decision-makers can get rewards or costs. In the learning process, the decision-makers learn how to maximize rewards or minimize costs through the interaction experience with the environment. The most important feature of RL is that its learning process is guided by reward signals or cost signals.

Example 1.1 (Maze) In Fig. 1.1, a robot in a maze observes its surroundings and tries to move. Stupid moves waste its time and energy, while smart moves lead it out of the maze. Here, the move is the action after its observation. The time and the energy it spends is the costs, which can also be viewed as negative rewards. In the learning process, the robot can only get reward or cost signals, but no one will tell it what are the correct moves.

© The Author(s), under exclusive license to Springer Nature Singapore Pte Ltd. 2024
Z. Xiao, *Reinforcement Learning*, https://doi.org/10.1007/978-981-19-4933-3_1

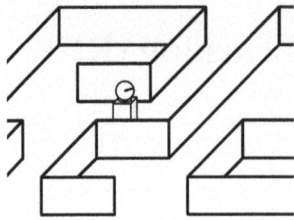

Fig. 1.1   Robot in a maze.

Interdisciplinary Reference 1.1
Behavior Psychology: Reinforcement Learning

Living beings learn to benefit themselves and avoid harm. For example, I made lots of decisions during my work. If some decision can help me get prompted and get more bonus, or help me avoid punished, I will make more similar decision afterward. The term "reinforcement" is used to depict the phenomenon that some inducements can make living beings tend to make some decisions, where the inducements can be termed "reinforcer" and the learning process can be termed "reinforcement learning" (Pavlov, 1928).

While positive reinforcement benefits the living beings and negative reinforcement harms the living beings, these two types of reinforcements are equivalent to each other (Michael, 1975). In the aforementioned example, getting prompted and getting more bonuses are positive reinforcement, and avoiding being fired and punished is negative reinforcement. Both positive reinforcement and negative reinforcement have similar effects.

Many problems in Artificial Intelligence (AI) try to enlarge benefits and avoid harm, so they adapt the term "reinforcement learning".

RL systems have the following important elements:

- **Reward** (or **cost** equivalently): The goal of RL is to maximize the long-term reward, or equivalently to minimize the long-term cost. In the example of the maze, the time and energy wasted generate negative rewards, and the result of going out of the maze generates positive rewards. The robot wants to get positive rewards while avoiding negative rewards as much as it can.
- **Policy**: Decision-makers will act according to their observations, and the way of making decisions is called policy. It is the policy that an RL algorithm learns. RL algorithm changes policy to maximize the total reward. In the example of the maze, the robot uses a policy to determine how to move.

RL systems adjust the policy to maximize the long-term reward. In the example of Maze, the robot adjusts its policy so that it can get out of the maze in a quicker and more efficient way.

RL differs essentially from supervised learning and unsupervised learning.

- The difference between RL and supervised learning is as follows: In supervised learning, learners know the correct actions, and can learn by comparing each decision with the correct decision. In RL, the correct action is unknown, so we can only learn from reward signals. RL wants to maximize the total reward for a whole period, and it usually seeks long-term superiority. Supervised learning compares the action for each decision, so it is myopic. Besides, supervised learning usually hopes that the learning results can be used in unknown data, and the results should generalize well. Contrarily, the policy obtained by RL can be used in the training environment. Therefore, supervised learning can be used for prediction or classification, such as predicting stock price, or identifying the contents of images, but RL does not excel in such tasks.
- The difference between RL and unsupervised learning is as follows: Unsupervised learning aims to discover the structure that data imply. RL has explicit quantitative goals, i.e. the reward. Therefore, unsupervised learning can be used in clustering, but RL does not excel in such tasks.

1.2 Applications of RL

RL has attained remarkable achievements in many applications. This section shares some success stories of RL to show its powerfulness.

- Video game: Video games include consoling games such as PacMan (Fig. 1.2), PC games such as StarCraft, and mobile games such as Flappy Bird. In a video game, players play according to the images on the screens to obtain high scores, or to win over other players. RL is usually more suitable for such game AI than supervised learning, since it is very difficult or uninteresting to point out the correct answer for every step in a game. RL-based AI can beat human players in many games, such as more than half of Atari 2600 games.

- Board game: Board games include the game of Go (Fig. 1.3), Reversi, and Gomoku. The player of a board game usually wants to beat the other player, and increases the chance of winning. However, it is difficult to obtain the correct answer for each step. That is exactly what RL is good at. DeepMind developed AI for the game of Go, and beaten Sedol Lee in Mar. 2016 and Jie Ke in May 2017, which drew exclusive attentions. DeepMind further developed more powerful board game AI such as AlphaZero and MuZero, and they can exceed human players by a large margin in many games such as the game of Go, Shogi, and chess.

- Robotics control: We can control robotics, such as androids, robot hands and robot feet to complete different tasks such as moving as far as possible (Fig. 1.4), grasping some objects, or keeping balance. Such a task can either be simulated

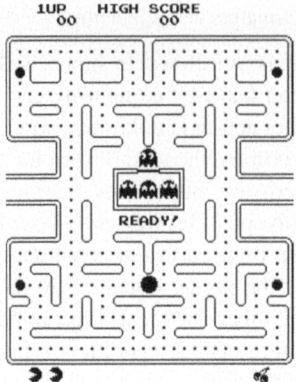

Fig. 1.2   PacMan in Atari 2600.
This figure is adapted from `https://en.wikipedia.org/wiki/Pac-Man#Gameplay`.

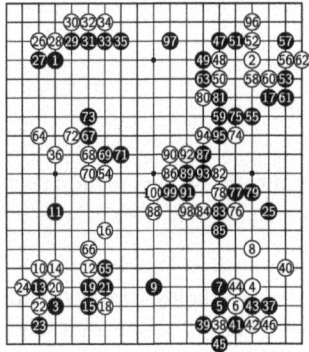

Fig. 1.3   A record for a game of Go.
The solid circles denote the stones of the black player, while the hollow circles denote the stones of
the white player. The number in the circle denotes the order in which stones have been placed on
the board. This picture is adapted from (Silver, 2017).

in a virtual environment (such as simulated using PC), and conducted in real
world. RL algorithms can help design control policies, and have achieved many
good results.

- Alignment of large models: Large models are machine learning models that
  have large amounts of parameters, and large models are expected to deal with
  complex and diverse tasks. Large models are usually trained by RL algorithms.
  For example, during the training process of the large language model ChatGPT,
  human evaluates the outputs of the model, and those feedbacks are fed into RL
  algorithms to adjust the model parameters (Fig. 1.5). In this way, RL-based
  training makes the model outputs align to our intentions.

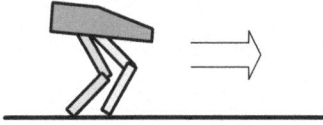

Fig. 1.4 Bipedal walker.
This figure is adapted from https://www.gymlibrary.dev/environments/box2d/bipedal_walker/.

Fig. 1.5 Large language models.

1.3 Agent–Environment Interface

Agent–environment interface (Fig. 1.6) is a powerful tool to analyze an RL system. It partitions the system into two parts: agent and environment.

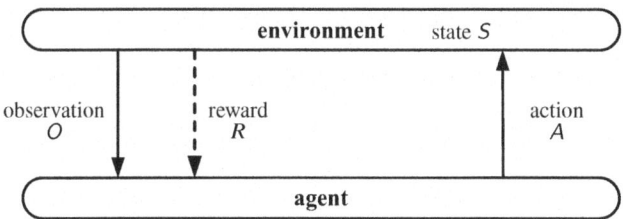

Fig. 1.6 Agent–environment interface.
The dash line of reward means that rewards are not available to agent during interactions.

- **Agent** is the decision-maker and learner. As a decision-maker, the agent can decide what action to take. As a learner, the agent can adjust its policy. There can be one or many agents in a RL system.
- **Environment** includes everything in RL system that agents do not include. It is what agents interact with.

The core idea of the agent–environment interface is to isolate the part that can be controlled subjectively and the part that can not be controlled subjectively.

Example 1.2 (Work) While I work, I can decide what to do, and learn to adjust the policy that determines what to do. Those parts belong to the agent. Meanwhile, my health, tiredness, and starvation are things that I can not directly control, so they belong to the environment. The agent can respond to the environment. Say, I may go to bed if I feel tired, or I may eat if I feel hungry.

> ! Note
>
> While RL systems are analyzed using agent–environment interface in most cases, it is not imperative to use agent–environment interface to analyze an RL system.

In an agent–environment interface, the interactions between agents and the environment involve the following two folds:

- Agents observe the environment and get the **observations** (denoted as O). The agents use some policy to determine what **actions** (denoted as A) they should take.
- After being impacted by the action, the environment signals **rewards** (denoted as R), and changes its **state** (denoted as S).

The agent can directly observe the observation O and the action A during interactions.

> ! Note
>
> States, observations, and actions are not necessarily numerical values. They can be stuffs such as feeling hungry and eating. This book uses a sans-serif typeface to denote such stuffs. Rewards are always numerical values (particularly, scalers). This book uses a serif typeface to denote numerical values, including scalers, vectors, and matrices.

Most RL systems are casual, which means what happened earlier can only impact what happens afterward. Such a problem is also called a sequential decision problem. For example, in a video game, every action a player takes may impact the scenarios afterward. For such systems, a time index t can be introduced. With the time index t, the state at time t can be denoted as S_t. The observation at time t can be denoted as O_t. The action at time t can be denoted as A_t. The reward at time t can be denoted as R_t.

> ! Note
>
> A system of agent–environment interface is not necessarily a sequential decision problem, and it does not necessarily require a timeline. Some problems are one-shot, which means that the agent only interacts with the environment once. For example, I may roll a die and get a reward equaling the outcome of the dice. This interaction does not need a time index.

When the interaction timing between the agent and the environment is countable, the agent–environment interface is called a discrete-time agent–environment interface. The interaction timing index can be mapped to $t = 0, 1, 2, 3, \ldots$ if the number of interactions is infinite or $t = 0, 1, \ldots, T$ if the number of interactions is finite, where T can be a random variable. Interactions at time t include the following two-fold:

- Environment signals the reward R_t and changes its state to S_t.
- Agent observes the observation O_t, and then decides to take the action A_t.

! Note

The intervals in the discrete-time index are not necessarily equal, or determined beforehand. We can always map a countable timing index to a non-negative integral set. Without loss of generality, we usually assume $t = 0, 1, 2, 3, \ldots$.

! Note

Different pieces of literature may use different notations. For example, some pieces of literature denote the reward signal after the action A_t as R_t, while this book denotes it as R_{t+1}. This book chooses this notation since R_{t+1} and S_{t+1} are determined in the same interaction step.

When agents learn, they not only know observations and actions, but also know rewards.

The system modeled as agent–environment interface may be further formulated to a more specific model, such as Markov decision process. Chapter 2 will introduce the Markov decision process formally.

1.4 Taxonomy of RL

There are various ways to classify RL (Fig. 1.7). The classification can be based on tasks, or based on algorithms.

1.4.1 Task-based Taxonomy

According to the task and the environment, an RL system can be classified in the following ways.

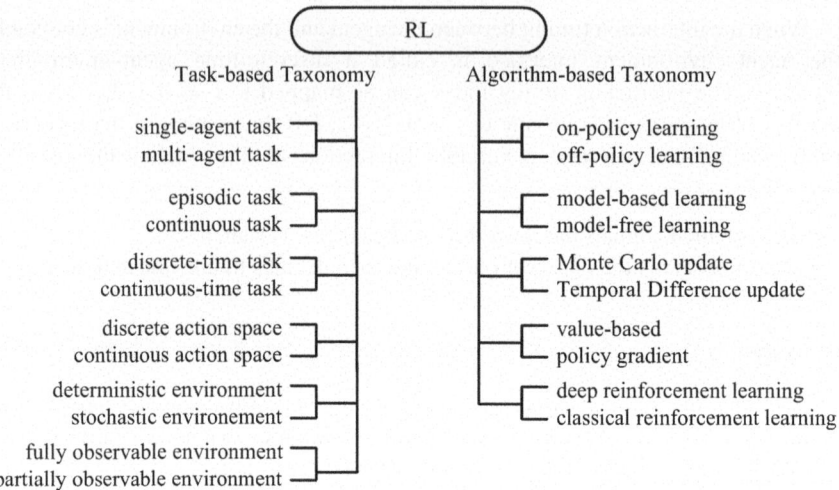

Fig. 1.7   Taxonomy of RL.
Each branch in the chart is not non-overlapping and exhaustive. See texts for details.

- **Single-agent RL** and **Multi-Agent RL (MARL)**. According to the number of agents in the RL task, an RL task can be a single-agent task or a multi-agent task. In a single-agent task, there is only one agent who can observe whatever it observes. In a multi-agent task, each agent may just observe its own observation. In a multi-agent task, agents may or may not communicate with each other. In a multi-agent task, different agents may have different goals, some of which may even conflict with each other. Without special notifications, this book considers single-agent tasks only.
- **Episodic task** and **sequential task**. An episodic task has start and end. For example, the game of Go is an episode task, where the board is empty in the beginning but full in the end. In the next round of the game, everything starts from empty. Contrarily, a sequential task has no beginning or no ends. For example, the scheduler of the public cloud services needs to process without interruption forever, and there are no ending situations designed for it.
- **Discrete-time task** and **continuous-time task**. If the time index of interactions between agents and the environment is countable, the timing index can be mapped to $t = 0, 1, 2, \ldots$ (if the number of interactions is countable infinite) or $t = 0, 1, \ldots, T - 1$ (if the number of interactions is finite, where T can be a random variable), and the task is a discrete-time task. If the time index of interactions between agents and the environment is uncountable, say, the set of non-negative real numbers or its non-empty continuous subset, the task is a continuous-time task. Note that this classification is not exclusive. Additionally, the intervals between two interactions can be stochastic, and the interval can further impact sequential evolution.

- **Discrete action space** and **continuous action space**. If the space is a countable set, the space is said to be discrete. Some pieces of literature imprecisely use the term discrete spaces to refer to finite spaces. If the space is a non-empty continuous interval (can be high-dimension intervals), the space is continuous. Taking the robot in the maze for example, if the robot has only 4 choices, i.e. east, north, west, and south, the action space is discrete; if the robot can move in a direction ranging from 0 to 360°, the action space is continuous. This classification is not exhaustive either.
- **Deterministic environment** or **stochastic environment**. The environment can be deterministic or stochastic. Taking the robot in the maze for example, if the maze is a particular fixed maze, the environment is deterministic. If the robot deploys a deterministic policy, the outcome will also be deterministic. However, if the maze changes randomly, the environment is stochastic.
- **Fully observable environment** or **partially observable environment**. Agents may observe the environment completely, partially, or not at all. If the agent can fully observe the environment, the environment is said to be fully observable. Otherwise, the environment is said to be partially observable. If the agents can observe nothing about the environment, we can not do anything meaningful, so the problem becomes trivial. For example, the game of Go can be viewed as a fully observable environment since players can observe the whole board. The poker is partially observable, since players do not know what cards do opponents have on their hands.

Some systems need to consider more than one task. Learning for multiple tasks is called **Multi-Task Reinforcement Learning** (MTRL). If many tasks differ only in the parameters of the reward function, such multi-task learning can be called **goal-conditioned reinforcement learning**. For example, in a multi-task system, each task wants the robot to go to a task-specific destination. The destination can be called the goal of the task, and those tasks are goal-conditional tasks. There is another type of system that we want to learn an agent such that the resulting agent can be used for some other different tasks. Such learning is called **transfer reinforcement learning**. If we want the resulting agent to be easily adapted to new tasks, we are conducting **meta reinforcement learning**. If the task is evolving, and our agent needs to adapt to the changing task, the learning is called **online reinforcement learning**, or **lifelong reinforcement learning**. Note that online reinforcement learning is not the antonym of offline reinforcement learning. **Offline reinforcement learning** is learning without interactions with environments. Offline reinforcement learning only learns from historical records of interactions, usually in batches, so it is also called **batch reinforcement learning**.

Additionally, **passive RL** is a kind of tasks that evaluate the performance of a fixed policy, while **active RL** tries to improve the policy. In this book, passive RL is called **policy evaluation**, while active RL is called **policy optimization**.

1.4.2 Algorithm-based Taxonomy

RL algorithms can be classified in the following ways.

- **On-policy** or **off-policy**. An on-policy algorithm learns from the policy that is used to play, while an off-policy algorithm learns from a different policy. For example, when learning how to play the game of Go, on-policy algorithms use the latest policy to play and learn from that, while off-policy algorithms can learn from records of games with old policies. In offline tasks, agents can not apply the latest action in the interaction, so only off-policy algorithms can be used in offline tasks. Please notice that online learning and on-policy learning are two different concepts.

- **Model-based** or **model-free**. Model-based algorithms use the model of environments, while model-free algorithms do not use the model of environments. Model-based algorithms may know the model of environments a priori, or they can learn the model from samples. Taking the game of Go for example, we know the rules of the game beforehand. The rule is the model. A model-based algorithm can use this model to play more games on other virtual boards, and learn from samples generated via the virtual playing. In the case that we do not know the model beforehand, the algorithm is model-based if it tries to learn the model of the environment, and it is model-free if it does not try to model the environment. Environments model, either given or learned, can be used to generate samples. If there are no environment models, agents can only get samples from actual interactions with the environment.

- **Monte Carlo update** or **Temporal Difference update**. Monte Carlo update is suitable for episode tasks, and it estimates the return at the end of each episode and learned from the estimates of episode returns. Temporal different update does not need to wait for finishing of episodes. It can use intermediate information to learn.

- **Valued-based**, **policy-based**, or **actor–critic**. Valued-based algorithms use state values and action values to denote the return we can get on the condition on a certain state and/or an action. (We will formally introduce these concepts in Sect. 2.2.) Policy-based algorithms change the policy directly without the help of values. Actor–critic methods combine value-based algorithms and policy-based methods. **Policy gradient** algorithms are the most important types of policy-based algorithms. Besides, there are also **gradient-free** algorithms.

- **Deep RL (DRL)** and classical RL. If an RL algorithm uses neural networks as components, it leverages Deep Learning (DL) method and is called Deep RL. Remarkably, RL and DL are two independent concepts (Fig. 1.8). An algorithm is an RL algorithm if it solves an RL task, while an algorithm is a DL algorithm if it uses neural networks. An RL algorithm may or may not be a DL algorithm, while a DL algorithm may or may not be an RL algorithm. If an algorithm is both a DL algorithm and an RL algorithm, it is a DRL algorithm. For example, video game AIs can use convolutional neural networks to process video images and solve RL tasks, and the algorithm is a DRL algorithm.

Fig. 1.8   Relationship among RL, DL, and DRL.

1.5 Performance Metrics

There are varied RL algorithms, each of which has its own limitations and can be applied to tasks with some properties. Some RL algorithms excel in some specific tasks, while some RL algorithms excel in other tasks. Therefore, the performance of RL algorithms is task-specific. When we want to use RL to solve a specific task, we need to focus on the performance of the designated task. In academia, researchers also use some generally accepted benchmark tasks, such as environments in the library Gym (will be introduced in detail in Sect. 1.6). Researchers also construct new tasks to demonstrate the superiority of some algorithms that are designed for a particular situation. Therefore, a saying goes, "1000 papers who claim to reach State-Of-The-Art (SOTA) actually reach 1000 different SOTAs."

Generally speaking, we care about the long-term reward and how it changes during the training. The most common way to demonstrate the performance is to show the long-term reward and its learn curves. Learning curve is a chart rather than a numerical value, but it can be further quantified. Some types of RL tasks, such as online RL, offline RL, and multi-task RL, may have their unique performance index.

In the sequel, we list the performance metrics of RL, starting from the most important ones to the least important ones in common cases.

- **Long-term reward**. This indicates the long-term reward that an algorithm can achieve when it is applied to the task. There are many ways to quantify it, such as return, and average reward. Sect. 15.1 compares different metrics. Many statistics derive from it, such as the mean and the standard deviation. We usually want larger returns and episode rewards. Secondly, we may also want small standard deviations.
- Optimality and convergence. Theorists care whether an RL algorithm converges to the optimal. We prefer the algorithms that can converge to the optimal.
- **Sample complexity**. It is usually costly and time-consuming to interact with environments, so we prefer to learn from as few samples as possible. Remarkably, some algorithms can make use of environment models or historical records to learn, so they have much smaller sample complexity. Some model-based algorithms can even solve the task without learning from samples. Strictly speaking, such algorithms are not RL algorithms. Algorithms for offline tasks learn from historical records. They do not interact with environments. Therefore, the "online" sample complexity of these offline RL algorithms is zero. However, offline RL tasks usually get a fixed number of samples, so the "offline" sample complexity is critical. We prefer small sample complexity.

- **Regret**. Regret is defined as the summation of episode regrets during the training process, where the episode regret is defined as the difference between the actual episode reward and the optimal episode reward after the training completes. We prefer small regret values. In some tasks, RL algorithms learn from scratch, and we start to accumulate regret values from the very beginning. In some other tasks, RL models are first pretrained in a pretraining environment, and then trained in another environment that is costly to interact with. We may only care about the regret in the costly environment. For example, an agent can be pre-trained in a simulation environment before it is formally trained in the real world. Failure in the real world is costly, so regrets in the real world are a matter of concern. For another example, a robot can be pre-trained in a lab before it is shown to clients. Sales who want to demonstrate the robots successfully to clients may care about the regret when the robot is trained in the venue of clients.
- Scalability. Some RL algorithms can use multiple workers to accelerate learning. Since RL tasks are usually time-consuming and computation-intensive, it is an advantage to have the scalability to speed up the learning.
- Convergence speed, training time, and time complexity. Theorists care whether an RL algorithm converges, and how fast it converges. Time complexity can also be used to indicate the speed of convergence. Convergence is a good property, and we prefer small time complexity.
- Occupied space and space complexity. We prefer small space occupation. Some use cases particularly care about the space for storing historical records, while some other use cases particularly care more about the storage on Graphics Random-Access Memory (RAM).

1.6 Case Study: Agent–Environment Interface in Gym

Gym (website: `https://www.gymlibrary.dev/`), created by the company OpenAI, is the most influential RL environment library. It implements hundreds of tasks with Python, which are modeled using discrete-time agent–environment interfaces. It is an open-source and free software. This section will install and use Gym, and write an example where an agent and an environment interact with each other.

Gym contains hundreds of tasks, including the following categories:

- Toy Text: This category includes some simple tasks that can be represented by strings.
- Classic Control: This category includes some simple kinetic tasks.
- Box2D: The tasks in this category use the library Box2D to create and visualize objects.
- Atari: This category includes video games that were invented for a game control called Atari. Players observe pixelized images and want to get high scores in those games.

Gym is an open-source library. Its codes can be accessible in the following GitHub URL:

```
https://github.com/openai/gym
```

1.6.1 Install Gym

Gym is compatible with all commonly-used operating systems, including Windows, Linux, and macOS. This section introduces how to install Gym with Anaconda 3.

When installing Gym, we can either install its minimum version, or install some of its extensions, such as its Toy Text, Classic Control, and Atari. You can input the following command as administrator (for Windows) or root (for Linux/macOS) in Anaconda Prompt to install them:

```
pip install gym[toy_text,classic_control,atari,accept-rom-license,other]
```

! Note

We will use Gym throughout the book (see Table 1.1 for details), so please make sure to install it. The aforementioned installation command can satisfy the usage of first 9 chapters of Gym in this book. The installation of other extensions will be introduced when they are used. Gym is evolving all the time, so it is recommended to install an environment when we indeed need it. It is unnecessary to install a complete version of Gym in the very beginning. The author of the book will update the installation method of Gym in GitHub repo of this book.

Table 1.1 Major Python modules that the codes in this book depend on.

Chapter	Primary library that agents depend on	Primary library that environments depend on
Chapter 1	Numpy	Gym with Classic Control (i.e. gym[classic_control])
Chapter 2–5	Numpy	Gym with Toy Text (i.e. gym[toy_text])
Chapter 6–9	TensorFlow or PyTorch	Gym with Classic Control (i.e. gym[classic_control])
Chapter 10	TensorFlow or PyTorch	Gym with Box2D (i.e. gym[box2d])
Chapter 11	Numpy	Gym with Box2D (i.e. gym[box2d])
Chapter 12	TensorFlow or PyTorch	Gym with Atari (i.e. gym[atari])
Chapter 13	Numpy	We will write custom environment to extend Gym
Chapter 14	TensorFlow or PyTorch	We will write custom environment to extend Gym
Chapter 15	Numpy	We will write custom environment to extend Gym
Chapter 16	TensorFlow or PyTorch	PyBullet (which extends the minimum version of Gym)

1.6.2 Use Gym

This section introduces how to use Gym APIs.

We need to import Gym if we want to use it. The following codes import Gym:

```
import gym
```

Upon Gym is imported, we can use the function `gym.make()` to get an environment object. Every task in Gym has an ID, which is a string in the form of `"Xxxxx-vd"`, such as `"CartPole-v0"`, and `"Taxi-v3"`. Note that the task ID also indicates the version of the task. The task of different versions may have different behaviors. The code to get an environment object of task `CartPole-v0` is:

```
env = gym.make('CartPole-v0')
```

We can use the following codes to see what tasks have been registered in Gym:

```
print(gym.envs.registry)
```

A task has its observation space and its action space. For an environment object env, `env.observation_space` records its observation space, while `env.action_space` records its action space. The class `gym.spaces.Box` can be used to represent a space of `np.array`'s. For a space with finite elements, we can also use the class `gym.spaces.Discrete` to represent a space of `int`'s. There are other space types. For example, the observation space of `CartPole-v0` is `Box(4,)`, so the observation is a `np.array` of the shape `(4,)`. The action space of `CartPole-v0` is `Discrete(2)`, so the action can be an int value within $\{0, 1\}$. A Box object uses members `low` and `high` to record the range of values. A Discrete object uses the member n to record the number of elements in the space.

Now we try to interact with the environment object env. Firstly, we need to initialize the environment. The code to initialize the object env is:

```
env.reset()
```

which returns the initial observation `observation` and an information variable `info` of the type `dict`. As we just mentioned, the type of `observation` is compatible with `env.observation_space`. For example, the `observation_space` of `'CartPole-v0'` object is `Box(4,)`, so the observation is a `np.array` object of the shape `(4,)`.

For each timestamp, we can use the member `env.step()` to interact with the environment object env. This member accepts a parameter, which is an action in the action space. This member returns the following five values:

- `observation`: a value sharing the same meaning with the first return value of `env.reset()`.
- `reward`: a `float` value.
- `terminated`: a `bool` value, indicating whether the episode has ended. The environments in Gym are mostly episodic. When `terminated` is `True`, it means that the episode has finished. In this case, we can call `env.reset()` to start the next episode.

- truncated: a bool value, indicating whether the episode has been truncated. We can limit the number of steps in an episode in both episodic tasks and sequential tasks, so that the tasks become episodic tasks with maximum number of steps. When an episode reaches its maximum step, the truncated indicator becomes True. Besides, there are other situation of truncated, especially the resources limitation such as the memory is not enough, or the data exceed valid range.
- info: a dict value, containing some optional debug information. It shares the same meaning with the second return value of env.reset().

Each time we call env.step(), we move the timeline for one step. In order to interact with the environment sequentially, we usually put it in a loop.

After env.reset() or env.step(), we may use the following code to visualize the environment:

```
env.render()
```

After using the environment, we can use the following code to close the environment:

```
env.close()
```

! Note

env.render() may render the environment in a new window. In this case, the best way to close the window is to call env.close(). Directly closing the window through the "X" button is not recommended and may lead to crashes.

When researchers use the environments in Gym as benchmark environments, they usually run 100 episodes successively and calculate the average episode rewards. They choose to run 100 episodes, rather than other numbers of episodes such as 128, just as a common practice. Some environments also have a reference threshold for average episode rewards. An algorithm is said to solve the task if and only if the average episode reward exceeds the reference threshold.

If a task has a reference episode reward threshold, the threshold is stored in

```
env.spec.reward_threshold
```

Online Contents

The GitHub repo of this book provides contents that guide advanced readers to understand the source codes of Gym library. Readers can learn the class gym.Env, the class gym.space.Space, the class gym.space.Box, the class gym.space.Discrete, the class gym.Wrapper, and the class gym.wrapper.TimeLimit here to better understand the details of Gym implementation.

1.6.3 Example: MountainCar

This section uses some self-contained examples to demonstrate the usage of environments in Gym. The examples consider a classical task called "Mountain Car" (Moore, 1990). This task has two versions: one version called MountainCar-v0 is with finite action space, another version called MountainCarContinuous-v0 is with continuous action space. We will use Gym APIs to interact with these two environments. Additionally, we will focus on the usage of Gym APIs, so we temporarily ignore the internals of the environments and the way to find the solutions. The details of the environments and solutions will be introduced in Sect. 6.5.

Firstly, we use MountainCar-v0, the version with finite action space. Every time we get a task, we need to know how to interact with the task. Especially, we need to know what is its observation space and its action space. We can use Code 1.1 to print its observation space and action space.

Remarkably, this book always uses the logging module to print out information, rather than calling the function print(). The logging can print out the time stamp along with the actual message, so we can better understand how long does it take to run the codes.

Code 1.1   Check the observation space and action space of the task MountainCar-v0.

MountainCar-v0_ClosedForm.ipynb

```
import gym
env = gym.make('MountainCar-v0')
for key in vars(env.spec):
    logging.info('%s: %s', key, vars(env.spec)[key])
for key in vars(env.unwrapped):
    logging.info('%s: %s', key, vars(env.unwrapped)[key])
```

The results are:

```
00:00:00 [INFO] entry_point: gym.envs.classic_control:MountainCarEnv
00:00:00 [INFO] reward_threshold: -110.0
00:00:00 [INFO] nondeterministic: False
00:00:00 [INFO] max_episode_steps: 200
00:00:00 [INFO] order_enforce: True
00:00:00 [INFO] kwargs: {}
00:00:00 [INFO] namespace: None
00:00:00 [INFO] name: MountainCar
00:00:00 [INFO] version: 0
00:00:00 [INFO] min_position: -1.2
00:00:00 [INFO] max_position: 0.6
00:00:00 [INFO] max_speed: 0.07
00:00:00 [INFO] goal_position: 0.5
00:00:00 [INFO] goal_velocity: 0
00:00:00 [INFO] force: 0.001
00:00:00 [INFO] gravity: 0.0025
00:00:00 [INFO] low: [-1.2  -0.07]
00:00:00 [INFO] high: [0.6  0.07]
00:00:00 [INFO] screen: None
00:00:00 [INFO] clock: None
00:00:00 [INFO] isopen: True
00:00:00 [INFO] action_space: Discrete(3)
00:00:00 [INFO] observation_space: Box([-1.2  -0.07], [0.6  0.07], (2,),
    float32)
```

```
24  00:00:00 [INFO] spec: EnvSpec(entry_point='gym.envs.classic_control:
        MountainCarEnv', reward_threshold=-110.0, nondeterministic=False,
        max_episode_steps=200, order_enforce=True, kwargs={}, namespace=None, name
        ='MountainCar', version=0)
25  00:00:00 [INFO] _np_random: RandomNumberGenerator(PCG64)
```

The results show that:

- The action space is `Discrete(3)`, so the action will be an `int` value within the set $\{0, 1, 2\}$.
- The observation space is `Box(2,)`, so the observation will be a `np.array` with the shape (2,).
- The maximum number of steps in an episode `max_episode_steps` is 200.
- Reference episode reward value `reward_threshold` is −110. An algorithm is said to solve the task if its average episode reward over 100 successive episodes exceeds this threshold.

Next, we prepare an agent for interacting with the environment. Generally, Gym does not include agents, so we need to implement an agent by ourselves. The class `ClosedFormAgent` in Code 1.2 can solve the task. It is a simplified agent that can only decide, but can not learn. Therefore, exactly speaking, this is not an RL agent. Nevertheless, it suffices to show the interaction between an agent and the environment.

Code 1.2   Closed-form agent for task the task `MountainCar-v0`.
`MountainCar-v0_ClosedForm.ipynb`

```
1   class CloseFormAgent:
2       def __init__(self, _):
3           pass
4
5       def reset(self, mode=None):
6           pass
7
8       def step(self, observation, reward, terminated):
9           position, velocity = observation
10          lb = min(-0.09 * (position + 0.25) ** 2 + 0.03,
11                  0.3 * (position + 0.9) ** 4 - 0.008)
12          ub = -0.07 * (position + 0.38) ** 2 + 0.07
13          if lb < velocity < ub:
14              action = 2  # push right
15          else:
16              action = 0  # push left
17          return action
18
19      def close(self):
20          pass
21
22
23  agent = CloseFormAgent(env)
```

Finally, we make the agent and the environment interact with each other. The function `play_episode()` in Code 1.3 plays an episode. It has the following five parameters:

- env: It is an environment object.
- agent: It is an agent object.

- seed: It can be None or an int value, which will be used as the seed of random number generator of the episode.
- mode: It can be None or the string "train". If it is the string "train", it tries to tell the agent to learn from interactions. However, if the agent can not learn, it is of no use.
- render: It is a bool value, indicating whether to visualize the interaction.

This function returns episode_reward and elapsed_step. episode_reward, which is a float, indicates the episode reward. elapsed_step, which is an int, indicates the number of steps in the episode.

Code 1.3   Play an episode.

MountainCar-v0_ClosedForm.ipynb

```
def play_episode(env, agent, seed=None, mode=None, render=False):
    observation, _ = env.reset(seed=seed)
    reward, terminated, truncated = 0., False, False
    agent.reset(mode=mode)
    episode_reward, elapsed_steps = 0., 0
    while True:
        action = agent.step(observation, reward, terminated)
        if render:
            env.render()
        if terminated or truncated:
            break
        observation, reward, terminated, truncated, _ = env.step(action)
        episode_reward += reward
        elapsed_steps += 1
    agent.close()
    return episode_reward, elapsed_steps
```

Leveraging the environment in Code 1.1, the agent in Code 1.2, and the interaction function in Code 1.3, the following codes make the agent and the environment interact for an episode with visualization. After the episode, they call env.close() to close the visualization window. At last, we show the results.

```
episode_reward, elapsed_steps = play_episode(env, agent, render=True)
env.close()
logging.info('episode reward = %.2f, steps = %d',
        episode_reward, elapsed_steps)
```

In order to evaluate the performance of the agent systematically, Code 1.4 calculates the average episode rewards over 100 successive episodes. The agent ClosedFormAgent can attain approximately -103, which exceeds the reward threshold -110. Therefore, the agent ClosedFormAgent solves this task.

Code 1.4   Test the performance by playing 100 episodes.

MountainCar-v0_ClosedForm.ipynb

```
episode_rewards = []
for episode in range(100):
    episode_reward, elapsed_steps = play_episode(env, agent)
    episode_rewards.append(episode_reward)
    logging.info('test episode %d: reward = %.2f, steps = %d',
            episode, episode_reward, elapsed_steps)
logging.info('average episode reward = %.2f ± %.2f',
        np.mean(episode_rewards), np.std(episode_rewards))
```

Next, we consider the task `MountainCarContinuous-v0`, the one with continuous action space. In order to import this task, we slightly modify Code 1.1, resulting in Code 1.5:

Code 1.5   Check the observation space and action space of the task `MountainCarContinuous-v0`.

`MountainCarContinuous-v0_ClosedForm.ipynb`

```
1  env = gym.make('MountainCarContinuous-v0')
2  for key in vars(env):
3      logging.info('%s: %s', key, vars(env)[key])
4  for key in vars(env.spec):
5      logging.info('%s: %s', key, vars(env.spec)[key])
```

The outputs are:

```
1   00:00:00 [INFO] entry_point: gym.envs.classic_control:Continuous_MountainCarEnv
2   00:00:00 [INFO] reward_threshold: 90.0
3   00:00:00 [INFO] nondeterministic: False
4   00:00:00 [INFO] max_episode_steps: 999
5   00:00:00 [INFO] order_enforce: True
6   00:00:00 [INFO] kwargs: {}
7   00:00:00 [INFO] namespace: None
8   00:00:00 [INFO] name: MountainCarContinuous
9   00:00:00 [INFO] version: 0
10  00:00:00 [INFO] min_action: -1.0
11  00:00:00 [INFO] max_action: 1.0
12  00:00:00 [INFO] min_position: -1.2
13  00:00:00 [INFO] max_position: 0.6
14  00:00:00 [INFO] max_speed: 0.07
15  00:00:00 [INFO] goal_position: 0.45
16  00:00:00 [INFO] goal_velocity: 0
17  00:00:00 [INFO] power: 0.0015
18  00:00:00 [INFO] low_state: [-1.2  -0.07]
19  00:00:00 [INFO] high_state: [0.6  0.07]
20  00:00:00 [INFO] screen: None
21  00:00:00 [INFO] clock: None
22  00:00:00 [INFO] isopen: True
23  00:00:00 [INFO] action_space: Box(-1.0, 1.0, (1,), float32)
24  00:00:00 [INFO] observation_space: Box([-1.2  -0.07], [0.6  0.07], (2,),
        float32)
25  00:00:00 [INFO] spec: EnvSpec(entry_point='gym.envs.classic_control:
        Continuous_MountainCarEnv', reward_threshold=90.0, nondeterministic=False,
         max_episode_steps=999, order_enforce=True, kwargs={}, namespace=None,
         name='MountainCarContinuous', version=0)
26  00:00:00 [INFO] _np_random: RandomNumberGenerator(PCG64)
```

The `action_space` of this environment is `Box(1,)`, and action is a `np.array` object with the shape (1,). The `observation_space` of this environment is `Box(2,)`, and observation is a `np.array` object with the shape (2,). The maximum number of steps in an episode becomes 999. The reward threshold for this task becomes 90, which means that an agent solves the task if the average episode rewards over 100 successive episodes exceed the threshold 90.

Different tasks need different agents. Code 1.6 provides an agent for the task `MountainCarContinuous-v0`. In member function `step()`, the observation is split into position and velocity, and the agent decides according to inequalities driven by these two components. We can use Codes 1.3 and 1.4 to test the performance of this agent. The test result shows that the average episode reward is around 93, which exceeds the reward threshold 90. Therefore, the agent in Code 1.6 solves the task `MountainCarContinuous-v0`.

Code 1.6   Closed-form agent for task `MountainCarContinous-v0`.
`MountainCarContinuous-v0_ClosedForm.ipynb`

```
class ClosedFormAgent:
    def __init__(self, _):
        pass

    def reset(self, mode=None):
        pass

    def step(self, observation, reward, terminated):
        position, velocity = observation
        if position > -4 * velocity or position < 13 * velocity - 0.6:
            force = 1.
        else:
            force = -1.
        action = np.array([force,])
        return action

    def close(self):
        pass


agent = ClosedFormAgent(env)
```

1.7 Summary

- Reinforcement Learning (RL) is a type of machine learning that improves policy using reward signals. Policy and reward are primary elements of an RL system. RL tries to maximize the long-term reward, or equivalently minimize the long-term cost.
- RL differs from supervised learning since RL does not require reference labels. RL differs from unsupervised learning since RL requires reward signals to learn.
- RL system is usually modeled as an agent–environment interface. The decision-maker and learner belong to agents, while other parts belong to the environment. Agents propose action to the environment, and get observation from environments.
- An RL task can be a single-agent RL task, or a multi-agent RL task. An RL task can be an episode task or a sequential task. The action space can be a discrete space or a continuous space. The environment can be deterministic or stochastic. The environment can be fully observable or partially observable.
- Multi-task RL considers multiple tasks, including goal-conditional RL, transfer RL, meta RL, and so on.
- Online RL, also known as lifelong learning, aims to adjust to changing environments. Offline RL, also known as batched RL, learns without direct interactions with the environments.

- An RL algorithm can be either on-policy or off-policy. On-policy algorithms learn using the policies that are interacting with the environments, while off-policy algorithms do not. An RL algorithm can be model-based or model-free. Model-based algorithms use the models of environments, either given or learned, but model-free algorithms do not. An RL algorithm can be Monte Carlo learning or Temporal-Difference learning. An RL algorithm can also be value-based and/or policy-based. Actor–critic methods combine the value-based method and policy-based method.
- An RL algorithm who uses deep learning is a Deep RL (DRL) algorithm.
- Gym is an open-source free library for RL environments. Its APIs include the following: (1) Get the environment: env = gym.make(TaskID); (2) initialize: env.reset(); (3) interact: env.step(action); (4) visualization: env.render(); (5) exit: env.close().

1.8 Exercises

1.8.1 Multiple Choices

1.1 On the goal of RL, the most accurate one is: (   )

A. RL tries to maximize the long-term reward or the long-term cost.
B. RL tries to minimize the long-term reward or the long-term cost.
C. RL tries to maximize the long-term reward or minimize the long-term cost.

1.2 On the agent–environment interface, the most accurate one is: (   )

A. RL system must be modeled as the agent–environment interface.
B. The intuition of the agent–environment interface is to separate the part that can be controlled directly and the part that can not be controlled directly.
C. The timing of interactions in the agent–environment interface must be determined a priori.

1.3 On the taxonomy of RL, the most accurate one is: (   )

A. An RL task can be episodic, or sequential.
B. On-policy learning means online learning, while off-policy learning means offline learning.
C. A discrete-time environment is an environment with discrete action space, while a continuous-time environment is an environment with continuous action space.

1.4 On the taxonomy of RL, the most accurate one is: (   )

A. An RL algorithm can be model-based or model-free.
B. We must use model-free algorithms if the model of the environment is unknown, while we must use model-based algorithms if the model of the environment is known.

C. Model-based algorithm is only suitable for deterministic environments. It is not suitable for stochastic environments.

1.8.2 Programming

1.5 Play with the task `'CartPole-v0'` by using Gym APIs. What are its observation space, action space, and the maximum number of steps? What is the type of observation and action? What is the reward threshold? Solve the task using the following agent: Denote the observation as x, v, θ, ω. Let the action be 1 if $3\theta + \omega > 0$. Otherwise, let the action be 0.

1.8.3 Mock Interview

1.6 How does RL differ from supervised learning and unsupervised learning?

1.7 Why do we usually use agent–environment interface to analyze RL systems?

Chapter 2
MDP: Markov Decision Process

This chapter covers

- Discrete-Time Markov Decision Process (DTMDP)
- discounted return
- values and their properties, including Bellman expectation equations
- partial order of policy
- policy improvement theorem
- discounted visitation frequency and discounted expectation, and their properties
- optimal policy, existence of optimal policy, and properties of optimal policy
- optimal values, existence of optimal values, and properties of optimal values, including Bellman optimal equations
- Linear Programming (LP) method to find optimal values
- finding optimal policy using optimal values

This chapter will introduce the most famous, most classical, and most important model in RL: Discrete-Time Markov Decision Process (DTMDP). We will first define DTMDP, introduce its properties, and introduce a way to find the optimal policy. This chapter is the most important chapter in this book, so I sincerely invite readers to understand it well.

2.1 MDP Model

2.1.1 DTMDP: Discrete-Time MDP

This subsection defines DTMDP. DTMDP is derived from the discrete-time agent–environment interface in the previous chapter.

Let us reframe how the agent and the environment interact with each other in a discrete-time agent–environment interface. In an agent–environment interface, the environment receives actions, and the agent receives observations during interactions. Mathematically, let the interaction time be $t = 0, 1, 2, \ldots$. At time t,

- The environment generates reward $R_t \in \mathcal{R}$ and reaches state $S_t \in \mathcal{S}$. Here, $\mathcal{R}$ is the **reward space** which is a subset of the real number set; $\mathcal{S}$ is the **state space**, which is the set of all possible states.
- The agent observes the environment and gets observation $O_t \in O$, where O is the **observation space**, the set of all possible observations. Then the agent decides to conduct action $A_t \in \mathcal{A}$, where $\mathcal{A}$ is the **action space**, the set of all possible actions.

> **! Note**
>
> The candidates of states, observations, actions, and rewards may differ for different t. For the convenience of notations, we usually use a large space to cover all possibilities, so that we can use the same notations for all time steps.

A discrete-time agent–environment interface has the following **trajectory**:

$$R_0, S_0, O_0, A_0, R_1, S_1, O_1, A_1, R_2, S_2, O_2, A_2, \ldots.$$

There is a terminal state for an episodic task. The terminal state, denoted as s_{end}, differs from other states: Episodes end at the terminal state, and there will be no further observations and actions afterwards. Therefore, the terminal state s_{end} is excluded from the state space $\mathcal{S}$ by default. We can use $\mathcal{S}^+$ to denote the state space with the terminal state.

The trajectory of an episodic task is in the form of

$$R_0, S_0, O_0, A_0, R_1, S_1, O_1, A_1, R_2, S_2, O_2, A_2, R_3, \ldots, R_T, S_T = s_{\text{end}},$$

where T is the number of steps to reach the terminal state.

> **! Note**
>
> The number of steps for an episode, denoted as T, can be a random variable. From the view of stochastic process, it can be viewed as a **stop time**.

In the discrete-time agent–environment interface, if agents can fully observe the state of the environment, we can assume $O_t = S_t$ ($t = 0, 1, 2, \ldots$) without the loss of generality. In such circumstance, the trajectory can be simplified to

$$R_0, S_0, A_0, R_1, S_1, A_1, R_2, S_2, A_2, R_3, \ldots, S_T = s_{\text{end}},$$

and the notations O_t and O are no longer needed.

> **! Note**
>
> The agent–environment interface does not require that the environment be fully observable. If the environment is not fully observable, the agent–environment interface can be modeled as other models such as Partially Observable Markov Decision Process (POMDP), which will be introduced in Sect. 15.5.

Furthermore, if the environment is probabilistic and Markovian, we will get DTMDP.

> **Interdisciplinary Reference 2.1**
> **Stochastic Process: Markov Process**
>
> Fix the time index set $\mathcal{T}$ (can be the set of natural numbers $\mathbb{N}$, or non-negative real numbers $[0, +\infty)$, etc). For a stochastic process $\{S_t : t \in \mathcal{T}\}$, if for any $i \in \mathbb{N}$, $t_0, t_1, \ldots, t_i, t_{i+1} \in \mathcal{T}$ (we may assume $t_0 < t_1 < \cdots < t_i < t_{i+1}$ without loss of generality), $s_0, s_1, \ldots, s_{i+1} \in \mathcal{S}$, we all have
>
> $$\Pr\left[S_{t_{i+1}} = s_{i+1} \middle| S_{t_0} = s_0, S_{t_1} = s_1, \ldots, S_{t_i} = s_i\right] = \Pr\left[S_{t_{i+1}} = s_{i+1} \middle| S_{t_i} = s_i\right],$$
>
> then we say the process $\{S_t : t \geq 0\}$ is a **Markov Process** (**MP**), or the process $\{S_t : t \geq 0\}$ is Markovian. Furthermore, if for any $t, \tau \in \mathcal{T}$ and $s, s' \in \mathcal{S}$, we have $t + \tau \in \mathcal{T}$ and
>
> $$\Pr\left[S_{t+\tau} = s' \middle| S_t = s\right] = \Pr\left[S_\tau = s' \middle| S_0 = s\right],$$
>
> we say $\{S_t : t \geq 0\}$ is a homogenous MP. In this book, MP by default refers to homogenous MP.
> Given $\tau \in \mathcal{T}$, we can further define transition probability over the time interval τ:
>
> $$p^{[\tau]}(s'|s) \stackrel{\text{def}}{=} \Pr\left[S_\tau = s' \middle| S_0 = s\right], \quad s, s' \in \mathcal{S}.$$
>
> It has the following properties:
>
> $$p^{[0]}(s'|s) = 1_{[s'=s]}, \qquad\qquad\qquad s, s' \in \mathcal{S},$$
> $$p^{[\tau'+\tau'']}(s'|s) = \sum_{s'' \in \mathcal{S}} p^{[\tau']}(s'|s'')p^{[\tau'']}(s'', s), \quad \tau', \tau'' \in \mathcal{T}, s, s' \in \mathcal{S},$$

where $1_{[\cdot]}$ is the indicator function. If we write the τ transition probability as a matrix $\mathbf{P}^{[\tau]}$ of shape $|\mathcal{S}| \times |\mathcal{S}|$ ($|\mathcal{S}|$ can be infinite), the above equations can be written as

$$\mathbf{P}^{[0]} = \mathbf{I},$$
$$\mathbf{P}^{[\tau'+\tau'']} = \mathbf{P}^{[\tau']}\mathbf{P}^{[\tau'']}, \quad \tau', \tau'' \in \mathcal{T},$$

where $\mathbf{I}$ is an identity matrix.

❗ Note

Here the matrices are denoted by upper case letters, although they are deterministic values. This book uses upper case letters to denote deterministic matrices.

If the time index set $\mathcal{T}$ is the set of natural numbers $\mathbb{N}$, the MP is a **Discrete-Time Markov Process (DTMP)**. If the time index set $\mathcal{T}$ is $[0, +\infty)$, the MP is a **Continuous-Time Markov Process (CTMP)**.

Example 2.1 A DTMP has two states $\mathcal{S} = \{\text{hungry}, \text{full}\}$, and its one-step transition probability is

$$\Pr\left[S_{t+1} = \text{hungry} \middle| S_t = \text{hungry}\right] = 1/2$$
$$\Pr\left[S_{t+1} = \text{full} \middle| S_t = \text{hungry}\right] = 1/2$$
$$\Pr\left[S_{t+1} = \text{hungry} \middle| S_t = \text{full}\right] = 0$$
$$\Pr\left[S_{t+1} = \text{full} \middle| S_t = \text{full}\right] = 1.$$

Its state transition graph is shown in Fig. 2.1.

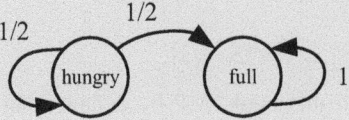

Fig. 2.1   State transition graph of the example.

If a fully-observable discrete-time agent–environment interface satisfies that, at time t, the probability to get reward $R_{t+1} = r$ and reach state $S_{t+1} = s'$ from state $S_t = s$ and action $A_t = a$, is

$$\Pr\left[S_{t+1} = s', R_{t+1} = r \middle| S_t = s, A_t = a\right],$$

the discrete-time agent–environment interface becomes a **Discrete-Time Markov Decision Process** (**DTMDP**). Remarkably, this definition assumes that the reward R_{t+1} and the next state S_{t+1} only depend on the current state S_t and the current action A_t, and are independent of earlier states and actions. That is what Markovian means. Markovian property is a very important feature of MDP, which requires that the state should include all historical information that can impact the future.

> ! **Note**
>
> Adding similar constraints on continuous-time agent–environment interface can lead to Continuous-Time MDP, which will be introduced in Sect. 15.2.

This book will focus on homogenous MDP, where the probability from state $S_t = s$ and action $A_t = a$ to reward $R_{t+1} = r$ and next state $S_{t+1} = s'$, i.e. $\Pr[S_{t+1} = s', R_{t+1} = r | S_t = s, A_t = a]$, does not change over time t. We can write it as

$$\Pr\left[S_{t+1} = s', R_{t+1} = r | S_t = s, A_t = a\right] = \Pr\left[S_1 = s', R_1 = r | S_0 = s, A_0 = a\right],$$
$$t \in \mathcal{T}, s \in \mathcal{S}, a \in \mathcal{A}, r \in \mathcal{R}, s' \in \mathcal{S}^+.$$

> ! **Note**
>
> MDP can be either homogenous or non-homogenous. This book focuses on homogenous MDP. Section 15.3 will discuss non-homogenous MDP.

In a DTMDP, the first reward R_0 is out of the control of the agent. Additionally, after the environment provides the first state S_0, all following stuffs will be independent of R_0. Therefore, we can exclude R_0 from the system. Therefore, a trajectory of DTMDP can be represented as

$$S_0, A_0, R_1, S_1, A_1, R_2, S_2, A_2, R_3, \ldots, S_T = s_{\text{end}}.$$

Actions and rewards are the important features of an MDP. If we delete all actions in a trajectory of an MDP, the trajectory becomes a trajectory of **Markov Reward Process** (**MRP**). Furthermore, if we delete all rewards in a trajectory of an MRP, the trajectory becomes a trajectory of MP. Figure 2.2 compares the trajectory of Discrete-Time MP (DTMP), Discrete-Time MRP (DTMRP), and DTMDP.

A DTMDP is a **Finite MDP** if and only if its state space $\mathcal{S}$, the action space $\mathcal{A}$, and the reward space $\mathcal{R}$ are all finite.

Example 2.2 (Feed and Full) This is an example of finite MDP. Its state space is $\mathcal{S} = \{\text{hungry, full}\}$, and its action space is $\mathcal{A} = \{\text{ignore, feed}\}$. And its reward space is $\mathcal{R} = \{-3, -2, -1, +1, +2, +3\}$. All these sets are finite, so this MDP is a finite MDP.

$$
\begin{aligned}
&\text{DTMP}: && S_0, && S_1, && S_2, && \dots \\
&\text{DTMRP}: && S_0, && R_1,S_1, && R_2,S_2, && R_3,\dots \\
&\text{DTMDP}: && S_0, A_0,R_1,S_1, A_1,R_2,S_2, A_2,R_3,\dots
\end{aligned}
$$

Fig. 2.2   Compare trajectories of DTMP, DTMRP, and DTMDP.

2.1.2 Environment and Dynamics

This section introduces the mathematical formulation of the environment in DTMDP.

The environment in a DTMDP can be characterized by initial state distribution and dynamics.

- **Initial state distribution** is denoted by p_{S_0}:

$$
p_{S_0}(s) \stackrel{\text{def}}{=} \Pr[S_0 = s], \quad s \in S.
$$

- **Dynamics**, also known as **transition probability**, is denoted by the function p:

$$
p(s', r \mid s, a) \stackrel{\text{def}}{=} \Pr\left[S_{t+1} = s', R_{t+1} = r \mid S_t = s, A_t = a\right],
$$
$$
s \in S, a \in \mathcal{A}(s), r \in \mathcal{R}, s' \in S^+,
$$

where the vertical bar "|" is adopted from the similar representation of conditional probability.

! Note

For simplicity, this book uses Pr [ ] to denote either probability or probability density. Please interpret the meaning of Pr [ ] according to the contexts. For example, in the definition of initial state distribution, if the state space S is a countable set, Pr [ ] should be interpreted as a probability. If the state space is a nonempty continuous subset of real numbers, Pr [ ] should be interpreted as a probability density. If you had difficulty in distinguish them, you may focus on the case of finite MDP, where the state space, action space, and reward space are all finite sets, so Pr [ ] is always a probability.

Example 2.3 (Feed and Full) We can further consider the example "Feed and Full" in the previous subsection. We can further define the initial state distribution as Table 2.1 and the dynamics as Table 2.2. The state transition graph Fig. 2.3 visualizes the dynamics.

We can derive other characteristics from the dynamics of the model, including:

Table 2.1   Example initial state distribution in the task "Feed and Full".

s	$p_{S_0}(s)$
hungry	1/2
full	1/2

Table 2.2   Example dynamics in the task "Feed and Full".

s	a	r	s'	$p(s',r\|s,a)$
hungry	ignore	−2	hungry	1
hungry	feed	−3	hungry	1/3
hungry	feed	+1	full	2/3
full	ignore	−2	hungry	3/4
full	ignore	+2	full	1/4
full	feed	+1	full	1
others				0

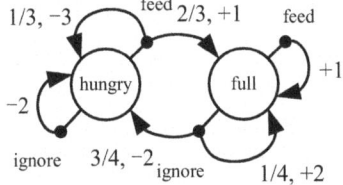

Fig. 2.3   State transition graph of the example "Feed and Full".

- transition probability

$$p(s'|s,a) \stackrel{\text{def}}{=} \Pr\left[S_{t+1} = s' | S_t = s, A_t = a\right], \quad s \in \mathcal{S}, a \in \mathcal{A}(s), s' \in \mathcal{S}^+.$$

We can prove that

$$p(s'|s,a) = \sum_{r \in \mathcal{R}} p(s',r|s,a), \quad s \in \mathcal{S}, a \in \mathcal{A}(s), s' \in \mathcal{S}^+.$$

- expected reward given the state–action pair

$$r(s,a) \stackrel{\text{def}}{=} \mathrm{E}[R_{t+1}|S_t = s, A_t = a], \quad s \in \mathcal{S}, a \in \mathcal{A}(s).$$

We can prove that

$$r(s,a) = \sum_{s \in \mathcal{S}^+, r \in \mathcal{R}} r p(s',r|s,a), \quad s \in \mathcal{S}, a \in \mathcal{A}(s).$$

- expected reward given state, action, and next state

$$r(s, a, s') \stackrel{\text{def}}{=} \mathrm{E}\left[R_{t+1} \middle| S_t = s, A_t = a, S_{t+1} = s'\right], \quad s \in S, a \in \mathcal{A}(s), s' \in S^+.$$

We can prove that

$$r(s, a, s') = \frac{\sum_{r \in \mathcal{R}} r\, p(s', r|s, a)}{p(s'|s, a)}, \quad s \in S, a \in \mathcal{A}(s), s' \in S^+.$$

> **! Note**
>
> In this book, the same letter can be used to denote different functions with different parameters. For example, the letter p in the form $p(s', r|s, a)$ and that in the form $p(s'|s, a)$ denote different conditional probability.

> **! Note**
>
> Summation may not converge when it is over infinite space, and the expectation may not exist when the random variable is over the infinite space. For the sake of simplicity, this book always assumes that the summation can converge and the expectation exists, unless specified otherwise. Summation over elements in a continuous space is essentially integration.

Example 2.4 (Feed and Full) We can further consider the example "Feed and Full" in the previous subsection. Based on the dynamics in Table 2.2, we can calculate the transition probability from state–action pair to the next state (Table 2.3), expected reward given a state–action pair (Table 2.4), and expected reward from a state–action pair to the next state (Table 2.5). Some calculation steps are shown as follows:

$$p(\text{hungry}|\text{hungry}, \text{ignore}) = \sum_{r \in \mathcal{R}} p(\text{hungry}, r|\text{hungry}, \text{ignore})$$
$$= p(\text{hungry}, -2|\text{hungry}, \text{ignore})$$
$$= 1$$
$$p(\text{hungry}|\text{hungry}, \text{feed}) = \sum_{r \in \mathcal{R}} p(\text{hungry}, r|\text{hungry}, \text{feed})$$
$$= p(\text{hungry}, -3|\text{hungry}, \text{feed})$$
$$= 1/3$$

$$r(\text{hungry}, \text{ignore}) = \sum_{s' \in S, r \in \mathcal{R}} r\, p(s', r|\text{hungry}, \text{ignore})$$

$$= -2p(\text{hungry}, -2|\text{hungry}, \text{ignore})$$
$$= -2$$
$$r(\text{hungry}, \text{feed}) = \sum_{s' \in S, r \in \mathcal{R}} r p(s', r|\text{hungry}, \text{feed})$$
$$= -3p(\text{hungry}, -3|\text{hungry}, \text{feed}) + 1 p(\text{hungry}, +1|\text{hungry}, \text{feed})$$
$$= -3 \cdot 1/3 + 1 \cdot 2/3$$
$$= -1/3$$

$$r(\text{hungry}, \text{ignore}, \text{hungry}) = \frac{\sum_{r \in \mathcal{R}} r p(\text{hungry}, r|\text{hungry}, \text{ignore})}{p(\text{hungry}|\text{hungry}, \text{ignore})}$$
$$= \frac{-2p(\text{hungry}, -2|\text{hungry}, \text{ignore})}{p(\text{hungry}|\text{hungry}, \text{ignore})}$$
$$= \frac{-2 \cdot 1}{1}$$
$$= -2$$
$$r(\text{hungry}, \text{feed}, \text{hungry}) = \frac{\sum_{r \in \mathcal{R}} r p(\text{hungry}, r|\text{hungry}, \text{feed})}{p(\text{hungry}|\text{hungry}, \text{feed})}$$
$$= \frac{-3p(\text{hungry}, -3|\text{hungry}, \text{ignore})}{p(\text{hungry}|\text{hungry}, \text{ignore})}$$
$$= \frac{-3 \cdot 1/3}{1/3}$$
$$= -3.$$

Table 2.3 Transition probability from state–action pair to the next state derived from Table 2.2.

s	a	s'	$p(s'|s, a)$
hungry	ignore	hungry	1
hungry	feed	hungry	1/3
hungry	feed	full	2/3
full	ignore	hungry	3/4
full	ignore	full	1/4
full	feed	full	1
	others		0

Table 2.4   Expected state–action reward derived from Table 2.2.

s	a	$r(s, a)$
hungry	ignore	−2
hungry	feed	−1/3
full	ignore	−1
full	feed	+1

Table 2.5   Expected reward from a state–action pair to the next state derived from Table 2.2.

s	a	s′	$r(s, a, s')$
hungry	ignore	hungry	−2
hungry	feed	hungry	−3
hungry	feed	full	+1
full	ignore	hungry	−2
full	ignore	full	+2
full	feed	full	+1
	others		N/A

2.1.3 Policy

Agents make decisions using a policy. The **policy** in MDP is formulated as the conditional probability from state to action, i.e.

$$\pi(a|s) \stackrel{\text{def}}{=} \Pr_{\pi}[A_t = a | S_t = s], \quad s \in \mathcal{S}, a \in \mathcal{A}.$$

! Note

The subscription in the probability operator $\Pr_{\pi}[\ ]$ means that the probability is over all trajectories whose actions are determined by policy π. From this sense, the subscription indicates the condition of probability. Similarly, we can define the expectation $E_{\pi}[\ ]$. The values of $\Pr_{\pi}[\ ]$ and $E_{\pi}[\ ]$ may be not only determined by the policy π, but also determined by the initial state distribution and dynamics of the environment. Since we usually consider a particular environment, we omit the initial state distribution and dynamics, and only write down the policy π explicitly. Besides, the policy π can also be omitted if the probability or expectation does not relate to the policy π.

Example 2.5 (Feed and Full) Continue the example. A possible policy is shown in Table 2.6.

Table 2.6   An example policy in the task "Feed and Full".

s	a	$\pi(a\|s)$
hungry	ignore	1/4
hungry	feed	3/4
full	ignore	5/6
full	feed	1/6

A policy is deterministic if, for every $s \in S$, there exists an action $a \in \mathcal{A}$ such that

$$\pi(a'|s) = 0, \quad a' \neq a.$$

In this case, the policy can be written as $\pi : S \to \mathcal{A}$, i.e. $\pi : s \mapsto \pi(s)$.

Example 2.6 (Feed and Full) Continue the example. Besides the stochastic policy in Table 2.6, the agent can also use the deterministic policy in Table 2.7. This deterministic policy can also be presented by Table 2.8.

Table 2.7   Another example policy in the task "Feed and Full".

s	a	$\pi(a\|s)$
hungry	ignore	0
hungry	feed	1
full	ignore	0
full	feed	1

Table 2.8   Alternative presentation of the deterministic policy in Table 2.7.

s	$\pi(s)$
hungry	feed
full	feed

We can further compute the following notations from the models of both environment and agent.

- Initial state–action distribution

$$p_{S_0, A_0, \pi}(s, a) \overset{\text{def}}{=} \Pr_\pi[S_0 = s, A_0 = a], \quad s \in S, a \in \mathcal{A}.$$

We can prove that

$$p_{S_0, A_0, \pi}(s, a) = p_{S_0}(s)\pi(a|s), \quad s \in S, a \in \mathcal{A}.$$

- Transition probability from state to next state

$$p_\pi(s'|s) \overset{\text{def}}{=} \Pr_\pi[S_{t+1} = s'|S_t = s], \quad s \in \mathcal{S}, s' \in \mathcal{S}^+.$$

We can prove that

$$p_\pi(s'|s) = \sum_a p(s'|s, a)\pi(a|s), \quad s \in \mathcal{S}, s' \in \mathcal{S}^+.$$

- Transition probability from state–action pair to next state–action pair

$$p_\pi(s', a'|s, a) \overset{\text{def}}{=} \Pr_\pi[S_{t+1} = s', A_{t+1} = a'|S_t = s, A_t = a]$$
$$s \in \mathcal{S}, a \in \mathcal{A}(s), s' \in \mathcal{S}^+, a' \in \mathcal{A}.$$

We can prove that

$$p_\pi(s', a'|s, a) = \pi(a'|s')p(s'|s, a), \quad s \in \mathcal{S}, a \in \mathcal{A}(s), s' \in \mathcal{S}^+, a' \in \mathcal{A}.$$

- Expected state reward

$$r_\pi(s) \overset{\text{def}}{=} \mathrm{E}_\pi[R_{t+1}|S_t = s], \quad s \in \mathcal{S}.$$

We can prove that

$$r_\pi(s) = \sum_a r(s, a)\pi(a|s), \quad s \in \mathcal{S}.$$

Example 2.7 (Feed and Full) Continue the example. From the initial state distribution in Table 2.1 and the policy in Table 2.6, we can calculate the initial state–action distribution (shown in Table 2.9). From the dynamics in Table 2.2 and the policy in Table 2.6, we can calculate transition probability from a state to the next state (shown in Table 2.10), transition probability from a state–action pair to the next state–action pair (shown in Table 2.11), and expected state reward (shown in Table 2.12).

Table 2.9   Initial state–action distribution derived from Tables 2.1 and 2.6.

s	a	$p_{S_0, A_0, \pi}(s, a)$
hungry	ignore	1/8
hungry	feed	3/8
full	ignore	5/12
full	feed	1/12

Table 2.10   Transition probability from a state to the next state derived from Tables 2.2 and 2.6.

s	s'	$p_\pi(s'|s)$
hungry	hungry	1/2
hungry	full	1/2
full	hungry	5/8
full	full	3/8

Table 2.11   Transition probability from a state–action pair to the next state–action pair derived from Tables 2.2 and 2.6.

s	a	s'	a'	$p_\pi(s', a'|s, a)$
hungry	ignore	hungry	ignore	1/4
hungry	ignore	hungry	feed	3/4
hungry	feed	hungry	ignore	1/12
hungry	feed	hungry	feed	1/4
hungry	feed	full	ignore	5/9
hungry	feed	full	feed	1/9
full	ignore	hungry	ignore	3/16
full	ignore	hungry	feed	9/16
full	ignore	full	ignore	5/24
full	ignore	full	feed	1/24
full	feed	full	ignore	5/6
full	feed	full	feed	1/6
others				0

Table 2.12   Expected state reward derived from Tables 2.2 and 2.6.

s	$r_\pi(s)$
hungry	$-3/4$
full	$-2/3$

2.1.4 Discounted Return

Chapter 1 told us that the reward is a core element in RL. The goal of RL is to maximize the total reward. This section formulates it mathematically.

Let us consider episodic tasks first. Let T denote the number of steps in an episode task. The **return** started from time t, denoted as G_t, can be defined as

$$G_t \overset{\text{def}}{=} R_{t+1} + R_{t+2} + \cdots + R_T, \quad \text{episodic task, without discount.}$$

The number of steps T can be a random variable. Therefore, in the definition of G_t, not only every term is random, but also the number of terms is random.

For sequential tasks, we also want to define their returns so that they can include all rewards after time t. However, sequential tasks have no ending time, so the aforementioned definition may not converge. In order to solve this problem, a concept called discount is introduced, and then discounted return is defined as

$$G_t \overset{\text{def}}{=} R_{t+1} + \gamma R_{t+2} + \gamma^2 R_{t+3} + \cdots = \sum_{\tau=0}^{+\infty} \gamma^\tau R_{t+\tau+1},$$

where $\gamma \in [0, 1]$ is called **discount factor**. The discount factor trades off between current rewards and future returns: Every unit of rewards after τ steps is equivalent to the γ^τ current rewards. In an extreme case, $\gamma = 0$ means the agent is myopic and ignores future rewards at all, and $\gamma = 1$ means that every future reward is as important as the current reward. Sequential tasks usually set $\gamma \in (0, 1)$. In this case, if the reward for every step is bounded, the return is bounded.

Now we have defined returns for both episodic tasks and sequential tasks. In fact, we can unify them to

$$G_t \overset{\text{def}}{=} \sum_{\tau=0}^{+\infty} \gamma^\tau R_{t+\tau+1}.$$

For episodic tasks, let $R_t = 0$ for $t > T$. In fact, the discount factor of an episodic task can also be < 1. Consequently, discount factor of an episodic task is usually $\gamma \in (0, 1]$, while discount factor of a sequential task is usually $\gamma \in (0, 1)$. We will use this unified notation throughout the book.

The performance of MDP can be other than metrics than the expectation of discounted return. For example, the average reward MDP in Sect. 15.1 defines average reward as the performance metrics of MDP.

Discounted return has the following recursive relationship:

$$G_t = R_{t+1} + \gamma G_{t+1}.$$

(Proof: This is due to $G_t = \sum_{\tau=0}^{+\infty} \gamma^\tau R_{t+\tau+1} = R_{t+1} + \sum_{\tau=1}^{+\infty} \gamma^\tau R_{t+\tau+1} = R_{t+1} + \sum_{\tau=0}^{+\infty} \gamma^\tau R_{(t+1)+\tau+1} = R_{t+1} + \gamma G_{t+1}.$)

This discounted return of the whole trajectory is G_0. Its expectation is called initial expected return, i.e.

$$g_\pi \overset{\text{def}}{=} \mathrm{E}_\pi[G_0].$$

The agent always wants to maximize g_π by choosing a smart policy. In order to emphasize the intention to maximize the expectation of discounted return, such MDP is sometimes called discounted MDP.

2.2 Value

Based on the definition of return, we can further define the concept of value, which is a very important concept in RL. This section will introduce the definition and properties of values.

2.2.1 Definition of Value

Fixed the dynamics of MDP, we can define **value** of a policy π in the following ways:

- **state value** (a.k.a. discounted state value) $v_\pi(s)$ is the expected return starting from the state s if all subsequent actions are decided by the policy π:

$$v_\pi(s) \stackrel{\text{def}}{=} \mathrm{E}_\pi[G_t|S_t = s], \quad s \in \mathcal{S}.$$

- **action value** (a.k.a. discounted action value) $q_\pi(s, a)$ is the expected return starting from the state–action pair (s, a) if all subsequent actions are decided by the policy π:

$$q_\pi(s, a) \stackrel{\text{def}}{=} \mathrm{E}_\pi[G_t|S_t = s, A_t = a], \quad s \in \mathcal{S}, a \in \mathcal{A}(s).$$

The terminal state s_{end} is not an ordinary state, and it has no subsequent actions. Therefore, we define $v_\pi(s_{\text{end}}) = 0$ and $q_\pi(s_{\text{end}}, a) = 0$ $(a \in \mathcal{A})$ for the sake of consistency.

Example 2.8 (Feed and Full) For the example in Tables 2.2 and 2.6, we have

$$
\begin{aligned}
v_\pi(\text{hungry}) &= \mathrm{E}_\pi\big[G_t|S_t = \text{hungry}\big] \\
v_\pi(\text{full}) &= \mathrm{E}_\pi[G_t|S_t = \text{full}] \\
q_\pi(\text{hungry}, \text{feed}) &= \mathrm{E}_\pi\big[G_t|S_t = \text{hungry}, A_t = \text{feed}\big] \\
q_\pi(\text{hungry}, \text{ignore}) &= \mathrm{E}_\pi\big[G_t|S_t = \text{hungry}, A_t = \text{ignore}\big] \\
q_\pi(\text{full}, \text{feed}) &= \mathrm{E}_\pi[G_t|S_t = \text{full}, A_t = \text{feed}] \\
q_\pi(\text{full}, \text{ignore}) &= \mathrm{E}_\pi\big[G_t|S_t = \text{full}, A_t = \text{ignore}\big].
\end{aligned}
$$

It is important to know the value of a policy. Calculating the values of a policy is called **policy evaluation**. RL researchers have spent lots of effects to find out good ways to evaluate a policy. In the sequential sections, we will learn some properties of values, which are helpful for the policy evaluation.

2.2.2 Properties of Value

This section introduces some relationships between values, including the famous Bellman expectation equations. These properties are useful for evaluating a policy.

Firstly, let us consider the relationship between state values and action values. State values can back up action values, and action values can back up state values.

- Use action values to back up state values:

$$v_\pi(s) = \sum_{a \in \mathcal{A}} \pi(a|s) q_\pi(s, a), \quad s \in \mathcal{S}.$$

(Proof: For an arbitrary state $s \in \mathcal{S}$, we have

$$
\begin{aligned}
v_\pi(s) &= \mathrm{E}_\pi[G_t | S_t = s] \\
&= \sum_{g \in \mathbb{R}} g \mathrm{Pr}_\pi[G_t = g | S_t = s] \\
&= \sum_{g \in \mathbb{R}} g \sum_{a \in \mathcal{A}} \mathrm{Pr}_\pi[G_t = g, A_t = a | S_t = s] \\
&= \sum_{g \in \mathbb{R}} g \sum_{a \in \mathcal{A}} \mathrm{Pr}_\pi[A_t = a | S_t = s] \mathrm{Pr}_\pi[G_t = g | S_t = s, A_t = a] \\
&= \sum_{a \in \mathcal{A}} \mathrm{Pr}_\pi[A_t = a | S_t = s] \sum_{g \in \mathbb{R}} g \mathrm{Pr}_\pi[G_t = g | S_t = s, A_t = a] \\
&= \sum_{a \in \mathcal{A}} \mathrm{Pr}_\pi[A_t = a | S_t = s] \mathrm{E}_\pi[G_t | S_t = s, A_t = a] \\
&= \sum_{a \in \mathcal{A}} \pi(a|s) q_\pi(s, a).
\end{aligned}
$$

The proof completes.) In this proof, the left-hand side of the equation is actually the state value at time t, and the right-hand side of the equation involves in the action values at time t. Therefore, this relationship uses action values at time t to back up state values at time t, and can be written as

$$v_\pi(S_t) = \mathrm{E}_\pi\big[q_\pi(S_t, A_t)\big].$$

This relationship can be illustrated by the **backup diagram** in Fig. 2.4(a), where hollow circles present states, and solid circles represent the state–action pairs.

- Use the state values to back up action values:

$$q_\pi(s, a) = r(s, a) + \gamma \sum_{s' \in \mathcal{S}} p(s'|s, a) v_\pi(s')$$

$$= \sum_{s' \in \mathcal{S}^+, r \in \mathcal{R}} p(s', r|s, a) [r + \gamma v_\pi(s')], \quad s \in \mathcal{S}, a \in \mathcal{A}.$$

(Proof: For an arbitrary state $s \in \mathcal{S}$ and action $a \in \mathcal{A}$, we have

$$E_\pi[G_t|S_t = s, A_t = a]$$

$$= \sum_{g \in \mathbb{R}} g \Pr_\pi[G_{t+1} = g|S_t = s, A_t = a]$$

$$= \sum_{g \in \mathbb{R}} g \sum_{s' \in \mathcal{S}^+} \Pr_\pi[S_{t+1} = s', G_{t+1} = g|S_t = s, A_t = a]$$

$$= \sum_{g \in \mathbb{R}} g \sum_{s' \in \mathcal{S}^+} \Pr_\pi[S_{t+1} = s'|S_t = s, A_t = a] \Pr_\pi[G_{t+1} = g|S_t = s, A_t = a, S_{t+1} = s']$$

$$= \sum_{g \in \mathbb{R}} g \sum_{s' \in \mathcal{S}^+} \Pr_\pi[S_{t+1} = s'|S_t = s, A_t = a] \Pr_\pi[G_{t+1} = g|S_{t+1} = s']$$

$$= \sum_{s' \in \mathcal{S}^+} \Pr[S_{t+1} = s'|S_t = s, A_t = a] \sum_{g \in \mathbb{R}} g \Pr_\pi[G_{t+1} = g|S_{t+1} = s']$$

$$= \sum_{s' \in \mathcal{S}^+} \Pr[S_{t+1} = s'|S_t = s, A_t = a] E_\pi[G_{t+1} = g|S_{t+1} = s']$$

$$= \sum_{s' \in \mathcal{S}^+} p(s'|s, a) v_\pi(s'),$$

where $\Pr_\pi[G_{t+1} = g|S_t = s, A_t = a, S_{t+1} = s'] = \Pr_\pi[G_{t+1} = g|S_{t+1} = s']$ uses the Markovian property of MDP. Therefore, we have

$$q_\pi(s, a) = E_\pi[G_t|S_t = s, A_t = a]$$

$$= E_\pi[R_{t+1} + \gamma G_{t+1}|S_t = s, A_t = a]$$

$$= E_\pi[R_{t+1}|S_t = s, A_t = a] + \gamma E_\pi[G_{t+1}|S_t = s, A_t = a]$$

$$= \sum_{s' \in \mathcal{S}^+, r \in \mathcal{R}} p(s', r|s, a) [r + \gamma v_\pi(s')].$$

The proof completes.) In this proof, the left-hand side of the equation is actually the action value at time t, and the right-hand side of the equation involves in the state values at time $t + 1$. Therefore, this relationship uses state values at time $t + 1$ to back up action values at time t, and can be written as

$$q_\pi(S_t, A_t) = E[R_{t+1} + \gamma v_\pi(S_{t+1})].$$

This relationship also has a vector form:

$$\mathbf{q}_\pi = \mathbf{r} + \gamma \mathbf{P}_{S_{t+1}|S_t, A_t}^\top \mathbf{v}_\pi,$$

where the column vector $\mathbf{v}_\pi = \left(v_\pi(s) : s \in \mathcal{S}\right)^\top$ has $|\mathcal{S}|$ elements, the column vector $\mathbf{q}_\pi = \left(q_\pi(s, a) : (s, a) \in \mathcal{S} \times \mathcal{A}\right)^\top$ has $|\mathcal{S}\|\mathcal{A}|$ elements, the column vector $\mathbf{r} = \left(r(s, a) : (s, a) \in \mathcal{S} \times \mathcal{A}\right)^\top$ has $|\mathcal{S}\|\mathcal{A}|$ elements, and the single-step transition matrix $\mathbf{P}_{S_{t+1}|S_t, A_t} = \left(p(s'|s, a) : s' \in \mathcal{S}, (s, a) \in \mathcal{S} \times \mathcal{A}\right)$ is a matrix of size $|\mathcal{S}| \times |\mathcal{S}\|\mathcal{A}|$, and its transpose $\mathbf{P}^\top_{S_{t+1}|S_t, A_t} = \left(p(s'|s, a) : (s, a) \in \mathcal{S} \times \mathcal{A}, s' \in \mathcal{S}\right)$ is a matrix of size $|\mathcal{S}\|\mathcal{A}| \times |\mathcal{S}|$. This relationship can be illustrated by the backup diagram Fig. 2.4(b).

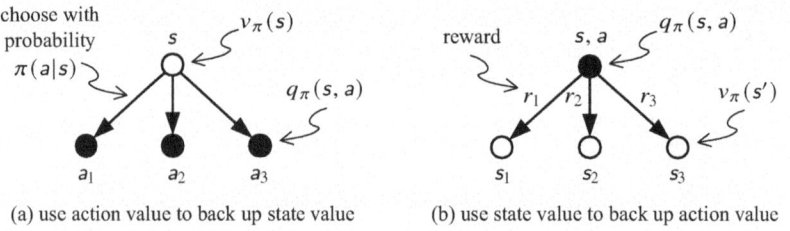

(a) use action value to back up state value   (b) use state value to back up action value

Fig. 2.4   Backup diagram that state values and action values represent each other.

> **! Note**
>
> Although the aforementioned rewritten, such as from $v_\pi(s) = \sum_a \pi(a|s) q_\pi(s, a)$ $(s \in \mathcal{S})$ to $v_\pi(S_t) = \mathrm{E}_\pi\left[q_\pi(S_t, A_t)\right]$, is not exactly the same if we investigate the notation rigidly, it still makes sense since the former equation can lead to the later equation. If $v_\pi(s) = \sum_a \pi(a|s) q_\pi(s, a)$ holds for all $s \in \mathcal{S}$, it also holds for $S_t \in \mathcal{S}$. Therefore, we can write $v_\pi(S_t) = \mathrm{E}_\pi\left[q_\pi(S_t, A_t)\right]$.

> **! Note**
>
> The vectors and transition matrix use the conventional of column vectors. The transition matrix is transposed here, because the transition matrix is defined for the transition from time t to time $t + 1$, but here we use the values at time $t + 1$ to backup values at time t.

> **! Note**
>
> This section only considers homogenous MDP, whose values do not vary with time. Therefore, the state values at time t are identical to the state values at time $t + 1$, and the action values at time t are identical to the action values at time $t + 1$.

Applying elimination on the above two relationships together, we can get **Bellman Expectation Equations** (Bellman, 1957), which has the following two forms.

- Use state values at time $t + 1$ to back up the state values at time t:

$$v_\pi(s) = r_\pi(s) + \gamma \sum_{s' \in S} p_\pi(s'|s) v_\pi(s'), \quad s \in S.$$

It can be rewritten as

$$v_\pi(S_t) = E_\pi\left[R_{t+1} + \gamma v_\pi(S_{t+1})\right].$$

It has vector form:

$$\mathbf{v}_\pi = \mathbf{r}_\pi + \gamma \mathbf{P}^\top_{S_{t+1}|S_t;\pi} \mathbf{v}_\pi,$$

where the column vector $\mathbf{v}_\pi = \left(v_\pi(s) : s \in S\right)^\top$ has $|S|$ elements, the column vector $\mathbf{r}_\pi = \left(r_\pi(s) : s \in S\right)^\top$ has $|S|$ elements, and transpose of the single-step transition matrix $\mathbf{P}^\top_{S_{t+1}|S_t;\pi}$ is an $|S| \times |S|$ matrix. The backup diagram is illustrated in Fig. 2.5(a).

- Use the action values at time $t + 1$ to represent the action values at time t:

$$q_\pi(s, a)$$
$$= r(s, a) + \gamma \sum_{s' \in S, a' \in \mathcal{A}} p_\pi(s', a'|s, a) q_\pi(s', a')$$
$$= \sum_{s' \in S, r \in \mathcal{R}} p(s', r|s, a)\left[r + \gamma \sum_{a' \in \mathcal{A}} \pi(a'|s') q_\pi(s', a')\right], \quad s \in S, a \in \mathcal{A}(s).$$

It can be rewritten as

$$q_\pi(S_t, A_t) = E_\pi\left[R_{t+1} + \gamma q_\pi(S_{t+1}, A_{t+1})\right].$$

Its vector form is

$$\mathbf{q}_\pi = \mathbf{r} + \gamma \mathbf{P}^\top_{S_{t+1}, A_{t+1}|S_t, A_t;\pi} \mathbf{q}_\pi,$$

where the column vector $\mathbf{q}_\pi = \left(q_\pi(s, a) : (s, a) \in S \times \mathcal{A}\right)^\top$ has $|S||\mathcal{A}|$ elements, the column vector $\mathbf{r} = \left(r(s, a) : (s, a) \in S \times \mathcal{A}\right)^\top$ has $|S||\mathcal{A}|$ elements, and transpose of the single-step transition matrix $\mathbf{P}^\top_{S_{t+1}, A_{t+1}|S_t, A_t;\pi}$ is an $|S||\mathcal{A}| \times |S||\mathcal{A}|$ matrix. The backup diagram is illustrated in Fig. 2.5(b).

! Note

Most part of the book will not use the vector representation. Sometimes we will still use it when it can make things much simpler. Ideally, we need to understand this representation.

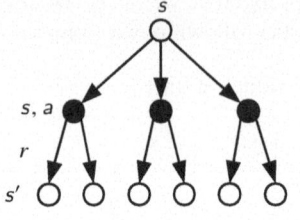

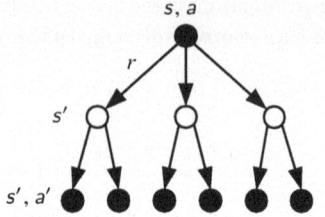

(a) use state value to back up state value   (b) use action value to back up action value

Fig. 2.5   State values and action values back up themselves.

2.2.3 Calculate Value

This section introduces the method to calculate values given the mathematical formulation of environment and policy. We will also prove that values do not depend on the initial state distribution.

For simplicity, this subsection only considers finite MDP.

In the previous section, we have known the relationship among values of a policy. We can use those relationships to get an equation system to calculate the values. The vector representations of Bellman expectation equations have the form $\mathbf{x}_\pi = \mathbf{r}_\pi + \gamma \mathbf{P}_\pi^\top \mathbf{x}_\pi$ for both state values and action values. Since $\mathbf{I} - \gamma \mathbf{P}_\pi^\top$ is usually invertible, we can solve the values using $\mathbf{x}_\pi = \left(\mathbf{I} - \gamma \mathbf{P}_\pi^\top\right)^{-1} \mathbf{r}_\pi$.

Approach 1: First plug the expected reward $r_\pi(s)$ $(s \in \mathcal{S})$, transition probability $p_\pi(s'|s)$ $(s, s' \in \mathcal{S})$ and discount factor γ into the state-value Bellman expectation equation and get the state value $v_\pi(s)$ $(s \in \mathcal{S})$. Then use the obtained state value $v_\pi(s)$ $(s \in \mathcal{S})$ and expected reward $r(s, a)$ $(s \in \mathcal{S}, a \in \mathcal{A})$, transition probability $p(s'|s, a)$ $(s \in \mathcal{S}, a \in \mathcal{A}, s' \in \mathcal{S})$, and discount factor γ to calculate action value $q_\pi(s, a)$ $(s \in \mathcal{S}, a \in \mathcal{A})$. This approach can be expressed in the vector form as follows: First apply $\mathbf{r}_\pi = \left(r_\pi(s) : s \in \mathcal{S}\right)^\top$ and $\mathbf{P}_{S_{t+1}|S_t;\pi}^\top = \left(p_\pi(s'|s) : s \in \mathcal{S}, s' \in \mathcal{S}\right)$ to $\mathbf{v}_\pi = \left(\mathbf{I} - \gamma \mathbf{P}_{S_{t+1}|S_t;\pi}^\top\right)^{-1} \mathbf{r}_\pi$ and obtain state value vector $\mathbf{v}_\pi = \left(v_\pi(s) : s \in \mathcal{S}\right)^\top$. Then apply $\mathbf{r} = \left(r(s, a) : (s, a) \in \mathcal{S} \times \mathcal{A}\right)^\top$, $\mathbf{P}_{S_{t+1}|S_t,A_t}^\top = \left(p(s'|s, a) : (s, a) \in \mathcal{S} \times \mathcal{A}, s' \in \mathcal{S}\right)$, and the state value vector to $\mathbf{q}_\pi = \mathbf{r} + \gamma \mathbf{P}_{S_{t+1}|S_t,A_t}^\top \mathbf{v}_\pi$ and obtain the action value vector $\mathbf{q}_\pi = \left(q_\pi(s, a) : (s, a) \in \mathcal{S} \times \mathcal{A}\right)^\top$.

Approach 2: First plug the expected reward $r(s, a)$ $(s \in \mathcal{S}, a \in \mathcal{A})$, transition probability $p_\pi(s', a'|s, a)$ $(s \in \mathcal{S}, a \in \mathcal{A}, s' \in \mathcal{S}, a' \in \mathcal{A})$, and discount factor γ to action-value Bellman expectation equation and get the action value $q_\pi(s, a)$ $(s \in \mathcal{S}, a \in \mathcal{A})$. Then use the obtained action value $q_\pi(s, a)$ $(s \in \mathcal{S})$ and policy $\pi(a|s)$ $(s \in \mathcal{S}, a \in \mathcal{A})$ to calculate state value $v_\pi(s)$ $(s \in \mathcal{S})$. This approach can be expressed in vector form as follows: First apply $\mathbf{r} = \left(r(s, a) : (s, a) \in \mathcal{S} \times \mathcal{A}\right)^\top$ and $\mathbf{P}_{S_{t+1},A_{t+1}|S_t,A_t;\pi}^\top = \left(p_\pi(s', a'|s, a) : (s, a) \in \mathcal{S} \times \mathcal{A}, (s', a') \in \mathcal{S} \times \mathcal{A}\right)$ to

$\mathbf{q}_\pi = \left(\mathbf{I} - \gamma \mathbf{P}^\top_{S_{t+1},A_{t+1}|S_t,A_t;\pi}\right)^{-1} \mathbf{r}$ and obtain the action value vector $\mathbf{q}_\pi = \left(q_\pi(s, a) : (s, a) \in \mathcal{S} \times \mathcal{A}\right)^\top$. Then apply $\pi(a|s)$ $(s \in \mathcal{S}, a \in \mathcal{A})$ and action values to $v_\pi(s) = \sum_a \pi(a|s)q_\pi(s, a)$ $(s \in \mathcal{S})$ and obtain the state values $v_\pi(s)$ $(s \in \mathcal{S})$.

There are other approaches as well. For example, we can plug the policy $\pi(a|s)$ $(s \in \mathcal{S}, a \in \mathcal{A})$, expected reward $r(s, a)$ $(s \in \mathcal{S}.a \in \mathcal{A})$, transition probability $p(s'|s, a)$ $(s \in \mathcal{S}, a \in \mathcal{A}, s' \in \mathcal{S})$ and discount factor γ all together to the relationship that state-value and action-value representing each other, and calculate both state value $v_\pi(s)$ $(s \in \mathcal{S})$ and action value $q_\pi(s, a)$ $(s \in \mathcal{S}, a \in \mathcal{A})$ from a linear equation system.

Example 2.9 (Feed and Full) In the example of Tables 2.2 and 2.6, set discount factor $\gamma = \frac{4}{5}$. We can get the values using the three approaches.

Approach 1: Plug expected rewards in Table 2.12, transition probabilities in Table 2.10, and discount factor into state-value Bellman expectation equations, we will have

$$v_\pi(\text{hungry}) = -\frac{3}{4} + \frac{4}{5}\left(\frac{1}{2}v_\pi(\text{hungry}) + \frac{15}{24}v_\pi(\text{full})\right)$$

$$v_\pi(\text{full}) = -\frac{2}{3} + \frac{4}{5}\left(\frac{1}{2}v_\pi(\text{hungry}) + \frac{9}{24}v_\pi(\text{full})\right).$$

Solving this equation set leads to state values

$$v_\pi(\text{hungry}) = -\frac{475}{132}, \quad v_\pi(\text{full}) = -\frac{155}{44}.$$

Plug in these state values, expected rewards in Table 2.4, transition probabilities in Table 2.3, and discount factor to the relationship that uses state values to back up action values, and we will get

$$q_\pi(\text{hungry, ignore}) = -2 + \frac{4}{5}\left(1\left(-\frac{475}{132}\right) + 0\left(-\frac{155}{44}\right)\right) = -\frac{161}{33},$$

$$q_\pi(\text{hungry, feed}) = -\frac{1}{3} + \frac{4}{5}\left(\frac{1}{3}\left(-\frac{475}{132}\right) + \frac{2}{3}\left(-\frac{155}{44}\right)\right) = -\frac{314}{99},$$

$$q_\pi(\text{full, ignore}) = -1 + \frac{4}{5}\left(\frac{3}{4}\left(-\frac{475}{132}\right) + \frac{1}{4}\left(-\frac{155}{44}\right)\right) = -\frac{85}{22},$$

$$q_\pi(\text{full, feed}) = +1 + \frac{4}{5}\left(0\left(-\frac{475}{132}\right) + 1\left(-\frac{155}{44}\right)\right) = -\frac{20}{11}.$$

The calculation steps can be written in vector forms as follows: The reward vector is

$$\mathbf{r}_\pi = \begin{bmatrix} r_\pi(\text{hungry}) \\ r_\pi(\text{full}) \end{bmatrix} = \begin{bmatrix} -3/4 \\ -2/3 \end{bmatrix}.$$

The transpose of transition matrix is

$$\mathbf{P}^\top_{S_{t+1}|S_t;\pi} = \begin{bmatrix} p_\pi(\text{hungry}|\text{hungry}) & p_\pi(\text{full}|\text{hungry}) \\ p_\pi(\text{hungry}|\text{full}) & p_\pi(\text{full}|\text{full}) \end{bmatrix} = \begin{bmatrix} 1/2 & 1/2 \\ 5/8 & 3/8 \end{bmatrix}.$$

The state value vector is

$$\mathbf{v}_\pi = \left(\mathbf{I} - \gamma \mathbf{P}^\top_{S_{t+1}|S_t;\pi}\right)^{-1} \mathbf{r}_\pi$$

$$= \left(\begin{bmatrix} 1 & 0 \\ 0 & 1 \end{bmatrix} - \frac{4}{5} \begin{bmatrix} 1/2 & 1/2 \\ 5/8 & 3/8 \end{bmatrix}\right)^{-1} \begin{bmatrix} -3/4 \\ -2/3 \end{bmatrix} = \begin{bmatrix} -475/132 \\ -155/44 \end{bmatrix}.$$

The action value vector is

$$\mathbf{q}_\pi = \mathbf{r} + \gamma \mathbf{P}^\top_{S_{t+1}|S_t,A_t} \mathbf{v}_\pi = \begin{bmatrix} -2 \\ -1/3 \\ -1 \\ +1 \end{bmatrix} + \frac{4}{5} \begin{bmatrix} 1 & 0 \\ 1/3 & 2/3 \\ 3/4 & 1/4 \\ 0 & 1 \end{bmatrix} \begin{bmatrix} -475/132 \\ -155/44 \end{bmatrix} = \begin{bmatrix} -161/33 \\ -314/99 \\ -85/22 \\ -20/11 \end{bmatrix}.$$

Approach 2: Plug expected rewards in Table 2.4 and transition probabilities in Table 2.3 into the action-value Bellman expectation equation, we have

$$q_\pi(\text{hungry, ignore}) = -2 + \frac{4}{5}\left(\frac{1}{4}q_\pi(\text{hungry, ignore}) + \frac{1}{12}q_\pi(\text{hungry, feed})\right.$$

$$\left. + \frac{3}{16}q_\pi(\text{full, ignore}) + 0q_\pi(\text{full, feed})\right)$$

$$q_\pi(\text{hungry, feed}) = -\frac{1}{3} + \frac{4}{5}\left(\frac{3}{4}q_\pi(\text{hungry, ignore}) + \frac{1}{4}q_\pi(\text{hungry, feed})\right.$$

$$\left. + \frac{9}{16}q_\pi(\text{full, ignore}) + 0q_\pi(\text{full, feed})\right)$$

$$q_\pi(\text{full, ignore}) = -1 + \frac{4}{5}\left(0q_\pi(\text{hungry, ignore}) + \frac{5}{9}q_\pi(\text{hungry, feed})\right.$$

$$\left. + \frac{5}{24}q_\pi(\text{full, ignore}) + \frac{5}{6}q_\pi(\text{full, feed})\right)$$

$$q_\pi(\text{full, feed}) = +1 + \frac{4}{5}\left(0q_\pi(\text{hungry, ignore}) + \frac{1}{9}q_\pi(\text{hungry, feed})\right.$$

$$\left. + \frac{1}{24}q_\pi(\text{full, ignore}) + \frac{1}{6}q_\pi(\text{full, feed})\right).$$

Solving this equality set and we will get

$$q_\pi(\text{hungry, ignore}) = -\frac{161}{33}, \quad q_\pi(\text{hungry, feed}) = -\frac{314}{99}$$

$$q_\pi(\text{full, ignore}) = -\frac{85}{22}, \quad q_\pi(\text{full, feed}) = -\frac{20}{11}.$$

Plug in these action values and policy in Table 2.6 into the relationship that uses state values to backup action values, and we will get state values as

$$v_\pi(\text{hungry}) = \frac{1}{4}\left(-\frac{161}{33}\right) + \frac{3}{4}\left(-\frac{314}{99}\right) = -\frac{475}{132}$$

$$v_\pi(\text{full}) = \frac{5}{6}\left(-\frac{85}{22}\right) + \frac{1}{6}\left(-\frac{20}{11}\right) = -\frac{155}{44}.$$

The calculation steps can be written in the following vector form: The reward vector is

$$\mathbf{r} = \begin{bmatrix} r(\text{hungry, ignore}) \\ r(\text{hungry, feed}) \\ r(\text{full, ignore}) \\ r(\text{full, feed}) \end{bmatrix} = \begin{bmatrix} -2 \\ -1/3 \\ -1 \\ +1 \end{bmatrix}.$$

The transpose of transition matrix is

$$\mathbf{P}^\mathsf{T}_{S_{t+1}, A_{t+1} | S_t, A_t; \pi} = \begin{bmatrix} 1/4 & 3/4 & 0 & 0 \\ 1/12 & 1/4 & 5/9 & 1/9 \\ 3/16 & 9/16 & 5/24 & 1/24 \\ 0 & 0 & 5/6 & 1/6 \end{bmatrix}.$$

The action value vector is

$$\mathbf{q}_\pi = \left(\mathbf{I} - \gamma\mathbf{P}^\mathsf{T}_{S_{t+1}, A_{t+1} | S_t, A_t; \pi}\right)^{-1}\mathbf{r}$$

$$= \left(\begin{bmatrix} 1 & 0 & 0 & 0 \\ 0 & 1 & 0 & 0 \\ 0 & 0 & 1 & 0 \\ 0 & 0 & 0 & 1 \end{bmatrix} - \frac{4}{5}\begin{bmatrix} 1/4 & 3/4 & 0 & 0 \\ 1/12 & 1/4 & 5/9 & 1/9 \\ 3/16 & 9/16 & 5/24 & 1/24 \\ 0 & 0 & 5/6 & 1/6 \end{bmatrix}\right)^{-1}\begin{bmatrix} -2 \\ -1/3 \\ -1 \\ +1 \end{bmatrix}$$

$$= \begin{bmatrix} -161/33 \\ -314/99 \\ -85/22 \\ -20/11 \end{bmatrix}.$$

Approach 3: Plug policy in Table 2.6, expected rewards in Table 2.4, and transition probabilities in Table 2.3 into the relationship among state values and action values, and we will get

$$v_\pi(\text{hungry}) = \frac{1}{4}q_\pi(\text{hungry, ignore}) + \frac{3}{4}q_\pi(\text{hungry, feed})$$

$$v_\pi(\text{full}) \qquad = \frac{5}{6}q_\pi(\text{full, ignore}) + \frac{1}{6}q_\pi(\text{full, feed})$$

$$q_\pi(\text{hungry, ignore}) = -2 + \frac{4}{5}\left(1v_\pi(\text{hungry}) + 0v_\pi(\text{full})\right)$$

$$q_\pi(\text{hungry, feed}) \quad = -\frac{1}{3} + \frac{4}{5}\left(\frac{1}{3}v_\pi(\text{hungry}) + \frac{2}{3}v_\pi(\text{full})\right)$$

$$q_\pi(\text{full, ignore}) \quad = -1 + \frac{4}{5}\left(\frac{3}{4}v_\pi(\text{hungry}) + \frac{1}{4}v_\pi(\text{full})\right)$$

$$q_\pi(\text{full, feed}) \quad = +1 + \frac{4}{5}\left(0v_\pi(\text{hungry}) + 1v_\pi(\text{full})\right).$$

Solving this equation set will lead to

$$v_\pi(\text{hungry}) \qquad = -\frac{475}{132}, \quad v_\pi(\text{full}) \qquad = -\frac{155}{44}$$

$$q_\pi(\text{hungry, ignore}) = -\frac{161}{33}, \quad q_\pi(\text{hungry, feed}) = -\frac{314}{99}$$

$$q_\pi(\text{full, ignore}) \quad = -\frac{85}{22}, \quad q_\pi(\text{full, feed}) \quad = -\frac{20}{11}.$$

All these three approaches lead to the state values in Table 2.13 and action values in Table 2.14.

Table 2.13 State values derived from Tables 2.2 and 2.6.

s	$v_\pi(s)$
hungry	−475/132
full	−155/44

Table 2.14 Action values derived from Tables 2.2 and 2.6.

s	a	$q_\pi(s, a)$
hungry	ignore	−161/33
hungry	feed	−314/99
ignore	ignore	−85/22
ignore	feed	−20/11

The whole calculation for values only uses the information of policy and dynamics. It does not use the information of initial state distributions. Therefore, initial state distribution has no impact on the value.

2.2.4 Calculate Initial Expected Returns using Values

This section shows how to use values to calculate the initial expected return.

The expected return at $t = 0$ can be calculated by

$$g_\pi = \mathrm{E}_{S_0 \sim p_{S_0}} \left[v_\pi(S_0) \right].$$

It can also be presented using vector forms:

$$g_\pi = \mathbf{p}_{S_0}^\top \mathbf{v}_\pi,$$

where $\mathbf{v}_\pi = \left(v_\pi(s) : s \in \mathcal{S} \right)^\top$ is the state value vector, and $\mathbf{p}_{S_0} = \left(p_{S_0}(s) : s \in \mathcal{S} \right)^\top$ is the initial state distribution vector.

Example 2.10 (Feed and Full) Starting from Tables 2.2 and 2.6, we can obtain the expected return at $t = 0$ as

$$g_\pi = \frac{1}{2}\left(-\frac{475}{132} \right) + \frac{1}{2}\left(-\frac{155}{44} \right) = -\frac{235}{66}.$$

2.2.5 Partial Order of Policy and Policy Improvement

We can define a partial order for policies based on the definition of values. A version of partial order is based on state values: For two policies p and π', if $v_\pi(s) \leq v_{\pi'}(s)$ holds for all state $s \in \mathcal{S}$, we say the policy π is not worse than the policy π', denoted as $\pi \preccurlyeq \pi'$.

Example 2.11 (Feed and Full) Consider the dynamics in Table 2.2 and discount factor $\gamma = \frac{4}{5}$. We have known that, for the policy in Table 2.6, the state values are shown in Table 2.13. We can use similar method to get the state values of the policy shown in Table 2.7 is Table 2.15. Comparing the state values of these two policies, we know that the policy shown in Table 2.7 is better than the policy shown in Table 2.6.

Table 2.15   State values derived from Tables 2.2 and 2.7.

s	$v_\pi(s)$
hungry	35/11
full	5

Now we are ready to learn the policy improvement theorem. Policy improvement theorem has many versions. One famous version is as follows:

Policy Improvement Theorem: For two policies π and π', if

$$v_\pi(s) \le \sum_a \pi'(a|s) q_\pi(s, a), \quad s \in S$$

i.e.

$$v_\pi(s) \le E_{A \sim \pi'(s)} \big[q_\pi(s, A) \big], \quad s \in S$$

then $\pi \preccurlyeq \pi'$, i.e.

$$v_\pi(s) \le v_{\pi'}(s), \quad s \in S.$$

Additionally, if there is a state $s \in S$ such that the former inequality holds, there is a state $s \in S$ such that the later inequality also holds. (Proof: Since

$$v_\pi(s) = E_{\pi'} \big[v_\pi(s) \big] = E_{\pi'} \big[v_\pi(S_t) | S_t = s \big], \qquad\qquad s \in S,$$

$$E_{A \sim \pi'(s)} \big[q_\pi(s, A) \big] = E_{A_t \sim \pi'(s)} \Big[E_{\pi'} \big[q_\pi(S_t, A_t) | S_t = s \big] \Big]$$
$$= E_{\pi'} \big[q_\pi(S_t, A_t) | S_t = s \big], \qquad\qquad s \in S,$$

the former inequality is equivalent to

$$v_\pi(s) = E_{\pi'} \big[v_\pi(S_t) | S_t = s \big] \le E_{\pi'} \big[q_\pi(S_t, A_t) | S_t = s \big], \quad s \in S,$$

where the expectation is over the trajectories starting from the state $S_t = s$ and subsequently driven by the policy π'. Furthermore, we have

$$E_{\pi'} \big[v_\pi(S_{t+\tau}) | S_t = s \big]$$
$$= E_{\pi'} \Big[E_{\pi'} \big[v_\pi(S_{t+\tau}) | S_{t+\tau} \big] \Big| S_t = s \Big]$$
$$\le E_{\pi'} \Big[E_{\pi'} \big[q_\pi(S_{t+\tau}, A_{t+\tau}) | S_{t+\tau} \big] \Big| S_t = s \Big]$$
$$= E_{\pi'} \big[q_\pi(S_{t+\tau}, A_{t+\tau}) | S_t = s \big], \quad s \in S, \tau = 0, 1, 2, \dots.$$

Considering

$$E_{\pi'} \big[q_\pi(S_{t+\tau}, A_{t+\tau}) | S_t = s \big] = E_{\pi'} \big[R_{t+\tau+1} + \gamma v_\pi(S_{t+\tau+1}) | S_t = s \big],$$
$$s \in S, \tau = 0, 1, 2, \dots$$

Therefore,

$$E_{\pi'} \big[v_\pi(S_{t+\tau}) | S_t = s \big] \le E_{\pi'} \big[R_{t+\tau+1} + \gamma v_\pi(S_{t+\tau+1}) | S_t = s \big],$$
$$s \in S, \tau = 0, 1, 2, \dots.$$

Hence,

$$v_\pi(s) = E_{\pi'} \big[v_\pi(S_t) | S_t = s \big]$$
$$\le E_{\pi'} \big[R_{t+1} + \gamma v_\pi(S_{t+1}) | S_t = s \big]$$
$$\le E_{\pi'} \Big[R_{t+1} + \gamma E_{\pi'} \big[R_{t+2} + \gamma v_\pi(S_{t+2}) | S_t = s \big] \Big| S_t = s \Big]$$

$$\leq E_{\pi'}\left[R_{t+1} + \gamma R_{t+2} + \gamma^2 R_{t+3} + \gamma^3 v_\pi(S_{t+3})\middle| S_t = s\right]$$

$$\cdots$$

$$\leq E_{\pi'}\left[R_{t+1} + \gamma R_{t+2} + \gamma^2 R_{t+3} + \gamma^3 R_{t+4} + \cdots \middle| S_t = s\right]$$
$$= E_{\pi'}[G_t | S_t = s]$$
$$= v_{\pi'}(s), \qquad\qquad s \in \mathcal{S}.$$

The proof for inequality is similar.)

> **! Note**
>
> The aforementioned proof uses the many properties of expectations. When reading the proof, make sure to understand what each expectation is over, and understand the applications of expectation. Similar proof methods are useful in RL.

Policy improvement theorem tells us, for an arbitrary policy π, if there exists a state–action pair (s', a') such that $\pi(a'|s') > 0$ and $q_\pi(s', a') < \max_a q_\pi(s', a)$, we can construct a new deterministic policy π' who takes action $\arg\max_a q_\pi(s', a)$ at state s' (if multiple actions reach maxim we can pick an arbitrary one) and share the same actions in states other than s. We can verify that,

$$v(s') = \sum_{a \in \mathcal{A}} \pi(a|s')q_\pi(s', a) < \sum_{a \in \mathcal{A}} \pi'(a|s')q_\pi(s', a)$$
$$v(s) = \sum_{a \in \mathcal{A}} \pi(a|s)q_\pi(s, a) = \sum_{a \in \mathcal{A}} \pi'(a|s)q_\pi(s, a), \quad s \neq s'.$$

So policy π and policy π' satisfy the conditions of the policy improvement theorem. In this way, we can improve policy π to a better policy π'.

Algorithm 2.1   Check whether the policy is optimal.

Input: action values of the policy q_π.
Output: an indicator to show whether the policy is optimal.

1. For each state–action pair $s \in \mathcal{S}, a \in \mathcal{A}(s)$,

    1.1. If $\pi(a|s) > 0$ and $q_\pi(s, a) < \max_{a' \in \mathcal{A}} q_\pi(s, a')$, it means the policy is not optimal, and can be further improved.

2. If we did not find a state–action pair in Step 1, the policy is optimal and can not be improved.

Algorithm 2.2 Policy improvement.

> **Input:** action values of the policy q_π.
> **Output:** improved policy π'.
>
> **1.** For each state $s \in S$, find an action a that maximizes $q_\pi(s, a)$. The new
>   policy $\pi'(s) \leftarrow \arg\max_{a \in \mathcal{A}} q_\pi(s, a)$.

Example 2.12 (Feed and Full) Consider the dynamics in Table 2.2 and discount factor $\gamma = \frac{4}{5}$. We have known that, for the policy in Table 2.6, the state values are shown in Table 2.13. Since $\pi(\text{ignore}|\text{hungry}) > 0$ and $q(\text{hungry}, \text{ignore}) < q(\text{hungry}, \text{feed})$ (or $\pi(\text{ignore}|\text{full}) > 0$ and $q(\text{full}, \text{ignore}) < q(\text{full}, \text{feed}))$, this policy can be improved. As for the policy shown in Table 2.7, its state values are shown in Table 2.15, and this policy can not be further improved.

Now we know that, for any policy π, if there exists $s \in S, a \in \mathcal{A}$ such that $q_\pi(s, a) > v_\pi(s)$, we can always construct a better policy π'. Repeating such constructions can lead to better policy. Such iterations end when $q_\pi(s, a) \leq v_\pi(s)$ holds for all $s \in S, a \in \mathcal{A}$.

The resulting policy is the maximal element in this partial order. We can prove that, any maximal element π_* in this partial order satisfying that

$$v_{\pi_*}(s) = \max_{a \in \mathcal{A}} q_{\pi_*}(s, a), \quad s \in S.$$

(Proof: Section 2.2.2 told us

$$v_{\pi_*}(s) = \sum_{a \in \mathcal{A}} \pi_*(a|s) q_{\pi_*}(s, a)$$

$$\leq \sum_{a \in \mathcal{A}} \pi_*(a|s) \max_{a' \in \mathcal{A}} q_{\pi_*}(s, a')$$

$$= \max_{a' \in \mathcal{A}} q_{\pi_*}(s, a'), \quad s \in S.$$

Therefore, at the end of iterations, we have

$$v_{\pi_*}(s) \leq \max_{a \in \mathcal{A}} q_{\pi_*}(s, a) \leq \max_{a \in \mathcal{A}} v_{\pi_*}(s) = v_{\pi_*}(s), \quad s \in S.$$

The proof completes.)

2.3 Visitation Frequency

The previous section covers values, a very important concept in RL. This section will cover another import concept called discounted visitation frequency, also known as discounted distribution, which is a dual quantity of values. Based on the discounted distribution, we can further define discounted expectations. They all play important roles in RL researches.

2.3.1 Definition of Visitation Frequency

Given the environment and the policy of an MDP, we can determine how many times a state or a state–action pair will be visited. Taking possible discounts into considerations, we can define a statistic called **discounted visitation frequency**, which is also known as **discounted distribution**. It has two forms:

- **Discounted state visitation frequency** (also known as discounted state distribution)

$$\text{episodic task:} \quad \eta_\pi(s) \stackrel{\text{def}}{=} \sum_{t=1}^{+\infty} \Pr_\pi[T = t] \sum_{\tau=0}^{t-1} \gamma^\tau \Pr_\pi[S_\tau = s], \quad s \in \mathcal{S},$$

$$\text{sequential task:} \quad \eta_\pi(s) \stackrel{\text{def}}{=} \sum_{\tau=1}^{+\infty} \gamma^\tau \Pr_\pi[S_\tau = s], \qquad\qquad s \in \mathcal{S}.$$

- **Discounted state–action visitation frequency** (also known as discounted state–action distribution)

episodic task:

$$\rho_\pi(s, a) \stackrel{\text{def}}{=} \sum_{t=1}^{+\infty} \Pr_\pi[T = t] \sum_{\tau=0}^{t-1} \gamma^\tau \Pr_\pi[S_\tau = s, A_\tau = a], \quad s \in \mathcal{S}, a \in \mathcal{A}(s),$$

sequential task:

$$\rho_\pi(s, a) \stackrel{\text{def}}{=} \sum_{\tau=1}^{+\infty} \gamma^\tau \Pr_\pi[S_\tau = s, A_\tau = a], \qquad\qquad s \in \mathcal{S}, a \in \mathcal{A}(s).$$

! Note

Although discounted visitation is also called discounted distribution, it may not be a probability distribution. We can verify that,

$$\text{episodic task:} \quad \sum_{s \in \mathcal{S}} \eta_\pi(s) = \sum_{s \in \mathcal{S}, a \in \mathcal{A}(s)} \rho_\pi(s, a) = \mathrm{E}_\pi\left[\frac{1 - \gamma^T}{1 - \gamma}\right],$$

$$\text{sequential task:} \quad \sum_{s \in \mathcal{S}} \eta_\pi(s) = \sum_{s \in \mathcal{S}, a \in \mathcal{A}(s)} \rho_\pi(s, a) = \frac{1}{1 - \gamma}.$$

Proof for the episodic tasks:

$$\sum_{s \in S} \eta_\pi(s) = \sum_{s \in S} \sum_{t=1}^{+\infty} \Pr_\pi[T = t] \sum_{\tau=0}^{t-1} \gamma^\tau \Pr_\pi[S_\tau = s]$$

$$= \sum_{t=1}^{+\infty} \Pr_\pi[T = t] \sum_{\tau=0}^{t-1} \gamma^\tau \sum_{s \in S} \Pr_\pi[S_\tau = s]$$

$$= \sum_{t=1}^{+\infty} \Pr_\pi[T = t] \sum_{\tau=0}^{t-1} \gamma^\tau$$

$$= \sum_{t=1}^{+\infty} \Pr_\pi[T = t] \frac{1 - \gamma^t}{1 - \gamma}$$

$$= E_\pi \left[\frac{1 - \gamma^T}{1 - \gamma} \right].$$

Apparently, this expectation does not always equal 1. Therefore, the discounted distribution is not always a distribution.

Example 2.13 (Feed and Full) For the example in Tables 2.2 and 2.6, we have

$$\eta_\pi(\text{hungry}) = \sum_{\tau=0}^{+\infty} \gamma^\tau \Pr_\pi[S_\tau = \text{hungry}]$$

$$\eta_\pi(\text{full}) = \sum_{\tau=0}^{+\infty} \gamma^\tau \Pr_\pi[S_\tau = \text{full}]$$

$$\rho_\pi(\text{hungry, ignore}) = \sum_{\tau=0}^{+\infty} \gamma^\tau \Pr_\pi[S_\tau = \text{hungry}, A_\tau = \text{ignore}]$$

$$\rho_\pi(\text{hungry, feed}) = \sum_{\tau=0}^{+\infty} \gamma^\tau \Pr_\pi[S_\tau = \text{hungry}, A_\tau = \text{feed}]$$

$$\rho_\pi(\text{full, ignore}) = \sum_{\tau=0}^{+\infty} \gamma^\tau \Pr_\pi[S_\tau = \text{full}, A_\tau = \text{ignore}]$$

$$\rho_\pi(\text{full, feed}) = \sum_{\tau=0}^{+\infty} \gamma^\tau \Pr_\pi[S_\tau = \text{full}, A_\tau = \text{feed}].$$

Rewards do not appear in the definition of discounted visitation frequency. Therefore, discounted visitation frequencies do not directly depend on rewards.

2.3.2 Properties of Visitation Frequency

Discounted state visitation frequency and discounted state–action visitation frequency have the following relationships.

- Use discounted state–action visitation frequency to back up discounted state visitation frequency

$$\eta_\pi(s) = \sum_{a \in \mathcal{A}(s)} \rho_\pi(s, a), \quad s \in S.$$

(Proof: For sequential tasks,

$$\eta_\pi(s) = \sum_{t=0}^{+\infty} \gamma^t \Pr_\pi[S_t = s]$$

$$= \sum_{t=0}^{+\infty} \gamma^t \sum_{a \in \mathcal{A}(s)} \Pr_\pi[S_t = s, A_t = a]$$

$$= \sum_{a \in \mathcal{A}(s)} \sum_{t=0}^{+\infty} \gamma^t \Pr_\pi[S_t = s, A_t = a]$$

$$= \sum_{a \in \mathcal{A}(s)} \rho_\pi(s, a).$$

We can prove for sequential tasks in a similar way.) This equation can be represented in vector form as follows:

$$\boldsymbol{\eta}_\pi = \mathbf{I}_{S|S,A} \boldsymbol{\rho}_\pi,$$

where the column vector $\boldsymbol{\eta}_\pi = (\eta_\pi(s) : s \in S)^\top$ has $|S|$ elements, column vector $\boldsymbol{\rho}_\pi = (\rho_\pi(s, a) : (s, a) \in S \times \mathcal{A})^\top$ has $|S\|\mathcal{A}|$ elements, and the matrix $\mathbf{I}_{S|S,A}$ has $|S| \times |S\|\mathcal{A}|$ elements.
- Use discounted state visitation frequency and policy to back up discounted state–action visitation frequency

$$\rho_\pi(s, a) = \eta_\pi(s)\pi(a|s), \quad s \in S, a \in \mathcal{A}(s).$$

(Proof: Taking sequential task as an example,

$$\rho_\pi(s, a) = \sum_{t=0}^{+\infty} \gamma^t \Pr_\pi[S_t = s, A_t = a]$$

$$= \sum_{t=0}^{+\infty} \gamma^t \Pr_\pi[S_t = s]\pi(a|s)$$

$$= \eta_\pi(s)\pi(a|s).$$

We can prove for episodic tasks in a similar way.)
- Use discounted state–action visitation frequency and dynamics to back up discounted state visitation frequency.

$$\eta_\pi(s') = p_{S_0}(s') + \gamma \sum_{s \in S, a \in \mathcal{A}(s)} p(s'|s, a)\rho_\pi(s, a), \quad s' \in S.$$

(Proof: For sequential tasks, taken the definition of $\rho_\pi(s, a)$ for examples, we have

$$\sum_{s \in S, a \in \mathcal{A}(s)} \gamma p(s'|s, a)\rho_\pi(s, a)$$

$$= \sum_{s \in S, a \in \mathcal{A}(s)} \gamma p(s'|s, a) \sum_{t=0}^{+\infty} \gamma^t \Pr_\pi[S_t = s, A_t = a]$$

$$= \sum_{s \in S, a \in \mathcal{A}(s)} \gamma p(s'|s, a) \sum_{s_0 \in S} p_{S_0}(s_0) \sum_{t=0}^{+\infty} \gamma^t \Pr_\pi[S_t = s, A_t = a|S_0 = s_0]$$

$$= \sum_{s_0 \in S} p_{S_0}(s_0) \sum_{s \in S, a \in \mathcal{A}(s)} \gamma^{t+1} \sum_{t=0}^{+\infty} p(s'|s, a)\Pr_\pi[S_t = s, A_t = a|S_0 = s_0]$$

$$= \sum_{s_0 \in S} p_{S_0}(s_0) \sum_{s \in S, a \in \mathcal{A}(s)} \gamma^{t+1} \sum_{t=0}^{+\infty} \Pr_\pi[S_{t+1} = s', S_t = s, A_t = a|S_0 = s_0]$$

$$= \sum_{s_0 \in S} p_{S_0}(s_0) \sum_{s \in S, a \in \mathcal{A}(s)} \gamma^{t+1} \sum_{t=0}^{+\infty} \Pr_\pi[S_{t+1} = s'|S_0 = s_0].$$

Let $1_{[\cdot]}$ be the indicator function. We have

$$\sum_{t=0}^{+\infty} \gamma^t \Pr_\pi[S_t = s'|S_0 = s_0]$$

$$= \Pr_\pi[S_0 = s'|S_0 = s_0] + \sum_{t=1}^{+\infty} \gamma^t \Pr_\pi[S_t = s'|S_0 = s_0]$$

$$= 1_{[s'=s_0]} + \sum_{t=0}^{+\infty} \gamma^{t+1} \Pr_\pi[S_{t+1} = s'|S_0 = s_0].$$

So

$$\sum_{t=0}^{+\infty} \gamma^{t+1} \Pr_\pi[S_{t+1} = s'|S_0 = s_0] = \left(\sum_{t=0}^{+\infty} \gamma^t \Pr_\pi[S_t = s'|S_0 = s_0]\right) - 1_{[s'=s_0]}.$$

Plugging in the aforementioned equation into the equation in the beginning of the proof, we have

$$\sum_{s \in \mathcal{S}, a \in \mathcal{A}(s)} \gamma p\big(s'\big|s, a\big) \rho_\pi(s, a)$$

$$= \sum_{s_0 \in \mathcal{S}} p_{S_0}(s_0) \left(\sum_{t=0}^{+\infty} \gamma^t \Pr_\pi \big[S_t = s' \big| S_0 = s_0 \big] - 1_{[s'=s_0]} \right)$$

$$= \sum_{s_0 \in \mathcal{S}} p_{S_0}(s_0) \sum_{t=0}^{+\infty} \gamma^t \Pr_\pi \big[S_t = s' \big| S_0 = s_0 \big] - p_{S_0}\big(s'\big)$$

$$= \eta_\pi\big(s'\big) - p_{S_0}\big(s'\big).$$

The proof completes.) Its vector form is

$$\boldsymbol{\eta}_\pi = \mathbf{p}_{S_0} + \gamma \mathbf{P}_{S_{t+1}|S_t, A_t} \boldsymbol{\rho}_\pi,$$

where the column vector $\boldsymbol{\eta}_\pi = \big(\eta_\pi(s) : s \in \mathcal{S}\big)^\top$ has $|\mathcal{S}|$ elements, the column vector $\mathbf{p}_{S_0} = \big(p_{S_0}(s) : s \in \mathcal{S}\big)^\top$ has $|\mathcal{S}|$ elements, the column vector $\boldsymbol{\rho}_\pi = \big(\rho_\pi(s, a) : (s, a) \in \mathcal{S} \times \mathcal{A}\big)^\top$ has $|\mathcal{S}||\mathcal{A}|$ elements, and the matrix $\mathbf{P}_{S_{t+1}|S_t, A_t} = \big(p(s'|s, a) : s' \in \mathcal{S}, (s, a) \in \mathcal{S} \times \mathcal{A}\big)$ has $|\mathcal{S}| \times |\mathcal{S}||\mathcal{A}|$ elements.

In the three aforementioned relationship, the first relationship uses neither the mathematical model of the environment nor the mathematical model of the agent. The second relationship uses the mathematical model of the agent, i.e. π. The third relationship uses the mathematical model of the environment, i.e. p_{S_0} and p.

The Bellman expectation equations among discounted visitation frequencies are as follows:

- Use the discounted state distribution at time t to back up the discounted state distribution at time $t + 1$:

$$\eta_\pi\big(s'\big) = p_{S_0}\big(s'\big) + \sum_{s \in \mathcal{S}} \gamma p_\pi\big(s'\big|s\big) \eta_\pi(s), \quad s' \in \mathcal{S}.$$

(Proof: Plug $\rho_\pi(s, a) = \eta_\pi(s)\pi(a|s)$ $(s \in \mathcal{S}, a \in \mathcal{A}(s))$ into $\eta_\pi(s') = p_{S_0}(s') + \sum_{s \in \mathcal{S}, a \in \mathcal{A}(s)} \gamma p(s'|s, a)\rho_\pi(s, a)$ $(s' \in \mathcal{S})$, and simplify using $p_\pi(s'|s) = \sum_s p(s'|s, a)\pi(a|s)$ $(s \in \mathcal{S}, s' \in \mathcal{S})$.) Its vector form is

$$\boldsymbol{\eta}_\pi = \mathbf{p}_{S_0} + \gamma \mathbf{P}_{S_{t+1}|S_t;\pi} \boldsymbol{\eta}_\pi,$$

where the column vector $\boldsymbol{\eta}_\pi = \big(\eta_\pi(s) : s \in \mathcal{S}\big)^\top$ has $|\mathcal{S}|$ elements the column vector $\mathbf{p}_{S_0} = \big(p_{S_0}(s) : s \in \mathcal{S}\big)^\top$ has $|\mathcal{S}|$ elements, and the matrix $\mathbf{P}_{S_{t+1}|S_t;\pi}$ has $|\mathcal{S}| \times |\mathcal{S}|$ elements.

- Use discounted state–action distribution at time t to back up the discounted state–action distribution at time $t + 1$:

$$\rho_\pi\big(s', a'\big) = p_{S_0, A_0; \pi}\big(s', a'\big) + \sum_{s \in \mathcal{S}, a \in \mathcal{A}(s)} \gamma p_\pi\big(s', a'\big|s, a\big) \rho_\pi(s, a),$$

$$s' \in \mathcal{S}, a' \in \mathcal{A}\big(s'\big).$$

(Proof: Multiply both sides of $\eta_\pi(s') = p_{S_0}(s') + \gamma \sum_{s \in S, a \in \mathcal{A}(s)} p(s'|s,a) \rho_\pi(s,a)$ $(s' \in S)$ by $\pi(a'|s')$, and simplify using $\rho_\pi(s',a') = \eta_\pi(s')\pi(a'|s')$ $(s' \in S, a' \in \mathcal{A}(s'))$, $p_{S_0,A_0;\pi}(a'|s') = \pi(a'|s')p_{S_0}(s')$ $(s' \in S, a' \in \mathcal{A}(s'))$, and $p_\pi(s',a'|s,a) = \pi(a'|s')p(s'|s,a)$ $(s \in S, a \in \mathcal{A}, s' \in S, a' \in \mathcal{A}).$) It has the vector form

$$\boldsymbol{\rho}_\pi = \mathbf{p}_{S_0,A_0;\pi} + \gamma \mathbf{P}_{S_{t+1},A_{t+1}|S_t,A_t;\pi} \boldsymbol{\rho}_\pi,$$

where the column vector $\boldsymbol{\rho}_\pi = \left(\rho_\pi(s,a) : (s,a) \in S \times \mathcal{A}\right)^\top$ has $|S\|\mathcal{A}|$ elements the column vector $\mathbf{p}_{S_0,A_0;\pi} = \left(p_{S_0,A_0;\pi} : (s,a) \in S \times \mathcal{A}\right)^\top$ has $|S\|\mathcal{A}|$ elements, and the matrix $\mathbf{P}_{S_{t+1},A_{t+1}|S_t,A_t;\pi} = \left(p_\pi(s',a'|s,a) : (s',a') \in S \times \mathcal{A},$ $(s,a) \in S \times \mathcal{A}\right)$ has $|S\|\mathcal{A}| \times |S\|\mathcal{A}|$ elements.

2.3.3 Calculate Visitation Frequency

Similar to the case of calculating values, there are many ways to calculate discounted visitation frequencies.

Approach 1: Plug the initial state distribution p_{S_0} $(s \in S)$, transition probability $p_\pi(s'|s)$ $(s \in S, s' \in S)$, and discount factor γ into the Bellman expectation equation of discounted state visitation frequencies to get the discounted state visitation frequencies. After that, use the obtained discounted state visitation frequencies and the policy $\pi(a|s)$ $(s \in S, a \in \mathcal{A})$ to calculate discounted state–action visitation frequencies.

Approach 2: Plug in the initial state–action distribution $p_{S_0,A_0;\pi}(s,a)$ $((s,a) \in S \times \mathcal{A})$, transition probability $p_\pi(s',a'|s,a)$ $((s,a) \in S \times \mathcal{A},$ $(s',a') \in S \times \mathcal{A})$ and discount factor γ into the Bellmen expectation equation of discounted state–action visitation frequencies to get the discounted state–action visitation frequencies. After that, sum up the obtained discounted state–action visitation frequency to get the discounted state visitation frequencies.

Example 2.14 (Feed and Full) In the example of Tables 2.2 and 2.6, set discount factor $\gamma = \frac{4}{5}$. We can get the discounted visitation frequencies in Tables 2.15 and 2.17.

Table 2.16   Discounted state visitation frequency derived from Tables 2.2 and 2.6.

s	$\eta_\pi(s)$
hungry	30/11
full	25/11

The calculation steps are as follows. Approach 1: The initial state distribution is

Table 2.17   Discounted state–action visitation frequency derived from Tables 2.2 and 2.6.

s	a	$\rho_\pi(s, a)$
hungry	ignore	15/22
hungry	feed	45/22
full	ignore	125/66
full	feed	25/66

$$\mathbf{p}_{S_0} = \begin{bmatrix} p_{S_0}(\text{hungry}) \\ p_{S_0}(\text{full}) \end{bmatrix} = \begin{bmatrix} 1/2 \\ 1/2 \end{bmatrix}.$$

The transition matrix is

$$\mathbf{P}_{S_{t+1}|S_t;\pi} = \begin{bmatrix} p_\pi(\text{hungry}|\text{hungry}) & p_\pi(\text{hungry}|\text{full}) \\ p_\pi(\text{full}|\text{hungry}) & p_\pi(\text{full}|\text{full}) \end{bmatrix} = \begin{bmatrix} 1/2 & 5/8 \\ 1/2 & 3/8 \end{bmatrix}.$$

Then the state visitation frequency is

$$\boldsymbol{\eta}_\pi = \left(\mathbf{I} - \gamma \mathbf{P}_{S_{t+1}|S_t;\pi}\right)^{-1} \mathbf{p}_{S_0} = \left(\begin{bmatrix} 1 & 0 \\ 0 & 1 \end{bmatrix} - \frac{4}{5} \begin{bmatrix} 1/2 & 5/8 \\ 1/2 & 3/8 \end{bmatrix} \right)^{-1} \begin{bmatrix} 1/2 \\ 1/2 \end{bmatrix} = \begin{bmatrix} 30/11 \\ 25/11 \end{bmatrix}.$$

And the state–action visitation frequency is

$$\rho_\pi(\text{hungry}, \text{ignore}) = \frac{30}{11} \times \frac{1}{4} = \frac{15}{22}$$

$$\rho_\pi(\text{hungry}, \text{feed}) = \frac{30}{11} \times \frac{3}{4} = \frac{45}{22}$$

$$\rho_\pi(\text{full}, \text{ignore}) = \frac{25}{11} \times \frac{5}{6} = \frac{125}{66}$$

$$\rho_\pi(\text{full}, \text{feed}) = \frac{25}{11} \times \frac{1}{6} = \frac{25}{66}.$$

Approach 2: The initial state–action distribution is

$$\mathbf{p}_{S_0, A_0;\pi} = \begin{bmatrix} p_{S_0, A_0;\pi}(\text{hungry}, \text{ignore}) \\ p_{S_0, A_0;\pi}(\text{hungry}, \text{feed}) \\ p_{S_0, A_0;\pi}(\text{full}, \text{ignore}) \\ p_{S_0, A_0;\pi}(\text{full}, \text{feed}) \end{bmatrix} = \begin{bmatrix} 1/8 \\ 3/8 \\ 5/12 \\ 1/12 \end{bmatrix}.$$

The transition matrix is

$$\mathbf{P}_{S_{t+1},A_{t+1}|S_t,A_t;\pi} = \begin{bmatrix} 1/4 & 1/12 & 3/16 & 0 \\ 3/4 & 1/4 & 9/16 & 0 \\ 0 & 5/9 & 5/24 & 5/6 \\ 0 & 1/9 & 1/24 & 1/6 \end{bmatrix}.$$

Then the state–action visitation frequency is

$$\boldsymbol{\rho}_\pi = \left(\mathbf{I} - \gamma\mathbf{P}_{S_{t+1},A_{t+1}|S_t,A_t;\pi}\right)^{-1} \mathbf{P}_{S_0,A_0;\pi}$$

$$= \left(\begin{bmatrix} 1 & 0 & 0 & 0 \\ 0 & 1 & 0 & 0 \\ 0 & 0 & 1 & 0 \\ 0 & 0 & 0 & 1 \end{bmatrix} - \frac{4}{5}\begin{bmatrix} 1/4 & 1/12 & 3/16 & 0 \\ 3/4 & 1/4 & 9/16 & 0 \\ 0 & 5/9 & 5/24 & 5/6 \\ 0 & 1/9 & 1/24 & 1/6 \end{bmatrix}\right)^{-1} \begin{bmatrix} 1/8 \\ 3/8 \\ 5/12 \\ 1/12 \end{bmatrix} = \begin{bmatrix} 15/22 \\ 45/22 \\ 125/66 \\ 25/66 \end{bmatrix}.$$

And the state visitation frequency is

$$\eta_\pi(\text{hungry}) = \frac{15}{22} + \frac{45}{22} = \frac{30}{11}$$
$$\eta_\pi(\text{full}) = \frac{125}{66} + \frac{25}{66} = \frac{25}{11}.$$

2.3.4 Equivalence between Visitation Frequency and Policy

Section 2.3.2 told us that the discounted visitation frequencies of an arbitrary policy π satisfy the following equation set:

$$\eta_\pi(s') = p_{S_0}(s') + \sum_{s\in S, a\in\mathcal{A}(s)} \gamma p(s'|s,a)\rho_\pi(s,a), \quad s' \in S$$

$$\eta_\pi(s) = \sum_{a\in\mathcal{A}(s)} \rho_\pi(s,a), \qquad\qquad\qquad s \in S$$

$$\rho_\pi(s,a) \geq 0, \qquad\qquad\qquad\qquad\qquad s \in S, a \in \mathcal{A}(s).$$

Note that the equation set does not include π explicitly. In fact, the solution of the above equation set is bijective with the policy. Exactly speaking, if a suite of $\eta(s)$ ($s \in S$) and $\rho(s,a)$ ($s \in S, a \in \mathcal{A}(s)$) satisfies the equation set,

$$\eta(s') = p_{S_0}(s') + \sum_{s\in S, a\in\mathcal{A}(s)} \gamma p(s'|s,a)\rho(s,a), \quad s' \in S$$

$$\eta(s) = \sum_{a\in\mathcal{A}(s)} \rho_\pi(s,a), \qquad\qquad\qquad s \in S$$

$$\rho(s,a) \geq 0, \qquad\qquad\qquad\qquad\qquad s \in S, a \in \mathcal{A}(s),$$

we can define a policy using

$$\pi(a|s) = \frac{\rho(s, a)}{\eta(s)}, \quad s \in S, a \in \mathcal{A}(s)$$

and the policy satisfies (1) $\eta_\pi(s) = \eta(s)$ ($s \in S$); (2) $\rho_\pi(s, a) = \rho(s, a)$ ($s \in S$, $a \in \mathcal{A}(s)$). (Proof: Consider sequential tasks. (1) For any $s' \in S$, taking the definition of policy π, we will know

$$\eta(s') = p_{S_0}(s') + \sum_{s \in S, a \in \mathcal{A}(s)} \gamma p(s'|s, a)\rho(s, a)$$

$$= p_{S_0}(s') + \sum_{s \in S, a \in \mathcal{A}(s)} \gamma p(s'|s, a)\pi(a|s)\eta(s)$$

$$= p_{S_0}(s') + \sum_{s \in S} \gamma p_\pi(s'|s)\eta(s).$$

Therefore, $\eta(s)$ ($s \in S$) satisfies the Bellman expectation equation. At the same time, $\eta_\pi(s)$ ($s \in S$) satisfies the same Bellman expectation equation, i.e.

$$\eta_\pi(s') = p_{S_0}(s') + \sum_{s \in S} \gamma p_\pi(s'|s)\eta_\pi(s), \quad s' \in S.$$

Since Bellmen expectation equations have a unique solution, we have $\eta_\pi(s) = \eta(s)$ ($s \in S$). (2) For any $s \in S, a \in \mathcal{A}(s)$, we have $\rho_\pi(s, a) = \eta_\pi(s)\pi(a|s)$, $\rho(s, a) = \eta(s)\pi(a|s)$, and $\eta_\pi(s) = \eta(s)$. Therefore, $\rho_\pi(s, a) = \rho(s, a)$.)

Furthermore, if a suite of $\rho(s, a)$ ($s \in S, a \in \mathcal{A}(s)$) satisfies

$$\sum_{a' \in \mathcal{A}(s')} \rho_\pi(s', a') = p_{S_0}(s') + \sum_{s \in S, a \in \mathcal{A}(s)} \gamma p(s'|s, a)\rho(s, a), \quad s' \in S$$

$$\rho(s, a) \geq 0, \qquad\qquad\qquad\qquad\qquad s \in S, a \in \mathcal{A}(s),$$

we can define a policy using

$$\pi(a|s) = \frac{\rho(s, a)}{\eta(s)}, \quad s \in S, a \in \mathcal{A}(s),$$

where $\eta(s) = \sum_{a \in \mathcal{A}(s)} \rho_\pi(s, a)$ and the policy satisfies (1) $\eta_\pi(s) = \eta(s)$ ($s \in S$); (2) $\rho_\pi(s, a) = \rho(s, a)$ ($s \in S, a \in \mathcal{A}(s)$).

2.3.5 Expectation over Visitation Frequency

Although discounted distribution is usually not a probability distribution, we can still define expectation over the discounted distribution. Mathematically speaking, given a deterministic function f, we can define the expectation over discounted distribution as

$$\mathrm{E}_{S\sim\eta_\pi}\left[f(S)\right] \overset{\text{def}}{=} \sum_{s\in S} \eta_\pi(s)f(s)$$

$$\mathrm{E}_{(S,A)\sim\rho_\pi}\left[f(S,A)\right] \overset{\text{def}}{=} \sum_{s\in S, a\in \mathcal{A}(s)} \rho_\pi(s,a)f(s,a).$$

Many statistics can be represented using the notation of expectation over discounted distribution.

Example 2.15 The expected discounted return at $t = 0$ can be represented as

$$g_\pi = \mathrm{E}_{(S,A)\sim\rho_\pi}\left[r(S,A)\right].$$

(Proof:

$$g_\pi = \mathrm{E}_\pi[G_0]$$

$$= \mathrm{E}_\pi\left[\sum_{t=0}^{+\infty}\gamma^t R_{t+1}\right]$$

$$= \sum_{t=0}^{+\infty}\gamma^t \mathrm{E}_\pi[R_{t+1}]$$

$$= \sum_{t=0}^{+\infty}\gamma^t \mathrm{E}_\pi\left[\mathrm{E}_\pi[R_{t+1}|S_t, A_t]\right] \qquad \text{due to law of total expectation}$$

$$= \sum_{t=0}^{+\infty}\gamma^t \mathrm{E}_\pi\left[r(S_t, A_t)\right] \qquad \text{due to definition of } r(S_t, A_t)$$

$$= \sum_{t=0}^{+\infty}\gamma^t \sum_{s\in S, a\in \mathcal{A}(s)} \mathrm{Pr}_\pi[S_t = s, A_t = a]r(s,a)$$

$$= \sum_{s\in S, a\in \mathcal{A}(s)} \left(\sum_{t=0}^{+\infty}\gamma^t \mathrm{Pr}_\pi[S_t = s, A_t = a]\right)r(s,a)$$

$$= \sum_{s\in S, a\in \mathcal{A}(s)} \rho_\pi(s,a)r(s,a) \qquad \text{due to definition of } \rho_\pi(s,a)$$

$$= \mathrm{E}_{(S,A)\sim\rho_\pi}\left[r(S,A)\right].$$

The proof completes.)

! **Note**

We have seen the way to convert between expectation over policy trajectory and the expectation over discounted distribution. This method will be repeatedly used in later chapters in the book.

Until now, we have known how to use discounted state–action visitation frequency and expected reward to calculate the expected discounted return at $t = 0$. It also has vector form. Let the vector $\mathbf{r} = \left(r(s, a) : s \in \mathcal{S}, a \in \mathcal{A} \right)^{\top}$ denote the expected rewards given state–action pairs, and use the vector $\boldsymbol{\rho} = \left(\rho_{\pi}(s, a) : s \in \mathcal{S}, a \in \mathcal{A} \right)^{\top}$ to denote the discounted state–action visitation frequencies. Then the expected discounted return at $t = 0$ can be represented as

$$g_{\pi} = \mathbf{r}^{\top} \boldsymbol{\rho}_{\pi}.$$

Example 2.16 (Feed and Full) Starting from Tables 2.4 and 2.17, we can obtain the expected return at $t = 0$ as

$$g_{\pi} = (-2) \cdot \frac{15}{22} + \left(-\frac{1}{3} \right) + (-1) \cdot \frac{125}{66} + 1 \cdot \frac{25}{66} = -\frac{235}{66}.$$

2.4 Optimal Policy and Optimal Value

2.4.1 From Optimal Policy to Optimal Value

We have learned the definition of values and partial order among values. Based on them, we can further define what is an optimal policy. The definition of optimal policy is as follows: Given an environment, if there exists a policy π_* such that any policy π satisfies $\pi \preceq \pi_*$, the policy π_* is called **optimal policy**.

The values of an optimal policy π_* have the following properties:

- The state values of optimal policy satisfy

$$v_{\pi_*}(s) = \max_{\pi} v_{\pi}(s), \quad s \in \mathcal{S}.$$

(Proof by contradiction: If the above equation does not hold, there exists a policy π and a state s such that $v_{\pi}(s) > v_{\pi_*}(s)$. Therefore, $v_{\pi} \preceq v_{\pi_*}$ does not hold, which conflicts with the assumption that the policy π_* is optimal.)

- The action values of the optimal policy satisfy

$$q_{\pi_*}(s, a) = \max_{\pi} q_{\pi}(s, a), \quad s \in \mathcal{S}, a \in \mathcal{A}(s).$$

Based on the aforementioned property of the values of optimal policy, we can define optimal values. **Optimal values** include the following two forms.

- **Optimal state value** is defined as

$$v_*(s) \overset{\text{def}}{=} \sup_{\pi} v_{\pi}(s), \quad s \in \mathcal{S}.$$

- **Optimal action value** is defined as

$$q_*(s, a) \stackrel{\text{def}}{=} \sup_{\pi} q_\pi(s, a), \quad s \in S, a \in \mathcal{A}(s).$$

Obviously, if there exists an optimal policy, the values of that optimal policy equal the optimal values.

Remarkably, not all environments have optimal values. Here is an example of an environment that has no optimal values: A one-shot environment has a singleton state space, and the action space is a close interval $\mathcal{A} = [0, 1]$. The episode reward R_1 is fully determined by the action A_0 in the following way: $R_1 = 0$ when $A_0 = 0$, and $R_1 = 1/A_0$ when $A_0 > 0$. Apparently, the reward is unbounded, and there does not exist an optimal policy.

2.4.2 Existence and Uniqueness of Optimal Policy

The previous section told us that not all environments have optimal values. For those environments that do not have optimal values, they do not have optimal policies either. Furthermore, even if an environment has optimal values, it does not necessarily have optimal policies. This section will examine an example that has optimal values but does not have optimal policies. We will also discuss the conditions that ensure the existence of an optimal policy. At last, we show that there may be multiple different optimal policies.

Firstly, let us examine an example of an environment that has optimal values, but has no optimal policies. Consider the following one-shot environment: the state space is a singleton $S = \{s\}$, and the action space is a bounded open interval $\mathcal{A} \subseteq \mathbb{R}$ (such as $(0, 1)$), and the episode reward equals the action $R_0 = A$. The state value for any policy π is $v_\pi(s) = \mathrm{E}_\pi[a] = \sum_{a \in \mathcal{A}} a\pi(a|s)$, and the action values are $q_\pi(s, a) = a$. Now we find the optimal values of the environment. Let us consider the optimal state values first. On one hand, the state values of an arbitrary policy satisfy $v_\pi(s) = \sum_{a \in \mathcal{A}} a\pi(a|s) \leq \sum_{a \in \mathcal{A}} (\sup \mathcal{A})\pi(a|s) = \sup \mathcal{A}$, so we have $v_*(s) = \sup_\pi v_\pi(s) \leq \sup_\pi (\sup \mathcal{A}) = \sup \mathcal{A}$. On the other hand, for any action $a \in \mathcal{A}$, we can construct a deterministic policy $\pi : s \mapsto a$, whose state value is $v_\pi(s) = a$. Therefore, $v_*(s) \geq a$. Since a can be any value inside the action space $\mathcal{A}$, so the optimal state value $v_\pi(s) = a$. Next, we check the optimal action values. When the action $a \in \mathcal{A}$ is fixed, no matter what that policy is, the action value is a. Therefore, the optimal action value is $q_*(s, a) = a$. In this example, if we further define the action space to be $\mathcal{A} = (0, 1)$, the optimal state value is $v_*(s) = \sup \mathcal{A} = 1$. However, the state value of any policy π satisfies $v_\pi(s) = \sum_{a \in \mathcal{A}} a\pi(a|s) < \sum_{a \in \mathcal{A}} \pi(a|s) = 1$, so $v_\pi(s) = v_*(s)$ does not hold. In this way, we prove that this environment has no optimal policy.

It is very complex to analyze when optimal policy exists. For example, an optimal policy exists when any one of the following conditions is met:

- The state space S is discrete (all finite sets are discrete), and the action spaces $\mathcal{A}(s)$ ($s \in S$) are finite.

- The state space S is discrete, and the action spaces $\mathcal{A}(s)$ $(s \in S)$ are compact (all bounded close intervals in real line are compact). And the transition probability $p(s'|s, a)$ $(s, s' \in S)$ is continuous over the action a.
- The state space S is Polish (examples of Polish spaces include n-dimension real space $\mathbb{R}^n$, and the close interval $[0, 1]$), and the action spaces $\mathcal{A}(s)$ $(s \in S)$ are finite.
- The state space S is Polish, and the action spaces $\mathcal{A}(s)$ $(s \in S)$ are compact metric spaces. And $r(s, a)$ is bounded.

These conditions and their proofs are all very complex. For simplicity, we usually incorrectly assume the existence of an optimal policy.

! **Note**

For the environment that has no optimal policy, we may also consider ε-optimal policies. The definition of ε-optimal policies is as follows: Given $\varepsilon > 0$, if a policy π_* satisfies

$$v_{\pi_*}(s) \quad > v_*(s) - \varepsilon, \qquad s \in S$$
$$q_{\pi_*}(s, a) > q_*(s, a) - \varepsilon, \quad s \in S, a \in \mathcal{A},$$

policy π_* is an ε-**optimal policy**. The concept of ε-optimal policy plays an important role in the theoretical analysis of RL algorithms, especially in the research of sample complexity. However, even when the optimal values exist, there may not exist ε-optimal policy.

The optimal policy may not be unique. Here is an example environment where there are multiple different optimal policies: Consider an environment that only has one-step interaction. Its state space has only one element, and its action space is $\mathcal{A} = (0, +\infty)$, and reward R_1 is a constant as $R_1 = 1$. In this example, all policies have the same values, so all policies are optimal policies. There are an infinite number of optimal policies.

2.4.3 Properties of Optimal Values

This section discusses the property of optimal values, under the assumption that both optimal values and optimal policy exist.

First, let us examine the relationship between optimal state values and optimal action values. It has the following twofold:

- Use optimal action values at time t to back up optimal state values at time t:

$$v_*(s) = \max_{a \in \mathcal{A}} q_*(s, a), \quad s \in S.$$

The backup diagram is depicted in Fig. 2.6(a). (Proof by contradiction: We first assume that the above equation does not hold. Then, there exists a state s' such that $v_*(s') < \max_{a \in \mathcal{A}} q_*(s', a)$. Additionally, there exists an action a' such that $q_*(s', a') = \max_{a \in \mathcal{A}} q_*(s', a)$. Therefore, $v_*(s') < q_*(s', a')$. Now we consider another policy π', which is defined as

$$
\pi'(a|s) = \begin{cases} 1, & s = s', a = a' \\ 0, & s = s', a \neq a' \\ \pi_*(a|s), & \text{otherwise.} \end{cases}
$$

We can verify that the old policy and the new policy π' satisfy the condition of policy improvement theorem. Therefore, $\pi_* \prec \pi'$, which conflicts with that π_* is the optimal policy. The proof completes.)

- Use optimal state values at time $t + 1$ to back up optimal action values at time t:

$$
q_*(s, a) = r(s, a) + \gamma \sum_{s' \in S} p(s'|s, a) v_*(s')
$$

$$
= \sum_{s' \in S^+, r \in \mathcal{R}} p(s', r|s, a)[r + \gamma v_*(s')], \quad s \in S, a \in \mathcal{A}.
$$

The backup diagram is depicted in Fig. 2.6(b). (Proof: Just plug in the optimal values into the relationship that uses state values to back up action values.) This relationship can be re-written as

$$
q_*(S_t, A_t) = E[R_{t+1} + \gamma v_*(S_{t+1})].
$$

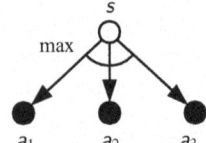

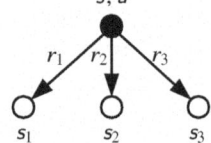

(a) use optimal action value to back up optimal state value

(b) use optimal state value to back up optimal action value

Fig. 2.6   Backup diagram for optimal state values and optimal action values backing up each other.

This section only considers homogenous MDP, whose optimal values do not vary with time. Therefore, the optimal state values at time t are identical to the optimal state values at time $t + 1$, and the optimal action values at time t are identical to the optimal action values at time $t + 1$.

Based on the relationship that optimal state values and optimal action values back up each other, we can further derive **Bellman Optimal Equation (BOE)**. It also has two forms.

- Use optimal state values at time $t + 1$ to back up optimal state values at time t:

$$v_*(s) = \max_{a \in \mathcal{A}} \left[r(s, a) + \gamma \sum_{s' \in S} p(s'|s, a) v_*(s') \right], \quad s \in S.$$

The backup diagram is depicted in Fig. 2.7(a). It can be re-written as

$$v_*(S_t) = \max_{a \in \mathcal{A}} E\left[R_{t+1} + \gamma v_*(S_{t+1}) \right].$$

- Use optimal action values at time $t + 1$ to back up optimal action values at time t:

$$q_*(s, a) = r(s, a) + \gamma \sum_{s' \in S} p(s'|s, a) \max_{a' \in \mathcal{A}} q_*(s', a'), \quad s \in S, a \in \mathcal{A}.$$

The backup diagram is depicted in Fig. 2.7(b). It can be re-written as

$$q_*(S_t, A_t) = E\left[R_{t+1} + \gamma \max_{a' \in \mathcal{A}} q_*(S_{t+1}, a') \right].$$

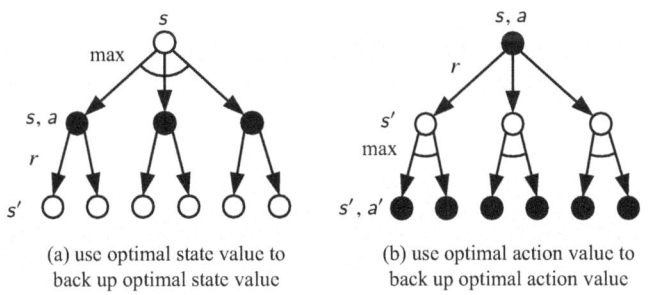

(a) use optimal state value to   (b) use optimal action value to
back up optimal state value   back up optimal action value

Fig. 2.7 **Backup diagram for optimal state values and optimal action values backing up themselves.**

Example 2.17 (Feed and Full) For the dynamics in Table 2.2 with discount factor $\gamma = \frac{4}{5}$, its optimal values satisfy

$$v_*(\text{hungry}) \qquad\qquad = \max\left\{q_*(\text{hungry}, \text{ignore}), q_*(\text{hungry}, \text{feed})\right\}$$

$$v_*(\text{full}) \qquad\qquad = \max\left\{q_*(\text{full}, \text{ignore}), q_*(\text{full}, \text{feed})\right\}$$

$$q_*(\text{hungry}, \text{ignore}) = -2 + \frac{4}{5}\left(1v_*(\text{hungry}) + 0v_*(\text{full})\right)$$

$$q_*(\text{hungry}, \text{feed}) \;\;= -\frac{1}{3} + \frac{4}{5}\left(\frac{1}{3}v_*(\text{hungry}) + \frac{2}{3}v_*(\text{full})\right)$$

$$q_*(\text{full}, \text{ignore}) \;\;= -1 + \frac{4}{5}\left(\frac{3}{4}v_*(\text{hungry}) + \frac{1}{4}v_*(\text{full})\right)$$

$$q_*(\text{full}, \text{feed}) \;\;= +1 + \frac{4}{5}\left(0v_*(\text{hungry}) + 1v_*(\text{full})\right).$$

Remarkably, the relationships among optimal values do not depend on the existences or uniqueness of optimal policy. When the optimal values exist but optimal policies are not existed or unique, those relationships always hold.

2.4.4 Calculate Optimal Values

This section introduces the methods to calculate optimal values. First, it shows how to use the relationship among optimal values to get optimal values, and then it shows how to use linear programming to get optimal values.

Example 2.18 (Feed and Full) The previous example has shown that, for the dynamics in Table 2.2 with discount factor $\gamma = \frac{4}{5}$, its optimal values satisfy

$$v_*(\text{hungry}) \qquad\qquad = \max\left\{q_*(\text{hungry}, \text{ignore}), q_*(\text{hungry}, \text{feed})\right\}$$

$$v_*(\text{full}) \qquad\qquad = \max\left\{q_*(\text{full}, \text{ignore}), q_*(\text{full}, \text{feed})\right\}$$

$$q_*(\text{hungry}, \text{ignore}) = -2 + \frac{4}{5}\left(1v_*(\text{hungry}) + 0v_*(\text{full})\right)$$

$$q_*(\text{hungry}, \text{feed}) \;\;= -\frac{1}{3} + \frac{4}{5}\left(\frac{1}{3}v_*(\text{hungry}) + \frac{2}{3}v_*(\text{full})\right)$$

$$q_*(\text{full}, \text{ignore}) \;\;= -1 + \frac{4}{5}\left(\frac{3}{4}v_*(\text{hungry}) + \frac{1}{4}v_*(\text{full})\right)$$

$$q_*(\text{full}, \text{feed}) \;\;= +1 + \frac{4}{5}\left(0v_*(\text{hungry}) + 1v_*(\text{full})\right).$$

This example solves this equation set directly to get the optimal values. This equation set includes the max operator, so it is a non-linear equation set. We can bypass the max operator by converting this equation set to several equation sets by discussion case by case.

Case I: $q_*(\text{hungry}, \text{ignore}) > q_*(\text{hungry}, \text{feed})$ and $q_*(\text{full}, \text{ignore}) > q_*(\text{full}, \text{feed})$. In this case, $v_*(\text{hungry}) = q_*(\text{hungry}, \text{ignore})$ and $v_*(\text{full}) = q_*(\text{full}, \text{ignore})$. So we get the following linear equation set

$$v_*(\text{hungry}) \qquad = q_*(\text{hungry, ignore})$$

$$v_*(\text{full}) \qquad\qquad = q_*(\text{full, ignore})$$

$$q_*(\text{hungry, ignore}) = -2 + \frac{4}{5}\Big(1v_*(\text{hungry}) + 0v_*(\text{full})\Big)$$

$$q_*(\text{hungry, feed}) \quad = -\frac{1}{3} + \frac{4}{5}\Big(\frac{1}{3}v_*(\text{hungry}) + \frac{2}{3}v_*(\text{full})\Big)$$

$$q_*(\text{full, ignore}) \qquad = -1 + \frac{4}{5}\Big(\frac{3}{4}v_*(\text{hungry}) + \frac{1}{4}v_*(\text{full})\Big)$$

$$q_*(\text{full, feed}) \qquad = +1 + \frac{4}{5}\Big(0v_*(\text{hungry}) + 1v_*(\text{full})\Big).$$

It leads to

$$v_*(\text{hungry}) = q_*(\text{hungry, ignore}) = \frac{6}{11}, \quad q_*(\text{hungry, feed}) = -\frac{23}{3},$$

$$v_*(\text{full}) = q_*(\text{full, ignore}) = -\frac{35}{4}, \qquad q_*(\text{full, feed}) = -6.$$

This solution set satisfies neither $q_*(\text{hungry, ignore}) > q_*(\text{hungry, feed})$ nor $q_*(\text{full, ignore}) > q_*(\text{full, feed})$, so it is not a valid solution.

Case II: $q_*(\text{hungry, ignore}) \leq q_*(\text{hungry, feed})$ and $q_*(\text{full, ignore}) > q_*(\text{full, feed})$. In this case, $v_*(\text{hungry}) = q_*(\text{hungry, feed})$ and $v_*(\text{full}) = q_*(\text{full, ignore})$. Solve the linear equation set, and we can get

$$q_*(\text{hungry, ignore}) = -\frac{22}{5}, \qquad v_*(\text{hungry}) = q_*(\text{hungry, feed}) = -3,$$

$$v_*(\text{full}) = q_*(\text{full, ignore}) = -\frac{7}{2}, \quad q_*(\text{full, feed}) = -\frac{9}{5}.$$

This solution set satisfies $q_*(\text{full, ignore}) > q_*(\text{full, feed})$, so it is not a valid solution.

Case III: $q_*(\text{hungry, ignore}) > q_*(\text{hungry, feed})$ and $q_*(\text{full, ignore}) \leq q_*(\text{full, feed})$. In this case, $v_*(\text{hungry}) = q_*(\text{hungry, ignore})$ and $v_*(\text{full}) = q_*(\text{full, feed})$. Solve the linear equation set, and we can get

$$v_*(\text{hungry}) = q_*(\text{hungry, ignore}) = -6, \quad q_*(\text{hungry, feed}) = -\frac{1}{3},$$

$$v_*(\text{full}) = q_*(\text{full, ignore}) = -10, \qquad q_*(\text{full, feed}) = 5.$$

This solution set does not satisfy $q_*(\text{hungry, ignore}) > q_*(\text{hungry, feed})$, so it is not a valid solution.

Case IV: $q_*(\text{hungry, ignore}) \leq q_*(\text{hungry, feed})$ and $q_*(\text{full, ignore}) \leq q_*(\text{full, feed})$. In this case, $v_*(\text{hungry}) = q_*(\text{hungry, feed})$ and $v_*(\text{full}) = q_*(\text{full, feed})$. Solve the linear equation set, and we can get

$$q_*(\text{hungry, ignore}) = \frac{6}{11}, \qquad v_*(\text{hungry}) = q_*(\text{hungry, feed}) = \frac{35}{11},$$

$$v_*(\text{full}) = q_*(\text{full, ignore}) = \frac{21}{11}, \quad q_*(\text{full, feed}) = 5.$$

This solution set satisfies both $q_*(\text{hungry, ignore}) \leq q_*(\text{hungry, feed})$ and $q_*(\text{full, ignore}) \leq q_*(\text{full, feed})$, so it is a valid solution set.

Therefore, among the four cases, only Case IV has a valid solution. In conclusion, we get the optimal state values in Table 2.18 and optimal action values in Table 2.19.

Table 2.18 Optimal state values derived from Table 2.2.

s	$v_*(s)$
hungry	35/11
full	5

Table 2.19 Optimal action values derived from Table 2.2.

s	a	$q_*(s, a)$
hungry	ignore	6/11
hungry	feed	35/11
full	ignore	21/11
full	feed	5

Remarkably, Bellman optimal equation only depends on the dynamics of the environment, and does not depend on the initial state distribution. Since Bellman optimal equation set fully determines the optimal values, the optimal values also only depend on the dynamics, and do not depend on the initial state distribution.

Such case-by-case analysis should consider $|\mathcal{A}|^{|S|}$ different cases, and the number of cases is too large. The remaining part of this subsection will introduce the linear programming method, which can reduce the computation complexity.

Consider the LP programming:

$$\underset{\rho(s,a):s\in S, a\in \mathcal{A}(s)}{\text{maximize}} \sum_{s\in S, a\in\mathcal{A}(s)} r(s, a)\rho(s, a)$$

$$\text{s.t.} \quad \sum_{a'\in\mathcal{A}(s')} \rho(s', a') - \gamma \sum_{s\in S, a\in\mathcal{A}(s)} p(s'|s, a)\rho(s, a) = p_{S_0}(s'), \ s' \in S$$

$$\rho(s, a) \geq 0, \qquad s \in S, a \in \mathcal{A}(s).$$

Recap that Sect. 2.3.4 introduced the equivalency between discounted distribution and policy. The equivalency tells us that, each feasible solution in this programming corresponds to a policy whose discounted distribution is exactly the solution, and every policy corresponds to a solution in the feasible region. In other word, the feasible region of this programming is exactly the policy space. Recap that Sect. 2.3.5 introduced how to use discounted distribution to express the expected discounted return. The expression tells us that the objective of this programming is exactly the expected initial return. Therefore, this programming maximizes the expected discounted return over all possible policies. That is exactly what we want to do. This programming can be written in the following vector form:

$$\underset{\rho}{\text{maximize}} \quad \mathbf{r}^\top \rho$$
$$\text{s.t.} \quad \mathbf{I}_{S|S,A}\,\rho - \gamma \mathbf{P}_{S_{t+1}|S_t,A_t}\,\rho = \mathbf{p}_{S_0}$$
$$\rho \geq \mathbf{0}.$$

Interdisciplinary Reference 2.2
Optimization: Duality in Linear Programming

The following two programmings are dual to each other.

- Primal:

$$\underset{\mathbf{x}}{\text{minimize}} \quad \mathbf{c}^\top \mathbf{x}$$
$$\text{s.t.} \quad \mathbf{A}\mathbf{x} \geq \mathbf{b}.$$

- Dual:

$$\underset{\mathbf{y}}{\text{maximize}} \quad \mathbf{b}^\top \mathbf{y}$$
$$\text{s.t.} \quad \mathbf{A}^\top \mathbf{y} = \mathbf{c}$$
$$\mathbf{y} \geq \mathbf{0}.$$

Using the duality of linear programming, we can view the aforementioned programming whose decision variables are state–action distribution as the dual form of the following linear programming:

$$\underset{v(s):s\in\mathcal{S}}{\text{minimize}} \quad \sum_{s\in\mathcal{S}} p_{S_0}(s)v(s)$$
$$\text{s.t.} \quad v(s) - \gamma \sum_{s'\in\mathcal{S}} p(s'|s,a)v(s'), \quad s \in \mathcal{S}, a \in \mathcal{A}(s).$$

This vector form of this primal programming is

$$\underset{\mathbf{v}}{\text{minimize}} \quad \mathbf{p}_{S_0}^\top \mathbf{v}$$
$$\text{s.t.} \quad \mathbf{I}_{S|S,A}^\top \mathbf{v} - \gamma \mathbf{P}_{S_{t+1}|S_t,A_t}^\top \mathbf{v} \geq \mathbf{r}.$$

Consider the following two expressions of expected initial return

$$g_\pi = \sum_{s \in S} p_{S_0}(s) v_\pi(s)$$

$$g_\pi = \sum_{s \in S, a \in \mathcal{A}(s)} r(s, a) \rho_\pi(s, a),$$

we know the primal programming and dual programming share the same optimal value. That is, the duality is strong-duality.

Section 2.4.3 told us that the optimal values do not depend on the initial state distribution. That is, the optimal values keep the same if we change the initial state distribution to an arbitrary distribution. Therefore, the optimal solution of the primal problem does not change if we replace the initial state distribution $p_{S_0}(s)$ with another arbitrary distribution, say $c(s)$ ($s \in S$), where $c(s)$ ($s \in S$) should satisfy $c(s) > 0$ ($s \in S$) and $\sum_{s \in S} c(s) = 1$. After the replacement, the primal programming becomes

$$\underset{v(s): s \in S}{\text{minimize}} \quad \sum_{s \in S} c(s) v(s)$$

s.t. $\quad v(s) - \gamma \sum_{s' \in S} p(s'|s, a) v(s') \geq r(s, a), \quad s \in S, a \in \mathcal{A}(s).$

The replacement will not change the solution of the programming. Furthermore, we can relax the constraint $\sum_{s \in S} c(s) = 1$ in $c(s)$ ($s \in S$) to $\sum_{s \in S} c(s) > 0$. Such relaxation does not change the optimal solution of the primal problem, although it indeed scales the optimal value by $\sum_{s \in S} c(s)$. Consequently, the primal problem can be further adjusted to

$$\underset{v(s): s \in S}{\text{minimize}} \quad \sum_{s \in S} c(s) v(s)$$

s.t. $\quad v(s) - \gamma \sum_{s' \in S} p(s'|s, a) v(s') \geq r(s, a), \quad s \in S, a \in \mathcal{A}(s),$

where $c(s) > 0$ ($s \in S$). For example, we can set $c(s) = 1$ ($s \in S$). This is the most common form of **Linear Programming** (**LP**) method.

People tend to use the primal programming whose decision variables are state values, rather than the duel programming whose decision variables are state–action distributions. The reasons are as follows: First, the optimal value is unique, but the optimal policies may not unique. Second, we can further change the primal programming such that it does not depend on the initial state distribution, while the duel programming always requires the knowledge of initial state distribution.

Next, let us go over an example that uses linear programming to find optimal values. This example will not use initial state distribution.

Example 2.19 (Feed and Full) Consider the dynamics in Table 2.2 with discount factor $\gamma = \frac{4}{5}$. Plug in Tables 2.3 and 2.4 to the linear programming and set $c(s) = 1$

$(s \in \mathcal{S})$, and we get

$$\underset{v(\text{hungry}), v(\text{full})}{\text{minimize}} \quad v(\text{hungry}) + v(\text{full})$$

$$\text{s.t.} \qquad v(\text{hungry}) - \frac{4}{5}\Big(1v(\text{hungry}) + 0v(\text{full})\Big) \geq -2$$

$$v(\text{hungry}) - \frac{4}{5}\Big(\frac{1}{3}v(\text{hungry}) + \frac{2}{3}v(\text{full})\Big) \geq -\frac{1}{3}$$

$$v(\text{full}) \quad - \frac{4}{5}\Big(\frac{3}{4}v(\text{hungry}) + \frac{1}{4}v(\text{full})\Big) \geq -1$$

$$v(\text{full}) \quad - \frac{4}{5}\Big(0v(\text{hungry}) + 1v(\text{full})\Big) \geq +1.$$

Solving this linear programming will lead to optimal state values in Table 2.18. Then we can further get optimal action values in Table 2.19.

Unfortunately, directly applying linear programming to find optimal values may encounter the following difficulties:

- It may be difficult to get all parameters for linear programming. Obtaining all parameters requires a complete understanding of the dynamics, which may be impossible for real-world tasks since real-world tasks are usually too complex to model accurately.
- It is highly complex to solve the linear programming. The primal programming has $|\mathcal{S}|$ decision variables and $|\mathcal{S}||\mathcal{A}|$ constraints. Real-world problems usually have very large state space, even infinite state space, and the combination of state space and action space may be even larger. In those case, solving those linear programming can be computationally infeasible. Therefore, people may consider some indirect method to find the optimal values, or even the sub-optimal values.

2.4.5 Use Optimal Values to Find Optimal Strategy

Previous sections cover the properties of optimal values and how to find optimal values. We can find optimal values from optimal policy, and vice visa. We will learn how to find an optimal policy from optimal values.

There may exist multiple optimal policies for a dynamics. However, the optimal values are unique. Therefore, all optimal policies share the same optimal values. Therefore, picking an arbitrary optimal policy is without loss of generality. One way to pick an optimal policy is to construct a deterministic policy in the following way:

$$\pi_*(s) = \underset{a \in \mathcal{A}}{\arg\max}\, q_*(s, a), \quad s \in \mathcal{S}.$$

Here, if multiple different actions a can attain the maximal values of $q_*(s, a)$, we can choose any one of them.

Example 2.20 (Feed and Full) For the dynamics in Table 2.2, we have found the optimal actions values as Table 2.19. Since $q_*(\text{hungry}, \text{ignore}) < q_*(\text{hungry}, \text{feed})$ and $q_*(\text{full}, \text{ignore}) < q_*(\text{full}, \text{feed})$, the optimal policy is $\pi_*(\text{hungry}) = \pi_*(\text{full}) = $ feed.

The previous section told us that the optimal values only depend on the dynamics of the environment, and do not depend on the initial state distribution. This section tells us the optimal policy, if exists, can be determined from optimal values. Therefore, optimal policy, if it exists, only depends on the dynamics of the environment, and does not depend on the initial state distribution.

2.5 Case Study: CliffWalking

This section considers the task `CliffWalking-v0`: As shown in Fig. 2.8, there is a 4×12 grid. In the beginning, the agent is at the left-bottom corner (state 36 in Fig. 2.8). In each step, the agent can move a step toward a direction selected from right, up, left, and down. And each move will be rewarded by -1. Additionally, the movement has the following constraints:

- The agent can not move out of the grid. If the agent wants to move out of the agent, the agent should be kept where it was. However, this attempt is still rewarded by -1.
- If the agent moves to the states 37–46 in the bottom row (which can be viewed as the cliff), the agent will be placed at the beginning state (state 36) and be rewarded -100.

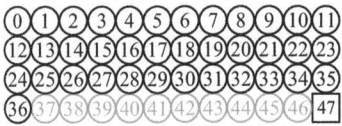

Fig. 2.8   Grid of the task `CliffWalking-v0`.
State 36 is the start point. States 37–46 are the cliff. State 47 is the goal.

Apparently, the optimal policy for this environment is: first move upward at the start state, and then move in the right directions all the way to the rightmost column, where the agent should move downward. The average episode reward is -13.

2.5.1 Use Environment

This section covers the usage of the environment. In fact, most of the contents in this section have been covered in Sect. 1.6, so readers should be able to finish them by themselves. For example, could you please try to explore what is the observation space, action space, and implement the agent with the optimal policy?

Code 2.1 imports this environment and checks the information of this environment.

Code 2.1 Import the environment `CliffWalking-v0` and check its information.

`CliffWalking-v0_Bellman_demo.ipynb`

```
import gym
import inspect
env = gym.make('CliffWalking-v0')
for key in vars(env):
    logging.info('%s: %s', key, vars(env)[key])
logging.info('type = %s', inspect.getmro(type(env)))
```

A state in this MDP is an int value among the state space $S = \{0, 1, \ldots, 46\}$, which indicates the position of the agent in Fig. 2.8. The state space including the terminal state is $S^+ = \{0, 1, \ldots, 46, 47\}$. An action is an int value among the action space $\mathcal{A} = \{0, 1, 2, 3\}$. Action 0 means moving upward. Action 1 means moving in the right direction. Action 2 means moving downward. Action 3 means moving in the right direction. The reward space is $\{-1, -100\}$. The reward is -100 when the agent reaches the cliff, otherwise -1.

The dynamic of the environment is saved in `env.P`. We can use the following codes to check the dynamic related to a state–action pair (for example, state 14, move down):

```
logging.info('P[14] = %s', env.P[14])
logging.info('P[14][2] = %s', env.P[14][2])
env.P[14][2]
```

It is a tuple, where each element consists of transition probability $p(s', r|s, a)$, next state s', reward r, and indicator of episode end $1_{[s'=s_{\text{end}}]}$. For example, `env.P[14][2]` is the list of a single tuple: `[(1.0, 26, -1, False)]`, meaning that

$$p(s_{26}, -1|s_{14}, a_2) = 1.$$

2.5.2 Policy Evaluation

This section considers policy evaluation. We use Bellman expectation equations to solve the state values and action values of a given policy. The Bellman equations that use state values to back up state values are:

$$v_\pi(s) = \sum_{a \in \mathcal{A}} \pi(a|s) \sum_{s' \in S^+, r \in \mathcal{R}} p(s', r|s, a)\left[r + \gamma v_\pi(s')\right], \quad s \in S.$$

This is a linear equation set, whose standard form is

$$v_\pi(s) - \gamma \sum_{a \in \mathcal{A}} \pi(a|s) \sum_{s' \in S^+, r \in \mathcal{R}} p(s', r|s, a) v_\pi(s')$$

$$= \sum_{a \in \mathcal{A}} \pi(a|s) \sum_{s' \in S^+, r \in \mathcal{R}} r p(s', r|s, a), \quad s \in S.$$

We can solve the standard form of the linear equation set and get the state values. After getting the state values, we can use the following relationship to find the action values.

$$q_\pi(s, a) = \sum_{s' \in S^+, r \in \mathcal{R}} p(s', r|s, a)\left[r + \gamma v_\pi(s')\right], \quad s \in S, a \in \mathcal{A}.$$

The function `evaluate_bellman()` in Code 2.2 implements the aforementioned functionality, including finding the state values by solving the linear equation set using the function `np.linalg.solve()` and finding the action values based on state values.

Code 2.2   Find states values and action values using Bellman expectation equations.
`CliffWalking-v0_Bellman_demo.ipynb`

```
def evaluate_bellman(env, policy, gamma=1., verbose=True):
    if verbose:
        logging.info('policy = %s', policy)
    a, b = np.eye(env.nS), np.zeros((env.nS))
    for state in range(env.nS - 1):
        for action in range(env.nA):
            pi = policy[state][action]
            for p, next_state, reward, terminated in env.P[state][action]:
                a[state, next_state] -= (pi * gamma * p)
                b[state] += (pi * reward * p)
    v = np.linalg.solve(a, b)
    q = np.zeros((env.nS, env.nA))
    for state in range(env.nS - 1):
        for action in range(env.nA):
            for p, next_state, reward, terminated in env.P[state][action]:
                q[state][action] += ((reward + gamma * v[next_state]) * p)
    if verbose:
        logging.info('state values = %s', v)
        logging.info('action values = %s', q)
    return v, q
```

Then we call the function `evaluate_bellman()` to evaluate a policy. Before that, we need to define a policy. For example, we can use the following line of code to define a policy that selects actions randomly:

```
policy = np.ones((env.nS, env.nA)) / env.nA
```

Or we can use the following lines of codes to define the optimal policy:

```
1   actions = np.ones(env.nS, dtype=int)
2   actions[36] = 0
3   actions[11::12] = 2
4   policy = np.eye(env.nA)[actions]
```

Or we can use the following lines of codes to generate a policy randomly:

```
1   policy = np.random.uniform(size=(env.nS, env.nA))
2   policy = policy / np.sum(policy, axis=1, keepdims=True)
```

After designating the policy, we use the following codes to evaluate the designated policy:

```
1   state_values, action_values = evaluate_bellman(env, policy)
```

2.5.3 Solve Optimal Values

This section will use LP method to find the optimal values and the optimal policy of the task.

The primal programming can be written as:

$$
\begin{aligned}
\operatorname*{minimize}_{v(s):s\in S} \quad & \sum_{s\in S} v(s) \\
\text{s.t.} \quad & v(s) - \gamma \sum_{s'\in S^{+}, r\in \mathcal{R}} p(s', r|s, a) v(s') \geq \sum_{s'\in S^{+}, r\in \mathcal{R}} r p(s', r|s, a), \\
& \hspace{6cm} s \in S, a \in \mathcal{A}
\end{aligned}
$$

where the coefficients in the objective function, $c(s)$ ($s \in S$), are fixed as 1. You can choose other positive numbers as well.

Code 2.3 uses the function scipy.optimize.linprog() to solve the linear programming. The 0th parameter of this function is the coefficients of the objective. We choose 1's here. The 1st and 2nd parameter are the values of **A** and **b**, where **A** and **b** are the coefficients of constraints in the form $\mathbf{Ax} \leq \mathbf{b}$. These values are calculated at the beginning of the function optimal_bellman(). Besides, the function scipy.optimize.linprog() has a compulsory keyword parameter bounds, which indicate the bounds of each decision variable. In our linear programming, decision variables have no bounds, but we still need to fill it. Additionally, the function has a keyword parameter method, which indicates the method of optimization. The default method is not able to deal with inequality constraints, so we choose the interior-point method, which is able to deal with inequality constraints.

Code 2.3   Find optimal values using LP method.
`CliffWalking-v0_Bellman_demo.ipynb`

```
import scipy.optimize

def optimal_bellman(env, gamma=1., verbose=True):
    p = np.zeros((env.nS, env.nA, env.nS))
    r = np.zeros((env.nS, env.nA))
    for state in range(env.nS - 1):
        for action in range(env.nA):
            for prob, next_state, reward, terminated in env.P[state][action]:
                p[state, action, next_state] += prob
                r[state, action] += (reward * prob)
    c = np.ones(env.nS)
    a_ub = gamma * p.reshape(-1, env.nS) - \
            np.repeat(np.eye(env.nS), env.nA, axis=0)
    b_ub = -r.reshape(-1)
    a_eq = np.zeros((0, env.nS))
    b_eq = np.zeros(0)
    bounds = [(None, None),] * env.nS
    res = scipy.optimize.linprog(c, a_ub, b_ub, bounds=bounds,
            method='interior-point')
    v = res.x
    q = r + gamma * np.dot(p, v)
    if verbose:
        logging.info('optimal state values = %s', v)
        logging.info('optimal action values = %s', q)
    return v, q
```

2.5.4 Solve Optimal Policy

This section will find the optimal policy using the optimal values that were found in the previous section.

Code 2.4 provides the codes to find an optimal policy from optimal action values. It applies the `argmax()` on the optimal action values, and gets an optimal deterministic policy.

Code 2.4   Find an optimal deterministic policy from optimal action values.
`CliffWalking-v0_Bellman_demo.ipynb`

```
optimal_actions = optimal_action_values.argmax(axis=1)
logging.info('optimal policy = %s', optimal_actions)
```

2.6 Summary

- In a Markov Decision Process (MDP), S is the state space. S^+ is the state space including the terminal state. $\mathcal{A}$ is the action space. $\mathcal{R}$ is the reward space.
- The environment in a Discrete-Time MDP (DTMDP) can be modeled using initial state distribution p_{S_0} and the dynamics p. $p(s', r|s, a)$ is the transition

probability from state s and action a to reward r and next state s'. Policy is denoted as π. $\pi(a|s)$ is the probability of action a given the state s.

- Initial distributions:

$$p_{S_0}(s) \overset{\text{def}}{=} \Pr[S_0 = s], \qquad\qquad s \in S$$

$$p_{S_0,A_0;\pi}(s, a) \overset{\text{def}}{=} \Pr_\pi[S_0 = s, A_0 = a], \quad s \in S, a \in \mathcal{A}(s).$$

Transition probabilities:

$$p(s'|s, a) \overset{\text{def}}{=} \Pr[S_{t+1} = s'|S_t = s, A_t = a], \qquad s \in S, a \in \mathcal{A}(s), s' \in S^+$$

$$p_\pi(s'|s) \overset{\text{def}}{=} \Pr_\pi[S_{t+1} = s'|S_t = s], \qquad\qquad s \in S, s' \in S^+$$

$$p_\pi(s', a'|s, a) \overset{\text{def}}{=} \Pr_\pi[S_{t+1} = s', A_{t+1} = a'|S_t = s, A_t = a],$$
$$s \in S, a \in \mathcal{A}(s), s' \in S^+, a' \in \mathcal{A}.$$

Expected rewards:

$$r(s, a) \overset{\text{def}}{=} \mathrm{E}[R_{t+1}|S_t = s, A_t = a], \quad s \in S, a \in \mathcal{A}(s)$$

$$r_\pi(s) \overset{\text{def}}{=} \mathrm{E}_\pi[R_{t+1}|S_t = s], \qquad\qquad s \in S.$$

- Return is the summation of future rewards, discounted by the discount factor γ.

$$G_t \overset{\text{def}}{=} \sum_{\tau=0}^{+\infty} \gamma^\tau R_{t+\tau+1}.$$

- Given a policy π, the values are defined as expected return starting from a state s or state–action pair (s, a):

$$\text{state value} \quad v_\pi(s) \overset{\text{def}}{=} \mathrm{E}_\pi[G_t|S_t = s], \qquad\qquad s \in S$$

$$\text{action value} \quad q_\pi(s, a) \overset{\text{def}}{=} \mathrm{E}_\pi[G_t|S_t = s, A_t = a], \quad s \in S, a \in \mathcal{A}(s).$$

- Relationships between state values and action values include

$$v_\pi(s) = \sum_{a \in \mathcal{A}} \pi(a|s)q_\pi(s, a), \qquad\qquad s \in S$$

$$q_\pi(s, a) = r(s, a) + \gamma \sum_{s' \in S} p(s'|s, a)v_\pi(s'), \quad s \in S, a \in \mathcal{A}(s).$$

- The visitation frequency of a policy π, a.k.a. discounted distributions, is defined as the discounted expectation of how many times a state or a state–action pair will be visited:

$$\eta_\pi(s) \overset{\text{def}}{=} \sum_{t=0}^{+\infty} \gamma^t \Pr_\pi[S_t = s], \qquad s \in S$$

$$\rho_\pi(s, a) \overset{\text{def}}{=} \sum_{t=0}^{+\infty} \gamma^t \Pr_\pi[S_t = s, A_t = a], \quad s \in S, a \in \mathcal{A}(s).$$

- The discounted visitation frequency of a policy π satisfies

$$\eta_\pi(s) = \sum_{a \in \mathcal{A}(s)} \rho_\pi(s, a) \qquad s \in S,$$

$$\rho_\pi(s, a) = \eta_\pi(s)\pi(a|s) \qquad s \in S, a \in \mathcal{A}(s),$$

$$\eta_\pi(s') = p_{S_0}(s') + \sum_{s \in S, a \in \mathcal{A}(s)} \gamma p(s'|s, a)\rho_\pi(s, a), \quad s' \in S.$$

- State values and action values do not depend on initial distributions. Discounted visitation frequencies do not depend on reward distributions.
- Expectation of initial return $g_\pi \overset{\text{def}}{=} \mathrm{E}_\pi[G_0]$ can be expressed as

$$g_\pi = \sum_{s \in S} p_{S_0}(s)v_\pi(s)$$

$$g_\pi = \sum_{s \in S, a \in \mathcal{A}(s)} r(s, a)\rho_\pi(s, a).$$

- We can define a partial order over the policy set using the definition of values. A policy is optimal when it is not worse than any policies.
- All optimal policies in an environment share the same state values and action values, which are equal to optimal state values (denoted as v_*) and optimal action values (denoted as q_*).
- Optimal state values and optimal action values satisfy

$$v_*(s) = \max_{a \in \mathcal{A}(s)} q_*(s, a), \qquad s \in S$$

$$q_*(s, a) = r(s, a) + \gamma \sum_{s' \in S} p(s'|s, a)v_*(s'), \quad s \in S, a \in \mathcal{A}(s).$$

- The vector representation of Bellman equation is in the form of

$$\mathbf{x} = \mathbf{y} + \gamma \mathbf{P} \mathbf{x},$$

where $\mathbf{x}$ can be state value vector, action value vector, discounted state distribution vector, or discounted state–action distribution vector.
- We can use the following Linear Programming (LP) to find the optimal state values,

$$\underset{v(s): s \in S}{\text{minimize}} \sum_{s \in S} c(s)v(s)$$

s.t. $\quad v(s) - \gamma \sum_{s' \in S} p(s'|s, a) v(s') \geq r(s, a), \quad s \in S, a \in \mathcal{A}(s)$

where $c(s) > 0$ $(s \in S)$.

- After finding the optimal action values, we can obtain an optimal deterministic policy using

$$\pi_*(s) = \arg\max_{a \in \mathcal{A}(s)} q_*(s, a), \quad s \in S,$$

if there is a state $s \in S$ such that there are many actions a that can maximize $q_*(s, a)$, we can pick an arbitrary action among them.

- Optimal values and optimal policy only depend on the dynamics of the environment, and do not depend on the initial state distribution.

2.7 Exercises

2.7.1 Multiple Choices

2.1 On the definition of state values, choose the correct one: (   )

A. $v(s) = E_\pi[G_t|S = s, A = a]$.
B. $v(s) = E_\pi[G_t|S = s]$.
C. $v(s) = E_\pi[G_t]$.

2.2 On the definition of action values, choose the correct ones: (   )

A. $q(s, a) = E_\pi[G_t|S = s, A = a]$.
B. $q(s, a) = E_\pi[G_t|S = s]$.
C. $q(s, a) = E_\pi[G_t]$.

2.3 On the discounted distribution of sequential tasks, choose the correct one: (   )

A. $\sum_{s \in S} \eta_\pi(s) = 1$.
B. $\sum_{a \in \mathcal{A}(s)} \rho_\pi(s, a) = 1$ holds for all $s \in S$.
C. $\rho_\pi(s, a) = \pi(a|s)\eta_\pi(s)$ holds for all $s \in S, a \in \mathcal{A}(s)$.

2.4 Which of the following equals to $E_\pi[G_0]$: (   )

A. $E_{S \sim \rho_\pi}[v(S)]$.
B. $E_{(S,A) \sim \rho_\pi}[q(S, A)]$.
C. $E_{(S,A) \sim \rho_\pi}[r(S, A)]$.

2.5 On the optimal values, choose the correct one: (   )

A. Optimal values do not depend on initial state distribution.
B. Optimal values do not depend on the dynamics.
C. Optimal values depend on neither initial state distributions nor dynamics.

2.6 On the values and optimal values, choose the correct one: (   )

A. $v_\pi(s) = \max_{a \in \mathcal{A}(s)} q_\pi(s, a)$.
B. $v_*(s) = \max_{a \in \mathcal{A}(s)} q_*(s, a)$.
C. $v_*(s) = \mathrm{E}_\pi[q_*(s, a)]$.

2.7 On DTMDP, choose the correct one: (   )

A. All environments have optimal policies.
B. Not all environments have optimal policies, but all environments have ε-optimal policies.
C. Not all environments have optimal policies. Not all environments have ε-optimal policies either.

2.7.2 Programming

2.8 Try to interact and solve the Gym environment `RouletteEnv-v0`. You can use whatever methods you can to do that

2.7.3 Mock Interview

2.9 What is MDP? What is DTMDP?

2.10 What are Bellman optimal equations? Why do not we always solve Bellman optimal equations directly to find the optimal policy?

Chapter 3
Model-Based Numerical Iteration

This chapter covers

- Bellman operators and their properties
- model-based numerical policy evaluation algorithm
- model-based numerical policy iterative algorithm
- model-based Value Iterative (VI) algorithm
- bootstrapping, and its advantages and disadvantages
- understanding model-based numerical iterative algorithm using Dynamic Programming (DP)

It is usually too difficult to directly solve Bellman equations. Therefore, this chapter considers an alternative method that solves Bellman equations. This method relies on the dynamics of the environment, and calculates iteratively. Since the iterative algorithms do not learn from data, they are not Machine Learning (ML) algorithms or RL algorithms.

3.1 Bellman Operators and Its Properties

This section will discuss the theoretical backgrounds of model-based numerical iterations. We will first prove that the value space is a complete metric space. Then we define two kinds of operators on the value space: Bellman expectation operators and Bellman optimal operators, and prove that both kinds of operators are contraction mapping. Then we show that values of policy and optimal values are the fixed points of the Bellman equation operators and Bellman optimal operators

Z. Xiao, *Reinforcement Learning*, https://doi.org/10.1007/978-981-19-4933-3_3

respectively. Finally, we show that we can use Banach fixed point theorem to find the values of a policy and the optimal values.

For simplicity, this chapter only considers finite MDP.

Interdisciplinary Reference 3.1
Functional Analysis: Metric and its Completeness

Metric (also known as distance), is a bivariate functional over a set. Given a set X and a functional $d : X \times X \rightarrow \mathbb{R}$ satisfying

- non-negative property: $d(x', x'') \geq 0$ holds for any $x', x'' \in X$;
- uniform property: $d(x', x'') = 0$ leads to $x' = x''$ for any $x', x'' \in X$;
- symmetric: $d(x', x'') = d(x'', x')$ holds for any $x', x'' \in X$;
- triangle inequality: $d(x', x''') \leq d(x', x'') + d(x', x''')$ holds for any $x', x'', x''' \in X$.

The pair (X, d) is called **metric space**.

A metric space is **complete** if all its Cauchy sequences converge within the space. For example, the set of real numbers $\mathbb{R}$ is complete. In fact, the set of real numbers can be defined using the completeness: The set of rational numbers is not complete, so mathematicians add irrational numbers to it to make it complete.

Let $\mathcal{V} \overset{\text{def}}{=} \mathbb{R}^{|S|}$ denote the set of all possible state values $v(s)$ ($s \in S$). Define a bivariate functional d_∞ as

$$d_\infty(v', v'') \overset{\text{def}}{=} \max_s |v'(s) - v''(s)|, \quad v', v'' \in \mathcal{V}.$$

We can prove that $(\mathcal{V}, d_\infty)$ is a metric space. (Proof: It is obviously non-negative, uniform, and symmetric. Now we prove it also satisfies the triangle inequality. For any $s \in S$, we have

$$\begin{aligned}
&|v'(s) - v'''(s)| \\
&= \left| [v'(s) - v''(s)] + [v''(s) - v'''(s)] \right| \\
&\leq |v'(s) - v''(s)| + |v''(s) - v'''(s)| \\
&\leq \max_s |v'(s) - v''(s)| + \max_s |v''(s) - v'''(s)|.
\end{aligned}$$

Therefore, the triangle inequality holds.)

The metric space $(\mathcal{V}, d_\infty)$ is complete. (Proof: Considering an arbitrary Cauchy sequence $\{v_k : k = 0, 1, 2, \ldots\}$. That is, for any $\varepsilon > 0$, there exists a positive integer κ such that, $d_\infty(v_{k'}, v_{k''}) < \varepsilon$ holds for all $k', k'' > \kappa$. Since $|v_k(s) - v_\infty(s)| < \varepsilon$ holds for any $s \in S$, $\{v_k(s) : k = 0, 1, 2, \ldots\}$ is a Cauchy sequence. Due to the completeness of the set of real numbers, we know that $\{v_k(s) : k = 0, 1, 2, \ldots\}$ converges to a real number, which can be denoted as $v_\infty(s)$. Now for any $\varepsilon > 0$, there exists a positive integer $\kappa(s)$ such that $|v_k(s) - v_\infty| < \varepsilon$ holds for all $k >$

$\kappa(s)$. Let $\kappa(\mathcal{S}) = \max_{s \in \mathcal{S}} \kappa(s)$, we have $d_\infty(v_k, v_\infty) < \varepsilon$ for all $k > \kappa(\mathcal{S})$. So $\{v_k : k = 0, 1, 2, \ldots\}$ converges to v_∞. Considering that $v_\infty \in \mathcal{V}$, we prove that the metric space is complete.)

Similarly, we can define action-value space $Q \overset{\text{def}}{=} \mathbb{R}^{|\mathcal{S}\|\mathcal{A}|}$ and the distance

$$d_\infty(q', q'') \overset{\text{def}}{=} \max_{s,a} |q'(s, a) - q''(s, a)|, \quad q', q'' \in Q.$$

And (Q, d_∞) is a complete metric space, too.

Now we define Bellman expectation operators and Bellman optimal operators.

Given the dynamics p and the policy π, we can define Bellman expectation operators of the policy π in the following two ways.

- Bellman expectation operator on the state-value space $\mathfrak{b}_\pi : \mathcal{V} \to \mathcal{V}$:

$$\mathfrak{b}_\pi(v)(s) \overset{\text{def}}{=} r_\pi(s) + \gamma \sum_{s'} p_\pi(s'|s) v(s'), \quad v \in \mathcal{V}, s \in \mathcal{S}.$$

- Bellman expectation operator on the action-value space $\mathfrak{b}_\pi : Q \to Q$:

$$\mathfrak{b}_\pi(q)(s, a) \overset{\text{def}}{=} r(s, a) + \gamma \sum_{s',a'} p_\pi(s', a'|s, a) q(s', a'), \quad q \in Q, s \in \mathcal{S}, a \in \mathcal{A}.$$

Given the dynamics p, we can define Bellman optimal operators in the following two forms.

- Bellman optimal operator on the state-value space $\mathfrak{b}_* : \mathcal{V} \to \mathcal{V}$:

$$\mathfrak{b}_*(v)(s) \overset{\text{def}}{=} \max_a \left[r(s, a) + \gamma \sum_{s'} p(s'|s, a) v(s') \right], \quad v \in \mathcal{V}, s \in \mathcal{S}.$$

- Bellman optimal operator on the action-value space $\mathfrak{b}_* : Q \to Q$:

$$\mathfrak{b}_*(q)(s, a) \overset{\text{def}}{=} r(s, a) + \gamma \sum_{s'} p(s'|s, a) \max_{a'} q(s', a'), \quad q \in Q, s \in \mathcal{S}, a \in \mathcal{A}.$$

We will prove a very important property of Bellman operators: they are all contraction mapping on the metric space $(\mathcal{V}, d_\infty)$ or (Q, d_∞).

Interdisciplinary Reference 3.2
Functional Analysis: Contraction Mapping

A functional $\mathfrak{f} : X \to X$ is called **contraction mapping** (or Lipschitzian mapping) over the metric space (X, d), if there exists a real number $\gamma \in (0, 1)$ such that

$$d\left(\mathfrak{f}(x'), \mathfrak{f}(x'')\right) < \gamma d(x', x''),$$

holds for any $x', x'' \in \mathcal{X}$. The positive real number γ is called Lipschitz constant.

This paragraph proves that the Bellman expectation operator on the state-value space $\mathfrak{b}_\pi : \mathcal{V} \to \mathcal{V}$ is a contraction mapping over the metric space $(\mathcal{V}, d_\infty)$. The definition of $\mathfrak{b}_\pi$ tells us, for any $v', v'' \in \mathcal{V}$, we have

$$\mathfrak{b}_\pi(v')(s) - \mathfrak{b}_\pi(v'')(s) = \gamma \sum_{s'} p_\pi(s'|s)\left[v'(s') - v''(s')\right], \quad s \in \mathcal{S}.$$

Therefore,

$$
\begin{aligned}
\left|\mathfrak{b}_\pi(v')(s) - \mathfrak{b}_\pi(v'')(s)\right| &\le \gamma \sum_{s'} p_\pi(s'|s) \max_{s''}\left|v'(s'') - v''(s'')\right| \\
&= \gamma \sum_{s'} p_\pi(s'|s) d_\infty(v', v'') \\
&= \gamma d_\infty(v', v''), \qquad s \in \mathcal{S}.
\end{aligned}
$$

Since s is an arbitrary state,

$$d_\infty\left(\mathfrak{b}_\pi(v'), \mathfrak{b}_\pi(v'')\right) = \max_s\left|\mathfrak{b}_\pi(v')(s) - \mathfrak{b}_\pi(v'')(s)\right| \le \gamma d_\infty(v', v'').$$

So $\mathfrak{b}_\pi$ is a contraction mapping when $\gamma < 1$.

This paragraph proves that the Bellman expectation operator on the action-value space $\mathfrak{b}_\pi : \mathcal{Q} \to \mathcal{Q}$ is a contraction mapping over the complete metric space $(\mathcal{Q}, d_\infty)$. The definition of $\mathfrak{b}_\pi$ tells us, for any $q', q'' \in \mathcal{Q}$, we have

$$\mathfrak{b}_\pi(q')(s,a) - \mathfrak{b}_\pi(q'')(s,a) = \gamma \sum_{s',a'} p_\pi(s',a'|s,a)\left[q'(s',a') - q''(s',a')\right],$$

$$s \in \mathcal{S}, a \in \mathcal{A}.$$

Therefore,

$$
\begin{aligned}
&\left|\mathfrak{b}_\pi(q')(s,a) - \mathfrak{b}_\pi(q'')(s,a)\right| \\
&\le \gamma \sum_{s',a'} p_\pi(s',a'|s,a) \max_{s'',a''}\left|q'(s'',a'') - q''(s'',a'')\right| \\
&= \gamma \sum_{s',a'} p_\pi(s',a'|s,a) d_\infty(q', q'') \\
&= \gamma d_\infty(q', q''), \qquad s \in \mathcal{S}, a \in \mathcal{A}.
\end{aligned}
$$

Since s is an arbitrary state and a is an arbitrary action,

$$d_\infty\left(\mathfrak{b}_\pi(q'), \mathfrak{b}_\pi(q'')\right) \le \gamma d_\infty(q', q'').$$

So $\mathfrak{b}_\pi$ is a contraction mapping when $\gamma < 1$.

This paragraph proves that the Bellman optimal operator on the state-value space $\mathfrak{b}_* : \mathcal{V} \to \mathcal{V}$ is a contraction mapping over the metric space $(\mathcal{V}, d_\infty)$. First, we prove the following auxiliary inequality:

$$\left| \max_a f'(a) - \max_a f''(a) \right| \le \max_a \left| f'(a) - f''(a) \right|$$

where f' and f'' are functions over a. (Proof: Let $a' = \arg\max_a f'(a)$. We have

$$\max_a f'(a) - \max_a f''(a) = f'(a') - \max_a f''(a) \le f'(a') - f''(a')$$

$$\le \max_a \left| f'(a) - f''(a) \right|.$$

Similarly, we can prove $\max_a f''(a) - \max_a f'(a) \le \max_a \left| f'(a) - f''(a) \right|$. Then the inequality is proved.) Due to this inequality, the following inequality holds for all $v', v'' \in \mathcal{V}$:

$$\mathfrak{b}_*(v')(s) - \mathfrak{b}_*(v'')(s)$$

$$= \max_a \left[r(s,a) + \gamma \sum_{s'} p(s'|s,a) v'(s') \right] - \max_a \left[r(s,a) + \gamma \sum_{s'} p(s'|s,a) v''(s') \right]$$

$$\le \max_a \left| \gamma \sum_{s'} p(s'|s,a) \left[v'(s') - v''(s') \right] \right|$$

$$\le \gamma \max_a \left| \sum_{s'} p(s'|s,a) \right| \max_{s'} \left| v'(s') - v''(s') \right|$$

$$\le \gamma d_\infty(v', v''), \qquad\qquad s \in S.$$

So it is obviously $\left| \mathfrak{b}_*(v')(s) - \mathfrak{b}_*(v'')(s) \right| \le \gamma d_\infty(v', v'')$ for all $s \in S$. Since s is an arbitrary state,

$$d_\infty\left(\mathfrak{b}_*(v'), \mathfrak{b}_*(v'') \right) = \max_s \left[\mathfrak{b}_*(v')(s) - \mathfrak{b}_*(v'')(s) \right] \le \gamma d_\infty(v', v'').$$

So $\mathfrak{b}_*$ is a contraction mapping when $\gamma < 1$.

This paragraph proves that the Bellman optimal operator on the action-value space $\mathfrak{b}_* : Q \to Q$ is a contraction mapping over the metric space (Q, d_∞). We once again use the inequality, $\left| \max_a f'(s) - \max_a f''(a) \right| \le \max_a \left| f'(a) - f''(a) \right|$, where f' and f'' are two arbitrary functions over a. We have

$$\left| \mathfrak{b}_*(q')(s,a) - \mathfrak{b}_*(q'')(s,a) \right|$$

$$= \gamma \sum_{s'} p(s'|s,a) \left| \max_{a'} q'(s',a') - \max_{a'} q''(s',a') \right|$$

$$\le \gamma \sum_{s'} p(s'|s,a) \max_{a'} \left| q'(s',a') - q''(s',a') \right|$$

$$\leq \gamma \sum_{s'} p(s'|s, a) d_\infty(q', q'')$$
$$= \gamma d_\infty(q', q''), \qquad s \in \mathcal{S}, a \in \mathcal{A}.$$

Since s is an arbitrary state and a is an arbitrary action,

$$d_\infty\Big(\mathfrak{b}_*(q'), \mathfrak{b}_*(q'')\Big) = \max_{s,a} \big|\mathfrak{b}_*(q')(s, a) - \mathfrak{b}_*(q'')(s, a)\big| \leq \gamma d_\infty(q', q'').$$

So $\mathfrak{b}_*$ is a contraction mapping when $\gamma < 1$.

> **! Note**
>
> In RL research, we often need to prove that an operator is a contraction mapping.

Interdisciplinary Reference 3.3
Functional Analysis: Fixed Point

Consider a functional $f : \mathcal{X} \rightarrow \mathcal{X}$ over the set $\mathcal{X}$. An element $x \in \mathcal{X}$ is called **fixed point** if it satisfies $f(x) = x$.

The state values of a policy satisfy the Bellman expectation equations that use state values to back up state values. Therefore, the state values are a fixed point of Bellman expectation operator on the state-value space. Similarly, the action values of a policy satisfy the Bellman expectation equations that use action values to back up action values. Therefore, the action values are a fixed point of Bellman expectation operator on the action-value space.

The optimal state values of a policy satisfy the Bellman optimal equations that use optimal state values to back up optimal state values. Therefore, the optimal state values are a fixed point of Bellman optimal operator on the state-value space. Similarly, the action values of a policy satisfy the Bellman optimal equations that use optimal action values to back up optimal action values. Therefore, the optimal action values are a fixed point of Bellman optimal operator on the action-value space.

Interdisciplinary Reference 3.4
Functional Analysis: Banach Fixed Point Theorem

Banach fixed point theorem, also known as contraction mapping theorem, is a very important property of a contraction mapping over a complete metric space. It provides a way to calculate fixed point for a contraction mapping over a complete metric space.

The content of **Banach fixed-point theorem** is: Let $(\mathcal{X}, d)$ be a non-empty complete metric space, and $f : \mathcal{X} \rightarrow \mathcal{X}$ is a contraction mapping over it. Then f

has a unique fixed point $x_{+\infty} \in X$. Furthermore, the fixed point $x_{+\infty}$ can be found in the following way: Starting from an arbitrary element $x_0 \in X$, iteratively define the sequence $x_k = f(x_{k-1})$ $(k = 1, 2, 3, \ldots)$. Then this sequence will converge to $x_{+\infty}$. (Proof: We can prove the sequence is a Cauchy sequence. For any k', k'' such that $k' < k''$, due to the triangle inequality of the metric, we know that

$$d(x_{k'}, x_{k''}) \le d(x_{k'}, x_{k'+1}) + d(x_{k'+1}, x_{k'+2}) + \cdots + d(x_{k''-1}, x_{k''}) \le \sum_{k=k'}^{+\infty} d(x_{k+1}, x_k).$$

Using the contraction mapping theorem over and over again, we will get $d(x_{k+1}, x_k) \le \gamma^k d(x_1, x_0)$ for any positive integer k. Plugging this into the above inequality will lead to

$$d(x_{k'}, x_{k''}) \le \sum_{k=k'}^{+\infty} d(x_{k+1}, x_k) \le \sum_{k=k'}^{+\infty} \gamma^k d(x_1, x_0) = \frac{\gamma^{k'}}{1-\gamma} d(x_1, x_0).$$

Since $\gamma \in (0, 1)$, the right-hand side of the above inequality can be arbitrarily small. Therefore, the sequence is a Cauchy sequence, and it converges to a fixed point in the space. Prove the uniqueness of the fixed point: Let x' and $expp$ be two fixed points. Then we have $d(x', x'') = d(f(x'), f(x'')) \le \gamma d(x', x'')$, so $d(x', x'') = 0$. Therefore, $x' = x''$.)

Banach fixed-point theorem tells us, starting from an arbitrary point and applying the contraction mapping iteratively will generate a sequence converging to the fixed point. The proof also provides a speed of convergence: the distance to the fixed point is in proportion to γ^k, where k is the number of iterations.

Now we have proved that Bellman expectation operator and Bellman optimal operator are contraction mappings over the complete metric space $(\mathcal{V}, d_\infty)$ or (Q, d_∞). What is more, Banach fixed point theorem tells us that we can find the fixed point of Bellman expectation operator and Bellman optimal operator. Since the values of the policy are the fixed point of the Bellman expectation operator, and optimal values are the fixed point of the Bellman optimal operator, we can find the values of the policy and optimal values iteratively. That is the theoretical foundation of model-based numerical iterations. We will see the detailed algorithms in the following sections.

3.2 Model-Based Policy Iteration

The previous section introduced the theoretical foundation of model-based numerical iterations. This section will introduce policy evaluation and policy iteration with known models of both environment and policy.

3.2.1 Policy Evaluation

Policy evaluation algorithm estimates the values of given policy. We have learned some approaches to calculate the values of given policy in Sect. 2.2.3. This subsection will introduce how to use numerical iterations to evaluate a policy.

Algorithm 3.1 is the algorithm that uses numerical iteration to estimate the state values of a policy. When both dynamic p and policy π are known, we can calculate $p_\pi(s'|s)$ ($s \in \mathcal{S}, s' \in \mathcal{S}^+$) and $r_\pi(s)$ ($s \in \mathcal{S}$), and then use them as inputs of the algorithm. Step 1 initializes the state values v_0. The state values of non-terminal states can be initialized to arbitrary values. For example, we can initialize all of them as zero. Step 2 iterates. Step 2.1 implements the Bellman expectation operator to update the state values. An iteration to update all state values is also called a sweep. The k-th sweep uses the values of v_{k-1} to get the updated values v_k ($k = 1, 2, \ldots$). In this way, we get a sequence of $v_0, v_1, v_2, \ldots$, which converges to the true state values. Step 2.2 checks the condition of breaking the iteration.

Algorithm 3.1   Model-based numerical iterative policy evaluation to estimate state values.

Inputs: $p_\pi(s'|s)$ ($s \in \mathcal{S}, s' \in \mathcal{S}^+$), and $r_\pi(s)$ ($s \in \mathcal{S}$).
Output: state value estimates.
Parameters: parameters that control the ending of iterations (such as the error tolerance $\vartheta_{\max}$, or the maximal number of iterations $k_{\max}$).

1. (Initialize) Set $v_0(s) \leftarrow$ arbitrary value ($s \in \mathcal{S}$). If there is a terminal state, set its value to 0, i.e. $v_0(s_{\text{end}}) \leftarrow 0$.
2. (Iterate) For $k \leftarrow 0, 1, 2, 3, \ldots$:

    2.1. (Update) For each state $s \in \mathcal{S}$, set $v_{k+1}(s) \leftarrow r_\pi(s) + \sum_{s'} p_\pi(s'|s) v_k(s')$.
    2.2. (Check and break) If the terminal condition for iterations is met (for example, $|v_{k+1}(s) - v_k(s)| < \vartheta_{\max}$ holds for all $s \in \mathcal{S}$, or the number of iterations reaches the maximum number $k = k_{\max}$), break the iterations.

The iterations can not loop forever. Therefore, we need to set a terminal condition for the iterations. Here are two common terminal conditions among varied conditions:

- The changes of all state values are less than a pre-designated tolerance $\vartheta_{\max}$, which is a small positive real number.
- The number of iterations reaches a pre-designated number $k_{\max}$, which is a large positive integer.

The tolerance and the maximum iterations numbers can be used either separately or jointly.

Algorithm 3.2 is the algorithm that uses numerical iteration to estimate the action values of a policy. When both dynamic p and policy π are known, we can calculate $p_\pi(s', a'|s, a)$ $(s \in S, a \in \mathcal{A}, s' \in S^+, a' \in \mathcal{A})$ and $r(s, a)$ $(s \in S, a \in \mathcal{A})$, and then use them as inputs of the algorithm. Step 1 initializes the state values. For example, we can initialize all action values as zero. Step 2 iterates. Step 2.1 implements the Bellman expectation operator to update the action values.

Algorithm 3.2 Model-based numerical iterative policy evaluation to estimate action values.

Inputs: $p_\pi(s', a'|s, a)$ $(s \in S, a \in \mathcal{A}, s' \in S^+, a' \in \mathcal{A})$, and $r(s, a)$ $(s \in S, a \in \mathcal{A})$.
Output: action value estimates.
Parameters: parameters that control the ending of iterations (such as the error tolerance ϑ_{max}, or the maximal number of iterations k_{max}).

1. (Initialize) Set $q_0(s, a) \leftarrow$ arbitrary value $(s \in S, a \in \mathcal{A})$. If there is a terminal state, set its value to 0, i.e. $q_0(s_{end}, a) \leftarrow 0$ $(a \in \mathcal{A})$.
2. (Iterate) For $k \leftarrow 0, 1, 2, 3, \ldots$:

   2.1. (Update) For $s \in S, a \in \mathcal{A}$, set $q_{k+1}(s, a) \leftarrow r(s, a) + \sum_{s', a'} p_\pi(s', a'|s, a) q_k(s', a')$.
   2.2. (Check and break) If the terminal condition for iterations is met (for example, $\left| q_{k+1}(s, a) - q_k(s, a) \right| < \vartheta_{max}$ holds for all $s \in S, a \in \mathcal{A}$, or the number of iterations reaches the maximum number $k = k_{max}$), break the iterations.

We usually use numerical iteration to calculate state values, but not action values, because action values can be easily obtained after we obtained the state values, and solving state values is less resource-consuming since the dimension of state-value space $|S|$ is smaller than the dimension of action-value space $|S \| \mathcal{A}|$.

We can improve Algo. 3.1 to make it more space-efficient. An idea of improvement is: Allocate two suites of storage spaces for odd iterations and even iterations respectively. In the beginning ($k = 0$, even), initialize the storage space for even iterations. During the k-th iteration, we use the values in even storage space to update odd storage space if k is odd, and use the values in odd storage space to update even storage space if k is even.

Algorithm 3.3 takes a step further by using one suite of storage space only. It always updates the state values inplace. Sometimes, some state values have been updated but some state values have not been updated. Consequently, the intermediate results of Algo. 3.3 may not match those of Algo. 3.1 exactly. Fortunately, Algo. 3.3 can also converge to the true state values.

Algorithm 3.3   Model-based numerical iterative policy evaluation to estimate action values (space-saving version).

Inputs: $p_\pi(s'|s)$ $(s \in \mathcal{S}, s' \in \mathcal{S}^+)$, and $r_\pi(s)$ $(s \in \mathcal{S})$.
Output: state value estimates $v(s)$ $(s \in \mathcal{S})$.
Parameters: parameters that control the ending of iterations (such as the error tolerance $\vartheta_{\max}$, or the maximal number of iterations $k_{\max}$).

1. (Initialize) Set $v(s) \leftarrow$ arbitrary value $(s \in \mathcal{S})$. If there is a terminal state, set its value to 0, i.e. $v(s_{\text{end}}) \leftarrow 0$.
2. (Iterate) For $k \leftarrow 0, 1, 2, 3, \ldots$:
    2.1. In the case that uses the maximum update difference as terminal condition, initialize the maximum update difference as 0, i.e. $\vartheta \leftarrow 0$.
    2.2. For each state $s \in \mathcal{S}$:
        2.2.1. Calculate new state value: $v_{\text{new}} \leftarrow r_\pi(s) + \sum_{s'} p_\pi(s'|s)v(s')$.
        2.2.2. In the case that uses the maximum update difference as terminal condition, update the maximum update difference in this sweep: $\vartheta \leftarrow \max\{\vartheta, |v_{\text{new}} - v(s)|\}$.
        2.2.3. Update the state value: $v(s) \leftarrow v_{\text{new}}$.
    2.3. If the terminal condition is met (for example, $\vartheta < \vartheta_{\max}$, or $k = k_{\max}$), break the loop.

Algorithm 3.4 is another equivalent algorithm. This algorithm does not pre-compute $p_\pi(s'|s)$ $(s \in \mathcal{S}, s' \in \mathcal{S}^+)$ and $r_\pi(s)$ $(s \in \mathcal{S})$ as inputs. Instead, it updates the state values using the following form of Bellmen expectation operator:

$$\mathfrak{b}_\pi(v)(s) = \sum_a \pi(a|s)\left[r(s, a) + \gamma \sum_{s'} p(s'|s, a)v(s')\right], \quad s \in \mathcal{S}.$$

Algorithms 3.3 and 3.4 are equivalent and they share the same results.

Algorithm 3.4   Model-based numerical iterative policy evaluation (space-saving version, alternative implementation).

Inputs: dynamics p, and the policy π.
Output: state value estimates $v(s)$ $(s \in \mathcal{S})$.
Parameters: parameters that control the ending of iterations (such as the error tolerance $\vartheta_{\max}$, or the maximal number of iterations $k_{\max}$).

1. (Initialize) Set $v(s) \leftarrow$ arbitrary value $(s \in \mathcal{S})$. If there is a terminal state, set its value to 0, i.e. $v(s_{\text{end}}) \leftarrow 0$.
2. (Iterate) For $k \leftarrow 0, 1, 2, 3, \ldots$:

2.1. In the case that uses the maximum update difference as terminal condition, initialize the maximum update difference as 0, i.e. $\vartheta \leftarrow 0$.

2.2. For each state $s \in \mathcal{S}$:

    2.2.1. Calculate action values that related to the new state: $q_{\text{new}}(a) \leftarrow r(s, a) + \sum_{s'} p(s'|s, a)v(s')$.

    2.2.2. Calculate new state value: $v_{\text{new}} \leftarrow \sum_a \pi(a|s)q_{\text{new}}(a)$.

    2.2.3. In the case that uses the maximum update difference as terminal condition, update the maximum update difference in this sweep: $\vartheta \leftarrow \max\left\{\vartheta, |v_{\text{new}} - v(s)|\right\}$.

    2.2.4. Update the state value: $v(s) \leftarrow v_{\text{new}}$.

2.3. If the terminal condition is met (for example, $\vartheta < \vartheta_{\text{max}}$, or $k = k_{\text{max}}$), break the loop.

Until now, we have learned the model-based numerical iterative policy evaluation algorithm. Section 3.2.2 will use this algorithm as a component of policy iteration algorithm to find the optimal policy. Section 3.3 will modify this algorithm to another numerical iterative algorithm that solves optimal values using value-based iterations.

3.2.2 Policy Iteration

Policy iteration combines policy evaluation and policy improvement to find optimal policy iteratively.

As shown in Fig. 3.1, starting from an arbitrary deterministic policy π_0, policy iteration evaluates policy and improves policy alternatively. Note that the policy improvement is a strict improvement, meaning that the improved policy differs from the old policy. For finite MDP, both state space and action space are finite, so possible policies are finite. Since possible policies are finite, the policy sequence $\pi_0, \pi_1, \pi_2, \ldots$ must converge. That is, there exists a positive integer k such that $\pi_{k+1} = \pi_k$, which means that $\pi_{k+1}(s) = \pi_k(s)$ holds for all $s \in \mathcal{S}$. Additionally, in the condition $\pi_k = \pi_{k+1}$, we have $\pi_k(s) = \pi_{k+1}(s) = \arg\max_a q_{\pi_k}(s, a)$, so $v_{\pi_k}(s) = \max_a q_{\pi_k}(s, a)$, which satisfies the Bellman optimal equations. Therefore, π_k is the optimal policy. In this way, we prove that the policy iteration converges to the optimal policy.

$$\pi_0 \xrightarrow{\text{evaluate policy}} v_{\pi_0}, q_{\pi_0} \xrightarrow{\text{improve policy}} \pi_1 \xrightarrow{\text{evaluate policy}} v_{\pi_1}, q_{\pi_1} \xrightarrow{\text{improve policy}} \cdots$$

Fig. 3.1 Policy improvement.

We have learned some methods to get the action values of a policy (such as Algo. 3.3), and also learned some algorithms to improve a policy given action values (such as Algo. 2.2). Now we iteratively apply these two kinds of algorithm, which leads to the policy iteration algorithm to find the optimal policy (Algo. 3.5).

Algorithm 3.5   Model-based policy iteration.

Input: dynamics p.
Output: optimal policy estimate.
Parameters: parameters that policy evaluation requires.

1. (Initialize) Initialize the policy π_0 as an arbitrary deterministic policy.
2. For $k \leftarrow 0, 1, 2, 3, \ldots$:

   2.1. (Evaluate policy) Calculate the values of the policy π_k using policy evaluation algorithm, and save them in q_{π_k}.
   2.2. (Improve policy) Use the action values q_{π_k} to improve the deterministic policy π_k, resulting in the improved deterministic policy π_{k+1}. If $\pi_{k+1} = \pi_k$, which means that $\pi_{k+1}(s) = \pi_k(s)$ holds for all $s \in S$, break the loop and return the policy π_k as the optimal policy.

We can also save the space usage of policy iteration algorithm by reusing the space. Algorithm 3.6 uses the same space $q(s, a)$ ($s \in S, a \in \mathcal{A}$) and the same $\pi(s)$ ($s \in S$) to store action values and deterministic policy respectively in all iterations.

Algorithm 3.6   Model-based policy iteration (space-saving version).

Input: dynamics p.
Output: optimal policy estimate π.
Parameters: parameters that policy evaluation requires.

1. (Initialize) Initialize the policy π as an arbitrary deterministic policy.
2. Do the following iteratively:

   2.1. (Evaluate policy) Use policy evaluation algorithm to calculate the values of the policy π and save the action values in q.
   2.2. (Improve policy) Use the action values q to improve policy, and save the updated policy in π. If the policy improvement algorithm indicates that the current policy π is optimal, break the loop and return the policy π as the optimal policy.

3.3 VI: Value Iteration

Value Iteration (**VI**) is a method to find optimal values iteratively. The policy evaluation algorithm in Sect. 3.2.1 uses Bellman expectation operator to find the state values of the given state. This section uses a similar structure, and uses Bellman optimal operator to find the optimal values and optimal policy.

Similar to policy evaluation algorithm, value iteration algorithm has parameters to control the terminal conditions of iterations, say the update tolerance $\vartheta_{\max}$ or maximum number of iterations $k_{\max}$.

Algorithm 3.7 shows the value iteration algorithm. Step 1 initializes the estimate of optimal state values, and Step 2 updates the estimate of optimal state values iteratively using Bellman optimal operator. According to Sect. 3.1, such iterations will converge to optimal state values. After obtaining the optimal values, we can obtain optimal policy easily.

Algorithm 3.7   Model-based VI.

Input: dynamics p.
Output: optimal policy estimate π.
Parameters: parameters that policy evaluation requires.

1. (Initialize) Set $v_0(s) \leftarrow$ arbitrary value ($s \in S$). If there is a terminal state, $v_0(s_{\text{end}}) \leftarrow 0$.
2. (Iterate) For $k \leftarrow 0, 1, 2, 3, \ldots$:

   2.1. (Update) For each state $s \in S$, set

   $$v_{k+1}(s) \leftarrow \max_a \left[r(s, a) + \gamma \sum_{s'} p(s'|s, a) v_k(s') \right].$$

   2.2. (Check and break) If a terminal condition is met, for example, the update tolerance is met ($\left| v_{k+1}(s) - v_k(s) \right| < \vartheta$ holds for all $s \in S$), or the maximum number of iterations are reached (i.e. $k = k_{\max}$), break the loop.

3. (Calculate optimal policy) For each state $s \in S$, calculate its action of the optimal deterministic policy:

   $$\pi(s) \leftarrow \arg\max_a \left[r(s, a) + \gamma \sum_{s'} p(s'|s, a) v_{k+1}(s') \right].$$

Similar to the case in numerical iterative policy evaluation algorithm, we can save space for value iteration algorithm. Algorithm 3.8 shows the space-saving version for the value iteration algorithm.

Algorithm 3.8   Model-based VI (space-saving version).

Input: dynamics p.
Output: optimal policy estimate π.

Parameters: parameters that policy evaluation requires.

1. (Initialize) Set $v(s) \leftarrow$ arbitrary value ($s \in S$). If there is a terminal state, $v(s_{\text{end}}) \leftarrow 0$.
2. (Iterate) For $k \leftarrow 0, 1, 2, 3, \ldots$:

   2.1. In the case that uses the maximum update difference as terminal condition, initialize the maximum update difference as 0, i.e. $\vartheta \leftarrow 0$.
   2.2. For each state $s \in S$:
   2.2.1. Calculate new state value

   $$v_{\text{new}} \leftarrow \max_{a} \left[r(s, a) + \gamma \sum_{s'} p(s'|s, a) v_k(s') \right].$$

   2.2.2. In the case that uses the maximum update difference as terminal condition, update the maximum update difference in this sweep: $\vartheta \leftarrow \max\{\vartheta, |v_{\text{new}} - v(s)|\}$.
   2.2.3. Update the state value $v(s) \leftarrow v_{\text{new}}$.
   2.3. If the terminal condition is met (for example, $\vartheta < \vartheta_{\text{max}}$, or $k = k_{\text{max}}$), break the loop.

3. (Calculate optimal policy) For each state $s \in S$, calculate its action of the optimal deterministic policy:

   $$\pi(s) \leftarrow \arg\max_{a} \left[r(s, a) + \gamma \sum_{s'} p(s'|s, a) v_{k+1}(s') \right].$$

3.4 Bootstrapping and Dynamic Programming

Both the model-based policy evaluation algorithm in Sect. 3.2.1 and the model-based value iteration algorithm in Sect. 3.3 use the ideas of bootstrapping and Dynamic Programming (DP). This section will introduce what is bootstrapping and DP. We will also discuss the demerits of vanilla DP and possible way to improve it.

Interdisciplinary Reference 3.5
Statistics: Bootstrap

The word "bootstrap" initially meant the strap in boot (see Fig. 3.2), which people can pull when they wear boots. Since the 19th century, the saying "to pull oneself up by one's bootstraps" is used to describe the self-reliant efforts in extreme difficult situation, and the word "bootstrap" gradually has the meaning of self-reliance without external inputs. In 1979, the statistics Bradley Efron used the name "bootstrap" to describe a non-parametric method that uses existing samples and their own statistics to generate new samples and their statistics. Now this term has been widely used in varied disciplines including computer sciences and financial mathematics.

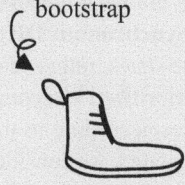

Fig. 3.2   Illustration of bootstrap.

Bootstrapping is a method to use existing estimates to estimate other things. In model-based numerical iterative algorithms, we use v_k to estimate v_{k+1} in k-th iteration. Here, estimates v_k are inaccurate and can be biased. Therefore, we may introduce the bias of v_k to the estimate v_{k+1} when we use v_k to calculate v_{k+1}. From this sense, a demerit of bootstrapping is that it propagates bias. The merit of bootstrapping is that it can make use of existing estimates for further estimation.

Bootstrapping can have obvious impacts on RL algorithms. We will see some algorithms without or with bootstrapping in the next few chapters.

The numerical iterative algorithms that solve Bellman expectation equations or Bellman optimal equations use the idea of DP, too.

Interdisciplinary Reference 3.6
Algorithm: Dynamic Programming

Dynamic Programming (DP) is a computer programming method (Bellman, 1957). The core idea of DP is twofold:

- Divide a complex problem into several easier sub-problems, while sub-problems can be further divided. In this way, we get lots of sub-problems.
- Although there are lots of sub-problems, lots of sub-problems are the same. So we can save lots of computation by re-using the results of sub-problems.

In k-th iteration ($k = 0, 1, 2, 3, \ldots$), we need to calculate $v_{k+1}(s)$ ($s \in \mathcal{S}$). In the context of DP, we can view getting $v_{k+1}(s)$ ($s \in \mathcal{S}$) as $|\mathcal{S}|$ complex problems. In order to solve each $v_{k+1}(s)$ ($s \in \mathcal{S}$), we need to use all elements in $\{v_k(s) : s \in \mathcal{S}\}$. Getting $v_k(s)$ ($s \in \mathcal{S}$) can be viewed as $|\mathcal{S}|$ sub-problems. Although we need to solve $|\mathcal{S}|$ complex problems, each of which requires of results of $|\mathcal{S}|$ sub-problems, we do not need to solve $|\mathcal{S}| \times |\mathcal{S}|$ sub-problems because all complex problems rely on the same $|\mathcal{S}|$ sub-problems. Therefore, we only need to solve $|\mathcal{S}|$ sub-problems. That is how DP applies in the numerical iterative algorithms.

Some real-world problems may encounter difficulty to use DP directly since they have a very large state space. For example, the game of Go has approximately $3^{19 \times 19} \approx 10^{172}$ states. It is even impossible to sweep all states. What is worse, most of updates may be meaningless when states the updates rely on have not been ever updated. For example, if we want to update a state s', which relies on some states $\{s : p(s'|s, a) > 0\}$. But if all states in $\{s : p(s'|s, a) > 0\}$ have not been updated, updating s' is meaningless. In this case, we may consider using **asynchronous DP** to avoid meaningless updates. Asynchronous DP only updates some states, rather than all states, in a sweep. Additionally, asynchronous DP can use prioritized sweeping to select what states to update. Prioritized sweeping selects states according to the Bellman error in the following way: After a state value has been updated, we consider all states who can be impacted, and calculate the Bellman error of those states. The Bellman error is defined as

$$\delta = \max_a \left[r(s, a) + \gamma \sum_{s'} p(s'|s, a) v(s') \right] - v(s).$$

Larger $|\delta|$ means that updating the state s will have larger impacts. Therefore, we may choose the state with the largest Bellman error. Practically, we may use a priority queue to maintain the values of $|\delta|$.

3.5 Case Study: FrozenLake

Task `FrozenLake-v1` is a text task in Gym. The task is as follows: A lake is represented by a 4×4 character grid (see below).

```
1   SFFF
2   FHFH
3   FFFH
4   HFFG
```

Some parts of the lake are frozen (denoted by the character 'F'), while other parts of the lake are not frozen and can be viewed as holes (denoted by the character 'H'). A player needs to move from the upper left start point (denoted by the character 'S') to the lower right goal (denoted by the character 'G'). In every iteration, each action chooses a direction from left, down, right, and up. Unfortunately, because the ice is slippery, the move is not always in the expected direction. For example, if an action wants to move left, the actual direction can also be down, right, or up. If the player reaches the cell 'G', the episode ends with reward 1. If the player reaches the cell 'H', the episode ends with reward 0. If the player reaches 'S' or 'F', the episode continues without additional reward.

The episode reward threshold of the task `FrozenLake-v1` is 0.70. A policy solves the problem when the average episode reward in 100 successive episodes exceeds this threshold. Theoretical results show that the average episode reward for optimal policy is around 0.74.

This section will solve the `FrozenLake-v1` using model-based numerical iterative algorithms.

3.5.1 Use Environment

This section introduces how to use the environment.

First, we use Code 3.1 to import the environment and check its basic information. This environment has 16 states $\{s_0, s_1, s_2, \ldots, s_{15}\}$ (including the terminal states), indicating what position is the player current at. The action space consists of 4 different actions $\{a_0, a_1, a_2, a_3\}$, indicating the direction "left", "down", "right", and "up" respectively. The threshold of episode reward is 0.7. The maximum number of steps in an episode is 100.

Code 3.1   Check the metadata of `FrozenLake-v1`.

`FrozenLake-v1_DP_demo.ipynb`

```
1  import gym
2  env = gym.make('FrozenLake-v1')
3  logging.info('observation space = %s', env.observation_space)
4  logging.info('action space = %s', env.action_space)
5  logging.info('number of states = %s', env.observation_space.n)
6  logging.info('number of actions = %s', env.action_space.n)
7  logging.info('reward threshold = %s', env.spec.reward_threshold)
8  logging.info('max episode steps = %s', env.spec.max_episode_steps)
```

The dynamic of the environment is saved in `env.P`. We can use the following codes to check the dynamic related to a state–action pair (for example, state 14, move right):

```
1  logging.info('P[14] = %s', env.P[14])
2  logging.info('P[14][2] = %s', env.P[14][2])
3  env.P[14][2]
```

It is a list of tuples. Each tuple element contains four elements: the transition probability $p(s', r|s, a)$, the next state s', reward r, and the indicator of episode end $1_{[s'=s_{end}]}$. For example, `env.P[14][2]` is the list of tuples: `[(0.3333333333333333, 14, 0.0, False), (0.3333333333333333, 15, 1.0, True), (0.3333333333333333, 10, 0.0, False)]`, meaning that

$$p(s_{14}, 0|s_{14}, a_2) = \frac{1}{3}$$

$$p(s_{15}, 1|s_{14}, a_2) = \frac{1}{3}$$

$$p(s_{10}, 0|s_{14}, a_2) = \frac{1}{3}.$$

In this environment, the number of steps is limited by the wrapper class `TimeLimitWrapper`. Therefore, strictly speaking, the location can not fully determine the state. Only the location and the current step together can determine the current state. For example, suppose the player is at the start location in the upper left corner. If there are only 3 steps remaining before the end of the episode, the player has no chances to reach the destination before the end of the episode. If there are hundreds of steps before the end of the episode, there are lots of opportunities to reach the destination. Therefore, strictly speaking, it is incorrect to view the location observation as the state. If the steps are not considered, the environment is in fact partially observed. However, intendedly incorrectly using a partially observable environment as a fully observable environment can often lead to meaningful results. All models are wrong, but some are useful. Mathematical models are all simplifications of complex problems. The modeling is successful if the model can identify the primary issue and solving the simplified model help solve the original complex problem. The simplification or misusing is acceptable and meaningful.

As for the task `FrozenLake-v1`, due to the stochasticity of the environment, even the optimal policy can not guarantee reaching the destination within 100 steps. From this sense, bounding the number of steps indeed impacts the optimal policy. However, the impact is limited. We may try to solve the problem with the incorrect assumption that there are no limits on step numbers, and then find a solution, which may be suboptimal for the environment with step number limit. When testing the performance of the policy, we need to test the policy with the environment with the step number limitation. More information about this can be found in Sect. 15.3.2.

This section uses Code 3.2 to interact with the environment. The function `play_policy()` accepts the parameter policy, which is a 16×4 `np.array` object representing the policy π. The function `play_policy()` returns a `float` value, representing the episode reward.

Code 3.2   Play an episode using the policy.
`FrozenLake-v1_DP_demo.ipynb`

```
1  def play_policy(env, policy, render=False):
2      episode_reward = 0.
3      observation, _ = env.reset()
4      while True:
5          if render:
6              env.render()   # render the environment
7          action = np.random.choice(env.action_space.n, p=policy[observation])
8          observation, reward, terminated, truncated, _ = env.step(action)
9          episode_reward += reward
10         if terminated or truncated:
11             break
12     return episode_reward
```

Next, we use the function `play_policy()` to test the performance of a random policy. Code 3.3 first constructs a random policy, whose $\pi(a|s) = \frac{1}{|\mathcal{A}|}$ for all $(s, a) \in \mathcal{S} \times \mathcal{A}$. Then we run 100 episodes and calculate the mean and the standard deviation. The mean is close to 0, meaning that the random policy can hardly reach the destination.

Code 3.3   Calculate the episode rewards of the random policy.
FrozenLake-v1_DP_demo.ipynb

```
1  random_policy = np.ones((env.observation_space.n, env.action_space.n)) / \
2         env.action_space.n
3
4  episode_rewards = [play_policy(env, random_policy) for _ in range(100)]
5  logging.info('average episode reward = %.2f ± %.2f',
6         np.mean(episode_rewards), np.std(episode_rewards))
```

3.5.2 Use Model-Based Policy Iteration

This section implements policy evaluation, policy improvements, and policy iterations.

First, let us consider policy evaluation. Code 3.4 first defines a function v2q(), which can calculates the action value based on the state value. Leveraging this function, the function evaluate_policy() calculates the state values of a given policy. This function uses a parameter tolerant to control of ending condition of iterations. Code 3.5 tests the function evaluate_policy(). It first finds the state values of the random policy, and then gets the action values using the function v2q().

Code 3.4   Implementation of Policy Evaluation.
FrozenLake-v1_DP_demo.ipynb

```
1  def v2q(env, v, state=None, gamma=1.):
2      # calculate action value from state value
3      if state is not None:  # solve for single state
4          q = np.zeros(env.action_space.n)
5          for action in range(env.action_space.n):
6              for prob, next_state, reward, terminated in env.P[state][action]:
7                  q[action] += prob * \
8                          (reward + gamma * v[next_state] * (1. - terminated))
9      else:  # solve for all states
10         q = np.zeros((env.observation_space.n, env.action_space.n))
11         for state in range(env.observation_space.n):
12             q[state] = v2q(env, v, state, gamma)
13     return q
14
15 def evaluate_policy(env, policy, gamma=1., tolerant=1e-6):
16     v = np.zeros(env.observation_space.n)  # initialize state values
17     while True:
18         delta = 0
19         for state in range(env.observation_space.n):
20             vs = sum(policy[state] * v2q(env, v, state, gamma))
21                     # update state value
22             delta = max(delta, abs(v[state]-vs))  # update max error
23             v[state] = vs
24         if delta < tolerant:  # check whether iterations can finish
25             break
26     return v
```

Code 3.5   Evaluate the random policy.

FrozenLake-v1_DP_demo.ipynb

```
1  v_random = evaluate_policy(env, random_policy)
2  logging.info('state value:\n%s', v_random.reshape(4, 4))
3
4  q_random = v2q(env, v_random)
5  logging.info('action value:\n%s', q_random)
```

Next, we consider policy improvement. The function `improve_policy()` in Code 3.6 realizes the policy improvement algorithm. A parameter of this function is `policy`, which will be rewritten by updated policy. This function will return a `bool` value `optimal` to show whether the input policy is optimal. Code 3.7 tests the function `improve_policy()`. It improves the random policy and gets the deterministic policy.

Code 3.6   Policy improvement.

FrozenLake-v1_DP_demo.ipynb

```
1  def improve_policy(env, v, policy, gamma=1.):
2      optimal = True
3      for state in range(env.observation_space.n):
4          q = v2q(env, v, state, gamma)
5          action = np.argmax(q)
6          if policy[state][action] != 1.:
7              optimal = False
8              policy[state] = 0.
9              policy[state][action] = 1.
10     return optimal
```

Code 3.7   Improve the random policy.

FrozenLake-v1_DP_demo.ipynb

```
1  policy = random_policy.copy()
2  optimal = improve_policy(env, v_random, policy)
3  if optimal:
4      logging.info('No update. Optimal policy is:\n%s', policy)
5  else:
6      logging.info('Updating completes. Updated policy is:\n%s', policy)
```

Now we implement policy iteration using the policy evaluation and policy improvement. The function `iterate_policy()` in Code 3.8 implements the policy iteration algorithm. Code 3.9 uses the function `iterate_policy()` to find the optimal policy, and tests the performance of this policy.

Code 3.8   Policy iteration.

FrozenLake-v1_DP_demo.ipynb

```
1  def iterate_policy(env, gamma=1., tolerant=1e-6):
2      policy = np.ones((env.observation_space.n,
3              env.action_space.n)) / env.action_space.n    # initialize
4      while True:
5          v = evaluate_policy(env, policy, gamma, tolerant)
6          if improve_policy(env, v, policy):
7              break
8      return policy, v
```

Code 3.9   Use policy iteration to find the optimal policy and test it.

FrozenLake-v1_DP_demo.ipynb

```
policy_pi, v_pi = iterate_policy(env)
logging.info('optimal state value =\n%s', v_pi.reshape(4, 4))
logging.info('optimal policy =\n%s',
        np.argmax(policy_pi, axis=1).reshape(4, 4))

episode_rewards = [play_policy(env, policy_pi) for _ in range(100)]
logging.info('average episode reward = %.2f ± %.2f',
        np.mean(episode_rewards), np.std(episode_rewards))
```

The optimal state value is:

$$\begin{bmatrix} 0.8235 & 0.8235 & 0.8235 & 0.8235 \\ 0.8235 & 0 & 0.5294 & 0 \\ 0.8235 & 0.8235 & 0.7647 & 0 \\ 0 & 0.8824 & 0.9412 & 0 \end{bmatrix}.$$

And the optimal policy is:

$$\begin{bmatrix} 0 & 3 & 3 & 3 \\ 0 & 0 & 0 & 0 \\ 3 & 1 & 0 & 0 \\ 0 & 2 & 1 & 0 \end{bmatrix}.$$

The result above shows that the optimal state value of the initial state is about 0.83. But if you use this policy to test, we can only get the average test episode rewards around 0.75, which is less than 0.83. The difference is because the calculation of optimal state values does not consider the upper bounds on the steps. Therefore, the estimated optimal state value of the initial state is smaller than the tested average episode rewards.

3.5.3 Use VI

This section tries to use VI to solve the task FrozenLake-v1. The function iterate_value() in Code 3.10 implements the value iteration algorithm. This function uses the parameter tolerant to control the end of iteration. Code 3.11 tests this function on the FrozenLake-v1 task.

Code 3.10   VI.

`FrozenLake-v1_DP_demo.ipynb`

```python
def iterate_value(env, gamma=1, tolerant=1e-6):
    v = np.zeros(env.observation_space.n)  # initialization
    while True:
        delta = 0
        for state in range(env.observation_space.n):
            vmax = max(v2q(env, v, state, gamma))  # update state value
            delta = max(delta, abs(v[state]-vmax))
            v[state] = vmax
        if delta < tolerant:  # check whether iterations can finish
            break

    # calculate optimal policy
    policy = np.zeros((env.observation_space.n, env.action_space.n))
    for state in range(env.observation_space.n):
        action = np.argmax(v2q(env, v, state, gamma))
        policy[state][action] = 1.
    return policy, v
```

Code 3.11   Find the optimal policy using the value iteration algorithm.

`FrozenLake-v1_DP_demo.ipynb`

```python
policy_vi, v_vi = iterate_value(env)
logging.info('optimal state value =\n%s', v_vi.reshape(4, 4))
logging.info('optimal policy = \n%s',
             np.argmax(policy_vi, axis=1).reshape(4, 4))
```

The optimal policy that is obtained by policy iteration is the same as the policy that is obtained by value iteration.

3.6 Summary

- Policy evaluation tries to estimate the values of the given policy. According to Banach fixed point theorem, we can iteratively solve Bellman expectation equation, and get the estimates of the values.
- We can use values of a policy to improve the policy. A method to improve the policy is to choose the action $\arg\max_a q_\pi(s, a)$ for every state s.
- The policy iteration algorithm finds an optimal policy by alternatively using the policy evaluation algorithm and the policy improvement algorithm.
- According to Banach fixed point theorem, Value Iteration (VI) algorithm can solve Bellman optimal equations and find the optimal values iteratively. The result optimal value estimates can be used to get an optimal policy estimate.
- Both the numerical iterative policy evaluation algorithm and value iterative algorithm use bootstrapping and Dynamic Programming (DP).

3.7 Exercises

3.7.1 Multiple Choices

3.1 On the model-based algorithms in this chapter, choose the correct one: (   )

A. The model-based policy evaluation algorithm can find an optimal policy of finite MDP.
B. The policy improvement algorithm can find an optimal policy of finite MDP.
C. The policy iteration algorithm can find an optimal policy of finite MDP.

3.2 On the model-based algorithms in this chapter, choose the correct one: (   )

A. Since Bellman expectation operator is a contraction mapping, according to Banach fixed point theorem, model-based policy evaluation algorithm can converge.
B. Since Bellman expectation operator is a contraction mapping, according to Banach fixed point theorem, model-based policy iteration algorithm can converge.
C. Since Bellman expectation operator is a contraction mapping, according to Banach fixed point theorem, model-based value iteration algorithm can converge.

3.3 On the model-based algorithms in this chapter, choose the correct one: (   )

A. The model-based policy evaluation algorithm uses bootstrapping, while the model-based value iteration algorithm uses dynamic programming.
B. The model-based policy evaluation algorithm uses bootstrapping, while the policy improvement algorithm uses dynamic programming.
C. The model-based value iteration algorithm uses bootstrapping, while the policy improvement algorithm uses dynamic programming.

3.7.2 Programming

3.4 Use the model-based value iteration algorithm to solve the task `FrozenLake8x8-v1`.

3.7.3 Mock Interview

3.5 Why model-based numerical iteration algorithms are not machine learning algorithms?

3.6 Compared to solve Bellman optimal equations directly, what is the advantage of model-based numerical iteration algorithms? What are the limitations of the model-based numerical iteration algorithms?

3.7 What is dynamic programming?

3.8 What is bootstrapping?

Chapter 4
MC: Monte Carlo Learning

This chapter covers

- Monte Carlo (MC) learning
- learning rate
- first visit and every visit
- MC update for policy evaluation
- MC learning for policy optimization
- exploring start
- soft policy
- importance sampling

We finally start to learn some real RL algorithms. RL algorithms can learn from the experience of interactions between agents and environments, and do not necessarily require the environment models. Since it is usually very difficult to build an environment model for a real-world task, the model-free algorithms, which do not require modeling the environment, have great advantages and are commonly used in practice.

Model-free learning can be further classified as Monte Carlo (MC) learning and Temporal Difference (TD) learning, which will be introduced in this chapter and the next chapter respectively. MC learning uses the idea of Monte Carlo method during the learning process. After the end of an episode, MC learning estimates the values of policy using the samples collected during the episodes. Therefore, MC learning can only be used in episodic tasks, since sequential tasks never end.

© The Author(s), under exclusive license to Springer Nature Singapore Pte Ltd. 2024
Z. Xiao, *Reinforcement Learning*, https://doi.org/10.1007/978-981-19-4933-3_4

Interdisciplinary Reference 4.1
Statistics: Monte Carlo Method

Monte Carlo method uses random samples to obtain numerical results for estimating deterministic values. It can be used for either stochastic problems or deterministic problems that have probabilistic interpretations.

Example 4.1 We want to find the area between $\frac{1-e^x}{\cos x}$ and x in the range $x \in [0, 1]$. There are no easy analytical solutions for this problem. But we can use Monte Carlo method to get a numerical solution. In Fig. 4.1, we uniformly sample a lot of (x, y) from the area $[0, 1] \times [0, 1]$, and calculate the percentage of samples that satisfy the relationship $\frac{1-e^x}{\cos x} \leq y < x$. The percentage is an estimate of the area we want to find. Since the percentage is a rational number while the area is an irrational number, the percentage will never equal the true area. However, their difference can be very small if the number of samples is sufficiently large.

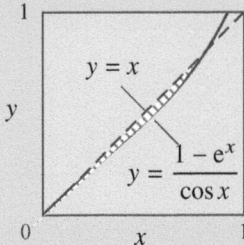

Fig. 4.1   An example task of Monte Carlo.

4.1 On-Policy MC Learning

This section introduces the on-policy MC learning algorithms. Similar to the previous chapter, we will first consider how to evaluate policy, and then consider how to find an optimal policy.

4.1.1 On-Policy MC Policy Evaluation

The basic idea of MC policy evaluation is as follows: Since state values and action values are the expectation of returns on the conditions of states and state–action pairs respectively, we can use Monte Carlo method to estimate the expectation. For example, among lots of trajectories, there are c trajectories that satisfy the condition that the trajectory had visited a given state (or a given state–action pair), and the returns of these trajectories are $g_1, g_2, \ldots, g_c$. Then, we can use Monte Carlo method to generate an estimate of the state value (or action value) as $\frac{1}{c} \sum_{i=1}^{c} g_i$.

This process is usually implemented incrementally in the following way: Suppose the first $c - 1$ return samples are $g_1, g_2, \ldots, g_{c-1}$, the estimate of value based on the first $c - 1$ samples is $\bar{g}_{c-1} = \frac{1}{c-1} \sum_{i=1}^{c-1} g_i$. Let g_c be the c-th sample. Then the estimate of values based on the first c samples is $\bar{g}_c = \frac{1}{c} \sum_{i=1}^{c} g_i$. We can prove that $\bar{g}_c = \bar{g}_{c-1} + \frac{1}{c}(g_c - \bar{g}_{c-1})$. Therefore, when we get the c-th return sample g_c, if we know the number c, we can use g_c to update the value estimate from the old estimate $\bar{g}_{c-1}$ to $\bar{g}_c$. Based on this method, incremental implementation counts the number of samples, and updates the value estimate upon it receives a new sample, without storing the sample itself. Such implementation obviously saves the space usage without any degradation on asymptotic timing complexity.

The incremental implementation of MC update can also be explained by **stochastic approximation**.

Interdisciplinary Reference 4.2
Stochastic Approximation: Robbins–Monro Algorithm

Robbins–Monro algorithm (Robbins, 1951) is one of the most important results in stochastic approximation.

Robbins–Monro algorithm tries to find a root of the equation $f(x) = 0$ with the limitation that we can only obtain the measurements of the random functions $F(x)$, where $f(x) = \mathrm{E}\big[F(x)\big]$.

Robbins–Monro algorithm solves the problem iteratively using

$$X_k = X_{k-1} - \alpha_k F(X_{k-1}),$$

where the **learning rate** sequence $\{\alpha_k : k = 1, 2, 3, \ldots\}$ satisfies

- (non-negative) $\alpha_k \geq 0$ $(k = 1, 2, 3, \ldots)$;
- (be able to converge to everywhere regardless of the start point) $\sum_{k=1}^{+\infty} \alpha_k = +\infty$; and
- (be able to converge regardless of noises) $\sum_{k=1}^{+\infty} \alpha_k^2 = +\infty$.

Robbins–Monro iteration converges to solution under some condition. (Proof: Here we prove the convergence in the mean square under the condition that

$$(x - x_*)f(x) \geq b(x - x_*)^2, \quad x \in \mathcal{X}$$
$$\big|F(x)\big|^2 \leq \zeta, \qquad\qquad x \in \mathcal{X},$$

where b and ζ are two positive real numbers, and x_* is the root of f. Since $X_k = X_{k-1} - \alpha_k F(X_{k-1})$, we have

$$\big|X_k - x_*\big|^2 = \big|X_{k-1} - \alpha_k F(X_{k-1}) - x_*\big|^2$$

$$= |X_{k-1} - x_*|^2 - 2\alpha_k(X_{k-1} - x_*)F(X_k) + \alpha_k^2|F(X_{k-1})|^2.$$

The sequence $|X_k - x_*|^2$ is called Lyapunov process. Its variation is

$$|X_k - x_*|^2 - |X_{k-1} - x_*|^2 = -2\alpha_k(X_{k-1} - x_*)F(X_k) + \alpha_k^2|F(X_{k-1})|^2.$$

Taking the expectation on the condition of X_{k-1} over the above equation and considering $\mathrm{E}\big[F(X_{k-1})|X_{k-1}\big]$, we have

$$\mathrm{E}\Big[|X_k - x_*|^2 \Big| X_{k-1}\Big] - |X_{k-1} - x_*|^2$$

$$= -2\alpha_k(X_{k-1} - x_*)f(X_{k-1}) + \alpha_k^2 \mathrm{E}\Big[|F(X_{k-1})|^2 \Big| X_{k-1}\Big].$$

Since $(X_{k-1} - x_*)f(X_{k-1}) \geq b|X_{k-1} - x_*|^2$ and $\mathrm{E}\Big[|F(X_{k-1})|^2 \Big| x_{k-1}\Big] \leq \zeta$, the above equation has the upper bound

$$\mathrm{E}\Big[|X_k - x_*|^2 \Big| X_{k-1}\Big] - |X_{k-1} - x_*|^2 = -2\alpha_k b|X_{k-1} - x_*|^2 + \alpha_k^2 \zeta.$$

Further taking expectation on X_{k-1}, we have

$$\mathrm{E}\Big[|X_k - x_*|^2\Big] - \mathrm{E}\Big[|X_{k-1} - x_*|^2\Big] = -2\alpha_k b|X_{k-1} - x_*|^2 + \alpha_k^2 \zeta,$$

or equivalently

$$2\alpha_k b|X_{k-1} - x_*|^2 = \mathrm{E}\Big[|X_{k-1} - x_*|^2\Big] - \mathrm{E}\Big[|X_k - x_*|^2\Big] + \zeta \alpha_k^2.$$

Summation over k leads to

$$2b\sum_{k=1}^{K} \alpha_k \mathrm{E}\Big[|X_{k-1} - x_*|^2\Big] \leq \mathrm{E}\Big[|X_0 - x_*|^2\Big] - \mathrm{E}\Big[|X_K - x_*|^2\Big] + \zeta \sum_{k=1}^{K} \alpha_k^2$$

$$\leq \mathrm{E}\Big[|X_0 - x_*|^2\Big] - 0 + \zeta \sum_{k=1}^{K} \alpha_k^2$$

$$< +\infty.$$

Therefore, $\sum_{k=1}^{K} \alpha_k \mathrm{E}\Big[|X_{k-1} - x_*|^2\Big]$ converges. Since $\sum_{k=1}^{K} \alpha_k = +\infty$, we have $\mathrm{E}\Big[|X_k - x_*|^2\Big] \rightarrow 0 \ (k \rightarrow +\infty)$. That is, the sequence $\{X_k : k = 0, 1, 2, \ldots\}$ converges in mean square.)

(Blum, 1954) proved that, when $f(x)$ is non-decreasing and has a unique root x_*, and $f'(x_*)$ exists and is positive, and $f'(x_*)$ is uniformly bounded, $F(x)$ with probability 1. (Nemirovski, 1983) showed the converge rate is $O\left(\frac{1}{\sqrt{k}}\right)$ when $f(x)$ is further restricted to be convex and α_k is carefully chosen.

Incremental implementation can be viewed from the aspect of Robbins–Monro algorithm in the following way: Consider estimating action values. Let $F(q) = G - q$, where q is the value we want to estimate. We observe many samples of returns, and update the value estimate using

$$q_k \leftarrow q_{k-1} + \alpha_k (g_k - q_{k-1}), \quad k = 1, 2, \ldots$$

where q_0 is an arbitrary initial value, and $\alpha_k = 1/k$ the sequence of **learning rates**. Learning rate sequence $\alpha_k = 1/k$ satisfies all conditions in the Robbins–Monro algorithm, which proves the convergence once again. There are other learning rate sequences that can ensure convergence. After the convergence, we have $\mathrm{E}\left[F(q(s, a))\right] = \mathrm{E}[G_t | S_t = s, A_t = a] - q(s, a) = 0$. We can analyze the case that estimates state values similarly by letting $F(v) = G - v$.

In the previous chapter, we have known that policy evaluation can either directly estimate state values or directly estimate action values. According to Bellman expectation equations, we can use state values to back up action values with the knowledge of the dynamics p, or use action values to back up state values with the knowledge of the policy π. When the dynamics of the task is given, states values and action values can back up each other. However, p is unknown in model-free learning, so we can only use action values to back up state values, but can not use state values to back up action values. Additionally, since we can use action values to improve policy, so learning action values is more important than learning state values.

A state (or a state–action pair) can be visited multiple times in an episode, and henceforth we can obtain inconsistent return samples in different visitation. **Every-visit Monte Carlo update** uses all return samples to update the value estimations, while **first-visit Monte Carlo update** only uses the sample when the state (or the state–action pair) is first visited. Both every-visit MC update and first-visit MC update converge to the true value, although they may have different estimates during the learning process.

Algorithm 4.1 shows every-visit MC update to evaluate action values. Step 1 initializes the action values $q(s, a)$ ($s \in \mathcal{S}, a \in \mathcal{A}$). The initialization values can be arbitrary, since they will be overwritten after the first update. If we use incremental implementation, we also need to initialize the counter with 0. Step 2 conducts MC update using incremental implementation. In every iteration of the loop, we first generate the trajectory of the episode, and then calculate the return sample and update $q(s, a)$ in reversed order. Here we use reversed order so that we can update G using the relationship $G_t = R_{t+1} + \gamma G_{t+1}$, which can reduce the computation complexity. If we use incremental implementation, the visitation number of the state–action pair (s, a) is stored in $c(s, a)$, and it will be increased by one every time the state–action pair (s, a) is visited. The breaking condition of the loop can be a maximal episode number k_{max} or a tolerance ϑ_{max} of updating precision, which is similar to the break condition of model-based numerical iterations in the previous chapter.

Algorithm 4.1   Evaluate action values using every-visit MC policy evaluation.

Inputs: environment (without mathematical model), and the policy π.
Output: action value estimates $q(s, a)$ ($s \in S, a \in \mathcal{A}$).

1. (Initialize) Initialize the action value estimates $q(s, a) \leftarrow$ arbitrary value ($s \in S, a \in \mathcal{A}$). If we use incremental implementation, initialize the counter $c(s, a) \leftarrow 0$ ($s \in S, a \in \mathcal{A}$).
2. (MC update) For each episode:
   2.1. (Sample trajectory) Use the policy π to generate a trajectory $S_0, A_0, R_1, S_1, \ldots, S_{T-1}, A_{T-1}, R_T, S_T$.
   2.2. (Initialize return) $G \leftarrow 0$.
   2.3. (Update) For $t \leftarrow T-1, T-2, \ldots, 0$:
      2.3.1. (Calculate return) $G \leftarrow \gamma G + R_{t+1}$.
      2.3.2. (Update action value) Update $q(S_t, A_t)$ to reduce $[G - q(S_t, A_t)]^2$. (For incremental implementation, set $c(S_t, A_t) \leftarrow c(S_t, A_t) + 1$, $q(S_t, A_t) \leftarrow q(S_t, A_t) + \frac{1}{c(S_t, A_t)}[G - q(S_t, A_t)]$.)

In Algo. 4.1, the update process is written as "Update $q(S_t, A_t)$ to reduce $[G - q(S_t, A_t)]^2$". It is a more general description compared to incremental implementation. We have known that, updating using

$$q(S_t, A_t) \leftarrow q(S_t, A_t) + \alpha [G - q(S_t, A_t)]$$

can reduce the difference between G and $q(S_t, A_t)$, which in fact reduces $[G - q(S_t, A_t)]^2$. Therefore, the updating in fact reduces $[G - q(S_t, A_t)]^2$. However, not all settings that reduce $[G - q(S_t, A_t)]^2$ are feasible. For example, it is not a good idea to set $G \leftarrow q(S_t, A_t)$, since $G \leftarrow q(S_t, A_t)$ is equivalent to setting $\alpha = 1$, which does not satisfy the convergence conditions of Robbins–Monro, so the algorithm may not converge if we set $G \leftarrow q(S_t, A_t)$.

After we obtain the action values, we can obtain state values using Bellman expectation equations. We can use MC updates to estimate state values directly.

Algorithm 4.2 is every-visit MC update to evaluate state values. Compared to Algo. 4.1, it just changes $q(s, a)$ to $v(s)$, and the counting will be revised.

Algorithm 4.2   Every-visit MC update to evaluate state values.

Inputs: environment (without mathematical model), and the policy π.
Output: state value estimates $v(s)$ ($s \in S$).

1. (Initialize) Initialize the state value estimates $v(s) \leftarrow$ arbitrary value ($s \in S$). If we use incremental implementation, initialize the counter $c(s) \leftarrow 0\,(s \in S)$.
2. (MC update) For each episode:

 2.1. (Sample trajectory) Use the policy π to generate a trajectory $S_0, A_0, R_1, S_1, \ldots, S_{T-1}, A_{T-1}, R_T, S_T$.
 2.2. (Initialize return) $G \leftarrow 0$.
 2.3. (Update) For $t \leftarrow T - 1, T - 2, \ldots, 0$:
  2.3.1. (Calculate return) $G \leftarrow \gamma G + R_{t+1}$.
  2.3.2. (Update action value) Update $v(S_t)$ to reduce $\left[G - v(S_t)\right]^2$. (For incremental implementation, set $c(S_t) \leftarrow c(S_t) + 1$, $v(S_t) \leftarrow v(S_t) + \frac{1}{c(S_t)}\left[G - v(S_t)\right]$.)

First-visit MC update is a more historical, more thoroughly researched algorithm. Algorithm 4.3 is the first-visit MC update to estimate action values. Compared to the every-visit version (Algo. 4.1), it first finds out when every state was visited for the first time and stores the results in $f(s, a) \leftarrow -1\,(s \in S, a \in \mathcal{A})$. When updating the value estimates, it only updates the estimates when the state was visited for the first time.

Algorithm 4.3   First-visit MC update to estimate action values.

Inputs: environment (without mathematical model), and the policy π.
Output: action value estimates $q(s, a)\,(s \in S, a \in \mathcal{A})$.

1. (Initialize) Initialize the action value estimates $q(s, a) \leftarrow$ arbitrary value ($s \in S, a \in \mathcal{A}$). If we use incremental implementation, initialize the counter $c(s, a) \leftarrow 0\,(s \in S, a \in \mathcal{A})$.
2. (MC update) For each episode:

 2.1. (Sample trajectory) Use the policy π to generate a trajectory $S_0, A_0, R_1, S_1, \ldots, S_{T-1}, A_{T-1}, R_T, S_T$.
 2.2. (Calculate steps that state–action pairs are first visited within the episode) For $t \leftarrow 0, 1, \ldots, T - 1$:
  2.2.1. If $f(S_t, A_t) < 0$, set $f(S_t, A_t) \leftarrow t$.
 2.3. (Initialize return) $G \leftarrow 0$.
 2.4. (Update) For $t \leftarrow T - 1, T - 2, \ldots, 0$:
  2.4.1. (Calculate return) $G \leftarrow \gamma G + R_{t+1}$.
  2.4.2. (Update when first visit) If $f(S_t, A_t) = t$, update $q(S_t, A_t)$ to reduce $\left[G - q(S_t, A_t)\right]^2$. (For incremental implementation, set $c(S_t, A_t) \leftarrow c(S_t, A_t) + 1$, $q(S_t, A_t) \leftarrow q(S_t, A_t) + \frac{1}{c(S_t, A_t)}\left[G - q(S_t, A_t)\right]$.)

Similar to the every-visit version, the first-visit version can directly estimate state vales (Algo. 4.4). We can also calculate state values from action values using Bellman expectation equations, too.

Algorithm 4.4   First-visit MC update to estimate state values.

Inputs: environment (without mathematical model), and the policy π.
Output: state value estimates $v(s)$ ($s \in S$).

1. (Initialize) Initialize the state value estimates $v(s) \leftarrow$ arbitrary value ($s \in S$). If we use incremental implementation, initialize the counter $c(s) \leftarrow 0$ ($s \in S$).
2. (MC update) For each episode:

   2.1. (Sample trajectory) Use the policy π to generate a trajectory $S_0, A_0, R_1, S_1, \ldots, S_{T-1}, A_{T-1}, R_T, S_T$.
   2.2. (Calculate steps that states are first visited within the episode) For $t \leftarrow 0, 1, \ldots, T - 1$:
     2.2.1. If $f(S_t) < 0$, set $f(S_t) \leftarrow t$.
   2.3. (Initialize return) $G \leftarrow 0$.
   2.4. (Update) For $t \leftarrow T - 1, T - 2, \ldots, 0$:
     2.4.1. (Calculate return) $G \leftarrow \gamma G + R_{t+1}$.
     2.4.2. (Update when first visit) If $f(S_t) = t$, update $v(S_t)$ to reduce $\left[G - v(S_t)\right]^2$. (For incremental implementation, set $c(S_t) \leftarrow c(S_t) + 1$, $v(S_t) \leftarrow v(S_t) + \frac{1}{c(S_t)}\left[G - v(S_t)\right]$.)

4.1.2 MC Learning with Exploration Start

This section introduces MC update algorithms to find the optimal policy.

The basic idea to find the optimal policy is as follows: In the previous section, we have known how to evaluate action values using MC updates. Upon getting estimates of action values, we can improve the policy, and get a new policy. Repeating such estimation and improvement may find the optimal policy.

When we use the aforementioned method, if we always start from the same start state, we may unfortunately not be able to find the optimal policy. It is because we may start from a bad policy, and then stuck into some bad states, and update values for those bad states. For example, In the MDP of Fig. 4.2, transferring from state s_{start} to the terminal state s_{end} can obtain reward 1, while transferring from state s_{middle} to state s_{end} can obtain reward 100. If MC updates always start from the state s_{start}, and the initial state values are all 0, and the initial policy is a deterministic policy such that $\pi(s_{\text{start}}) = \pi(s_{\text{middle}}) = a_{\text{to end}}$, we can get the trajectory $s_{\text{start}}, a_{\text{to end}}, +1, s_{\text{end}}$ (Note that the state s_{middle} has not been visited), and the updated action values are $q(s_{\text{start}}, a_{\text{to end}}) \leftarrow 1$, and $q(s_{\text{start}}, a_{\text{to middle}})$ keeps as 0 without updating. After that, we update the policy, but the policy does not change. Henceforth, no matter how many iterations we conduct, the trajectory will not change, and the policy will not change. We will not find the optimal policy $\pi_*(s_{\text{start}}) = a_{\text{to middle}}$.

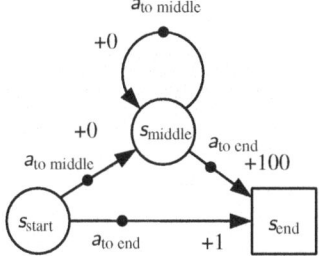

Fig. 4.2 An example where the optimal policy may not be found without exploring start.

Exploring start is a possible solution to deal with this problem. Exploring start changes the initial state distribution so that an episode can start with any one of state–action pairs. After applying exploration start, all state–action pairs will be visited.

Algorithm 4.5 shows the MC update algorithm with exploring start. This algorithm also has an every-visit version and a first-visit version. These two versions differ in Steps 2.3 and 2.5.2: The every-visit version does not need Step 2.3, and it always updates the action values in Step 2.5.2; the first-visit version uses Step 2.3 to know when every state–action value is first visited, and it updates the actions only when the state–action pair is first visited in Step 2.5.2.

Algorithm 4.5 MC update with exploring start (maintaining policy explicitly).

---

1. (Initialize) Initialize the action value estimates $q(s, a) \leftarrow$ arbitrary value ($s \in \mathcal{S}$, $a \in \mathcal{A}$). If we use incremental implementation, initialize the counter $c(s, a) \leftarrow 0$ ($s \in \mathcal{S}$, $a \in \mathcal{A}$).
   Initialize the deterministic policy $\pi(s) \leftarrow$ arbitrary action ($s \in \mathcal{S}$).
2. (MC update) For each episode:

   2.1. (Initialize episode start) Choose the start state–action pair $S_0 \in \mathcal{S}, A_0 \in \mathcal{A}$, such that all state–action pairs can be chosen as the start state–action pair.
   2.2. (Sample trajectory) Starting from (S_0, A_0), use the policy π to generate a trajectory $S_0, A_0, R_1, S_1, \ldots, S_{T-1}, A_{T-1}, R_T, S_T$.
   2.3. (For the first-visit version, find when a state–action pair is first visited) Initialize $f(s, a) \leftarrow -1$ ($s \in \mathcal{S}$, $a \in \mathcal{A}$). Then for each $t \leftarrow 0, 1, \ldots, T - 1$, if $f(S_t, A_t) < 0$, set $f(S_t, A_t) \leftarrow t$.

2.4. (Initialize return) $G \leftarrow 0$.

2.5. (Update) For $t \leftarrow T - 1, T - 2, \ldots, 0$:

    2.5.1. (Calculate return) $G \leftarrow \gamma G + R_{t+1}$.

    2.5.2. (Update action value estimate) Update $q(S_t, A_t)$ to reduce $\left[G - q(S_t, A_t)\right]^2$. (For incremental implementation, set $c(S_t, A_t) \leftarrow c(S_t, A_t) + 1$, $q(S_t, A_t) \leftarrow q(S_t, A_t) + \frac{1}{c(S_t, A_t)}\left[G - q(S_t, A_t)\right]$.) For the first-visit version, update the counter and action estimates only when $f(S_t, A_t) = t$.

    2.5.3. (Improve policy) $\pi(S_t) \leftarrow \arg\max_a q(S_t, a)$. (If there are multiple actions a that can maximize $q(S_t, a)$, choose an arbitrary action.)

We can choose to maintain the policy implicitly, since the greedy policy can be calculated from the action values. Specifically speaking, given the action values $q(S_t, \cdot)$ under a state S_t, we pick the action $\arg\max_a q(S_t, a)$ as A_t. Algo. 4.6 shows the version that maintains the policy implicitly. Maintaining the policy implicitly can save space.

Algorithm 4.6 MC update with exploring start (maintaining policy implicitly).

1. (Initialize) Initialize the action value estimates $q(s, a) \leftarrow$ arbitrary value ($s \in \mathcal{S}, a \in \mathcal{A}$). If we use incremental implementation, initialize the counter $c(s, a) \leftarrow 0$ ($s \in \mathcal{S}, a \in \mathcal{A}$).

2. (MC update) For each episode:

   2.1. (Initialize episode start) Choose the start state–action pair $S_0 \in \mathcal{S}$, $A_0 \in \mathcal{A}$, such that all state–action pairs can be chosen as the start state–action pair.

   2.2. (Sample trajectory) Starting from (S_0, A_0), use the policy derived from the action values q to generate a trajectory $S_0, A_0, R_1, S_1, \ldots, S_{T-1}, A_{T-1}, R_T, S_T$. (That is, choose the action that maximize the action value.)

   2.3. (For the first-visit version, find when a state–action pair is first visited) Initialize $f(s, a) \leftarrow -1$ ($s \in \mathcal{S}, a \in \mathcal{A}$). Then for e very $t \leftarrow 0, 1, \ldots, T - 1$, if $f(S_t, A_t) < 0$, set $f(S_t, A_t) \leftarrow t$.

   2.4. (Initialize return) $G \leftarrow 0$.

   2.5. (Update) For $t \leftarrow T - 1, T - 2, \ldots, 0$:

       2.5.1. (Calculate return) $G \leftarrow \gamma G + R_{t+1}$.

       2.5.2. (Update action value estimate) Update $q(S_t, A_t)$ to reduce $\left[G - q(S_t, A_t)\right]^2$. (For incremental implementation, set $c(S_t, A_t) \leftarrow c(S_t, A_t) + 1$, $q(S_t, A_t) \leftarrow q(S_t, A_t) + \frac{1}{c(S_t, A_t)}\left[G - q(S_t, A_t)\right]$.) For the first-visit version, update the counter and action estimates only when $f(S_t, A_t) = t$.

4.1.3 MC Learning on Soft Policy

This section considers MC update with soft policy. It can explore without exploring start.

First, let us see what is a soft policy. A policy π is a **soft policy**, if and only if $\pi(a|s) > 0$ holds for every $s \in S, a \in \mathcal{A}$. A soft policy can choose all possible actions. Therefore, starting a state s, it can reach all possible state–action pairs that we can reach from the state s. From this sense, using soft policy can help to explore more states or state–action pairs.

A policy π is called ε-soft policy, if and only if there exists an $\varepsilon > 0$ such that $\pi(a|s) > \varepsilon / |\mathcal{A}(s)|$ holds for all $s \in S, a \in \mathcal{A}$.

ε-soft policies are all soft policies. All soft policies in finite MDPs are ε-soft policies.

Given an environment and a deterministic policy, the ε-soft policy that is the closest to the deterministic policy is called ε-**greedy policy** of the deterministic policy. Specifically, the ε-soft policy of the deterministic policy

$$\pi(a|s) = \begin{cases} 1, & s \in S, a = a^* \\ 0, & s \in S, a \neq a^* \end{cases}$$

is

$$\pi(a|s) = \begin{cases} 1 - \varepsilon + \frac{\varepsilon}{|\mathcal{A}(s)|}, & s \in S, a = a^* \\ \frac{\varepsilon}{|\mathcal{A}(s)|}, & s \in S, a \neq a^*. \end{cases}$$

This ε-greedy policy assigns probability ε equally to all actions, and assigns the remaining $(1 - \varepsilon)$ probability to the action a^*.

ε-greedy policies are all ε soft policies. Furthermore, they are all soft policies.

MC update with soft policy uses ε-soft policy during iterations. Particularly, the policy improvement updates an old ε-soft policy to a new ε-greedy policy, which can be explained by the policy improvement theorem, too. (Proof: Consider updating from a ε soft policy π to the following ε-greedy policy

$$\pi'(a|s) = \begin{cases} 1 - \varepsilon + \frac{\varepsilon}{|\mathcal{A}(s)|}, & a = \arg\max_{a'} q_\pi(s, a') \\ \frac{\varepsilon}{|\mathcal{A}(s)|}, & a \neq \arg\max_{a'} q_\pi(s, a'). \end{cases}$$

According to the policy improvement theorem, we have $\pi \leqslant \pi'$ if

$$\sum_a \pi'(a|s) q_\pi(s, a) \geq v_\pi(s), \quad s \in S.$$

Now we verify the aforementioned inequality. Considering

$$\sum_a \pi'(a|s) q_\pi(s, a) = \frac{\varepsilon}{|\mathcal{A}(s)|} \sum_a q_\pi(s, a) + (1 - \varepsilon) \max_a q_\pi(s, a),$$

and noticing $1 - \varepsilon > 0$ and

$$1 - \varepsilon = \sum_a \left(\pi(a|s) - \frac{\varepsilon}{|\mathcal{A}(s)|} \right),$$

we have

$$(1 - \varepsilon)\max_a q_\pi(s, a) = \sum_a \left(\pi(a|s) - \frac{\varepsilon}{|\mathcal{A}(s)|} \right) \max_a q_\pi(s, a)$$

$$\geq \sum_a \left(\pi(a|s) - \frac{\varepsilon}{|\mathcal{A}(s)|} \right) q_\pi(s, a)$$

$$= \sum_a \pi(a|s) q_\pi(s, a) - \frac{\varepsilon}{|\mathcal{A}(s)|} \sum_a q_\pi(s, a).$$

Therefore,

$$\sum_a \pi'(a|s) q_\pi(s, a)$$

$$= \frac{\varepsilon}{|\mathcal{A}(s)|} \sum_a q_\pi(s, a) + (1 - \varepsilon)\max_a q_\pi(s, a)$$

$$\geq \frac{\varepsilon}{|\mathcal{A}(s)|} \sum_a q_\pi(s, a) + \sum_a \pi(a|s) q_\pi(s, a) - \frac{\varepsilon}{|\mathcal{A}(s)|} \sum_a q_\pi(s, a)$$

$$= \sum_a \pi(a|s) q_\pi(s, a)$$

$$= v_\pi(s).$$

So the condition of policy improvement theorem has been verified.)

Algorithm 4.7 shows the MC update algorithm that uses soft policy. This algorithm also has an every-visit version and a first-visit version, which differ in Step 2.2 and Step 2.4.2. Step 1 not only initializes the action values, but also initializes the policy as an ε-soft policy. In order to ensure that the condition of policy improvement theorem is always met, we need to ensure that policy is always an ε-soft policy. Therefore, Step 1 needs to initialize the policy as an ε-soft policy, and Step 2 should always update the policy to an ε-soft policy. Since we use ε-soft policy to generate trajectories, all reachable states or state–action pairs will be visited. Therefore, we can explore more thoroughly, and we are more likely to find the global optimal ε-soft policy.

Algorithm 4.7   MC update with soft policy (maintaining policy explicitly).

---

1. (Initialize) Initialize the action value estimates $q(s, a)$ ← arbitrary value ($s \in S, a \in \mathcal{A}$). If we use incremental implementation, initialize the counter $c(s, a) \leftarrow 0$ ($s \in S, a \in \mathcal{A}$). Initialize the policy $\pi(\cdot|\cdot)$ to an arbitrary ε-soft policy.

2. (MC update) For each episode:

   2.1. (Sample trajectory) Use the policy π to generate a trajectory $S_0, A_0, R_1, S_1, \ldots, S_{T-1}, A_{T-1}, R_T, S_T$.

   2.2. (For the first-visit version, find when a state–action pair is first visited) Initialize $f(s, a) \leftarrow -1$ ($s \in S, a \in \mathcal{A}$). Then for e very $t \leftarrow 0, 1, \ldots, T - 1$, if $f(S_t, A_t) < 0$, set $f(S_t, A_t) \leftarrow t$.

   2.3. (Initialize return) $G \leftarrow 0$.

   2.4. (Update) For $t \leftarrow T - 1, T - 2, \ldots, 0$:

   2.4.1. (Calculate return) $G \leftarrow \gamma G + R_{t+1}$.

   2.4.2. (Update action value estimate) Update $q(S_t, A_t)$ to reduce $[G - q(S_t, A_t)]^2$. (For incremental implementation, set $c(S_t, A_t) \leftarrow c(S_t, A_t) + 1$, $q(S_t, A_t) \leftarrow q(S_t, A_t) + \frac{1}{c(S_t, A_t)}[G - q(S_t, A_t)]$.) For the first-visit version, update the counter and action estimates only when $f(S_t, A_t) = t$.

   2.4.3. (Improve policy) $A^* \leftarrow \arg\max_a q(S_t, a)$. (If there are multiple actions a that can maximize $q(S_t, a)$, choose an arbitrary action.) Update $\pi(\cdot|S_t)$ to the ε-greedy policy of the deterministic policy $\pi(a|S_t) = 0$ ($a \neq A^*$) (For example, $\pi(a|S_t) \leftarrow \frac{\varepsilon}{|\mathcal{A}(s)|}$ ($a \in \mathcal{A}(s)$), $\pi(A^*|S_t) \leftarrow \pi(A^*|S_t) + (1 - \varepsilon)$.)

We can also maintain the policy implicitly, since the ε-greedy policy can be calculated from the action values. Specifically speaking, given the action values $q(S_t, \cdot)$ under a state S_t, we can determine an action using the ε-greedy policy derived from $q(S_t, \cdot)$ in the following way: First draw a random number X from the uniform distribution ranging $[0, 1]$. If $X < \varepsilon$, we explore, and choose an action within $\mathcal{A}(s)$ as A_t. Otherwise, choose the optimal action $\arg\max_a q(S_t, a)$ as A_t. In this way, we do not need to store and maintain the policy π explicitly.

Algorithm 4.8   MC update with soft policy (maintaining policy explicitly).

1. (Initialize) Initialize the action value estimates $q(s, a) \leftarrow$ arbitrary value ($s \in S, a \in \mathcal{A}$). If we use incremental implementation, initialize the counter $c(s, a) \leftarrow 0$ ($s \in S, a \in \mathcal{A}$).

2. (MC update) For each episode:

   2.1. (Sample trajectory) Use ε-greedy policy derived from q to generate a trajectory $S_0, A_0, R_1, S_1, \ldots, S_{T-1}, A_{T-1}, R_T, S_T$.

   2.2. (For the first-visit version, find when a state–action pair is first visited) Initialize $f(s, a) \leftarrow -1$ ($s \in S, a \in \mathcal{A}$). Then for each $t \leftarrow 0, 1, \ldots, T - 1$, if $f(S_t, A_t) < 0$, set $f(S_t, A_t) \leftarrow t$.

2.3. (Initialize return) $G \leftarrow 0$.
2.4. (Update) For $t \leftarrow T - 1, T - 2, \ldots, 0$:
    2.4.1. (Calculate return) $G \leftarrow \gamma G + R_{t+1}$.
    2.4.2. (Update action value estimate) Update $q(S_t, A_t)$ to reduce $\left[G - q(S_t, A_t) \right]^2$. (For incremental implementation, set $c(S_t, A_t) \leftarrow c(S_t, A_t) + 1$, $q(S_t, A_t) \leftarrow q(S_t, A_t) + \frac{1}{c(S_t,A_t)} \left[G - q(S_t, A_t) \right]$.) For the first-visit version, update the counter and action estimates only when $f(S_t, A_t) = t$.

Besides, immediate policy improvement after each value estimation updates is not obligatory. In the case that an episode has much more steps than the number of states, we can also update the policy after the episode ends to reduce computation.

4.2 Off-Policy MC Learning

This section considers off-policy MC learning. In an off-policy algorithm, the policy that is updated and the policy that generates samples can be different. This section will leverage a technique of Monte Carlo method called importance sampling into MC learning, and use it to evaluate policy and find an optimal policy.

4.2.1 Importance Sampling

This section considers off-policy RL based on importance sampling. Importance sampling can promote exploration, too. For example, given a policy, we may need to estimate the state value of a particular state. Unfortunately, if we use the policy to generate trajectories, it is very unlikely that the particular state is visited. Therefore, there are few samples that can be used for estimating that state value, so the variance of the state value estimate can be large. Importance sampling considers using another policy to generate samples so that the particular state can be more frequently visited, and henceforth the samples are more efficiently used.

Interdisciplinary Reference 4.3
Statistics: Importance Sampling

Importance sampling is a technique to reduce the variance in MC algorithm. It changes the sampling probability distributions so that the sampling can be more efficient. The ratio of new probability to old probability is called importance sampling ratio.

Example 4.2 This example revisits the problem that finds the area between $\frac{1-e^x}{\cos x}$ and x in the range $x \in [0, 1]$ (Fig. 4.1). Previously, we draw samples from $[0, 1] \times [0, 1]$, whose area is 1. But it is not efficient, since most of samples do not satisfy the target relationship $\frac{1-e^x}{\cos x} \leq y < x$. In this case, we can use importance sampling to improve the sample efficiency. Since $\frac{1-e^x}{\cos x} > x - 0.06$ for all $x \in [0, 1]$, we can draw samples uniformly from the range $\{(x, y) : x \in [0, 1], y \in [x - 0.06, x]\}$, whose area is 0.06, and calculate the percentage of samples that satisfy $\frac{1-e^x}{\cos x} \leq y < x$. Then the area estimate is the percentage multiplied by 0.06. Apparently, the probability that new samples satisfy the target relationship is $\frac{1}{0.06}$ times the probability of the old samples. Therefore, we can estimate the area using much fewer new samples. The benefit can be quantified by the importance sampling ratio $\frac{1}{0.06}$.

This section considers off-policy RL using importance sampling. We call the policy to update as the **target policy** (denoted as π), and call the policy to generate samples as the **behavior policy** (denoted as b). For on-policy algorithms, these two policies are the same. But for off-policy algorithms based on importance sampling, these two policies are different. That is, we use a different behavior policy to generate samples, and then use the generated samples to update the statistics about the target policy.

Given a state S_t at time t, we can generate a trajectory $S_t, A_t, R_{t+1}, S_{t+1}, A_{t+1}, \ldots,$ $S_{T-1}, A_{T-1}, R_T, S_T$ using either policy π or policy b. The probabilities of this trajectory generated by the policy π and policy b are

$$\Pr_\pi[A_t, R_{t+1}, S_{t+1}, A_{t+1}, \ldots, S_{T-1}, A_{T-1}, R_T, S_T | S_t]$$
$$= \pi(A_t | S_t) p(S_{t+1}, R_{t+1} | S_t, A_t) \pi(A_{t+1} | S_{t+1}) \cdots p(S_T, R_T | S_{T-1}, A_{T-1})$$
$$= \prod_{\tau=t}^{T-1} \pi(A_\tau | S_\tau) \prod_{\tau=t}^{T-1} p(S_{\tau+1}, R_{\tau+1} | S_\tau, A_\tau),$$
$$\Pr_b[A_t, R_{t+1}, S_{t+1}, A_{t+1}, \ldots, S_{T-1}, A_{T-1}, R_T, S_T | S_t]$$
$$= b(A_t | S_t) p(S_{t+1}, R_{t+1} | S_t, A_t) b(A_{t+1} | S_{t+1}) \cdots p(S_T, R_T | S_{T-1}, A_{T-1})$$
$$= \prod_{\tau=t}^{T-1} b(A_\tau | S_\tau) \prod_{\tau=t}^{T-1} p(S_{\tau+1}, R_{\tau+1} | S_\tau, A_\tau),$$

respectively. The ratio of these two probabilities is called the **importance sample ratio**:

$$\rho_{t:T-1} = \frac{\Pr_\pi[A_t, R_{t+1}, S_{t+1}, A_{t+1}, \ldots, S_{T-1}, A_{T-1}, R_T, S_T | S_t]}{\Pr_b[A_t, R_{t+1}, S_{t+1}, A_{t+1}, \ldots, S_{T-1}, A_{T-1}, R_T, S_T | S_t]} = \prod_{\tau=t}^{T-1} \frac{\pi(A_\tau | S_\tau)}{b(A_\tau | S_\tau)}.$$

This ratio depends on the policies, but it does not depend on the dynamics. To ensure the ratio is always well defined, we require the policy π is absolutely continuous with respect to b (denoted as $\pi \ll b$), meaning that for all $(s, a) \in \mathcal{S} \times \mathcal{A}$ such that $\pi(a|s) > 0$, we have $b(a|s) > 0$. We also have the following convention: if $\pi(a|s) = 0$, $\frac{\pi(a|s)}{b(a|s)}$ is defined as 0, regardless of the values of $b(a|s)$. With such convention and the condition of $\pi \ll b$, $\frac{\pi(a|s)}{b(a|s)}$ is always well-defined.

Given a state–action pair (S_t, A_t), the probabilities to generate the trajectory $S_t, A_t, R_{t+1}, S_{t+1}, A_{t+1}, \ldots, S_{T-1}, A_{T-1}, R_T, S_T$ using the policy π and the policy b are

$$\text{Pr}_\pi[R_{t+1}, S_{t+1}, A_{t+1}, \ldots, S_{T-1}, A_{T-1}, R_T, S_T | S_t, A_t]$$
$$= p(S_{t+1}, R_{t+1}|S_t, A_t)\pi(A_{t+1}|S_{t+1}) \cdots p(S_T, R_T|S_{T-1}, A_{T-1})$$
$$= \prod_{\tau=t+1}^{T-1} \pi(A_\tau|S_\tau) \prod_{\tau=t}^{T-1} p(S_{\tau+1}, R_{\tau+1}|S_\tau, A_\tau),$$
$$\text{Pr}_b[A_t, R_{t+1}, S_{t+1}, A_{t+1}, \ldots, S_{T-1}, A_{T-1}, R_T, S_T | S_t]$$
$$= p(S_{t+1}, R_{t+1}|S_t, A_t)b(A_{t+1}|S_{t+1}) \cdots p(S_T, R_T|S_{T-1}, A_{T-1})$$
$$= \prod_{\tau=t+1}^{T-1} b(A_\tau|S_\tau) \prod_{\tau=t}^{T-1} p(S_{\tau+1}, R_{\tau+1}|S_\tau, A_\tau),$$

respectively. The importance sample ratio is:

$$\rho_{t+1:T-1} = \prod_{\tau=t+1}^{T-1} \frac{\pi(A_\tau|S_\tau)}{b(A_\tau|S_\tau)}.$$

In on-policy MC update, after getting the return samples $g_1, g_2, \ldots, g_c$, we use the simple average $\frac{1}{c} \sum_{i=1}^{c} g_i$ as the value estimates. This implicitly assumes that samples $g_1, g_2, \ldots, g_c$ are with equal probabilities. Similarly, if we use the behavior policy b to generate return samples $g_1, g_2, \ldots, g_c$, we can view those return samples are with equal probability with regard to the behavior policy b. However, these return samples are not with equal probabilities with regard to the target policy π. For the target policy π, the probabilities of those samples are proportion to the important sample ratios. Therefore, we can use weighted average of these samples to estimate. Specifically, let ρ_i $(1 \le i \le c)$ be the importance sample ratio of the sample g_i, which is also the weights for the weighted average. Then the weighted average of those samples is

$$\frac{\sum_{i=1}^{c} \rho_i g_i}{\sum_{i=1}^{c} \rho_i}.$$

MC update with importance sampling can be implemented incrementally, too. However, we no longer record the number of samples for each state or state–action pair. Instead, we record the summation of weights. Each time we update incrementally, we first update the summations of weights, and then update the estimate. For example, updating state values can be written as

$$c \leftarrow c + \rho$$

$$v \leftarrow v + \frac{\rho}{c}(g - v)$$

where c is the summation of weights with abuse of notation usage, and updating action values can be written as

$$c \leftarrow c + \rho$$
$$q \leftarrow q + \frac{\rho}{c}(g - q).$$

4.2.2 Off-Policy MC Policy Evaluation

Section 4.1.1 introduced on-policy algorithms to evaluate policies. They first use the target policy to generate samples, and then use the samples to update value estimates. In those algorithms, the policy to generate samples and the policy to update are the same, so those algorithms are on-policy. This section will introduce off-policy MC policy evaluation based on importance sampling, which uses another behavior policy to generate samples.

Algorithm 4.9 is the weighted importance sample off-policy MC policy evaluation. As usual, the first-visit version and the every-visit version are shown together here. Step 1 initializes the action value estimates, while Step 2 conducts off-policy update. The off-policy update needs to determine a behavior policy b for importance sampling. The behavior policy b can be either different or the same for different episodes, but they should always satisfy $\pi \ll b$. All soft policies satisfy this condition. Then we use the behavior policy to generate samples. Using these samples, we calculate return and update value estimates and ratio in reversed order. In the beginning, the ratio ρ is set to 1. It will decrease. If the ratio ρ becomes 0, which is usually due to $\pi(A_t|S_t) = 0$, the ratio will always be 0 afterward, so it is meaningless to continue iterating. Therefore, Step 2.5.4 checks whether $\pi(A_t|S_t) = 0$. This check is necessary, since it ensures that all updates will observe $c(s, a) > 0$. If there were no such checks, $c(s, a)$ can be 0 both before the update and after the update, which further leads to the "division by 0" error when we update $q(s, a)$. This check avoids such errors.

Algorithm 4.9   Evaluate action values using off-policy MC update based on importance sampling.

---

1. (Initialize) Initialize the action value estimates $q(s, a) \leftarrow$ arbitrary value ($s \in \mathcal{S}, a \in \mathcal{A}$). If we use incremental implementation, initialize the counter $c(s, a) \leftarrow 0$ ($s \in \mathcal{S}, a \in \mathcal{A}$).
2. (MC update) For each episode:

   2.1. (Designate behavior policy) Designate a behavior policy b such that $\pi \ll b$.

2.2. (Sample trajectory) Use the policy b to generate a trajectory $S_0, A_0, R_1, S_1, \ldots, S_{T-1}, A_{T-1}, R_T, S_T$.

2.3. (For the first-visit version, find when a state–action pair is first visited) Initialize $f(s, a) \leftarrow -1$ ($s \in \mathcal{S}, a \in \mathcal{A}$). Then for each $t \leftarrow 0, 1, \ldots, T - 1$, if $f(S_t, A_t) < 0$, set $f(S_t, A_t) \leftarrow t$.

2.4. (Initialize return and ratio) $G \leftarrow 0$ and $\rho \leftarrow 1$.

2.5. (Update) For $t \leftarrow T - 1, T - 2, \ldots, 0$:

    2.5.1. (Calculate return) $G \leftarrow \gamma G + R_{t+1}$.

    2.5.2. (Update action value estimate) Update $q(S_t, A_t)$ to reduce $\rho \big[G - q(S_t, A_t)\big]^2$. (For incremental implementation, set $c(S_t, A_t) \leftarrow c(S_t, A_t) + \rho$, $q(S_t, A_t) \leftarrow q(S_t, A_t) + \frac{\rho}{c(S_t, A_t)}\big[G - q(S_t, A_t)\big]$.) For the first-visit version, update the counter and action estimates only when $f(S_t, A_t) = t$.

    2.5.3. (Update importance sampling ratio) $\rho \leftarrow \rho \frac{\pi(A_t|S_t)}{b(A_t|S_t)}$.

    2.5.4. (Check early stop condition) If $\rho = 0$, break the loop of Step 2.5.

4.2.3 Off-Policy MC Policy Optimization

Sections 4.1.2 and 4.1.3 introduced on-policy algorithms to find optimal policies. They use the latest estimates of optimal policy to generate samples, and then use the samples to update optimal policy. In those algorithms, the policy to generate samples and the policy to be updated are the same, so those algorithms are on-policy. This section will introduce off-policy MC update to find the optimal policy, based on importance sampling, which uses another behavior policy to generate samples.

Algorithm 4.10 shows an off-policy MC update algorithm to find an optimal policy based on importance sampling. As usual, the every-visit version and the first-visit version are shown together. Furthermore, the version that maintains the policy explicitly and the version that maintains the policy implicitly are shown together. These two versions differ in Step 1, Step 2.5.3, and Step 2.5.4. Step 1 initializes the action value estimates. If we want to maintain the policy explicitly, we need to initialize the policy, too. Step 2 conducts MC update. Since the target policy π changes during the updates, we usually set the behavior policy b to be a soft policy so that $\pi \ll b$ can always hold. We can use either different behavior policies or the same behavior policy in different episodes. Besides, we also limit our target policy π to be a deterministic policy. That is, for every state $s \in \mathcal{S}$, there exists an action $a \in \mathcal{A}(s)$ such that $\pi(a|s) = 1$, while other actions $\pi(\cdot|s) = 0$. Step 2.5.4 and Step 2.5.5 use this property to check the early stop condition and update the importance sampling ratio. If $A_t \neq \pi(S_t)$, we have $\pi(A_t|S_t) = 0$, and the importance sampling ratio after the update will be 0. Therefore, we need to exit the loop to avoid the "division by 0" error. If $A_t = \pi(S_t)$, we have $\pi(A_t|S_t) = 1$. So the updating statement $\rho \leftarrow \rho \frac{\pi(A_t|S_t)}{b(A_t|S_t)}$ can be simplified to $\rho \leftarrow \rho \frac{1}{b(A_t|S_t)}$.

Algorithm 4.10   Find an optimal policy using off-policy MC update based on importance sampling.

1. (Initialize) Initialize the action value estimates $q(s, a)$ ← arbitrary value ($s \in S$, $a \in \mathcal{A}$). If we use incremental implementation, initialize the counter $c(s, a) \leftarrow 0$ ($s \in S$, $a \in \mathcal{A}$). If the policy is maintained explicitly, initialize the policy $\pi(s) \leftarrow \arg\max_a q(s, a)$ ($s \in S$).

2. (MC update) For each episode:

    2.1. (Designate behavior policy) Designate a behavior policy b such that $\pi \ll b$.

    2.2. (Sample trajectory) Use the policy b to generate a trajectory $S_0, A_0, R_1, S_1, \ldots, S_{T-1}, A_{T-1}, R_T, S_T$.

    2.3. (For the first-visit version, find when a state–action pair is first visited) Initialize $f(s, a) \leftarrow -1$ ($s \in S$, $a \in \mathcal{A}$). Then for each $t \leftarrow 0, 1, \ldots, T - 1$, if $f(S_t, A_t) < 0$, set $f(S_t, A_t) \leftarrow t$.

    2.4. (Initialize return and ratio) $G \leftarrow 0$ and $\rho \leftarrow 1$.

    2.5. (Update) For $t \leftarrow T - 1, T - 2, \ldots, 0$:

        2.5.1. (Calculate return) $G \leftarrow \gamma G + R_{t+1}$.

        2.5.2. (Update action value estimate) Update $q(S_t, A_t)$ to reduce $\rho[G - q(S_t, A_t)]^2$. (For incremental implementation, set $c(S_t, A_t) \leftarrow c(S_t, A_t) + \rho$, $q(S_t, A_t) \leftarrow q(S_t, A_t) + \frac{\rho}{c(S_t, A_t)}[G - q(S_t, A_t)]$.) For the first-visit version, update the counter and action estimates only when $f(S_t, A_t) = t$.

        2.5.3. (If we maintain the policy explicitly, update the policy) If the policy is maintained explicitly, $\pi(S_t) \leftarrow \arg\max_a q(S_t, a)$.

        2.5.4. (Check early stop condition) If $A_t \neq \pi(S_t)$, break the loop of Step 2.5. (We can conduct similar judgments if we maintain the policy implicitly.)

        2.5.5. (Update importance sampling ratio) $\rho \leftarrow \rho \frac{\pi(A_t|S_t)}{b(A_t|S_t)}$.

This section applies a variance reduction technique in Monte Carlo method, importance sampling, to the MC update algorithm. Other variance reduction methods, such as control variate and antithetic variable, can also be used in MC update to boost the performance in some tasks.

4.3 Case Study: Blackjack

This section considers the comparing card game `Blackjack-v1`.

The rule of Blackjack is as follows: A player competes against with a dealer. In each episode, either the player or the dealer can win, or they may tie. At the beginning of an episode, both the player and the dealer have two cards. The player can see two cards at the player's hand, and one card within the two at the dealer's hand. Then the player can choose to "stand" or "hit". If the player chooses to hit, the player will have one more card. Then we sum up the values of all cards in the player's hand, where the value of each card is shown in Table 4.1. Especially, the Ace can be counted as either 1 or 11. Then the player repeats this process: if the total value of the player exceeds 21, the player "busts" and loses the game. Otherwise, the player can choose to hit or stand again. If the player stands before it gets bust, the total value now is the final value of the player. Then the dealer shows both of their cards. If the total value of the dealer is smaller than 17, it hits. Otherwise, it stands. Whenever the total value of the dealer exceeds 21, the dealer busts and loses the game. If the dealer stands before it busts, we compare the total values of the player and the total values of the dealer. Whoever has larger values wins. If the player and the dealer have the same total value, they tie.

Table 4.1   Value of the cards in Blackjack.

card	value
A	1 or 11
2	2
3	3
. . .	. . .
9	9
10, J, Q, K	10

4.3.1 Use Environment

The task `Blackjack-v1` in Gym implements this game. The action space of this environment is `Discrete(2)`. The action is an int value, either 0 or 1. 0 means that the player stands, while 1 means that the player hits. The observation space is `Tuple(Discrete(32), Discrete(11), Discrete(2))`. The observation is a `tuple` consisting of three elements. The three elements are:

- The total value of the player, which is an `int` value ranging from 4 to 21. When the player has an Ace, we will use the following rule to determine the value of the Ace: try to maximize the total reward under the condition that the total value of the player does not exceed 21. (Therefore, at most one ace will be calculated as 11.)
- The face-up card of the dealer, which is an `int` value ranging from 1 to 10. The Ace is shown as 1 here.
- Whether an Ace of player is calculated as 11 when we calculate the total value of the player, which is a `bool` value.

Online Contents

Advanced readers can check the explanation of the class `gym.spaces.Tuple` in GitHub repo of this book.

Code 4.1 shows the function `play_policy()`, which plays an episode using a given policy. The parameter `policy` is a `np.array` object with the shape (22,11,2,2), storing the probability of all actions under the condition of all states. The function `ob2state()` is an auxiliary function that converts an observation to a state. We know the observation is a tuple, whose last element is a bool value. Such value format is inconvenient for indexing the policy object. Therefore, the function `ob2state()` refactors the observation tuple to a state such that it can be used as an index to obtain probability from the `np.array` object `policy`. After obtaining the probability, we can use the function `np.random.choice()` to choose an action. The 0-th parameter of the function `np.random.choice()` indicates the output candidates. If it is an `int` value a, the output is from `np.arange(a)`. The function `np.random.choice()` has a keyword parameter p. indicating the probability of each outcome. Besides, if we do not assign a `np.array` object for the parameter policy of the function `play_policy()`, it chooses actions randomly.

Code 4.1 Play an episode.
Blackjack-v1_MonteCarlo_demo.ipynb

```
 1  def ob2state(observation):
 2      return observation[0], observation[1], int(observation[2])
 3
 4  def play_policy(env, policy=None, verbose=False):
 5      observation, _ = env.reset()
 6      reward, terminated, truncated = 0., False, False
 7      episode_reward, elapsed_steps = 0., 0
 8      if verbose:
 9          logging.info('observation = %s', observation)
10      while True:
11          if verbose:
12              logging.info('player = %s, dealer = %s', env.player, env.dealer)
13          if policy is None:
14              action = env.action_space.sample()
15          else:
16              state = ob2state(observation)
17              action = np.random.choice(env.action_space.n, p=policy[state])
18          if verbose:
19              logging.info('action = %s', action)
20          observation, reward, terminated, truncated, _ = env.step(action)
21          if verbose:
22              logging.info('observation = %s', observation)
23              logging.info('reward = %s', reward)
24              logging.info('terminated = %s', terminated)
25              logging.info('truncated = %s', truncated)
26          episode_reward += reward
27          elapsed_steps += 1
28          if terminated or truncated:
29              break
30      return episode_reward, elapsed_steps
31
32  episode_reward, elapsed_steps = play_policy(env, verbose=True)
33  logging.info("episode reward: %.2f", episode_reward)
```

4.3.2 On-Policy Policy Evaluation

The task `Blackjack-v1` has the following properties:

- There are no duplicate states in each trajectory. Since the player always has more cards than the previous steps, one of the following holds: The total value increases, or the value of an Ace changes from 1 to 11.
- Only the reward of the last step is non-zero.

Using these two properties, we can simplify the MC policy evaluations in the following ways:

- Since every state will be visited only once, there is no need to distinguish between every-visit and first-visit.
- When the discount factor $\gamma = 1$, the reward in the last step is the episode return. So we do not need to calculate the episode return in reversed order.

Leveraging the aforementioned simplifications, Code 4.2 shows the on-policy MC policy evaluation algorithm. The function `evaluate_action_monte_carlo()` accepts the environment env and the policy `policy`, and calculates and returns action value q. This function does not distinguish between the first-visit version and the every-visit version. This function uses the latest reward as the episode return, and update value estimates in normal order.

Code 4.2 On-Policy MC evaluation.
`Blackjack-v1_MonteCarlo_demo.ipynb`

```
 1  def evaluate_action_monte_carlo(env, policy, episode_num=500000):
 2      q = np.zeros_like(policy)  # action value
 3      c = np.zeros_like(policy)  # count
 4      for _ in range(episode_num):
 5          # play an episode
 6          state_actions = []
 7          observation, _ = env.reset()
 8          while True:
 9              state = ob2state(observation)
10              action = np.random.choice(env.action_space.n, p=policy[state])
11              state_actions.append((state, action))
12              observation, reward, terminated, truncated, _ = env.step(action)
13              if terminated or truncated:
14                  break  # end of episode
15          g = reward  # return
16          for state, action in state_actions:
17              c[state][action] += 1.
18              q[state][action] += (g - q[state][action]) / c[state][action]
19      return q
```

The following codes show how to use the function `evaluate_action_monte_carlo()`. It evaluates the deterministic policy `policy`, who hits when < 20, and stands otherwise. The function `evaluate_action_monte_carlo()` returns the estimate of action values q. Then we obtain the state value estimates v using the action value estimates q.

```
 1  policy = np.zeros((22, 11, 2, 2))
 2  policy[20:, :, :, 0] = 1  # stand when >=20
```

```
3   policy[:20, :, :, 1] = 1  # hit when <20
4
5   q = evaluate_action_monte_carlo(env, policy)  # action value
6   v = (q * policy).sum(axis=-1)  # state value
```

Now we try to visualize the value estimates. Since q is a 4-dimension np.array, while v is a 3-dimension np.array, v is easier to be visualized compared to q. Code 4.3 shows a function plot() to visualize a 3-dimension array that can be indexed by a state. There are two subplots. The left subplot is for the case where the last component of the state is 0, while the right subplot is for the cases where the last component of the state is 1. For each subplot, X-axis is the total value of the player, and Y-axis is the exposed card of the dealer. Note, that we only show the range from 12 to 21 for the total reward of the player, since this range interests us most. If the total value of the player is ≤ 11, the player will always get larger total value without worrying get busted by hitting one more time. Therefore, the player will always hit when its total value ≤ 11. Therefore, we are more interested in the policy when the total value of the player ranges from 12 to 21.

Code 4.3   Visualize a 3-dimension np.array, which can be indexed by a state.
Blackjack-v1_MonteCarlo_demo.ipynb

```
1   def plot(data):
2       fig, axes = plt.subplots(1, 2, figsize=(9, 4))
3       titles = ['without ace', 'with ace']
4       have_aces = [0, 1]
5       extent = [12, 22, 1, 11]
6       for title, have_ace, axis in zip(titles, have_aces, axes):
7           dat = data[extent[0]:extent[1], extent[2]:extent[3], have_ace].T
8           axis.imshow(dat, extent=extent, origin='lower')
9           axis.set_xlabel('player sum')
10          axis.set_ylabel('dealer showing')
11          axis.set_title(title)
```

We can plot the state values by calling the function plot() using the following codes, resulting in Fig. 4.3. Due to the randomness of the environment, the figure may differ in your own machine. Increasing the number of iterations may reduce the difference.

```
1   plot(v)
```

4.3.3 On-Policy Policy Optimization

This section considers on-policy MC update to find optimal values and optimal policies.

Code 4.4 shows the on-policy MC with exploring start. Implementing exploring start needs to change the initial state distribution, so it requires us to understand how the environment is implemented. The implementation of the environment Blackjack-v1 can be found in:

```
1   https://github.com/openai/gym/blob/master/gym/envs/toy_text/blackjack.py
```

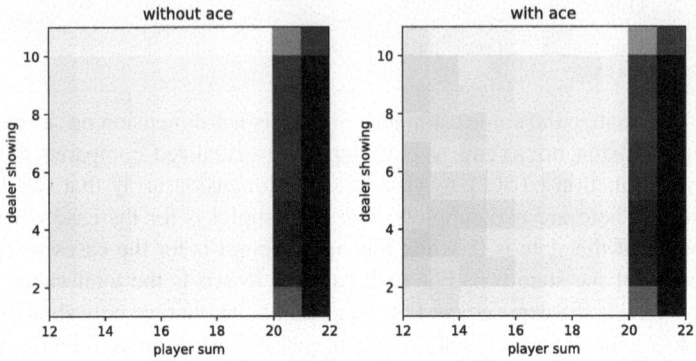

Fig. 4.3   State value estimates obtained by policy evaluation algorithm.
Darker means larger values.

After reading the source codes, we know that the cards of the player and the dealer are stored in env.player and env.dealer respectively, both of them are lists consisting of two int values. We can change the initial state by changing these two members. The first few lines within the for loop try to change the initial state distribution. In the previous section, we have known that we are more interested when the total value of the player ranges from 12 to 21. Therefore, we only cover this range for exploring start. Using the generated state, we can calculate the probability of player's card and the dealer's exposed card. Although there may be multiple card combinations for the same state, picking an arbitrary combination is without loss of generality since different combinations with the same state have the same effects. After calculating the card, we assign the cards to env.player and env.dealer[0], overwriting the initial state. Therefore, the episode can start with our designated state.

Code 4.4   On-policy MC update with exploring start.
Blackjack-v1_MonteCarlo_demo.ipynb

```
 1  def monte_carlo_with_exploring_start(env, episode_num=500000):
 2      policy = np.zeros((22, 11, 2, 2))
 3      policy[:, :, :, 1] = 1.
 4      q = np.zeros_like(policy)  # action values
 5      c = np.zeros_like(policy)  # counts
 6      for _ in range(episode_num):
 7          # choose initial state randomly
 8          state = (np.random.randint(12, 22), np.random.randint(1, 11),
 9                   np.random.randint(2))
10          action = np.random.randint(2)
11          # play an episode
12          env.reset()
13          if state[2]:  # has ace
14              env.unwrapped.player = [1, state[0] - 11]
15          else:  # no ace
16              if state[0] == 21:
17                  env.unwrapped.player = [10, 9, 2]
18              else:
19                  env.unwrapped.player = [10, state[0] - 10]
```

```
20      env.unwrapped.dealer[0] = state[1]
21      state_actions = []
22      while True:
23          state_actions.append((state, action))
24          observation, reward, terminated, truncated, _ = env.step(action)
25          if terminated or truncated:
26              break  # end of episode
27          state = ob2state(observation)
28          action = np.random.choice(env.action_space.n, p=policy[state])
29      g = reward  # return
30      for state, action in state_actions:
31          c[state][action] += 1.
32          q[state][action] += (g - q[state][action]) / c[state][action]
33          a = q[state].argmax()
34          policy[state] = 0.
35          policy[state][a] = 1.
36  return policy, q
```

The following codes use the function `monte_carlo_with_exploring_start()` to get optimal policy estimates and optimal value estimates, and use the function `plot()` to visualize the optimal policy estimates and optimal value estimates as Figs. 4.4 and 4.5 respectively.

```
1  policy, q = monte_carlo_with_exploring_start(env)
2  v = q.max(axis=-1)
3  plot(policy.argmax(-1))
4  plot(v)
```

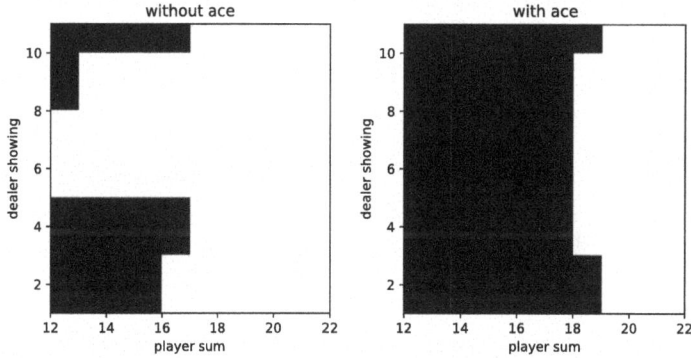

Fig. 4.4 Optimal policy estimates.
Black means hit, while white means stand.

Code 4.5 implements the on-policy MC update with soft policy to find optimal policy and optimal values. The function `monte_carlo_with_soft()` uses a parameter epsilon to designate ε in ε-soft policy. At the initialization step, we initialize the policy as $\pi(a|s) = 0.5$ ($s \in \mathcal{S}, a \in \mathcal{A}$) to make sure the policy before iterations is ε-soft policy.

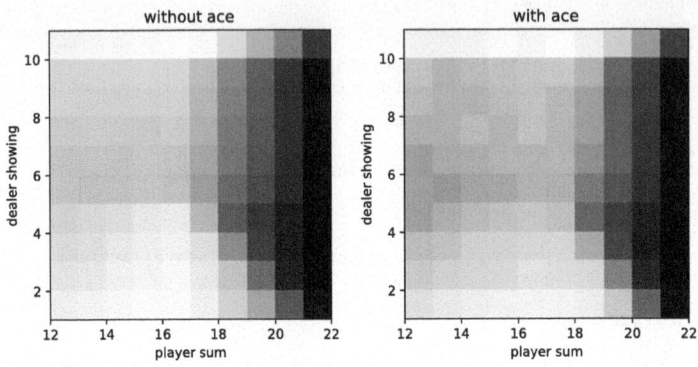

Fig. 4.5   Optimal state value estimates.
Darker means larger values.

Code 4.5   MC update with soft policy.

`Blackjack-v1_MonteCarlo_demo.ipynb`

```
def monte_carlo_with_soft(env, episode_num=500000, epsilon=0.1):
    policy = np.ones((22, 11, 2, 2)) * 0.5  # soft policy
    q = np.zeros_like(policy)  # action values
    c = np.zeros_like(policy)  # counts
    for _ in range(episode_num):
        # play an episode
        state_actions = []
        observation, _ = env.reset()
        while True:
            state = ob2state(observation)
            action = np.random.choice(env.action_space.n, p=policy[state])
            state_actions.append((state, action))
            observation, reward, terminated, truncated, _ = env.step(action)
            if terminated or truncated:
                break  # end of episode
        g = reward  # return
        for state, action in state_actions:
            c[state][action] += 1.
            q[state][action] += (g - q[state][action]) / c[state][action]
            # soft update
            a = q[state].argmax()
            policy[state] = epsilon / 2.
            policy[state][a] += (1. - epsilon)
    return policy, q
```

The following codes use the function `monte_carlo_with_soft()` to get the optimal policy estimates and the optimal value estimates, and visualize them as Fig. 4.4 and Fig. 4.5 respectively.

```
policy, q = monte_carlo_with_soft(env)
v = q.max(axis=-1)
plot(policy.argmax(-1))
plot(v)
```

4.3.4 Off-Policy Policy Evaluation

This section implements off-policy algorithms based on importance sampling.

Code 4.6 implements the off-policy policy evaluation to estimate action values. The parameters of the function `evaluate_monte_carlo_importance_sample()` include not only the target policy `policy`, but also the behavior policy `behavior_policy`. Within an episode, we update the importance sampling ratio in reversed order to make the updates more efficient.

Code 4.6 Policy evaluation based on importance sampling.
`Blackjack-v1_MonteCarlo_demo.ipynb`

```
1   def evaluate_monte_carlo_importance_sample(env, policy, behavior_policy,
2           episode_num=500000):
3       q = np.zeros_like(policy)  # action values
4       c = np.zeros_like(policy)  # counts
5       for _ in range(episode_num):
6           # play episode using behavior policy
7           state_actions = []
8           observation, _ = env.reset()
9           while True:
10              state = ob2state(observation)
11              action = np.random.choice(env.action_space.n,
12                      p=behavior_policy[state])
13              state_actions.append((state, action))
14              observation, reward, terminated, truncated, _ = env.step(action)
15              if terminated or truncated:
16                  break  # finish the episode
17          g = reward  # return
18          rho = 1.  # importance sampling ratio
19          for state, action in reversed(state_actions):
20              c[state][action] += rho
21              q[state][action] += (rho / c[state][action] * (g -
22                      q[state][action]))
23              rho *= (policy[state][action] / behavior_policy[state][action])
24              if rho == 0:
25                  break  # early stop
26      return q
```

The following codes call the function `evaluate_monte_carlo_importance_sample()` with the behavior function $\pi(a|s) = 0.5$ ($s \in \mathcal{S}, a \in \mathcal{A}$). The results are the same as that of on-policy algorithms (Fig. 4.3).

```
1   policy = np.zeros((22, 11, 2, 2))
2   policy[20:, :, :, 0] = 1  # stand when >=20
3   policy[:20, :, :, 1] = 1  # hit when <20
4   behavior_policy = np.ones_like(policy) * 0.5
5   q = evaluate_monte_carlo_importance_sample(env, policy, behavior_policy)
6   v = (q * policy).sum(axis=-1)
```

4.3.5 Off-Policy Policy Optimization

Finally, let us implement the off-policy MC update to find an optimal policy. Code 4.7 shows how to find the optimal policy using importance sampling. Once again, the soft policy $\pi(a|s) = 0.5$ ($s \in \mathcal{S}, a \in \mathcal{A}$) is used as the behavior policy. And reversed order is used to update the importance sampling ratio more efficiently.

Code 4.7   Importance sampling policy optimization with soft policy.
`Blackjack-v1_MonteCarlo_demo.ipynb`

```python
def monte_carlo_importance_sample(env, episode_num=500000):
    policy = np.zeros((22, 11, 2, 2))
    policy[:, :, :, 0] = 1.
    behavior_policy = np.ones_like(policy) * 0.5  # soft policy
    q = np.zeros_like(policy)  # action values
    c = np.zeros_like(policy)  # counts
    for _ in range(episode_num):
        # play using behavior policy
        state_actions = []
        observation, _ = env.reset()
        while True:
            state = ob2state(observation)
            action = np.random.choice(env.action_space.n,
                    p=behavior_policy[state])
            state_actions.append((state, action))
            observation, reward, terminated, truncated, _ = env.step(action)
            if terminated or truncated:
                break  # finish the episode
        g = reward  # return
        rho = 1.  # importance sampling ratio
        for state, action in reversed(state_actions):
            c[state][action] += rho
            q[state][action] += (rho / c[state][action] *
                    (g - q[state][action]))
            # improve the policy
            a = q[state].argmax()
            policy[state] = 0.
            policy[state][a] = 1.
            if a != action:  # early stop
                break
            rho /= behavior_policy[state][action]
    return policy, q
```

The following codes call the function `monte_carlo_importance_sample()` to get the optimal policy estimates and optimal value estimates. The results are the same as those of on-policy algorithms (Figs. 4.4 and 4.5).

```python
policy, q = monte_carlo_importance_sample(env)
v = q.max(axis=-1)
plot(policy.argmax(-1))
plot(v)
```

4.4 Summary

- Model-free RL and its convergence are based on stochastic approximation theory, especially the Robbins–Monro algorithm.
- Monte Carlo (MC) update uses the Monte Carlo method to estimate the expectation of returns. The incremental implementation uses the estimate from the first $c - 1$ samples and the counter c to get the estimate from the first c samples.

- The same state (or the same state–action pair) can be visited multiple times in an episode. Every-visit algorithm uses the return samples every time it is visited, while the first-visit algorithm only uses the return samples the first time it is visited.
- MC policy evaluation updates the value estimates at the end of each episode. MC update to find the optimal policy update the value estimates and improve the policy at the end of each episode.
- Exploring start, soft policy, and importance sampling are ways of explorations.
- On-policy MC uses the policy to update to generate trajectory samples, while off-policy MC uses a different policy to generate trajectory samples.
- Importance sampling uses a behavior policy b to update the target policy π. The importance sampling ratio is defined as

$$\rho_{t:T-1} \stackrel{\text{def}}{=} \prod_{\tau=t}^{T-1} \frac{\pi(A_\tau|S_\tau)}{b(A_\tau|S_\tau)}.$$

4.5 Exercises

4.5.1 Multiple Choices

4.1 Which of the following learning sequences satisfies the condition of the Robbins–Monro algorithm: (   )

A. $\alpha_k = 1, k = 1, 2, 3, \ldots$.
B. $\alpha_k = \frac{1}{k}, k = 1, 2, 3, \ldots$.
C. $\alpha_k = \frac{1}{k^2}, k = 1, 2, 3, \ldots$.

4.2 On exploration and exploitation, choose the correct one: (   )

A. Exploration uses the current information to maximize the reward, while exploitation tries to understand more about the environment.
B. Exploration tries to understand more about the environment, while exploitation uses the current information to maximize the reward.
C. Exploration uses the current information to maximize the reward, while exploitation tries to use the optimal policy.

4.3 On exploration, choose the correct one: (   )

A. Exploration start can not be applied to all tasks since it needs to modify the environment.
B. Soft policy can not be applied to all tasks since it needs to modify the environment.
C. Importance sampling can not be applied to all tasks since it needs to modify the environment.

4.4 On RL based on importance sampling, choose the correct one: (   )

A. The policy to generate samples is called behavior policy, and the policy to update is called the target policy.
B. The policy to generate samples is called behavior policy, and the policy to update is called the evaluation policy.
C. The policy to generate samples is called evaluation policy, and the policy to update is called the target policy.

4.5 On RL importance sampling, choose the correct one: (   )

A. The target policy π should be absolutely continuous to the behavior policy b, which can be denoted as $\pi \ll b$.
B. The behavior policy b should be absolutely continuous to the target policy π, which can be denoted as $b \ll \pi$.
C. The target policy π should be absolutely continuous to the behavior policy b, which can be denoted as $b \ll \pi$.

4.5.2 Programming

4.6 Solve `CliffWalking-v0` using the MC update algorithms in this chapter.

4.5.3 Mock Interview

4.7 Why Monte Carlo update algorithms in RL are called "Monte Carlo"?

4.8 Why importance sampling RL algorithms are off-policy algorithms? Why do we use importance sampling in RL?

Chapter 5
TD: Temporal Difference Learning

This chapter covers

- Temporal Difference (TD) return
- TD learning for policy evaluation
- TD learning for policy optimization
- SARSA algorithm
- Expected SARSA algorithm
- TD learning with importance sampling
- Q learning algorithm
- Double Q learning algorithm
- λ return
- eligibility trace
- TD(λ) algorithm

This chapter discusses another RL algorithm family called Temporal Difference (TD) learning. Both MC learning and TD learning are model-free learning algorithms, which means that they can learn from samples without an environment model. The difference between MC learning and TD learning is, TD learning uses bootstrapping, meaning that it uses existing value estimates to update value estimates. Consequently, TD learning can update value estimates without requiring that an episode finishes. Therefore, TD learning can be used for either episodic tasks or sequential tasks.

© The Author(s), under exclusive license to Springer Nature Singapore Pte Ltd. 2024
Z. Xiao, *Reinforcement Learning*, https://doi.org/10.1007/978-981-19-4933-3_5

5.1 TD return

MC learning in previous chapter uses the following way to estimate values: First sample a trajectory starting from a state s or a state–action pair (s, a) to the end of the episode to get the return sample G_t, and then estimate the values according to $v_\pi(s) = \mathrm{E}_\pi[G_t|S_t = s]$ or $q_\pi(s, a) = \mathrm{E}_\pi[G_t|S_t = s, A_t = a]$. This section will introduce a new statistics call TD return (denoted as U_t). We do not need to sample to the end of episode to get the TD return sample, and the TD return U_t also satisfies $v_\pi(s) = \mathrm{E}_\pi[U_t|S_t = s]$ and $q_\pi(s, a) = \mathrm{E}_\pi[U_t|S_t = s, A_t = a]$, so that we can estimate values according to the samples of U_t.

Given a positive integer n ($n = 1, 2, 3, \ldots$), we can define n-**step TD return** as follows:

- n-step TD return bootstrapped from state value:

$$
U_{t:t+n}^{(v)} \overset{\text{def}}{=} \begin{cases} R_{t+1} + \gamma R_{t+2} + \cdots + \gamma^{n-1} R_{t+n} + \gamma^n v(S_{t+n}), & t + n < T \\ R_{t+1} + \gamma R_{t+2} + \cdots + \gamma^{T-t-1} R_T & t + n \geq T, \end{cases}
$$

where the superscript (v) in $U_{t:t+n}^{(v)}$ means that the return is based on state values, and the subscript $t : t + n$ means that we use $v(S_{t+n})$ to estimate the TD target when $S_{t+n} \neq s_{\text{end}}$. We can write $U_{t:t+n}^{(v)}$ as U_t when there are no confusions.

- n-step TD return bootstrapped from action value:

$$
U_{t:t+n}^{(q)} \overset{\text{def}}{=} \begin{cases} R_{t+1} + \gamma R_{t+2} + \cdots + \gamma^{n-1} R_{t+n} + \gamma^n q(S_{t+n}, A_{t+n}), & t + n < T \\ R_{t+1} + \gamma R_{t+2} + \cdots + \gamma^{T-t-1} R_T & t + n \geq T, \end{cases}
$$

We can write $U_{t:t+n}^{(q)}$ as U_t when there are no confusions.

We can prove that

$$
\begin{aligned}
v_\pi(s) &= \mathrm{E}_\pi[U_t|S_t = s], & s \in \mathcal{S} \\
q_\pi(s, a) &= \mathrm{E}_\pi[U_t|S_t = s, A_t = a], & s \in \mathcal{S}, a \in \mathcal{A}(s).
\end{aligned}
$$

(Proof: When $t + n \geq T$, G_t is exactly U_t, so the equations obviously hold. When $t + n < T$, $v_\pi(s) = \mathrm{E}_\pi\left[U_{t:t+n}^{(v)}\middle|S_t = s\right]$ ($s \in \mathcal{S}$) can be proved by

$$
\begin{aligned}
v_\pi(s) &= \mathrm{E}_\pi[G_t|S_t = s] \\
&= \mathrm{E}_\pi\left[R_{t+1} + \gamma R_{t+2} + \cdots + \gamma^{n-1} R_{t+n} + \gamma^n G_{t+n}\middle|S_t = s\right] \\
&= \mathrm{E}_\pi\left[R_{t+1} + \gamma R_{t+2} + \cdots + \gamma^{n-1} R_{t+n} + \gamma^n \mathrm{E}_\pi[G_{t+n}|S_{t+n}]\middle|S_t = s\right] \\
&= \mathrm{E}_\pi\left[R_{t+1} + \gamma R_{t+2} + \cdots + \gamma^{n-1} R_{t+n} + \gamma^n v(S_{t+n})\middle|S_t = s\right] \\
&= \mathrm{E}_\pi[U_t|S_t = s].
\end{aligned}
$$

Other cases can be proved similarly.)

The most common value of n is 1, where n-step TD return degrades to **one-step TD return**:

- 1-step TD return bootstrapped from state value:

$$U_{t:t+1}^{(v)} = \begin{cases} R_{t+1} + \gamma v(S_{t+1}), & S_{t+1} \neq s_{\text{end}}, \\ R_{t+1}, & S_{t+1} = s_{\text{end}}. \end{cases}$$

- 1-step TD return bootstrapped from action value:

$$U_{t:t+1}^{(q)} = \begin{cases} R_{t+1} + \gamma q(S_{t+1}, A_{t+1}), & S_{t+1} \neq s_{\text{end}}, \\ R_{t+1}, & S_{t+1} = s_{\text{end}}. \end{cases}$$

Some environments can provide an indicator D_t to show whether a state S_t is a terminal state (The letter D stands for "done"), defined as

$$D_t \overset{\text{def}}{=} \begin{cases} 1, & S_t = s_{\text{end}}, \\ 0, & S_t \neq s_{\text{end}}. \end{cases}$$

We can use this indicator to simplify the formulation of the TD return. 1-step TD return can be simplified to

$$U_{t:t+1}^{(v)} = R_{t+1} + \gamma(1 - D_{t+1})v(S_{t+1})$$

$$U_{t:t+1}^{(q)} = R_{t+1} + \gamma(1 - D_{t+1})q(S_{t+1}, A_{t+1}).$$

Remarkably, the multiplication between $(1 - D_{t+1})$ and $v(S_{t+1})$ (or $q(S_{t+1}, A_{t+1})$) should be understood as follows: If $1 - D_{t+1} = 0$, the results of multiplication is 0, no matter whether $v(S_{t+1})$ (or $q(S_{t+1}, A_{t+1})$) is well defined. n-step TD returns ($n = 1, 2, \ldots$) can be similarly represented as

$$U_{t:t+n}^{(v)} = R_{t+1} + \gamma(1 - D_{t+1})R_{t+2} + \cdots$$
$$+ \gamma^{n-1}(1 - D_{t+n-1})R_{t+n} + \gamma^n(1 - D_{t+n})v(S_{t+n}),$$

$$U_{t:t+n}^{(q)} = R_{t+1} + \gamma(1 - D_{t+1})R_{t+2} + \cdots$$
$$+ \gamma^{n-1}(1 - D_{t+n-1})R_{t+n} + \gamma^n(1 - D_{t+n})q(S_{t+n}, A_{t+n}).$$

The multiplications should be understood in a similar way.

Figure 5.1 compares the backup diagram of return samples based on TD and MC. Recall that we use hollow circles to denote states, and solid circles to denote state–action pairs. Figure 5.1(a) is the backup diagram of action values. If we use the state–action pair that is one-step away in the future to estimate the action value of the current state–action pair, the return sample is one-step TD return. If we use state–action pair that is two-step away in the future to estimate the action value of the current state–action pair, the return sample is two-step TD return. So on and so forth.

Finally, we consider many, many steps, and then the return is without bootstrapping. Figure 5.1(b) is the backup diagram of state values. They can be analyzed similarly.

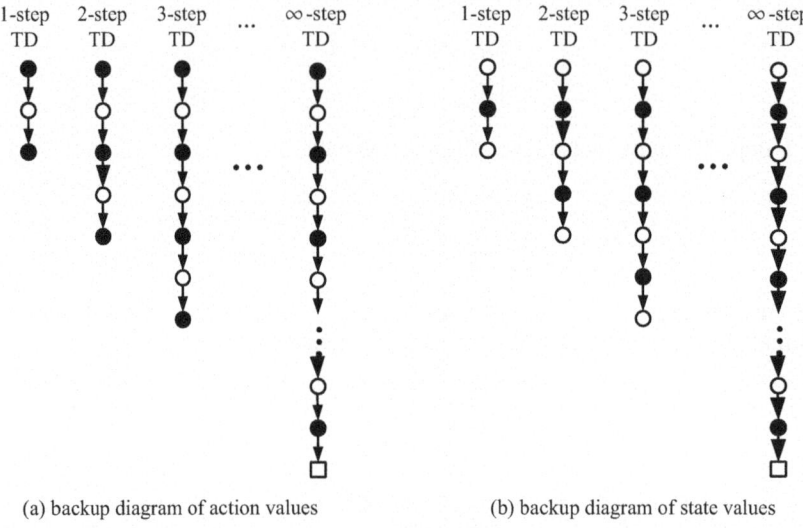

(a) backup diagram of action values   (b) backup diagram of state values

Fig. 5.1   Backup diagram of TD return and MC return.
This figure is adapted from (Sutton, 1998).

The greatest advantage of TD return over return is: We only need to interact n step to get the n-step TD return, rather than proceeding to the end of episode. Therefore, TD return can be used in both episodic tasks and sequential tasks.

5.2 On-Policy TD Learning

5.2.1 TD Policy Evaluation

This section uses TD return to evaluate a policy.

Similar to the cases of MC update, for model-free algorithms, action values are more important than state values, since action values can be used to determine policy and state values, but state values can not determine action values.

In the previous chapter, on-policy MC algorithms try to reduce $\left[G_t - q(S_t, A_t)\right]^2$ when updating $q(S_t, A_t)$. If the incremental implementation is used, the updating is of the form

$$q(S_t, A_t) \leftarrow q(S_t, A_t) + \alpha \left[G_t - q(S_t, A_t)\right],$$

where G_t is the return sample. The return sample is G_t for MC algorithms, while it is U_t for TD algorithms. We can obtain TD policy evaluation by replacing G_t in MC policy evaluation by U_t.

Define **TD error** as

$$\Delta_t \stackrel{\text{def}}{=} U_t - q(S_t, A_t).$$

The update can also be expressed as

$$q(S_t, A_t) \leftarrow q(S_t, A_t) + \alpha \Delta_t.$$

Updating values also uses a parameter called learning rate α, which is a positive number. In the MC algorithms in the previous chapter, the learning rate is usually $\frac{1}{c(S_t, A_t)}$, which relates to the state–action pair, and is monotonically decreasing. We can use such decreasing learning rate in TD learning, too. Meanwhile, in TD learning, the estimates will be increasingly accurate. Since TD uses bootstrapping, the return samples will be increasingly trustful. Therefore, we may put more weights on recent samples than historical samples, so it also makes sense to make the learning rate not decrease that fast, or even keep the learning rate constant. However, the learning rate should be always within $(0, 1]$. Designing a smart schedule for learning rate may benefit the learning.

The TD return can be either one-step TD return or multi-step TD return. Let us consider one-step TD return first.

Algorithm 5.1 shows the one-step TD return policy evaluation algorithm. The parameters of the algorithm include an optimizer, the discounted rate, and so on. The optimizer contains the learning rate α implicitly. Besides the learning rate α and the discount factor γ, there are also parameters to control the number of episodes and the maximum steps in an episode. We have known that TD learning can be used in either episodic tasks or sequential tasks. For sequential tasks, we can pick some ranges of steps as an episode, or we can update them as a very long episode. Step 1 initializes the action value estimates, while Step 2 conducts TD update. Each time an action is executed, we can get a new state. If the state is not a terminal state, we use the policy to decide an action, and calculate TD return. If the state is a terminal state, we calculate TD return directly. Especially, when the state is a terminal state, 1-step TD return equals the latest reward. Then we use the TD return sample to update action value.

Algorithm 5.1   One-step TD policy evaluation to estimate action values.

Inputs: environment (without mathematical model), and the policy π.
Output: action value estimates $q(s, a)$ ($s \in \mathcal{S}, a \in \mathcal{A}$).
Parameters: optimizer (containing learning rate α), discount factor γ, and parameters to control the number of episodes and maximum steps in an episode.

1. (Initialize) Initialize the action value estimates $q(s, a) \leftarrow$ arbitrary value ($s \in \mathcal{S}, a \in \mathcal{A}$).

---

2. (TD update) For each episode:

    2.1. (Initialize state–action pair) Select the initial state S, and then use the policy $\pi(\cdot|S)$ to determine the action A.

    2.2. Loop until the episode ends (for example, reach the maximum step, or S is the terminal state):

        2.2.1. (Sample) Execute the action A, and then observe the reward R and the next state S'.

        2.2.2. (Decide) If the state S' is not a terminal state, use the input policy $\pi(\cdot|S')$ to determine the action A'.

        2.2.3. (Calculate TD return) If the state S' is not a terminal state, $U \leftarrow R + \gamma q(S', A')$. If the state S' is the terminal state, $U \leftarrow R$.

        2.2.4. (Update value estimate) Update $q(S, A)$ to reduce $\left[U - q(S, A)\right]^2$. (For example, $q(S, A) \leftarrow q(S, A) + \alpha\left[U - q(S, A)\right]$.)

        2.2.5. $S \leftarrow S'$, $A \leftarrow A'$.

For the environments that provide indicators to show whether the episode has ended, such as the episodic environments in Gym, using the formulation with the indicators will simplify the control process of the algorithm. Algorithm 5.2 shows how to use this indicator to simplify implementation. Specifically, Step 2.2.1 obtains not only the new state S', but also the indicator D' to show whether the episode has ended. When Step 2.2.3 calculates TD return, if the new state S' is the terminal state, this indicator is 1, which leads to $U \leftarrow R$. In this case, the action value estimate $q(S', A')$ has no impacts. Therefore, in Step 2.2.2, no matter whether the new state S' is the terminal state, we can pick an action. Besides, in Step 1, we not only initialize the action value estimates for all states in the state space, but also initialize the action value estimates related to the terminal state arbitrarily. We know that action values related to the terminal state should be 0, so arbitrary initialization is incorrect. However, it does not matter since the action value estimates related to the terminal states are not actually used.

Algorithm 5.2   One-step TD policy evaluation to estimate action values with an indicator of episode end.

Inputs: environment (without mathematical model), and the policy π.

Output: action value estimates $q(s, a)$ ($s \in \mathcal{S}, a \in \mathcal{A}$).

Parameters: optimizer (containing learning rate α), discount factor γ, and parameters to control the number of episodes and maximum steps in an episode.

1. (Initialize) Initialize the action value estimates $q(s, a) \leftarrow$ arbitrary value ($s \in \mathcal{S}^+, a \in \mathcal{A}$).

2. (TD update) For each episode:

    2.1. (Initialize state–action pair) Select the initial state S, and then use the policy $\pi(\cdot|S)$ to determine the action A.

    2.2. Loop until the episode ends (for example, reach the maximum step, or S is the terminal state):

        2.2.1. (Sample) Execute the action A, and then observe the reward R, the next state S', and the indicator of episode end D'.

        2.2.2. (Decide) Use the policy $\pi(\cdot|S')$ to determine the action A'. (The action can be arbitrarily chosen if $D' = 1$.)

        2.2.3. (Calculate TD return) $U \leftarrow R + \gamma(1 - D')q(S', A')$.

        2.2.4. (Update value estimate) Update $q(S, A)$ to reduce $[U - q(S, A)]^2$. (For example, $q(S, A) \leftarrow q(S, A) + \alpha[U - q(S, A)]$.)

        2.2.5. $S \leftarrow S', A \leftarrow A'$.

Similarly, Algo. 5.3 shows the one-step TD policy evaluation to estimate state values.

Algorithm 5.3   One-step TD policy evaluation to estimate state values.

Inputs: environment (without mathematical model), and the policy π.

Output: state value estimates $v(s)$ ($s \in S$).

Parameters: optimizer (containing learning rate α), discount factor γ, and parameters to control the number of episodes and maximum steps in an episode.

1. (Initialize) Initialize the state value estimates $v(s) \leftarrow$ arbitrary value ($s \in S^+$).

2. (TD update) For each episode:

    2.1. (Initialize state) Select the initial state S.

    2.2. Loop until the episode ends (for example, reach the maximum step, or S is the terminal state):

        2.2.1. (Decide) Use the policy $\pi(\cdot|S)$ to determine the action A.

        2.2.2. (Sample) Execute the action A, and then observe the reward R, the next state S', and the indicator of episode end D'.

        2.2.3. (Calculate TD return) $U \leftarrow R + \gamma(1 - D')v(S')$.

        2.2.4. (Update value estimate) Update $v(S)$ to reduce $[U - v(S)]^2$. (For example, $v(S) \leftarrow v(S) + \alpha[U - v(S)]$.)

        2.2.5. $S \leftarrow S'$.

Since the indicator of episode end D' is usually used in the form of $1 - D'$, some implementations calculate the mask $1 - D'$ or the discounted mask $\gamma(1 - D')$ as intermediate results.

This book will always assume that the environment can provide the episode-end indicator along with the next state. For the environment that can not show such indicator, we can judge whether the state is a terminal state to get this indicator.

Both MC policy evaluation and TD policy evaluation can get the true values asymptotically, and each of them has its own advantage. We can not say that one method is always better than the other. In lots of tasks, TD learning with a constant learning rate may converge faster than MC learning. However, TD learning is more demanding to the Markov property of the environment.

Here is an example to demonstrate the difference between MC learning and TD learning. For example, a DTMRP has the following five trajectory samples (only show states and rewards):

$$s_A, 0$$

$$s_B, 0, s_A, 0$$

$$s_A, 1$$

$$s_B, 0, s_A, 0$$

$$s_A, 1$$

MC updating yields state value estimates $v(s_A) = \frac{2}{5}$ and $v(s_B) = 0$, while TD updating yields state value estimates $v(s_A) = v(s_B) = \frac{2}{5}$. These two methods yield the same estimates for $v(s_A)$, but quite different estimates for $v(s_B)$. The reason is: MC updating only uses the two trajectories that containing s_B to estimate $v(s_B)$. However, TD updating finds that every trajectory that has ever reached state s_B will definitely reach state s_A, so TD updating uses all trajectories to estimate $v(s_A)$, and then uses $v(s_A)$ to estimate $v(s_B)$. Which one is better? If the process is indeed MRP and we have correctly identified the state space $S = \{s_A, s_B\}$, TD updating can use more trajectory samples to estimate the state values of s_B, so the estimate is more accurate. However, if the environment is actually not an MRP, or $\{s_A, s_B\}$ is actually not the state space, the reward after visiting s_A may depend on whether we have visited s_B. For example, if a trajectory has visited s_B, it will have 0 reward after visiting state s_A. In this case, MC updating is more robust to such errors. This example compares the merit and demerit of MC updating and TD updating.

Now we consider multi-step TD policy evaluation. Multi-step TD policy evaluation is the tradeoff between single-step TD policy evaluation and MC policy evaluation. Algorithms 5.4 and 5.5 show the algorithm that uses multi-step TD updating to estimate action values and state values respectively. When this algorithm is implemented, (S_t, A_t, R_t, D_t) and $(S_{t+n+1}, A_{t+n+1}, R_{t+n+1}, D_{t+n+1})$ can share the same space, so it only needs $n + 1$ copies of space.

Algorithm 5.4 n-step TD policy evaluation to estimate action values.

Inputs: environment (without mathematical model), and the policy π.
Output: action value estimates $q(s, a)$ ($s \in S, a \in \mathcal{A}$).

Parameters: step n, optimizer (containing learning rate α), discount factor γ, and parameters to control the number of episodes and maximum steps in an episode.

1. (Initialize) Initialize the action value estimates $q(s, a) \leftarrow$ arbitrary value $(s \in \mathcal{S}^+, a \in \mathcal{A})$.
2. (TD update) For each episode:
   2.1. (Sample first n steps) Use the policy π to generate the first n steps of the trajectory $S_0, A_0, R_1, \ldots, R_n, S_n$. (If the terminal state is encountered, set subsequent rewards to 0 and subsequent states to s_{end}.) Each state S_t $(1 \leq t \leq n)$ is accompanied by an indicator of episode end D_t.
   2.2. For $t = 0, 1, 2, \ldots$ until $S_t = s_{\text{end}}$:
       2.2.1. (Decide) Use the policy $\pi(\cdot|S_{t+n})$ to determine the action A_{t+n}. (The action can be chosen arbitrarily chosen if $D_{t+n} = 1$.)
       2.2.2. (Calculate TD return) $U \leftarrow R_{t+1} + \gamma(1 - D_{t+1})R_{t+2} + \cdots + \gamma^{n-1}(1 - D_{t+n-1})R_{t+n} + \gamma^n(1 - D_{t+n})q(S_{t+n}, A_{t+n})$.
       2.2.3. (Update value estimate) Update $q(S_t, A_t)$ to reduce $\left[U - q(S_t, A_t)\right]^2$. (For example, $q(S_t, A_t) \leftarrow q(S_t, A_t) + \alpha\left[U - q(S_t, A_t)\right]$.)
       2.2.4. (Sample) If S_{t+n} is not the terminal state, execute A_{t+n}, and observe the reward R_{t+n+1}, the next state S_{t+n+1}, and the indicator of episode end D_{t+n+1}. If S_{t+n} is the terminal state, set $R_{t+n+1} \leftarrow 0$, $S_{t+n+1} \leftarrow s_{\text{end}}$, and $D_{t+n+1} \leftarrow 1$.

Algorithm 5.5 n**-step TD policy evaluation to estimate state values.**

Inputs: environment (without mathematical model), and the policy π.
Output: state value estimates $v(s)$ $(s \in \mathcal{S})$.
Parameters: step n, optimizer (containing learning rate α), discount factor γ, and parameters to control the number of episodes and maximum steps in an episode.

1. (Initialize) Initialize the state value estimates $v(s) \leftarrow$ arbitrary value $(s \in \mathcal{S}^+)$.
2. (TD update) For each episode:
   2.1. (Sample first n steps) Use the policy π to generate the first n steps of the trajectory $S_0, A_0, R_1, \ldots, R_n, S_n$. (If the terminal state is encountered, set subsequent rewards to 0 and subsequent states to s_{end}.) Each state S_t $(1 \leq t \leq n)$ is accompanied by an indicator of episode end D_t.

2.2. For $t = 0, 1, 2, \ldots$ until $S_t = s_{\text{end}}$:

    2.2.1. (Calculate TD return) $U \leftarrow R_{t+1} + \gamma(1 - D_{t+1})R_{t+2} + \cdots + \gamma^{n-1}(1 - D_{t+n-1})R_{t+n} + \gamma^n(1 - D_{t+n})v(S_{t+n})$.

    2.2.2. (Update value estimate) Update $v(S_t)$ to reduce $\left[U - v(S_t)\right]^2$. (For example, $v(S_t) \leftarrow v(S_t) + \alpha\left[U - v(S_t)\right]$.)

    2.2.3. (Decide) Use the policy $\pi(\cdot|S_{t+n})$ to determine the action A_{t+n}. (The action can be chosen arbitrarily if $D_{t+n} = 1$.)

    2.2.4. (Sample) If S_{t+n} is not the terminal state, execute A_{t+n}, and observe the reward R_{t+n+1}, the next state S_{t+n+1}, and the indicator of episode end D_{t+n+1}. If S_{t+n} is the terminal state, set $R_{t+n+1} \leftarrow 0$, $S_{t+n+1} \leftarrow s_{\text{end}}$, and $D_{t+n+1} \leftarrow 1$.

5.2.2 SARSA

This section considers an on-policy TD policy optimization algorithm. This algorithm is called **SARSA** algorithm, which is named after the random variable $(S_t, A_t, R_{t+1}, S_{t+1}, A_{t+1})$ "State–Action–Reward–State–Action". This algorithm uses $U_t = R_{t+1} + \gamma(1 - D_{t+1})q_t(S_{t+1}, A_{t+1})$ to get the one-step TD return U_t, and then update $q(S_t, A_t)$ using (Rummery, 1994)

$$q(S_t, A_t) \leftarrow q(S_t, A_t) + \alpha\left[U_t - q(S_t, A_t)\right],$$

where α is the learning rate.

Algorithm 5.6 shows the SARSA algorithm. SARSA algorithm is similar to the one-step action value policy evaluation algorithm, and it updates the policy after each value estimate update. In Step 2.2, after the optimal action value estimate q is updated, we improve the policy by modifying the optimal policy estimate π. The policy improvement can follow the ε-greedy approach so that the optimal policy estimate π is always a soft policy. When the algorithm finishes, we obtain the estimates for optimal action values and optimal policy.

Algorithm 5.6   SARSA (maintaining the policy explicitly).

Inputs: environment (without mathematical model).
Output: optimal action value estimates $q(s, a)$ ($s \in \mathcal{S}, a \in \mathcal{A}$), and optimal policy estimate $\pi(a|s)$ ($s \in \mathcal{S}, a \in \mathcal{A}$).
Parameters: optimizer (containing learning rate α), discount factor γ, and parameters to control the number of episodes and maximum steps in an episode.

1. (Initialize) Initialize the action value estimates $q(s, a) \leftarrow$ arbitrary value $(s \in \mathcal{S}^+, a \in \mathcal{A})$.
   Use the action value estimates $q(s, a)$ $(s \in \mathcal{S}, a \in \mathcal{A})$ to determine the policy π (for example, use the ε-greedy policy).
2. (TD update) For each episode:

   2.1. (Initialize state–action pair) Select the initial state S, and then use the policy $\pi(\cdot|S)$ to determine the action A.
   2.2. Loop until the episode ends (for example, reach the maximum step, or S is the terminal state):
   2.2.1. (Sample) Execute the action A, and then observe the reward R, the next state S', and the indicator of episode end D'.
   2.2.2. (Decide) Use the input policy $\pi(\cdot|S')$ to determine the action A'. (The action can be arbitrarily chosen if $D' = 1$.)
   2.2.3. (Calculate TD return) $U \leftarrow R + \gamma(1 - D')q(S', A')$.
   2.2.4. (Update value estimate) Update $q(S, A)$ to reduce $\left[U - q(S, A)\right]^2$. (For example, $q(S, A) \leftarrow q(S, A) + \alpha\left[U - q(S, A)\right]$.)
   2.2.5. (Improve policy) Use $q(S, \cdot)$ to modify $\pi(\cdot|S)$ (say, using ε-greedy policy).
   2.2.6. $S \leftarrow S', A \leftarrow A'$.

We can also implement SARSA without maintaining the optimal policy estimate explicitly, which is shown in Algo. 5.7. When the optimal policy estimate is not maintained explicitly, it has been implicitly maintained in the action values estimates. A common way is to assume the policy is the ε-greedy policy, so we can determine the policy from the action value estimates. When we need to determine the action A on a state S, we can generate a random variance X, which is uniformly distributed in the range $[0, 1]$. If $X < \varepsilon$, we explore and randomly pick an action in the action space. Otherwise, choose an action that maximizes $q(S, \cdot)$.

Algorithm 5.7   SARSA (maintaining the policy implicitly).

Input: environment (without mathematical model).
Output: optimal action value estimates $q(s, a)$ $(s \in \mathcal{S}, a \in \mathcal{A})$. We can use these optimal action value estimates to generate optimal policy estimate $\pi(a|s)$ $(s \in \mathcal{S}, a \in \mathcal{A})$ if needed.
Parameters: optimizer (containing learning rate α), discount factor γ, and parameters to control the number of episodes and maximum steps in an episode.

1. (Initialize) Initialize the action value estimates $q(s, a) \leftarrow$ arbitrary value $(s \in \mathcal{S}^+, a \in \mathcal{A})$.
2. (TD update) For each episode:

2.1. (Initialize state–action pair) Select the initial state S, and then use the policy $\pi(\cdot|S)$ to determine the action A.

2.2. Loop until the episode ends (for example, reach the maximum step, or S is the terminal state):

    2.2.1. (Sample) Execute the action A, and then observe the reward R, the next state S', and the indicator of episode end D'.

    2.2.2. (Decide) Use the policy derived from the action values $q(S', \cdot)$ (say ε-greedy policy) to determine the action A'. (The action can be arbitrarily chosen if $D' = 1$.)

    2.2.3. (Calculate TD return) $U \leftarrow R + \gamma(1 - D')q(S', A')$.

    2.2.4. (Update value estimate) Update $q(S, A)$ to reduce $\left[U - q(S, A)\right]^2$. (For example, $q(S, A) \leftarrow q(S, A) + \alpha\left[U - q(S, A)\right]$.)

    2.2.5. (Improve policy) Use $q(S, \cdot)$ to modify $\pi(\cdot|S)$ (say, using ε-greedy policy).

    2.2.6. $S \leftarrow S', A \leftarrow A'$.

If we replace the one-step TD return by n-step TD return, we obtain the n-step SARSA algorithm (shown in Algo. 5.8). Compared to n-step TD policy evaluation, it decides using the policy derived from the latest action value estimates.

Algorithm 5.8 n-step SARSA.

Inputs: environment (without mathematical model).

Output: optimal action value estimates $q(s, a)$ ($s \in \mathcal{S}, a \in \mathcal{A}$). We can use these optimal action value estimates to generate optimal policy estimate $\pi(a|s)$ ($s \in \mathcal{S}, a \in \mathcal{A}$) if needed.

Parameters: step n, optimizer (containing learning rate α), discount factor γ, and parameters to control the number of episodes and maximum steps in an episode.

1. (Initialize) Initialize the action value estimates $q(s, a) \leftarrow$ arbitrary value ($s \in \mathcal{S}^+, a \in \mathcal{A}$).

2. (TD update) For each episode:

    2.1. (Sample first n steps) Use the policy derived from the action values q (say ε-greedy policy) to generate the first n steps of the trajectory $S_0, A_0, R_1, \ldots, R_n, S_n$. (If the terminal state is encountered, set subsequent rewards to 0 and subsequent states to s_{end}.) Each state S_t ($1 \le t \le n$) is accompanied by an indicator of episode end D_t.

    2.2. For $t = 0, 1, 2, \ldots$, until $S_t = s_{\text{end}}$:

        2.2.1. (Decide) Use the policy derived from the action values $q(S_{t+n}, \cdot)$ (say, ε-greedy policy) to determine the action A_{t+n}. (The action can be arbitrarily chosen if $D_{t+n} = 1$.)

2.2.2. (Calculate TD return) $U \leftarrow R_{t+1} + \gamma(1 - D_{t+1})R_{t+2} + \cdots + \gamma^{n-1}(1 - D_{t+n-1})R_{t+n} + \gamma^n(1 - D_{t+n})q(S_{t+n}, A_{t+n})$.

2.2.3. (Update value estimate) Update $q(S_t, A_t)$ to reduce $[U - q(S_t, A_t)]^2$. (For example, $q(S_t, A_t) \leftarrow q(S_t, A_t) + \alpha[U - q(S_t, A_t)]$.)

2.2.4. (Sample) If S_{t+n} is not the terminal state, execute A_{t+n}, and observe the reward R_{t+n+1}, the next state S_{t+n+1}, and the indicator of episode end D_{t+n+1}. If S_{t+n} is the terminal state, set $R_{t+n+1} \leftarrow 0$, $S_{t+n+1} \leftarrow s_{\text{end}}$, and $D_{t+n+1} \leftarrow 1$.

5.2.3 Expected SARSA

SARSA algorithm has a variant: **expected SARSA**. Expected SARSA differs from SARSA in a way that expected SARSA uses state-value one-step TD return $U_{t:t+1}^{(v)} = R_{t+1} + \gamma(1 - D_{t+1})v(S_{t+1})$ rather than action-value one-step TD return $U_{t:t+1}^{(q)} = R_{t+1} + \gamma(1 - D_{t+1})q(S_{t+1}, A_{t+1})$. According to the relationship between state values and action values, the state-value one-step TD return can be written as (Sutton, 1998)

$$U_t = R_{t+1} + \gamma(1 - D_{t+1}) \sum_{a \in \mathcal{A}(S_{t+1})} \pi(a|S_{t+1})q(S_{t+1}, a).$$

Compared to SARSA, expected SARSA needs to calculate $\sum_a \pi(a|S_{t+1})q(S_{t+1}, a)$, so it needs more computation resources. Meanwhile, such computation reduces some negative impacts due to some bad actions at the later stage of learning. Therefore, expected SARSA may be more stable than SARSA, and expected SARSA usually uses a larger learning rate compared to SARSA.

Algorithm 5.9 shows the expected SARSA algorithm. It is modified from one-step TD state-value policy evaluation algorithm. Although expected SARSA and SARSA may share similar ways to control the episode numbers and the step number in an episode, expected SARSA has a simpler looping structure since expected SARSA does not need A_{t+1} when updating $q(S_t, A_t)$. We can always let the policy π to be a ε-greedy policy. If ε is very small, the ε-greedy policy will be very close to a deterministic policy. In this case, $\sum_a \pi(a|S_{t+1})q(S_{t+1}, a)$ in expected SARSA will be very close to $q(S_{t+1}, A_{t+1})$ in SARSA.

Algorithm 5.9   Expected SARSA.

1. (Initialize) Initialize the action value estimates $q(s, a) \leftarrow$ arbitrary value ($s \in \mathcal{S}^+, a \in \mathcal{A}$).
If the policy is maintained explicitly, use the action value estimates q to determine the policy π (for example, use the ε-greedy policy).

2. (TD update) For each episode:

    2.1. (Initialize state) Select the initial state S.

    2.2. Loop until the episode ends (for example, reach the maximum step, or S is the terminal state):

        2.2.1. (Decide) Use the policy $\pi(\cdot|S)$ to determine the action A. If the policy is not maintained explicitly, use the policy derived from the action value q (say, ε-greedy policy).

        2.2.2. (Sample) Execute the action A, and then observe the reward R, the next state S', and the indicator of episode end D'.

        2.2.3. (Calculate TD return) $U \leftarrow R + \gamma(1 - D') \sum_{a \in \mathcal{A}(S')} \pi(a|S')q(S', a)$.

        2.2.4. (Update value estimate) Update $q(S, A)$ to reduce $[U - q(S, A)]^2$. (For example, $q(S, A) \leftarrow q(S, A) + \alpha[U - q(S, A)]$.)

        2.2.5. (Improve policy) For the situation that the policy is maintained explicitly, use action value estimates $q(S, \cdot)$ to modify $\pi(\cdot|S)$.

        2.2.6. $S \leftarrow S'$.

There is also n-step expected SARSA, whose TD return is

$$
\begin{aligned}
U_t = & R_{t+1} + \gamma(1 - D_{t+1})R_{t+2} + \cdots \\
& + \gamma^{n-1}(1 - D_{t+n-1})R_{t+n} + \gamma^n(1 - D_{t+n}) \sum_{a \in \mathcal{A}(S_{t+n})} \pi(a|S_{t+n})q(S_{t+n}, a).
\end{aligned}
$$

Algorithm 5.10 shows the policy optimization algorithm using multi-step expected SARSA algorithm.

Algorithm 5.10 n-step expected SARSA.

1. (Initialize) Initialize the action value estimates $q(s, a) \leftarrow$ arbitrary value $(s \in \mathcal{S}^+, a \in \mathcal{A})$.
If the policy is maintained explicitly, use the action value estimates q to determine the policy π (for example, use the ε-greedy policy).

2. (TD update) For each episode:

    2.1. (Sample first n steps) Use the policy π to generate the first n steps of the trajectory $S_0, A_0, R_1, \ldots, R_n, S_n$. If the policy is not explicitly maintained, use the policy derived from the action values q (say ε-greedy policy). (If the terminal state is encountered, set subsequent rewards to 0 and subsequent states to s_{end}.) Each state S_t $(1 \leq t \leq n)$ is accompanied by an indicator of episode end D_t.

    2.2. For $t = 0, 1, 2, \ldots$, until $S_t = s_{\text{end}}$:

2.2.1. (Calculate TD return) $U \leftarrow R_{t+1} + \sum_{k=1}^{n-1} \gamma^k (1 - D_{t+k}) R_{t+k+1} + \gamma^n (1 - D_{t+n}) \sum_{a \in \mathcal{A}(S_{t+n})} \pi(a|S_{t+n}) q(S_{t+n}, a)$. If the policy is not explicitly maintained, it can be derived from action value q.

2.2.2. (Update value estimate) Update $q(S, A)$ to reduce $[U - q(S_t, A_t)]^2$. (For example, $q(S_t, A_t) \leftarrow q(S_t, A_t) + \alpha [U - q(S_t, A_t)]$.)

2.2.3. (Improve policy) For the situation that the policy is maintained explicitly, use action value estimates $q(S_t, \cdot)$ to update the policy estimate $\pi(\cdot|S_t)$.

2.2.4. (Decide and Sample) If the state S_{t+n} is not the terminal state, use the policy $\pi(\cdot|S_{t+n})$ or the policy derived form action values $q(S_{t+n}, \cdot)$ to determine the action A_{t+n}. Execute A_{t+n}, and observe the reward R_{t+n+1}, the next state S_{t+n+1}, and the indicator of episode end D_{t+n+1}. If S_{t+n} is the terminal state, set $R_{t+n+1} \leftarrow 0$, $S_{t+n+1} \leftarrow s_{\text{end}}$, and $D_{t+n+1} \leftarrow 1$.

5.3 Off-Policy TD Learning

This section introduces off-policy TD learning. Off-policy TD algorithms are usually more popular than on-policy TD algorithms. Especially, Q learning has become one of the most fundamental RL algorithms.

5.3.1 Off-Policy Algorithm based on Importance Sampling

TD policy evaluation and policy optimization can leverage importance sampling, too. For n-step TD policy evaluation or n-step off-policy SARSA, the n-step TD return $U_{pq:t:t+n}$ relies on the trajectory $S_t, A_t, R_{t+1}, S_{t+1}, A_{t+1}, \ldots, S_{t+n}, A_{t+n}$. Given the state–action pair (S_t, A_t), the target policy π and the behavior policy b generates the trajectory with the probabilities:

$$\Pr_\pi[R_{t+1}, S_{t+1}, A_{t+1}, \ldots, S_{t+n}|S_t, A_t] = \prod_{\tau=t+1}^{t+n-1} \pi(A_\tau|S_\tau) \prod_{\tau=t}^{t+n-1} p(S_{\tau+1}, R_{\tau+1}|S_\tau, A_\tau)$$

$$\Pr_b[R_{t+1}, S_{t+1}, A_{t+1}, \ldots, S_{t+n}|S_t, A_t] = \prod_{\tau=t+1}^{t+n-1} b(A_\tau|S_\tau) \prod_{\tau=t}^{t+n-1} p(S_{\tau+1}, R_{\tau+1}|S_\tau, A_\tau).$$

The ratio of the aforementioned two probabilities is the importance sampling ratio:

$$\rho_{t+1:t+n-1} = \frac{\mathrm{Pr}_\pi[R_{t+1}, S_{t+1}, A_{t+1}, \ldots, S_{t+n}|S_t, A_t]}{\mathrm{Pr}_b[R_{t+1}, S_{t+1}, A_{t+1}, \ldots, S_{t+n}|S_t, A_t]} = \prod_{\tau=t+1}^{t+n-1} \frac{\pi(A_\tau|S_\tau)}{b(A_\tau|S_\tau)}.$$

That is, for the sample generated by the behavior policy b, the probability that the policy π generates it is $\rho_{t+1:t+n-1}$ times of the probability that the policy b generates it. Therefore, the TD return has the weight $\rho_{t+1:t+n-1}$. Taken this weight into consideration, we get the TD policy evaluation or SARSA with importance sampling.

Algorithm 5.11 shows the n-step version. You can sort out the one-step version by yourself.

Algorithm 5.11 n-step TD policy evaluation of SARSA with importance sampling.

Inputs: environment (without mathematical model). For policy evaluation, input policy π, too.
Output: action value estimates $q(s, a)$ $(s \in S, a \in \mathcal{A})$. For policy optimization, output the optimal policy estimates π if it is maintained explicitly.
Parameters: step n, optimizer (containing learning rate α), discount factor γ, and parameters to control the number of episodes and maximum steps in an episode.

1. (Initialize) Initialize the action value estimates $q(s, a)$ ← arbitrary value $(s \in S^+, a \in \mathcal{A})$.
   For policy optimization that maintains the policy explicitly, use the action value q to determine π (say, ε-greedy policy).
2. (TD update) For each episode:
   2.1. (Designate behavior policy) Designate a behavior policy b such that $\pi \ll b$.
   2.2. (Sample first n steps) Use the behavior policy b to generate the first n steps of the trajectory $S_0, A_0, R_1, \ldots, R_n, S_n$. (If the terminal state is encountered, set subsequent rewards to 0 and subsequent states to s_{end}.) Each state S_t $(1 \le t \le n)$ is accompanied by an indicator of episode end D_t.
   2.3. For $t = 0, 1, 2, \ldots$, until $S_t = s_{\mathrm{end}}$:
       2.3.1. (Decide) Use the policy derived from the action values $b(\cdot|S_{t+n})$ to determine the action A_{t+n}.
       2.3.2. (Calculate TD return) $U \leftarrow R_{t+1} + \gamma(1 - D_{t+1})R_{t+2} + \cdots + \gamma^{n-1}(1 - D_{t+n-1})R_{t+n} + \gamma^n(1 - D_{t+n})q(S_{t+n}, A_{t+n})$.
       2.3.3. (Calculate importance sampling ratio) ρ ← $\prod_{\tau=t+1}^{\min\{t+n, T\}-1} \frac{\pi(A_\tau|S_\tau)}{b(A_\tau|S_\tau)}$. For policy optimization, the policy can be derived from the action value estimates if it is maintained implicitly.

2.3.4. (Update value estimate) Update $q(S_t, A_t)$ to reduce $[U - q(S_t, A_t)]^2$. (For example, $q(S_t, A_t) \leftarrow q(S_t, A_t) + \alpha[U - q(S_t, A_t)]$.)

2.3.5. (Improve policy) For policy optimization and the policy is maintained explicitly, use action value estimates $q(S, \cdot)$ to update the policy estimate $\pi(\cdot|S)$.

2.3.6. (Sample) If S_{t+n} is not the terminal state, execute A_{t+n}, and observe the reward R_{t+n+1}, the next state S_{t+n+1}, and the indicator of episode end D_{t+n+1}. If S_{t+n} is the terminal state, set $R_{t+n+1} \leftarrow 0$, $S_{t+n+1} \leftarrow s_{\text{end}}$, and $D_{t+n+1} \leftarrow 1$.

We can use a similar method to apply importance sampling to TD policy evaluation to estimate state values and to the expected SARSA algorithm. Specifically, starting from a state S_t, the n-step trajectory $S_t, A_t, R_{t+1}, S_{t+1}, A_{t+1}, \ldots, S_{t+n}$ is generated by the policy π and the behavior policy b with the probabilities:

$$\text{Pr}_\pi[A_t, R_{t+1}, S_{t+1}, A_{t+1}, \ldots, S_{t+n}|S_t] = \prod_{\tau=t}^{t+n-1} \pi(A_\tau|S_\tau) \prod_{\tau=t}^{t+n-1} p(S_{\tau+1}, R_{\tau+1}|S_\tau, A_\tau)$$

$$\text{Pr}_b[A_t, R_{t+1}, S_{t+1}, A_{t+1}, \ldots, S_{t+n}|S_t] = \prod_{\tau=t}^{t+n-1} b(A_\tau|S_\tau) \prod_{\tau=t}^{t+n-1} p(S_{\tau+1}, R_{\tau+1}|S_\tau, A_\tau).$$

The ratio of the aforementioned two probabilities is the importance sampling ratio of TD policy evaluation for state value and expected SARSA:

$$\rho_{t:t+n-1} = \frac{\text{Pr}_\pi[A_t, R_{t+1}, S_{t+1}, A_{t+1}, \ldots, S_{t+n}|S_t]}{\text{Pr}_b[A_t, R_{t+1}, S_{t+1}, A_{t+1}, \ldots, S_{t+n}|S_t]} = \prod_{\tau=t+1}^{t+n-1} \frac{\pi(A_\tau|S_\tau)}{b(A_\tau|S_\tau)}.$$

5.3.2 Q Learning

Among one-step TD on-policy policy optimization algorithms, SARSA algorithm uses $U_t = R_{t+1} + \gamma(1 - D_{t+1})q(S_{t+1}, A_{t+1})$ to generate return sample, while expected SARSA algorithm uses $U_t = R_{t+1} + \gamma(1 - D_{t+1})v(S_{t+1})$ to generate return sample. In both algorithms, the return samples are generated using the policy that is currently being maintained. After the action values are updated, the policy is updated either explicitly or implicitly. Q learning tries to take a step further: It uses an improved deterministic policy to generate action sample, and calculate the TD return using the samples generated by the improved deterministic policy. Henceforth, the update of Q learning is (Watkin, 1989):

$$U_t = R_{t+1} + \gamma(1 - D_{t+1}) \max_{a \in \mathcal{A}(S_{t+1})} q(S_{t+1}, a).$$

The rationale is, when we use S_{t+1} to back up U_t, we may use the improved policy according to $q(S_{t+1}, \cdot)$, rather than the original $q(S_{t+1}, A_{t+1})$ or $v(S_{t+1})$, so that the value is closer to the optimal values. Therefore, we can use a deterministic policy after policy improvement to generate samples. However, since the deterministic policy to generate the sample is not the maintained policy, which is usually a ε-soft policy rather than a deterministic policy, Q learning is off-policy.

Algorithm 5.12 shows the Q learning algorithm. Q learning shares the same control structure with expected SARSA, but it uses a different TD return to update the optimal action value estimates $q(S_t, A_t)$. Implementation of Q learning usually does not maintain the policy explicitly.

Algorithm 5.12   Q learning.

Inputs: environment (without mathematical model). For policy evaluation, input policy π, too.

Outputs: optimal action value estimates $q(s, a)$ $(s \in \mathcal{S}, a \in \mathcal{A})$. We can use these optimal action value estimates to generate optimal policy estimate $\pi(a|s)$ $(s \in \mathcal{S}, a \in \mathcal{A})$ if needed.

1. (Initialize) Initialize the action value estimates $q(s, a) \leftarrow$ arbitrary value $(s \in \mathcal{S}^+, a \in \mathcal{A})$.
2. (TD update) For each episode:

   2.1. (Initialize state) Select the initial state S.
   2.2. Loop until the episode ends (for example, reach the maximum step, or S is the terminal state):

   2.2.1. (Decide) Use the policy derived from the action value q (say, ε-greedy policy) to determine the action A.
   2.2.2. (Sample) Execute the action A, and then observe the reward R, the next state S', and the indicator of episode end D'.
   2.2.3. (Calculate   TD   return) U $\leftarrow$ R $+$ $\gamma(1 - D') \max_{a \in \mathcal{A}(S')} q(S', a)$.
   2.2.4. (Update value estimate) Update $q(S, A)$ to reduce $\left[U - q(S, A)\right]^2$. (For example, $q(S, A) \leftarrow q(S, A) + \alpha\left[U - q(S, A)\right]$.)
   2.2.5. (Improve policy) Use $q(S, \cdot)$ to modify $\pi(\cdot|S)$ (say, using ε-greedy policy).
   2.2.6. $S \leftarrow S'$.

There is also multi-step Q learning, whose TD return is

$$
\begin{aligned}
U_t =& R_{t+1} + \gamma(1 - D_{t+1})R_{t+2} + \cdots \\
&+ \gamma^{n-1}(1 - D_{t+n-1})R_{t+n} + \gamma^n(1 - D_{t+n}) \max_{a \in \mathcal{A}(S_{t+n})} q(S_{t+n}, a).
\end{aligned}
$$

The algorithm has a similar control structure with Algo. 5.10, which is omitted here.

5.3.3 Double Q Learning

The Q learning algorithm bootstraps on $\max_a q(S_{t+1}, a)$ to obtain the TD return for updating. This may make the optimal action value estimate over-biased, which is called "maximization bias".

Example 5.1 (Maximization Bias) Consider the episodic DTMDP in Fig. 5.2. The state space is $\mathcal{S} = \{s_{\text{start}}, s_{\text{middle}}\}$. The initial state is always s_{start}, whose action space is $\mathcal{A}(s_{\text{start}}) = \{a_{\text{to middle}}, a_{\text{to end}}\}$. The action $a_{\text{to middle}}$ will lead to the state s_{middle} with the reward is 0, and the action $a_{\text{to end}}$ will lead to the terminal state s_{end} with the reward $+1$. There are lots of different actions (say, 1000 different actions) starting from the state s_{middle}, all of which lead to the terminal state s_{end} with all rewards being normal distributed random variables with mean 0 and standard deviation 100. Theoretically, the optimal value of this DTMDP is $v_*(s_{\text{middle}}) = q_*(s_{\text{middle}}, \cdot) = 0$ and $v_*(s_{\text{start}}) = q_*(s_{\text{start}}, a_{\text{to end}}) = 1$, and the optimal policy is $\pi_*(s_{\text{start}}) = a_{\text{to end}}$. However, Q learning suffers the following setback: We may observe that some trajectories that have visited s_{middle} may have large rewards when some actions are selected. This is just because the reward has a very large standard deviation, and each action may have too few samples to correctly estimate the expectation. However, $\max_{a \in \mathcal{A}(s_{\text{middle}})} q(s_{\text{middle}}, a)$ will be larger than actual values, which makes the policy tends to choose $a_{\text{to middle}}$ at the initial state s_{start}. This error will take lots of samples to remedy.

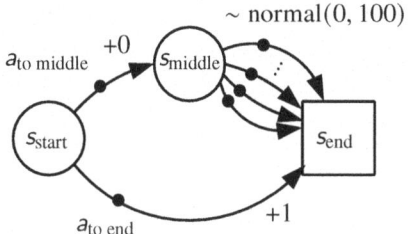

Fig. 5.2   Maximization bias in Q learning.

Double Q Learning (Hasselt, 2010) reduces the impact of maximization bias by using two independent estimates of action value $q^{(0)}$ and $q^{(1)}$. It replaces $\max_a q(S_{t+1}, a)$ in Q learning by $q^{(0)}\left(S_{t+1}, \arg\max_a q^{(1)}(S_{t+1}, a)\right)$ or $q^{(1)}\left(S_{t+1}, \arg\max_a q^{(0)}(S_{t+1}, a)\right)$. If $q^{(0)}$ and $q^{(1)}$ are completely independent estimates, $\mathrm{E}\left[q^{(0)}(S_{t+1}, A^*)\right] = q(S_{t+1}, A^*)$ where $A^* = \arg\max_a q^{(1)}(S_{t+1}, a)$, and the bias is totally eliminated here. Note that Double Q Learning needs to update both $q^{(0)}$ and $q^{(1)}$. Although $q^{(0)}$ and $q^{(1)}$ may not be totally independent, the maximization bias is much less than using a single action value estimate.

In Double Q Learning, each learning step updates the action value estimates in one of the following two ways:

- Calculate TD return $U_t^{(0)} = R_{t+1} + \gamma(1 - D_{t+1})q^{(1)}\Big(S_{t+1}, \arg\max_a q^{(0)}(S_{t+1}, a)\Big)$, and update $q^{(0)}(S_t, A_t)$ to minimize the difference between $q^{(0)}(S_t, A_t)$ and $U_t^{(0)}$. For example, we may use the loss $\Big[U_t^{(0)} - q^{(0)}(S_t, A_t)\Big]^2$, or update using $q^{(0)}(S_t, A_t) \leftarrow q^{(0)}(S_t, A_t) + \alpha\Big[U_t^{(0)} - q^{(0)}(S_t, A_t)\Big]$.

- Calculate TD return $U_t^{(1)} = R_{t+1} + \gamma(1 - D_{t+1})q^{(1)}\Big(S_{t+1}, \arg\max_a q^{(0)}(S_{t+1}, a)\Big)$, and update $q^{(0)}(S_t, A_t)$ to minimize the difference between $q^{(0)}(S_t, A_t)$ and $U_t^{(1)}$. For example, we may use the loss $\Big[U_t^{(1)} - q^{(1)}(S_t, A_t)\Big]^2$, or update using $q^{(1)}(S_t, A_t) \leftarrow q^{(1)}(S_t, A_t) + \alpha\Big[U_t^{(1)} - q^{(1)}(S_t, A_t)\Big]$.

Algorithm 5.13 shows the Double Q Learning algorithm. The output of this algorithm is the action value estimates $\frac{1}{2}\Big(q^{(0)} + q^{(1)}\Big)$, the average of $q^{(0)}$ and $q^{(1)}$. During iterations, we can also use $q^{(0)} + q^{(1)}$ to replace $\frac{1}{2}\Big(q^{(0)} + q^{(1)}\Big)$ to slightly save the computation, since the policy derived from $q^{(0)} + q^{(1)}$ is the same as the policy derived from $\frac{1}{2}\Big(q^{(0)} + q^{(1)}\Big)$.

Algorithm 5.13   Double Q Learning.

Input: environment (without mathematical model).

Outputs: optimal   action   value   estimates $\frac{1}{2}\Big(q^{(0)} + q^{(1)}\Big)(s, a)$ $(s \in \mathcal{S}, a \in \mathcal{A})$. We can use these optimal action value estimates to generate optimal policy estimate $\pi(a|s)$ $(s \in \mathcal{S}, a \in \mathcal{A})$ if needed.

1. (Initialize) Initialize the action value estimates $q^{(i)}(s, a) \leftarrow$ arbitrary value $(s \in \mathcal{S}^+, a \in \mathcal{A}, i \in \{0, 1\})$.
2. (TD update) For each episode:

    2.1. (Initialize state) Select the initial state S.
    2.2. Loop until the episode ends (for example, reach the maximum step, or S is the terminal state):

    2.2.1. (Decide) Use the policy derived from the action value $\Big(q^{(0)} + q^{(1)}\Big)$ (say, ε-greedy policy) to determine the action A.

    2.2.2. (Sample) Execute the action A, and then observe the reward R, the next state S', and the indicator of episode end D'.

    2.2.3. (Choose between two action value estimates) Choose a value I from the two-element set $\{0, 1\}$ with equal probability. Then $q^{(I)}$ will be the action value estimate to be updated.

    2.2.4. (Calculate   TD   return) $U \leftarrow R +$ $\gamma(1 - D')q^{(1-I)}\Big(S', \arg\max_a q^{(I)}(S', a)\Big)$.

> **2.2.5.** (Update value estimate) Update $q^{(I)}(S,A)$ to reduce $\left[U - q^{(I)}(S,A)\right]^2$. (For example, $q^{(I)}(S,A) \leftarrow q^{(I)}(S,A) + \alpha\left[U - q(S,A)\right]$.)
> **2.2.6.** $S \leftarrow S'$.

Double learning has a variant that maintains two suites of action value estimates: $q^{(0)}$ and $q^{(1)}$. During training, each time we use min $\left\{q^{(0)}, q^{(1)}\right\}$ to update TD return sample, and randomly pick one value estimate to update it. This variant can be further extended to n-ple learning ($n = 2, 3, \ldots$), which maintains n suites of action value estimates $q^{(0)}, q^{(1)}, \ldots, q^{(n-1)}$ and use $\min_{0 \le i < n} q^{(i)}$ to update TD return sample. Details are omitted here.

5.4 Eligibility Trace

Eligibility trace is a mechanism that trades off between MC learning and TD learning. It can improve the learning performance with simple implementation.

5.4.1 λ Return

Before we formally learn the eligibility trace algorithm, let us learn the definition of λ return and an offline λ-return algorithm.

Given $\lambda \in [0, 1]$, λ-**return** is the weighted average of TD returns $U_{t:t+1}, U_{t:t+2}, U_{t:t+3}, \ldots$ with the weights $(1 - \lambda), (1 - \lambda)\lambda, (1 - \lambda)\lambda^2, \ldots$:

$$\text{episodic task:} \quad U_t^{\langle\lambda\rangle} \overset{\text{def}}{=} (1 - \lambda) \sum_{n=1}^{T-t-1} \lambda^{n-1} U_{t:t+n} + \lambda^{T-t-1} G_t,$$

$$\text{sequential task:} \quad U_t^{\langle\lambda\rangle} \overset{\text{def}}{=} (1 - \lambda) \sum_{n=1}^{+\infty} \lambda^{n-1} U_{t:t+n}.$$

The λ return $U_t^{\langle\lambda\rangle}$ can be viewed as a tradeoff between the MC return G_t and one-step TD return $U_{t:t+1}$: When $\lambda = 1$, $U_t^{\langle 1 \rangle} = G_t$ is the MC return; when $\lambda = 0$, $U_t^{\langle 0 \rangle} = U_{t:t+1}$ is the single-step TD return. The backup diagram of λ return is shown in Fig. 5.3.

Offline λ-return algorithm uses λ return to update the value estimates, either action value estimates or state value estimates. Compared to MC update, it only changes the return samples from G_t to $U_t^{\langle\lambda\rangle}$. For episodic tasks, offline λ return algorithm calculates $U_t^{\langle\lambda\rangle}$ for each $t = 0, 1, 2, \ldots$, and update all action value

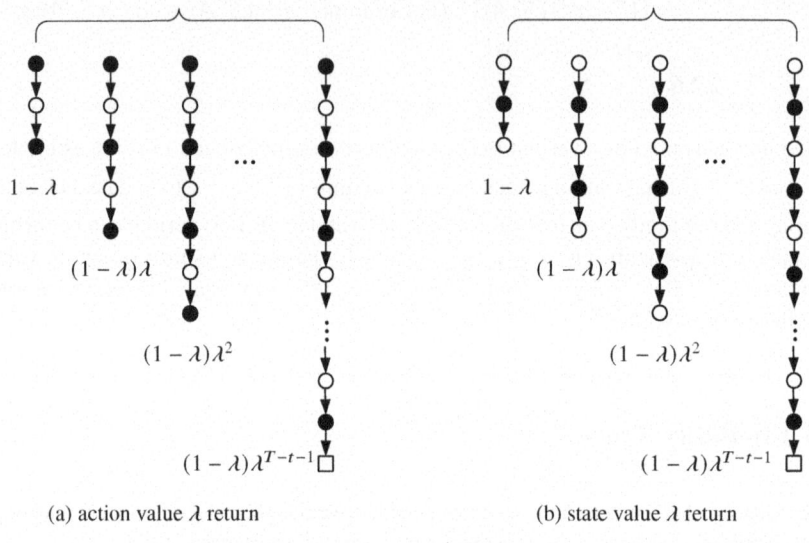

(a) action value λ return   (b) state value λ return

Fig. 5.3   Backup diagram of λ return.
This figure is adapted from (Sutton, 1998).

estimates. Sequential tasks can not use offline λ return algorithm, since $U_t^{\langle \lambda \rangle}$ can not be computed.

The name of "offline λ return algorithm" contains the word "offline", since it only updates at the end of episodes, like MC algorithms. Such algorithms are suitable for offline tasks, but can also be used for other tasks. Therefore, from the modern view, offline λ return algorithm is not an offline algorithm that is specially designed for offline tasks.

Since λ return trades off between G_t and $U_{t:t+1}$, offline λ return algorithm may perform better than both MC learning and TD learning in some tasks. However, offline λ return algorithm has obvious shortcomings: First, it can only be used in episodic tasks. It can not be used in sequential tasks. Second, at the end of each episode, it needs to calculate $U_t^{\langle \lambda \rangle}$ ($t = 0, 1, \ldots, T - 1$), which requires lots of computations. The eligibility trace algorithms in the next section will deal with these shortcomings.

5.4.2 TD(λ)

TD(λ) algorithm is one of the most influential RL algorithms in history.

TD(λ) is an improved algorithm based on offline λ return algorithm. The intuition of the improvement is as follows: When offline λ return algorithm updates an optimal value estimate $q(S_{t-n}, A_{t-n})$ (or $v(S_{t-n})$) for $n = 1, 2, \ldots$, the weight of TD return $U_{t-n:t}$ is $(1 - \lambda)\lambda^{n-1}$. Although we can calculate $U_{t-n}^{\langle \lambda \rangle}$ only at the end of the episode, we can calculate $U_{t-n:t}$ upon we get (S_t, A_t) (or S_t). Therefore, we can partially update $q(S_{t-n}, A_{t-n})$ immediately after getting (S_t, A_t). Since this holds for all $n = 1, 2, \ldots$, we can update $q(S_\tau, A_\tau)$ ($\tau = 0, 1, \ldots, t - 1$) immediately after getting $q(S_t, A_t)$, and the weight of updating $q(S_\tau, A_\tau)$ is in proportion to $\lambda^{t-\tau}$.

Accordingly, given the trajectory $S_0, A_0, R_1, S_1, A_1, R_2, \ldots$, we may introduce a concept called the eligibility trace $e_t(s, a)$ ($s \in \mathcal{S}, a \in \mathcal{A}(s)$) to represent the weight of using the one-step TD return $U_{t:t+1} = R_{t+1} + \gamma q(S_{t+1}, A_{t+1})(1 - D_{t+1})$ based on the state–action pair (S_t, A_t) to update the state–action pair (s, a). Mathematically, eligibility trace is defined recursively as follows: When $t = 0$, define $e_0(s, a) = 0$ ($s \in \mathcal{S}, a \in \mathcal{A}(s)$); when $t > 0$,

$$e_t(s, a) = \begin{cases} 1 + \beta\gamma\lambda e_{t-1}(s, a), & S_t = s, A_t = a, \\ \gamma\lambda e_{t-1}(s, a), & \text{otherwise,} \end{cases}$$

where $\beta \in [0, 1]$ is a pre-defined parameter. The definition of eligibility trace should be understood in the following way: For a state–action pair (S_τ, A_τ) in history, it has $t - \tau$ steps away from the time t, so the weight of $U_{\tau:t}$ in the λ return $U_t^{\langle \lambda \rangle}$ is $(1 - \lambda)\lambda^{t-\tau-1}$. Noticing $U_{\tau:t} = R_{\tau+1} + \cdots + \gamma^{t-\tau-1}U_{t-1:t}$, $U_{t-1:t}$ should be added up to $U_\tau^{\langle \lambda \rangle}$ with the discount $(1 - \lambda)(\lambda\gamma)^{t-\tau-1}$. For every step further away, the eligibility trace should be decayed using the factor $\lambda\gamma$. For the latest state–action pair (S_τ, A_τ), we should amplify its weight. The strength of such amplification can be

- $\beta = 1$. Such eligibility trace is also called **accumulating trace**.
- $\beta = 1 - \alpha$ (where α is the learning rate). Such eligibility trace is called **dutch trace**.
- $\beta = 0$. Such eligibility trace is called **replacing trace**.

If $\beta = 1$, the eligibility trace is directly added by 1. If $\beta = 0$, the trace is set to be 1. Since the eligibility trace is always within $[0, 1]$, so we amplify the eligibility trace when the trace is set to be 1. If $\beta \in (0, 1)$, the amplification is between that of $\beta = 0$ and $\beta = 1$.

Combining eligibility trace with single-step TD policy evaluation algorithm can lead to eligibility trace policy evaluation algorithm. Combining eligibility trace with single-step TD policy optimization algorithm such as SARSA can lead to eligibility trace policy optimization algorithm such as SARSA(λ). Algorithm 5.14 shows the TD(λ) policy evaluation algorithm and the SARSA(λ) algorithm.

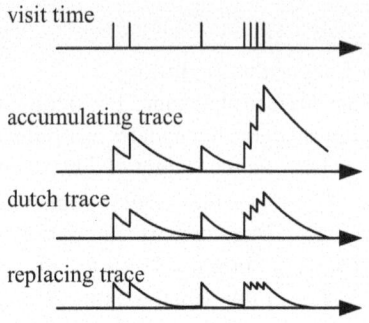

Fig. 5.4   Compare different eligibility traces.
The figure is adapted from (Singh, 1996).

Algorithm 5.14   TD(λ) policy evaluation or SARSA(λ).

Inputs: environment (without mathematical model). For policy evaluation, we also need to input the policy π.
Output: optimal action value estimates $q(s, a)$ ($s \in \mathcal{S}, a \in \mathcal{A}$).
Parameters: eligibility trace parameter λ and β, optimizer (containing learning rate α), discount factor γ, and parameters to control the number of episodes and maximum steps in an episode.

1. (Initialize) Initialize the action value estimates $q(s, a) \leftarrow$ arbitrary value ($s \in \mathcal{S}^+, a \in \mathcal{A}$).
2. (TD update) For each episode:
    2.1. (Initialize eligibility trace) $e(s, a) \leftarrow 0$ ($s \in \mathcal{S}, a \in \mathcal{A}$).
    2.2. (Initialize state–action pair) Select the initial state S, and then use the policy $\pi(\cdot|S)$ to determine the action A.
    2.3. Loop until the episode ends (for example, reach the maximum step, or S is the terminal state):
        2.3.1. (Sample) Execute the action A, and then observe the reward R, the next state S', and the indicator of episode end D'.
        2.3.2. (Decide) Use the input policy $\pi(\cdot|S')$ or the policy derived from the optimal action value estimate $q(S', \cdot)$ to determine the action A'. (The action can be arbitrarily chosen if $D' = 1$.)
        2.3.3. (Update eligibility traces) $e(s, a) \leftarrow \gamma \lambda e(s, a)$ ($s \in \mathcal{S}$, $a \in \mathcal{A}(s)$), and then $e(S, A) \leftarrow 1 + \beta e(S, A)$.
        2.3.4. (Calculate TD return) $U \leftarrow R + \gamma(1 - D')q(S', A')$.
        2.3.5. (Update value estimates) $q(s, a) \leftarrow q(s, a) + \alpha e(s, a)[U - q(S, A)]$ ($s \in \mathcal{S}, a \in \mathcal{A}(s)$).

2.3.6. $S \leftarrow S', A \leftarrow A'$.

There are also eligibility traces for state values. Given the trajectory $S_0, A_0, R_1, S_1, A_1, R_1, \ldots$, the eligibility trace $e_t(s)$ $(s \in S)$ shows the weight of one-step TD return $U_{t:t+1} = R_{t+1} + \gamma(1 - D_{t+1})v(S_{t+1})$ when it is used to update the state s $(s \in S)$. Mathematically, the eligibility trace of the state s $(s \in S)$ is defined as follows: When $t = 0$, $e_o(s) = 0$ $(s \in S)$. When $t > 0$,

$$
e_t(s) = \begin{cases} 1 + \beta\gamma\lambda e_{t-1}(s), & S_t = s, \\ \gamma\lambda e_{t-1}(s), & \text{otherwise.} \end{cases}
$$

Algorithm 5.15 shows the eligibility trace algorithm to evaluate state values.

Algorithm 5.15 TD(λ) policy evaluation to estimate state values.

Inputs: environment (without mathematical model), and the policy π.
Output: optimal state value estimates $v(s)$ $(s \in S)$.
Parameters: eligibility trace parameter λ and β, optimizer (containing learning rate α), discount factor γ, and parameters to control the number of episodes and maximum steps in an episode.

1. (Initialize) Initialize the state value estimates $v(s) \leftarrow$ arbitrary value $(s \in S^+)$.
2. (TD update) For each episode:
   2.1. (Initialize eligibility trace) $e(s) \leftarrow 0$ $(s \in S)$.
   2.2. (Initialize state) Select the initial state S.
   2.3. Loop until the episode ends (for example, reach the maximum step, or S is the terminal state):
   2.3.1. (Decide) Use the policy $\pi(\cdot|S)$ to determine the action A.
   2.3.2. (Sample) Execute the action A, and then observe the reward R, the next state S', and the indicator of episode end D'.
   2.3.3. (Update eligibility traces) $e(s) \leftarrow \gamma\lambda e(s)$ $(s \in S)$, and then $e(S) \leftarrow 1 + \beta e(S)$.
   2.3.4. (Calculate TD return) $U \leftarrow R + \gamma(1 - D')v(S')$.
   2.3.5. (Update value estimates) $v(s) \leftarrow v(s) + \alpha e(s)\left[U - v(S)\right]$ $(s \in S)$.
   2.3.6. $S \leftarrow S'$.

Compared to offline λ return algorithm, TD(λ) algorithm has the following advantages:

- TD(λ) can be used in both episodic tasks and sequential tasks.
- TD(λ) updates the value estimates after each step, so the updating is timelier.
- TD(λ) has little computation after each step. The computation is always not heavy.

5.5 Case Study: Taxi

This section considers the task Taxi-v3, which dispatches a taxi (Dietterich, 2000). As shown in Fig. 5.5, there are four taxi stands in a 5×5 grid. At the beginning of each episode, a passenger randomly appears in one taxi stand, and wants to go to another random taxi stand. At the same time, the taxi is randomly located at one of 25 locations. The taxi needs to move to the passenger, pick the passenger up, move to the destination, and drop the passenger. The taxi can move one grid at each step in a direction that does not have solid fence, at the cost of reward -1. Each time the taxi picks up or drops off the passenger in an incorrect way, it will be penalized for the reward -10. Successfully finishing the task can obtain reward 20. We want to maximize the total episode rewards.

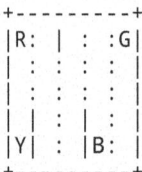

Fig. 5.5   ASCII map of the task Taxi-v3.
B, G, R, and Y are four taxi stands.
This figure is adapted from (Dietterich, 2000).

5.5.1 Use Environment

The environment of the task Taxi-v3 can be initialized by using env.reset() and be stepped by using env.step(). env.render() can visualize the task as an ASCII map (Fig. 5.5). In this map, the location of passenger and destination are shown in colors, while the location of the taxi is highlighted. Specifically, if the passenger is not in the taxi, the location of the passenger is shown in blue. The destination is shown in red. If the passenger is not in the taxi, the location of the taxi is highlighted in yellow. If the passenger is in the taxi, the location of the taxi is highlighted in green.

The observation space of the environment is Discrete(500), which means the observation is an int value within the range $[0, 500)$. We can use the function env.decode() to convert the int value into a tuple of length 4. The meaning of elements in the tuple (taxirow, taxicol, passloc, destidx) is as follows:

- (taxirow, taxicol) is the current location of the taxi. Both of them are int values ranging $\{0, 1, 2, 3, 4\}$, which shows the location of the taxi.

- passloc is the location of the passenger. It is an int value ranging $\{0, 1, 2, 3, 4\}$. 0, 1, 2, and 3 mean that the passenger is waiting at the taxi stand shown in Fig. 5.5, and 4 means the passenger is in the taxi.
- destidx is the destination. It is an int value ranging $\{0, 1, 2, 3\}$. The meaning of the number is shown in Fig. 5.5, too.

The total number of states is $(5 \times 5) \times 5 \times 4 = 500$.

Table 5.1 Taxi stands in the task Taxi-v3.

passloc or destidx	letter in ASCII map	coordinate in the grid
0	R	$(0, 0)$
1	G	$(0, 4)$
2	Y	$(4, 0)$
3	B	$(4, 3)$

The action space is Discrete(6), which means that the action is an int value within $\{0, 1, 2, 3, 4, 5\}$. The meaning of the action is shown in Table 5.2. Table 5.2 also shows the hint provided by env.render() and the possible reward after the action is executed.

Table 5.2 Actions in the task Taxi-v3.

action	meaning	hint in env.render()	reward after the action
0	move down	South	-1
1	move up	North	-1
2	move right	East	-1
3	move left	West	-1
4	try to pick up the passager	Pickup	-1 or -10
5	try to drop off the passager	Dropoff	$+20$ or -10

Code 5.1 instantiates the environment using the function gym.make(), where the parameter render_mode is designated as "ansi" so that env.render() will show the environment using texts. Then it gets the locations of taxi, passenger, and destination using env.decode(). Finally, it uses env.render() to get the string that represents the map, and use the function print() to visualizes the map.

Code 5.1   Initialize and visualize the task.

Taxi-v3_SARSA_demo.ipynb

```
1  import gym
2  env = gym.make('Taxi-v3', render_mode="ansi")
3  state, _ = env.reset()
4  taxirow, taxicol, passloc, destidx = env.unwrapped.decode(state)
5  logging.info('location of taxi = %s', (taxirow, taxicol))
6  logging.info('location of passager = %s', env.unwrapped.locs[passloc])
7  logging.info('location of destination = %s', env.unwrapped.locs[destidx])
8  print(env.render())
```

Now we have known how to use this environment.

5.5.2 On-Policy TD

This section uses SARSA algorithm and expected SARSA algorithm to find the optimal policy.

The class SARSAAgent in Code 5.2 implements the SARSA algorithm. At the start of an episode, we will call its member reset() of the agent object agent. agent.reset() has a parameter mode, indicating whether the upcoming episode is for training or for testing. If the upcoming episode is for training, the parameter is set to be "train". If the upcoming episode is for testing, the parameter is set to be "test". Agents behave differently between training episodes and testing episodes. Agents need to update value estimates or optimal policy estimates during training, but they do not during testing. Agents need to trade off between exploration and exploitation during training, but they simply exploit during testing. Every time we call env.step(), we can get an observation, a reward, and an indicator to show whether the episode finishes. All this information will be passed to agent.step(). The logic of agent.step() has two components: deciding and learning, which are also the two roles of the agent. In the part of deciding, the agent first checks whether it needs to consider exploration. If the current episode is for training, the agent can explore using ε-greedy policy. If the current episode is for testing, the agent will not explore. If the agent uses ε-greedy policy, it first draws a random number uniformly between 0 and 1. If this random number is smaller than ε, the agent explores by picking a random action. Otherwise, the agent exploits by choosing an action that maximizes the action value estimates. Next, the agent learns. Before the core learning logics, it first saves the observation, reward, indicator of episode end, and the action into a trajectory. The reason to save them is, the learning process not only needs the most recent observation, reward, indicator, and action, but also needs previous observation and so on. For example, in SARSA algorithm, each update involves two state–action pairs in successive steps. Therefore, we need to save trajectory during training. SARSA algorithm can update the action value estimates upon it collects two steps. The member learn() implements the core logic of learning. It first obtains

suitable values from the trajectory, then computes TD return, and updates action
value estimates.

Code 5.2   SARSA agent.

Taxi-v3_SARSA_demo.ipynb

```python
class SARSAAgent:
    def __init__(self, env):
        self.gamma = 0.9
        self.learning_rate = 0.2
        self.epsilon = 0.01
        self.action_n = env.action_space.n
        self.q = np.zeros((env.observation_space.n, env.action_space.n))

    def reset(self, mode=None):
        self.mode = mode
        if self.mode == 'train':
            self.trajectory = []

    def step(self, observation, reward, terminated):
        if self.mode == 'train' and np.random.uniform() < self.epsilon:
            action = np.random.randint(self.action_n)
        else:
            action = self.q[observation].argmax()
        if self.mode == 'train':
            self.trajectory += [observation, reward, terminated, action]
            if len(self.trajectory) >= 8:
                self.learn()
        return action

    def close(self):
        pass

    def learn(self):
        state, _, _, action, next_state, reward, terminated, next_action = \
                self.trajectory[-8:]

        target = reward + self.gamma * \
                self.q[next_state, next_action] * (1. - terminated)
        td_error = target - self.q[state, action]
        self.q[state, action] += self.learning_rate * td_error


agent = SARSAAgent(env)
```

Code 5.3 shows the codes to train SARSA agent. It repeatedly calls the function
play_episiode() in Code 1.3 in Sect. 1.6.3 to interact with the environment.
Code 5.3 controls the ending of training in the following way: If the average episode
reward of recent 200 episodes exceeds the reward threshold, finish training. Here,
the number of episodes and the threshold can be adjusted for different environments
and agents.

Code 5.3   Train the agent.

Taxi-v3_SARSA_demo.ipynb

```python
episode_rewards = []
for episode in itertools.count():
    episode_reward, elapsed_steps = play_episode(env, agent, seed=episode,
            mode='train')
    episode_rewards.append(episode_reward)
    logging.info('train episode %d: reward = %.2f, steps = %d',
            episode, episode_reward, elapsed_steps)
```

```
8      if np.mean(episode_rewards[-200:]) > env.spec.reward_threshold:
9          break
10  plt.plot(episode_rewards)
```

Code 1.4 in Sect. 1.6.3 tests the trained agent. It calculates the average episode reward in 100 successive episodes. If the average episode reward exceeds the threshold, which is 8 for the task Taxi-v3, the agent solves the task.

We can use the following code to show the optimal action value estimates

```
1  pd.DataFrame(agent.q)
```

We can use the following codes to show the optimal policy estimates:

```
1  policy = np.eye(agent.action_n)[agent.q.argmax(axis=-1)]
2  pd.DataFrame(policy)
```

Next, we consider expected SARSA algorithm. The class ExpectedSARSAAgent in Code 5.4 implements the expected SARSA agent. The only difference between this class and the class SARSAAgent in Code 5.3 is in the function learn().

Code 5.4   Expected SARSA agent.
Taxi-v3_ExpectedSARSA.ipynb

```
1   class ExpectedSARSAAgent:
2       def __init__(self, env):
3           self.gamma = 0.99
4           self.learning_rate = 0.2
5           self.epsilon = 0.01
6           self.action_n = env.action_space.n
7           self.q = np.zeros((env.observation_space.n, env.action_space.n))
8
9       def reset(self, mode=None):
10          self.mode = mode
11          if self.mode == 'train':
12              self.trajectory = []
13
14      def step(self, observation, reward, terminated):
15          if self.mode == 'train' and np.random.uniform() < self.epsilon:
16              action = np.random.randint(self.action_n)
17          else:
18              action = self.q[observation].argmax()
19          if self.mode == 'train':
20              self.trajectory += [observation, reward, terminated, action]
21              if len(self.trajectory) >= 8:
22                  self.learn()
23          return action
24
25      def close(self):
26          pass
27
28      def learn(self):
29          state, _, _, action, next_state, reward, terminated, _ = \
30                  self.trajectory[-8:]
31
32          v = (self.q[next_state].mean() * self.epsilon + \
33                  self.q[next_state].max() * (1. - self.epsilon))
34          target = reward + self.gamma * v * (1. - terminated)
35          td_error = target - self.q[state, action]
36          self.q[state, action] += self.learning_rate * td_error
37
38
39  agent = ExpectedSARSAAgent(env)
```

We use Codes 5.3 and 1.4 to train and test the agent, respectively. For this task, expected SARSA performs slightly better than SARSA.

5.5.3 Off-Policy TD

This section implements Q Learning algorithm and Double Q Learning algorithm.

First, let us consider Q Learning algorithm (shown in Code 5.5). The class QLearningAgent and the class ExpectedSARSAAgent differ in the bootstrapping logics in learn(). The agent is still trained using Code 5.3, and tested using Code 1.4.

Code 5.5 Q Learning agent.
Taxi-v3_QLearning.ipynb

```
class QLearningAgent:
    def __init__(self, env):
        self.gamma = 0.99
        self.learning_rate = 0.2
        self.epsilon = 0.01
        self.action_n = env.action_space.n
        self.q = np.zeros((env.observation_space.n, env.action_space.n))

    def reset(self, mode=None):
        self.mode = mode
        if self.mode == 'train':
            self.trajectory = []

    def step(self, observation, reward, terminated):
        if self.mode == 'train' and np.random.uniform() < self.epsilon:
            action = np.random.randint(self.action_n)
        else:
            action = self.q[observation].argmax()
        if self.mode == 'train':
            self.trajectory += [observation, reward, terminated, action]
            if len(self.trajectory) >= 8:
                self.learn()
        return action

    def close(self):
        pass

    def learn(self):
        state, _, _, action, next_state, reward, terminated, _ = \
                self.trajectory[-8:]

        v = reward + self.gamma * self.q[next_state].max() * (1. - terminated)
        target = reward + self.gamma * v * (1. - terminated)
        td_error = target - self.q[state, action]
        self.q[state, action] += self.learning_rate * td_error


agent = QLearningAgent(env)
```

The class DoubleQLearningAgent in Code 5.6 implements the Double Q Learning agent. Since Double Q Learning involves two sets of action value estimates, the class DoubleQLearningAgent and the class QLearningAgent have different constructors and learn() function. During learning, Double Q Learning agent swaps two action value estimates with the probability 0.5. The agent is still trained and tested by Codes 5.3 and 1.4 respectively. Since the maximization bias is not significant in this task, Double Q Learning shows no advantages over the vanilla Q Learning algorithm.

Code 5.6 Double Q Learning agent.

Taxi-v3_DoubleQLearning.ipynb

```
 1  class DoubleQLearningAgent:
 2      def __init__(self, env):
 3          self.gamma = 0.99
 4          self.learning_rate = 0.1
 5          self.epsilon = 0.01
 6          self.action_n = env.action_space.n
 7          self.qs = [np.zeros((env.observation_space.n, env.action_space.n)) for
 8                  _ in range(2)]
 9
10      def reset(self, mode=None):
11          self.mode = mode
12          if self.mode == 'train':
13              self.trajectory = []
14
15      def step(self, observation, reward, terminated):
16          if self.mode == 'train' and np.random.uniform() < self.epsilon:
17              action = np.random.randint(self.action_n)
18          else:
19              action = (self.qs[0] + self.qs[1])[observation].argmax()
20          if self.mode == 'train':
21              self.trajectory += [observation, reward, terminated, action]
22              if len(self.trajectory) >= 8:
23                  self.learn()
24          return action
25
26      def close(self):
27          pass
28
29      def learn(self):
30          state, _, _, action, next_state, reward, terminated, _ = \
31                  self.trajectory[-8:]
32
33          if np.random.randint(2):
34              self.qs = self.qs[::-1] # swap two elements
35          a = self.qs[0][next_state].argmax()
36          v = reward + self.gamma * self.qs[1][next_state, a] * (1. - terminated)
37          target = reward + self.gamma * v * (1. - terminated)
38          td_error = target - self.qs[0][state, action]
39          self.qs[0][state, action] += self.learning_rate * td_error
40
41
42  agent = DoubleQLearningAgent(env)
```

5.5.4 Eligibility Trace

This section considers SARSA(λ) algorithm. The class SARSALambdaAgent in Code 5.7 implements the SARSA(λ) agent. Compared to the class SARSAAgent, it has more parameters for eligibility traces, including lambd for controlling decay speed and beta for controlling enhancement amplitude. Note that we use the parameter name `lambd`, which originates from removing the ending letter in the word "lambda", since `lambda` is a Python keyword. The class `SARSALambdaAgent` is also trained by Code 5.3 and tested by Code 1.4. Due to the usage of eligibility trace, SARSA(λ) usually performs better than one-step SARSA in this task.

Code 5.7 SARSA(λ) agent.

`Taxi-v3_SARSALambda.ipynb`

```python
class SARSALambdaAgent:
    def __init__(self, env):
        self.gamma = 0.99
        self.learning_rate = 0.1
        self.epsilon = 0.01
        self.lambd = 0.6
        self.beta = 1.
        self.action_n = env.action_space.n
        self.q = np.zeros((env.observation_space.n, env.action_space.n))

    def reset(self, mode=None):
        self.mode = mode
        if self.mode == 'train':
            self.trajectory = []
            self.e = np.zeros(self.q.shape)

    def step(self, observation, reward, terminated):
        if self.mode == 'train' and np.random.uniform() < self.epsilon:
            action = np.random.randint(self.action_n)
        else:
            action = self.q[observation].argmax()
        if self.mode == 'train':
            self.trajectory += [observation, reward, terminated, action]
            if len(self.trajectory) >= 8:
                self.learn()
        return action

    def close(self):
        pass

    def learn(self):
        state, _, _, action, next_state, reward, terminated, next_action = \
                self.trajectory[-8:]

        # update eligibility trace
        self.e *= (self.lambd * self.gamma)
        self.e[state, action] = 1. + self.beta * self.e[state, action]

        # update value
        target = reward + self.gamma * \
                self.q[next_state, next_action] * (1. - terminated)
        td_error = target - self.q[state, action]
        self.q += self.learning_rate * self.e * td_error


agent = SARSALambdaAgent(env)
```

We trained and tested lots of agents in this section. Some agents perform better than others. Overall speaking, the reason why one algorithm outperforms the other is complex. Maybe an algorithm is suitable for the task, or a suite of parameters is more suitable for the algorithm. There are no algorithms such that it can always outperform other algorithms in all tasks. An algorithm that performs excellently on one task may underperform other algorithms in another task.

5.6 Summary

- 1-step Temporal Difference return (TD return) is defined as

$$U_{t:t+n}^{(v)} \stackrel{\text{def}}{=} R_{t+1} + (1 - D_{t+1})\gamma v(S_{t+1}),$$

$$U_{t:t+n}^{(q)} \stackrel{\text{def}}{=} R_{t+1} + (1 - D_{t+1})\gamma q(S_{t+1}, \mathcal{A}_{t+1}),$$

and n-step TD return is defined as:

$$U_{t:t+n}^{(v)} \stackrel{\text{def}}{=} R_{t+1} + \gamma(1 - D_{t+1})R_{t+2} + \cdots$$
$$+ \gamma^{n-1}(1 - D_{t+n-1})R_{t+n} + \gamma^n(1 - D_{t+n})v(S_{t+n}),$$

$$U_{t:t+n}^{(q)} \stackrel{\text{def}}{=} R_{t+1} + \gamma(1 - D_{t+1})R_{t+2} + \cdots$$
$$+ \gamma^{n-1}(1 - D_{t+n-1})R_{t+n} + \gamma^n(1 - D_{t+n})q(S_{t+n}, A_{t+n}),$$

where the indicator

$$D_t \stackrel{\text{def}}{=} \begin{cases} 1, & S_t = s_{\text{end}}, \\ 0, & S_t \neq s_{\text{end}}. \end{cases}$$

- TD return uses bootstrapping.
- TD error is defined as

$$\varDelta_t \stackrel{\text{def}}{=} U_t - v(S_t),$$

$$\varDelta_t \stackrel{\text{def}}{=} U_t - q(S_t, A_t).$$

- TD return in SARSA algorithm is $U_t = R_{t+1} + \gamma(1 - D_{t+1})q(S_{t+1}, A_{t+1})$.
- TD return in expected SARSA algorithm is $U_t = R_{t+1} + \gamma(1 - D_{t+1})$ $\sum_a \pi(a|S_{t+1})q(S_{t+1}, a)$.
- TD return in Q learning algorithm is $U_t = R_{t+1} + \gamma(1 - D_{t+1})\max_a q(S_{t+1}, a)$.
- In order to resist maximization bias, Double Q Learning maintains two independent action value estimates $q^{(0)}$ and $q^{(1)}$. Each update randomly updates one of them. TD return for updating $q^{(i)}$ ($i \in \{0, 1\}$) is $U_t^{(i)} = R_{t+1} + \gamma(1 - D_{t+1})q^{(1-i)}(S_{t+1}, \arg\max_a(S_{t+1}, a))$.
- λ return is the weighted average of multiple TD returns, where the weights are decayed at the speed of $\lambda \in [0, 1]$. It is the trade-off between vanilla return G_t and one-step TD return $U_{t:t+1}$.
- Offline λ-return algorithm is only suitable for episodic tasks. It updates using the λ-return U_t^λ at the end of each episode.
- SARSA(λ) uses TD return $U_t = R_{t+1} + \gamma(1 - D_{t+1})q(S_{t+1}, A_{t+1})$ with the weight called eligibility traces. Eligibility traces is decayed at the speed of $\gamma\lambda$, and be amplified at the current state–action pair.

5.7 Exercises

5.7.1 Multiple Choices

5.1 Choose the correct one: (   )

A. n-step TD update with $n = 0$ is equivalent to MC update.
B. n-step TD update with $n = 1$ is equivalent to MC update.
C. n-step TD update with $n \to +\infty$ is equivalent to MC update.

5.2 Choose the correct one: (   )

A. Model-based dynamic programming does NOT use bootstrapping.
B. MC learning does NOT use bootstrapping.
C. TD learning does NOT use bootstrapping.

5.3 On Q learning, the letter "Q" stands for: (   )

A. state values.
B. action values.
C. advantages.

5.4 Choose the correct one: (   )

A. SARSA algorithm is off-policy.
B. Expected SARSA algorithm is off-policy.
C. Q learning algorithm is off-policy.

5.5 On Double Q Learning, choose the correct one: (   )

A. Double learning can reduce bias.
B. Double learning can reduce variance.
C. Double learning can reduce both bias and variance.

5.6 On SARSA(λ) algorithm, choose the correct one: (   )

A. SARSA(λ) with $\lambda = 0$ is equivalent to one-step SARSA.
B. SARSA(λ) with $\lambda = 1$ is equivalent to one-step SARSA.
C. SARSA(λ) with $\lambda \to +\infty$ is equivalent to one-step SARSA.

5.7.2 Programming

5.7 Solve the task CliffWalking-v0 using Q learning.

5.7.3 Mock Interview

5.8 What is the difference between Monte Carlo learning and Temporal Difference learning?

5.9 Is Q learning an on-policy algorithm or an off-policy algorithm? Why?

5.10 Why does Double Q Learning use double learning?

5.11 What is eligibility trace?

Chapter 6
Function Approximation

This chapter covers

- function approximation
- parametric function approximation
- non-parametric function approximation
- stochastic gradient descent algorithm
- semi-gradient descent algorithm
- eligibility trace semi-gradient descent algorithm
- convergence of function approximation
- Baird's counterexample
- Deep Q Network (DQN) algorithm
- experience replay
- target network
- Double DQN algorithm
- Dueling DQN algorithm

For algorithms in Chaps. 3–5, each update can only update the value estimate for one state or one state–action pair. Unfortunately, state spaces or action spaces can be very large in some tasks, even can be infinitely large. In those tasks, it is impossible to update all states or all state–action pairs one by one. Function approximation tries to deal with this problem by approximating all these values using a parametric function, so updating the parameters of this parametric function can update the value estimates of lots of states or state–action pairs. Particularly, when we update the parameters according to the experience that visits a state or a state–action pair, the value estimates of states or state–action pairs that we have not visited yet can also be updated.

© The Author(s), under exclusive license to Springer Nature Singapore Pte Ltd. 2024

Z. Xiao, *Reinforcement Learning*, https://doi.org/10.1007/978-981-19-4933-3_6

This chapter will introduce the general method of function approximation, for both policy evaluation and policy optimization. We will also introduce some common function forms, including linear models and neural networks. Furthermore, we will introduce some RL algorithms, including DRL algorithms that use neural networks.

6.1 Basic of Function Approximation

Function approximation uses a mathematical model to estimate the real value. For MC algorithms and TD algorithms, the target to approximate are values (including optimal values), such as v_π, q_π, v_*, and q_*. We will consider other things to approximate, such as policy and dynamic, in remaining chapters of this book.

Interdisciplinary Reference 6.1
Machine Learning: Parametric Model and Nonparametric Model

A math model (for supervised learning, non-supervised learning, or reinforcement learning) can be classified as parametric or non-parametric.

Parametric model predefines the functional form of the model. For example, consider feature vector $\mathbf{x}$ and target vector $\mathbf{y}$, and their relationship is determined by a random mapping $\mathbf{F}$, i.e. $\mathbf{y} = \mathbf{F}(\mathbf{x})$. We need to model the random mapping. A parametric model assumes that the model has some form $\mathbf{y} = \mathbf{f}(\mathbf{x}; \mathbf{w}) + \mathbf{n}$, where $\mathbf{f}$ is a predefined function form, $\mathbf{w}$ are parameters to learn, and $\mathbf{n}$ are noises.

Example 6.1 (Linear Model) Linear model uses the linear combination of feature vectors as target, and its mathematical form is of

$$f(\mathbf{x}; \mathbf{w}) = \mathbf{x}^\top \mathbf{w} = \sum_{j \in \mathcal{J}} x_j w_j.$$

Its form is simple and easy to understand. However, the quality of features makes a significant difference on performance. People usually need to construct features for better performance. There are many existing algorithms on how to construct features.

Example 6.2 (Neural Network) Neural network is a powerful model that has universally expression ability. It is the most popular feature construction method in recent years. Users need to specify the network structures and parameters. It is difficult to explain.

Parametric models have the following advantages:

- Parametric models are easy to train: Training is fast and need few data, especially when the number of parameters is small.
- Parametric models can be interpreted easily, especially when the function form is simple.

Parametric models have the following disadvantages:

- Parametric models limit the function form, but the optimal solution may not have this form. Therefore, the parametric models will not be optimal. And it is easier to get underfitting.
- Parametric models require a priori knowledge to determine the function form. For example, the neural network model needs some a priori knowledge for designing suitable network structure and parameters.

A successful parametric model requires both suitable function forms and suitable parameters.

Nonparametric models directly learn the model from data without assumption of specific function form. Examples of nonparametric model include nearest neighbor, decision tree, kernel methods.

> **❗ Note**
>
> Nonparametric models can have many parameters, and the number of parameters may increase to arbitrarily large with the increase of data. Nonparametric models differ from parametric models in that they do not have presupposed function forms, rather than they do not have parameters.

Nonparametric models have the following advantages:

- Nonparametric models are more adaptable, since they are not constraints by some function forms. Therefore, they can fit the data better.
- They require little a priori knowledge.

Nonparametric models have the following disadvantages:

- They are more difficult to train. They tend to need more data, and involve more parameters when the models are complex, and the training is slower.
- It is more likely to have overfitting.

When the feature set is a finite set, we can generate an estimate for every feature. Such method is called tabular method. Since tabular method does not assume a function form, it is a nonparametric model. However, since the number of parameters is fixed and limited, it can be viewed as a parametric model as well. Furthermore, it can be viewed as a special case of linear model. To be specific, consider a feature space $\mathcal{X}$, who has $|\mathcal{X}|$ elements. We can map every feature $x \in \mathcal{X}$ to a vector $\left(1_{[x=x']} : x' \in \mathcal{X}\right)^{\top}$, whose element is 1 at the position of x but 0 at other $|\mathcal{X}| - 1$ positions. This vector of length $|\mathcal{X}|$ is equivalent to the feature vector in the linear model.

In RL, function approximation uses mathematical models to estimate functions such as value functions. Similar to the case of general machine learning, it can be classified to either parametric or nonparametric. Parametric models include linear model and neural networks, while nonparametric models include kernel method. This section will focus on parametric model.

The mathematical expression of parametric value models are as follows:

- Parametric model of state values can be denoted as $v(s; \mathbf{w})$ $(s \in S)$, where $\mathbf{w}$ is the parameters. It can be used to approximate either state values of a policy or the optimal state values.
- Parametric model of action values can be denoted as $q(s, a; \mathbf{w})$ $(s \in S, a \in \mathcal{A}(s))$, where $\mathbf{w}$ is the parameters. If the action space is finite, we can also use a vector function $\mathbf{q}(s; \mathbf{w}) = \left(q(s, a; \mathbf{w}) : a \in \mathcal{A} \right)$ $(s \in S)$ to estimate action values. Each element in the vector function $\mathbf{q}(s; \mathbf{w})$ corresponds to an action, and the input of the vector function is the state. They can be used to approximate either action values of a policy or the optimal action values.

There are few constraints on the forms of the function $v(s; \mathbf{w})$ $(s \in S)$, $q(s, a; \mathbf{w})$ $(s \in S, a \in \mathcal{A}(s))$, and $\mathbf{q}(s; \mathbf{w})$ $(s \in S)$. They can be linear models, neural networks, or anything else. We need to designate the form beforehand. Given the form of the function and the function parameters $\mathbf{w}$, all values of the function will be fully determined.

Linear models and artificial neural networks are the most popular forms of functions.

- Linear approximation uses the linear combination of multiple vectors to approximate values. The vectors, usually called features, depend on the inputs, such as states or state–action pairs. Taking the action value for example, we can define a vector as feature for each state–action pair, i.e. $\mathbf{x}(s, a) = \left(x_j(s, a) : j \in \mathcal{J} \right)$ and then further define the linear combination of those features as the approximation function:

$$q(s, a; \mathbf{w}) = \left[\mathbf{x}(s, a) \right]^\top \mathbf{w} = \sum_{j \in \mathcal{J}} x_j(s, a) w_j, \quad s \in S, a \in \mathcal{A}(s).$$

There is linear approximation for state values, too:

$$v(s; \mathbf{w}) = \left[\mathbf{x}(s) \right]^\top \mathbf{w} = \sum_{j \in \mathcal{J}} x_j(s) w_j, \quad s \in S.$$

- Artificial neural networks are another type of commonly-used functions. Since artificial neural networks are very expressive, and can find features automatically, so we can directly input state–action pairs or states into the neural networks. The RL algorithms using artificial neural networks are both DL algorithms and RL algorithms, so they are DRL algorithms.

Remarkably, in Chaps. 3–5, we estimate a value estimate for each state or each state–action value. This method is the **tabular method**. The tabular method can be viewed as a special case of linear approximation. Taking the action value as example, we can view the tabular methods as the special case of linear approximation with features in the form of

$$
\left(0,\ldots,0,\underset{\underset{s,\,a}{\uparrow}}{1},0,\ldots,0\right)^{\top},
$$

which has an element 1 at some state–action pairs, and has elements 0 elsewhere. From this aspect, the combination of all features are all action values, and weights of the linear combination exactly equal the action values.

6.2 Parameter Update using Gradient

Finding the best parameters is an optimization problem. We can use either gradient-based algorithms or non-gradient methods to solve an optimization problem. Gradient-based methods include gradient descent algorithms. Such methods require that the function is subdifferentiable with respect to the parameters. Both linear function and neural networks meet this requirement. There are also methods that do not require gradients, such as evolution strategy. Gradient-based algorithms are generally believed to be faster to find the optimal parameters, so they are most commonly used.

This section discusses how to update the parameters **w** using gradient-based methods, including the stochastic gradient descent for MC learning, semi-gradient descent for TD learning, and the semi-gradient descent with eligibility trace with eligibility trace learning. These methods can be used for both policy evaluation and policy optimization.

6.2.1 SGD: Stochastic Gradient Descent

This section considers the Stochastic Gradient Descent algorithm, which is suitable for MC update. After using function approximation, we need to learn the parameters of the function, rather than the value estimate of single state or single state–action pair. Stochastic method applies the Stochastic Gradient Descent algorithm to RL algorithms. It can be used in both policy evaluation and policy optimization.

Interdisciplinary Reference 6.2
Stochastic Optimization: Stochastic Gradient Descent

Stochastic Gradient Descent (SGD) is a first-order iterative optimization algorithm to find the optimal point for a subdifferentiable objective. It is a stochastic approximation of the gradient descent optimization.

The idea of this algorithm is as follows: Consider a differentiable and convex function $f(\mathbf{x}) = E[F(\mathbf{x})]$. We want to find the root of $\nabla f(\mathbf{x}) = \mathbf{0}$, or equivalently the root of $E[\nabla F(\mathbf{x})] = \mathbf{0}$. According to the Robbins–Monro algorithm, we may consider the following update:

$$\mathbf{x}_{k+1} = \mathbf{x}_k - \alpha_k \nabla F(\mathbf{x}_k),$$

where k is the index of iteration, and α_k is the learning rate. Furthermore, if we move the parameters against the direction of gradient, it is more likely that $\nabla f(\mathbf{x})$ will be close to $\mathbf{0}$. Such stochastic gradient descent can be more specifically called Steepest Gradient Descent.

SGD algorithm has many extensions and variants, such as momentum, RMSProp, and Adam. You can refer to Sect. 4.2 of the book (Xiao, 2018) for more details. SGD and its variants have been implemented by lots of software packages, such as TensorFlow and PyTorch.

Besides the Steepest Gradient Descent, there are other stochastic optimization methods, such as Natural Gradient, which will be covered in Sect. 8.4 of this book.

Algorithm 6.1 shows the SGD algorithm that evaluates policy using MC update. Unlike the MC policy evaluation algorithms in Chap. 4 (that is, Algos. 4.1–4.4), which updates the value estimate of single state or single state–action pair in each update, Algo. 6.1 updates the function parameter $\mathbf{w}$ during the update. In Step 2.3.2, the updating tries to minimize the difference between the return G_t and the action value estimate $q(S_t, A_t; \mathbf{w})$ (or the state value estimate $v(S_t; \mathbf{w})$). Therefore, we can define the sample loss as $[G_t - q(S_t, A_t; \mathbf{w})]^2$ (or $[G_t - v(S_t; \mathbf{w})]^2$ for state value), and define the total loss of an episode as $\sum_{t=0}^{T-1} [G_t - q(S_t, A_t; \mathbf{w})]^2$ (or $\sum_{t=0}^{T-1} [G_t - v(S_t; \mathbf{w})]^2$ for state values). Changing the parameter $\mathbf{w}$ in the opposite direction of the gradient of loss with respect to the parameter $\mathbf{w}$ may reduce the loss. Therefore, we can first calculate the gradient $\nabla q(S_t, A_t; \mathbf{w})$ (or $\nabla v(S_t; \mathbf{w})$), and then updating using

$$\mathbf{w} \leftarrow \mathbf{w} - \frac{1}{2}\alpha_t \nabla [G_t - q(S_t, A_t; \mathbf{w})]^2$$
$$= \mathbf{w} + \alpha_t [G_t - q(S_t, A_t; \mathbf{w})] \nabla q(S_t, A_t; \mathbf{w}), \quad \text{update action value}$$
$$\mathbf{w} \leftarrow \mathbf{w} - \frac{1}{2}\alpha_t \nabla [G_t - v(S_t; \mathbf{w})]^2$$
$$= \mathbf{w} + \alpha_t [G_t - v(S_t; \mathbf{w})] \nabla v(S_t; \mathbf{w}), \qquad \text{update state value.}$$

Algorithm 6.1   Policy evaluation with function approximation and SGD.

---

1. (Initialize parameters) $\mathbf{w} \leftarrow$ arbitrary values.
2. For each episode:

    2.1. (Sample) Use the policy π to generate the trajectory $S_0, A_0, R_1, S_1, A_1, R_2, \ldots, S_{T-1}, A_{T-1}, R_T, S_T$.

    2.2. (Initialize return) $G \leftarrow 0$.

    2.3. For each $t \leftarrow T - 1, T - 2, \ldots, 0$:

        2.3.1. (Update return) $G \leftarrow \gamma G + R_{t+1}$.

        2.3.2. (Update value parameter) If we are to estimate action values, update the parameter $\mathbf{w}$ to reduce $\left[G - q(S_t, A_t; \mathbf{w}) \right]^2$ (for example, $\mathbf{w} \leftarrow \mathbf{w} + \alpha \left[G - q(S_t, A_t; \mathbf{w}) \right] \nabla q(S_t, A_t; \mathbf{w})$). If we are to estimate state values, update the parameter $\mathbf{w}$ to reduce $\left[G - v(S_t; \mathbf{w}) \right]^2$ (for example, $\mathbf{w} \leftarrow \mathbf{w} + \alpha \left[G - v(S_t; \mathbf{w}) \right] \nabla v(S_t; \mathbf{w})$).

We can use software such as TensorFlow and PyTorch to calculate gradients and update the parameters.

SGD policy optimization can be obtained by combining policy evaluation with SGD policy evaluation and policy improvement. Algorithm 6.2 shows the SGD policy optimization algorithm. The difference between this algorithm and the policy optimization algorithms in Chap. 4 is that this algorithm updates the parameter $\mathbf{w}$ rather than the value estimates of single state or single state–action pair. Policy optimization algorithm with function approximation seldom maintains the policy explicitly.

Algorithm 6.2   Policy optimization with function approximation and SGD.

---

1. (Initialize parameters) $\mathbf{w} \leftarrow$ arbitrary values.
2. For each episode:

    2.1. (Sample) Use the policy derived from current optimal action value estimates $q(\cdot, \cdot; \mathbf{w})$ (say, ε-greedy policy) to generate the trajectory $S_0, A_0, R_1, S_1, A_1, R_2, \ldots, S_{T-1}, A_{T-1}, R_T, S_T$.

    2.2. (Initialize return) $G \leftarrow 0$.

    2.3. For each $t \leftarrow T - 1, T - 2, \ldots, 0$:

        2.3.1. (Update return) $G \leftarrow \gamma G + R_{t+1}$.

        2.3.2. (Update optimal action value estimate) Update the parameter $\mathbf{w}$ to reduce $\left[G - q(S_t, A_t; \mathbf{w}) \right]^2$ (for example, $\mathbf{w} \leftarrow \mathbf{w} + \alpha \left[G - q(S_t, A_t; \mathbf{w}) \right] \nabla q(S_t, A_t; \mathbf{w})$).

6.2.2 Semi-Gradient Descent

TD learning uses bootstrapping to generate return samples, so the return samples depend on the parameter $\mathbf{w}$. Taking the action values as an example, one-step TD return is $U_t = R_{t+1} + \gamma q(S_{t+1}, A_{t+1}; \mathbf{w})$, which depends on the parameter $\mathbf{w}$. When we update to reduce the difference between U_t and the action value estimate $q(S_t, A_t; \mathbf{w})$, we can define the sample loss as $\left[U_t - q(S_t, A_t; \mathbf{w})\right]^2$. When the TD learning algorithms in Chap. 5 try to update the value estimates, they will not change the target U_t. Similarly, when a TD learning algorithm with function approximation tries to update the function parameter $\mathbf{w}$, it should not try to change the target U_t, either. Therefore, during the updating, we need to calculate the gradients of value estimates $q(S_t, A_t; \mathbf{w})$ with respect to the parameter $\mathbf{w}$, but we shall not calculate the gradients of the target $U_t = R_{t+1} + \gamma q(S_{t+1}, A_{t+1}; \mathbf{w})$ with respect to the parameter $\mathbf{w}$. That is the principle of the **semi-gradient descent**. We have similar analysis for state values.

Semi-gradient descent can be used for policy evaluation and policy optimization (Algos. 6.3 and 6.4). These algorithms use the indicator of episode ending in Algo. 5.2. Algorithms with function approximation tends to use such indicator. This is because, when we design the form of approximation function, we seldom consider the terminal states, so the designated form of function will be directly applied to the state space with the terminal state, i.e. S^+. However, the form of the function can not guarantee that the values of the function on the terminal state is 0. Therefore, it is convenient to use this indicator of episode ends to calculate TD return. For sequential tasks, there are no terminal states, so this indicator is always 0, and there is no need to maintain it.

Algorithm 6.3   Semi-gradient descent policy evaluation to estimate action values or SARSA policy optimization.

---

1. (Initialize parameters) $\mathbf{w} \leftarrow$ arbitrary values.
2. For each episode:

   2.1. (Initialize state–action pair) Choose the initial state S.
        For policy evaluation, use the policy $\pi(\cdot|S)$ to determine the action A.
        For policy optimization, use the policy derived from current optimal action value estimates $q(S, \cdot; \mathbf{w})$ (such as ε-greedy policy) to determine the action A.
   2.2. Loop until the episode ends:
        2.2.1. (Sample) Execute the action A, and then observe the reward R, the next state S', and the indicator of episode end D'.
        2.2.2. (Decide) For policy evaluation, use the policy $\pi(\cdot|S')$ to determine the action A'. For policy optimization, use the policy derived from the current optimal action value estimates

---

$q(S', \cdot; \mathbf{w})$ (for example, ε-greedy policy) to determine the action A'. (The action can be arbitrarily chosen if $D' = 1$.)

2.2.3. (Calculate TD return) $U \leftarrow R + \gamma(1 - D')q(S', A'; \mathbf{w})$.

2.2.4. (Update action value parameter) Update the parameter $\mathbf{w}$ to reduce $\left[U - q(S, A; \mathbf{w})\right]^2$ (For example, $\mathbf{w} \leftarrow \mathbf{w} + \alpha\left[U - q(S, A; \mathbf{w})\right]\nabla q(S, A; \mathbf{w})$). Note, you should not re-calculate U, and you should not calculate the gradient of U with respect to $\mathbf{w}$.

2.2.5. $S \leftarrow S', A \leftarrow A'$.

Algorithm 6.4 Semi-gradient descent policy evaluation to estimate state values, or expected SARSA policy optimization, or Q learning.

1. (Initialize parameters) $\mathbf{w} \leftarrow$ arbitrary values.

2. For each episode:

2.1. (Initialize state) Choose the initial state S.

2.2. Loop until the episode ends:

2.2.1. (Decide) For policy evaluation, use the policy $\pi(\cdot|S)$ to determine the action A.

For policy optimization, use the policy derived from the current optimal action value estimates $q(S, \cdot; \mathbf{w})$ (for example, ε-greedy policy) to determine the action A.

2.2.2. (Sample) Execute the action A, and then observe the reward R, the next state S', and the indicator of episode end D'.

2.2.3. (Calculate TD return) For policy evaluation for state values, set $U \leftarrow R + \gamma(1 - D')v(S'; \mathbf{w})$.

For expected SARSA, set $U \leftarrow R + \gamma(1 - D')\sum_a \pi(a|S'; \mathbf{w})q(S', a; \mathbf{w})$, where $\pi(\cdot|S'; \mathbf{w})$ is the policy derived from $q(S', \cdot; \mathbf{w})$ (for example, ε-greedy policy).

For Q learning, set $U \leftarrow R + \gamma(1 - D')\max_a q(S', a; \mathbf{w})$.

2.2.4. (Update value parameter) For policy evaluation for estimate state values, update the parameter $\mathbf{w}$ to reduce $\left[U - v(S; \mathbf{w})\right]^2$ (For example, $\mathbf{w} \leftarrow \mathbf{w} + \alpha\left[U - v(S; \mathbf{w})\right]\nabla v(S; \mathbf{w})$).

For expected SARSA or Q learning, update the parameter $\mathbf{w}$ to reduce $\left[U - q(S, A; \mathbf{w})\right]^2$ (For example, $\mathbf{w} \leftarrow \mathbf{w} + \alpha\left[U - q(S, A; \mathbf{w})\right]\nabla q(S, A; \mathbf{w})$).

Note, you should not re-calculate U, and you should not calculate the gradient of U with respect to $\mathbf{w}$.

2.2.5. $S \leftarrow S'$.

If we use soft packages such as TensorFlow or PyTorch to calculate gradients, we must make sure not to calculate the gradients of the target with respect to the parameter **w**. The software package may have some functions to stop the gradient calculation (such as `stop_gradient()` in TensorFlow or `detach()` in PyTorch), and we can use them to stop the gradient calculation. Another way is to make another copy of the parameter, say $\mathbf{w}_{target} \leftarrow \mathbf{w}$. Then use $\mathbf{w}_{target}$ to calculate the return sample, and calculate the loss gradient only with respect to the original parameter. In this way, we can avoid calculating gradients of target with respect to the original parameter.

6.2.3 Semi-Gradient Descent with Eligibility Trace

We have learned the eligibility trace algorithms in Sect. 5.4. Eligibility trace can trade off between MC update and one-step TD update, so it may perform better than both MC update and TD update. Eligibility trace algorithm maintains an eligibility trace for each value estimate to indicate weight for updating. Recently visited states or state–action pairs have larger weights. The states or state–action pairs that were visited long ago have small weights. The states or state–action pairs that have never been visited have no weights. Each updating will update all eligibility traces of the whole trajectory, and we update the value estimates for the whole trajectory with eligibility traces as weights.

Function approximation can be also applied to eligibility trace. In such cases, the eligibility trace will correspond to the value parameters **w**. To be more specific, the eligibility trace parameter **z** will have the same shape with the value estimate **w**, and they have mapping one-to-one: Each element in the eligibility trace parameter indicates the weight that we need to use to update the value parameters. For example, an element z in eligibility trace parameter **z** is the weight of the element w in the value parameter **w**, so we need to update **w** using

$$\text{update action value:} \quad w \leftarrow w + \alpha \big[U - q(S_t, A_t; \mathbf{w})\big] z$$
$$\text{update state value:} \quad w \leftarrow w + \alpha \big[U - v(S_t; \mathbf{w})\big] z.$$

For the whole value parameter, we have

$$\text{update action value:} \quad \mathbf{w} \leftarrow \mathbf{w} + \alpha \big[U - q(S_t, A_t; \mathbf{w})\big] \mathbf{z}$$
$$\text{update state value:} \quad \mathbf{w} \leftarrow \mathbf{w} + \alpha \big[U - v(S_t; \mathbf{w})\big] \mathbf{z}.$$

When selecting accumulative trace as the eligibility trace, the updating is defined as follows: When $t = 0$, set $\mathbf{z}_0 = \mathbf{0}$. When $t > 0$, set

$$\text{update traces of action values:} \quad \mathbf{z}_t \leftarrow \gamma \lambda \mathbf{z}_{t-1} + \nabla q(S_t, A_t; \mathbf{w})$$
$$\text{update traces of state values:} \quad \mathbf{z}_t \leftarrow \gamma \lambda \mathbf{z}_{t-1} + \nabla v(S_t; \mathbf{w}).$$

Algorithms 6.5 and 6.6 show the algorithms to use eligibility trace in policy evaluation and policy optimization. Both of them use the accumulative trace.

Algorithm 6.5 TD(λ) policy evaluation for action values or SARSA.

1. (Initialize parameters) $\mathbf{w} \leftarrow$ arbitrary values.
2. For each episode:
   2.1. (Initialize eligibility trace) $\mathbf{z} \leftarrow \mathbf{0}$.
   2.2. (Initialize state–action pair) Choose the initial state S.
        For policy evaluation, use the policy $\pi(\cdot|S)$ to determine the action A.
        For policy optimization, use the policy derived from current optimal action value estimates $q(S, \cdot; \mathbf{w})$ (such as ε-greedy policy) to determine the action A.
   2.3. Loop until the episode ends:
        2.3.1. (Sample) Execute the action A, and then observe the reward R, the next state S', and the indicator of episode end D'.
        2.3.2. (Decide) For policy evaluation, use the policy $\pi(\cdot|S')$ to determine the action A'. For policy optimization, use the policy derived from the current optimal action value estimates $q(S', \cdot; \mathbf{w})$ (for example, ε-greedy policy) to determine the action A'. (The action can be arbitrarily chosen if $D' = 1$.)
        2.3.3. (Calculate TD return) $U \leftarrow R + \gamma(1 - D')q(S', A'; \mathbf{w})$.
        2.3.4. (Update eligibility trace) $\mathbf{z} \leftarrow \gamma\lambda\mathbf{z} + \nabla q(S, A; \mathbf{w})$.
        2.3.5. (Update action value parameter) $\mathbf{w} \leftarrow \mathbf{w} + \alpha[U - q(S, A; \mathbf{w})]\mathbf{z}$.
        2.3.6. $S \leftarrow S', A \leftarrow A'$.

Algorithm 6.6 TD(λ) policy evaluation for state values, or expected SARSA, or Q learning.

1. (Initialize parameters) $\mathbf{w} \leftarrow$ arbitrary values.
2. For each episode:
   2.1. (Initialize eligibility trace) $\mathbf{z} \leftarrow \mathbf{0}$.
   2.2. (Initialize state) Choose the initial state S.
   2.3. Loop until the episode ends:
        2.3.1. (Decide) For policy evaluation, use the policy $\pi(\cdot|S)$ to determine the action A.
               For policy optimization, use the policy derived from the current optimal action value estimates $q(S, \cdot; \mathbf{w})$ (for example, ε-greedy policy) to determine the action A.
        2.3.2. (Sample) Execute the action A, and then observe the reward R, the next state S', and the indicator of episode end D'.

2.3.3. (Calculate TD return) For policy evaluation for state values, set $U \leftarrow R + \gamma(1 - D')v(S'; \mathbf{w})$.

For expected SARSA, set $U \leftarrow R + \gamma(1 - D')$ $\sum_a \pi(a|S'; \mathbf{w})q(S', a; \mathbf{w})$, where $\pi(\cdot|S'; \mathbf{w})$ is the policy derived from $q(S', \cdot; \mathbf{w})$ (for example, ε-greedy policy).

For Q learning, set $U \leftarrow R + \gamma(1 - D')\max_a q(S', a; \mathbf{w})$.

2.3.4. (Update eligibility trace) $\mathbf{z} \leftarrow \gamma\lambda\mathbf{z} + \nabla q(S, A; \mathbf{w})$.

2.3.5. (Update value parameter) For policy evaluation for estimate state values, set $\mathbf{w} \leftarrow \mathbf{w} + \alpha[U - v(S; \mathbf{w})]\mathbf{z}$.

For expected SARSA or Q learning, set $\mathbf{w} \leftarrow \mathbf{w} + \alpha[U - q(S, A; \mathbf{w})]\mathbf{z}$.

2.3.6. $S \leftarrow S'$.

6.3 Convergence of Function Approximation

This section considers the convergence of RL algorithms.

6.3.1 Condition of Convergence

Tables 6.1 and 6.2 summarize the convergence of policy evaluation algorithms and policy optimization algorithms respectively. In these two tables, the tabular algorithms refer to the algorithms in Chap. 4–5, which do not use function approximation. Those tabular algorithms can converge to true values or optimal values in general. However, for function approximation methods, the convergence can be guaranteed for SGD with MC learning, but can not be guaranteed for the semi-gradient descent with TD learning.

Table 6.1 Convergence of policy evaluation algorithms.

algorithm	tabular	linear approximation	non-linear approximation
on-policy MC	converge	converge	converge
on-policy TD	converge	converge	not always converge
off-policy MC	converge	converge	converge
off-policy TD	converge	not always converge	not always converge

Linear approximation has a simple structure among all forms of function approximations, so linear approximation has extra convergence opportunity than general cases.

Table 6.2   Convergence of policy optimization algorithms.

algorithm	tabular	linear approximation	non-linear approximation
MC	converge	converge or swing around optimal solution	not always converge
SARSA	converge	converge or swing around optimal solution	not always converge
Q learning	converge	not always converge	not always converge

All convergence in the table, of course, is guaranteed only when the learning rate satisfies the condition of Robbins–Monro algorithm. For those convergence cases, their convergence is usually proved by verifying the condition of Robbins–Monro algorithm.

6.3.2 Baird's Counterexample

Remarkably, for off-policy Q learning, the convergence can not be guaranteed even when the function approximation uses the linear model. Researchers noticed that, we can not guarantee the convergence of an algorithm that satisfies the three conditions (1) off-policy (2) bootstrapping, and (3) function approximation. Baird's counterexample is a famous example to demonstrate that.

Baird's counterexample considers the following MDP (Baird, 1995): As Fig. 6.1, the state space is $S = \left\{s^{(0)}, s^{(1)}, \ldots, s^{(5)}\right\}$, and action space is $\mathcal{A} = \left\{a^{(0)}, a^{(1)}\right\}$. The initial state distribution is the uniform distribution over all states in the state space. At time t ($t = 0, 1, \ldots$), no matter what state is agent at, action $a^{(0)}$ will lead to next state $a^{(0)}$ and get reward 0, while action $a^{(1)}$ will lead to next state uniformly selected from $S \setminus \left\{s^{(0)}\right\}$ with reward 0, too. The discount factor γ is very close to 1 (say 0.99). Obviously, the state values and the action values of an arbitrary policy are 0, so the optimal state values and the optimal action values are all 0. Every policy is optimal.

In order to prove that the algorithm with off-policy, bootstrapping, and function approximation at the same time may not converge, we will designate a policy and try to evaluate its state values using off-policy TD learning algorithm with function approximation. And we will prove that the learning diverges.

- The policy to evaluate is the deterministic policy $\pi(a|s) = \begin{cases} 1, & s \in S, a = a^{(0)} \\ 0, & s \in S, a = a^{(1)}. \end{cases}$

  It always selects the action $a^{(0)}$ and its next state is always $s^{(0)}$. The state values of this policy is $v_\pi(s) = 0$ ($s \in S$).
- Function approximation: We designate the following linear approximation: $v\left(s^{(i)}; \mathbf{w}\right) = \left(\mathbf{g}^{(i)}\right)^\top \mathbf{w}$ ($i = 0, 1, \ldots, |S| - 1$), where $\mathbf{w} = \left(w^{(0)}, \ldots, w^{(|S|)}\right)^\top$ is the parameter to learn, and $\mathbf{g}^{(i)}$ is

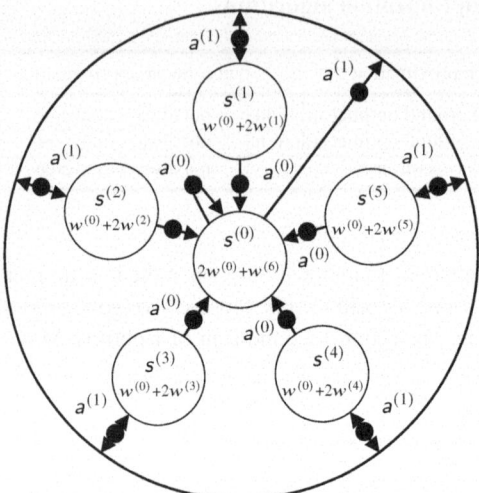

Fig. 6.1   MDP in Baird's counterexample.

$$
\begin{bmatrix}
\left(\mathbf{g}^{(0)}\right)^{\mathsf{T}} \\
\left(\mathbf{g}^{(1)}\right)^{\mathsf{T}} \\
\left(\mathbf{g}^{(2)}\right)^{\mathsf{T}} \\
\vdots \\
\left(\mathbf{g}^{(|\mathcal{S}|-1)}\right)^{\mathsf{T}}
\end{bmatrix}
=
\begin{bmatrix}
2 & 0 & 0 & \cdots & 1 \\
1 & 2 & 0 & \cdots & 0 \\
1 & 0 & 2 & \cdots & 0 \\
\vdots & \vdots & \vdots & \ddots & \vdots \\
1 & 0 & 2 & \cdots & 0
\end{bmatrix}.
$$

Obviously, the gradient of the approximation function with respect to the parameter $\mathbf{w}$ is $\nabla v\left(s^{(i)}, \mathbf{w}\right) = \mathbf{g}^{(i)}$ $(i = 0, 1, \ldots, |\mathcal{S}| - 1)$.

- Bootstrapping: We use one-step TD return: For time t, if the next state is $S_{t+1} = s^{(i)}$, the TD return is $U_t = \gamma v(S_{t+1}; \mathbf{w})$, and the TD error is $\delta_t = U_t - v(S_t; \mathbf{w}) = \gamma v(S_{t+1}; \mathbf{w}) - v(S_t; \mathbf{w})$.

- Off-policy: We use the behavior policy $b(a|s) = \begin{cases} 1/|\mathcal{S}|, & s \in \mathcal{S}, a = a^{(0)} \\ 1 - 1/|\mathcal{S}|, & s \in \mathcal{S}, a = a^{(1)}. \end{cases}$

  At time t, if we choose the action $A_t = a^{(0)}$, the importance sampling ratio with respect to the target policy is $\rho_t = \dfrac{\pi\left(a^{(0)}|S_t\right)}{b\left(a^{(0)}|S_t\right)} = \dfrac{1}{1/|\mathcal{S}|} = |\mathcal{S}|$, and the next state will be $s^{(0)}$. If we choose the action $A_t = a^{(1)}$, the importance sampling ratio with respect to the target policy is $\rho_t = \dfrac{\pi\left(a^{(1)}|S_t\right)}{b\left(a^{(1)}|S_t\right)} = \dfrac{0}{1/|\mathcal{S}|} = 0$, and the next state will be selected from $\mathcal{S} \setminus \left\{s^{(0)}\right\}$ with equal probability. We can verify that, using this behavior policy, the state of any time is always uniformed distributed among the state space S.

The learning process updates parameters using

$$\mathbf{w}_{t+1} \leftarrow \mathbf{w}_t + \alpha \rho_t \delta_t \nabla v(S_t; \mathbf{w}),$$

where the learning rate α is a small positive number (say $\alpha = 0.01$). Remarkably, if the behavior policy chooses the action $A_t = a^{(1)}$ at time t, the updating is simplified to $\mathbf{w}_{t+1} \leftarrow \mathbf{w}_t$ due to $\rho_t = 0$, which means that the parameter does not change. If the behavior policy chooses the action $A_t = a^{(0)}$, the importance sampling ratio is $\rho_t = |S|$.

We can prove that, if the initial parameters $\mathbf{w}_0$ are not all zeros (for example, initial parameters are $\mathbf{w}_0 = (1, 0, \ldots, 0)^\top$), the aforementioned learning diverges. In fact, the elements in the parameters will increase to infinity, with the ratio approximated to $5 : 2 : \cdots : 2 : 10(\gamma - 1)$. As shown in Fig. 6.2, $w^{(0)}$ increases to positive infinity with the fastest speed, and $\left\{ w^{(i)} : i = 1, 2, \ldots, 5 \right\}$ will increase to positive infinity too. $w^{(6)}$ will decrease to negative infinity slowly. The trend will continue forever, so the learning will not converge to some values.

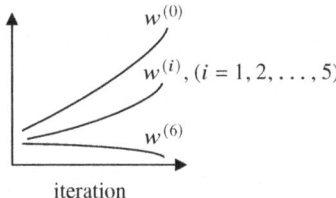

Fig. 6.2   Trend of parameters with iterations.

We can use mathematical induction to roughly verify this trend. Suppose in some iteration, the elements of $\mathbf{w}_t$ is approximately $\left(5\chi, 2\chi, \ldots, 2\chi, 10(\gamma - 1)\chi \right)$. Now we will discuss how the values of $\mathbf{w}_t$ change afterwards. Previously analysis has told use the parameters can only change when $A_t = a^{(0)}$. We can consider the following cases.

- If $S_t = s^{(0)}$ and $A_t = a^{(0)}$, we have

$$\delta_t = .\gamma \left(\mathbf{g}^{(0)} \right)^\top \mathbf{w}_t - \left(\mathbf{g}^{(0)} \right)^\top \mathbf{w}_t$$
$$= (\gamma - 1)\left(2w_t^{(0)} + w_t^{(6)} \right)$$
$$= (\gamma - 1)\left(2 \cdot 5\chi + 10(\gamma - 1)\chi \right)$$
$$\approx 10(\gamma - 1)\chi.$$

Since $\mathbf{g}^{(0)} = (2, 0, \ldots, 0, 1)^{\top}$, the parameter $w^{(0)}$ and $w^{(6)}$ will increase by $\alpha\rho_t$ times of $20(\gamma - 1)\chi$ and $10(\gamma - 1)\chi$ respectively.

- If $S_t = s^{(i)}$ ($i > 0$) and $A_t = a^{(0)}$, we have

$$
\begin{aligned}
\delta_t &= \gamma\left(\mathbf{g}^{(0)}\right)^{\top}\mathbf{w}_t - \left(\mathbf{g}^{(0)}\right)^{\top}\mathbf{w}_t \\
&= \gamma\left(2w_t^{(0)} + w_t^{(6)}\right) - \left(w_t^{(0)} + 2w_t^{(i)}\right) \\
&= \gamma\left(2 \cdot 5\chi + 10(\gamma - 1)\chi\right) - (5\chi + 2 \cdot 2\chi) \\
&= \left(10\gamma^2 - 9\right)\chi \\
&\approx \chi.
\end{aligned}
$$

Consider the expression of $\mathbf{g}^{(i)}$, the parameter $w^{(0)}$ and $w^{(i)}$ will be increased by $\alpha\rho_t$ times χ and 2χ respectively.

Since the state is uniformly distributed over the state space, the increase of elements in the parameters are:

$$
\begin{aligned}
w^{(0)} &: & 20(\gamma - 1)\chi + 5\chi \approx 5\chi \\
w^{(i)} \ (i = 1, \ldots, 5) &: & 2\chi \\
w^{(6)} &: & 10(\gamma - 1)\chi.
\end{aligned}
$$

The ratio of these increments is approximately $5 : 2 : \cdots : 2 : 10(\gamma - 1)$. The verification completes.

6.4 DQN: Deep Q Network

This section considers a very hot function approximation policy optimization algorithm: **Deep Q Network (DQN)**. The primary intuition of DQN is to use an artificial neural network to approximate the action values. Since the artificial neural networks are very expressive and can find features automatically, they have much larger potential than classic feature engineering methods.

Section 6.3.2 showed that we can not guarantee the convergence of an algorithm that uses off-policy, bootstrapping, and function approximation at the same time. The training may be very unstable and difficult. Therefore, researchers come up with some tricks, including:

- **Experience replay**: Store the transition experience (including the state, action, reward in the historical trajectory), and sample afterward from the storage according to some rules.
- **Target network**: Change the way to update the networks so that we do not immediately use the parameter we just learn for bootstrapping.

6.4.1 Experience Replay

The key idea of experience replay is to store the experiment so that it can be repeatedly used (Lin, 1993). Experience replay consists of two steps: storing and replaying.

- Store: Store the experience, which is a part of the trajectory such as $(S_t, A_t, R_{t+1}, S_{t+1}, D_{t+1})$, in the storage.
- Replay: Randomly select some experiences from the storage, according to some selection rule.

Experience replay has the following advantages:

- The same experience can be used multiple times, so the sample efficiency is improved. It is particularly useful when it is difficult or expensive to obtain data.
- It re-arranges the experiences, so the relationship among adjacent experiences is minimized. Consequently, the distribution of data is more stable, which makes the training of neural networks easier.

Shortcomings for experience replay include:

- It takes some space to store experiences.
- The length of experience is limited, so it can not be used for MC update when episodes have infinite steps. Experience replay is usually used for TD updating. For one-step TD learning, the experience can be in the form of $(S_t, A_t, R_{t+1}, S_{t+1}, D_{t+1})$. For n-step TD learning, the experience can be in the form of $(S_t, A_t, R_{t+1}, S_{t+1}, D_{t+1}, \ldots, S_{t+n}, D_{t+n})$.

Algorithm 6.7 shows the DQN algorithm with experience replay. Step 1.1 initializes the network parameter $\mathbf{w}$, which should follow the convention in deep learning. Step 1.2 initializes the storage for experience. Step 2.2.1 saves the interaction experience between the agent and the environment. Step 2.2.2 replays the experience, and conducts TD update using the replayed experience. Specially, Step 2.2.2.2 calculates the TD return for every experience entry. We usually calculate all entries in batch. In order to calculate in batch, we usually explicitly maintain the indicators of episode ends for episodic tasks. Therefore, when collecting experiences, we also obtain and save the indicators of episode ends. Step 2.2.2.3 uses the whole batch to update network parameters.

Algorithm 6.7   DQN policy optimization with experience replay (loop over episodes).

Parameters: parameters for experience replay (for example, the capacity of the storage, number of replayed samples in each batch), and other parameters (such as the optimizer, and the discount factor).

1. Initialize:

   1.1. (Initialize network parameter) Initialize $\mathbf{w}$.

1.2. (Initialize experience storage) $\mathcal{D} \leftarrow \varnothing$.

2. For each episode:

    2.1. (Initialize state) Choose the initial state S.

    2.2. Loop until the episode ends:

        2.2.1. (Collect experiences) Do the following once or multiple times:

            2.2.1.1. (Decide) Use the policy derived from $q(S, \cdot; \mathbf{w})$ (for example, ε-greedy policy) to determine the action A.

            2.2.1.2. (Sample) Execute the action A, and then observe the reward R, the next state S', and the indicator of episode end D'.

            2.2.1.3. Save the experience (S, A, R, S', D') in the storage $\mathcal{D}$.

            2.2.1.4. $S \leftarrow S'$.

        2.2.2. (Use experiences) Do the following once or multiple times:

            2.2.2.1. (Replay) Sample a batch of experience $\mathcal{B}$ from the storage $\mathcal{D}$. Each entry is in the form of (S, A, R, S', D').

            2.2.2.2. (Calculate TD return) $U \leftarrow R + \gamma(1 - D')\max_a q(S', a; \mathbf{w})$ $((S, A, R, S', D') \in \mathcal{B})$.

            2.2.2.3. (Update value parameter) Update $\mathbf{w}$ to reduce $\frac{1}{|\mathcal{B}|}\sum_{(S,A,R,S',D')\in\mathcal{B}}\left[U - q(S, A; \mathbf{w})\right]^2$. (For example, $\mathbf{w} \leftarrow \mathbf{w} + \alpha\frac{1}{|\mathcal{B}|}\sum_{(S,A,R,S',D')\in\mathcal{B}}\left[U - q(S, A; \mathbf{w})\right]\nabla q(S, A; \mathbf{w})$.)

We can also implement the algorithm in a way that does not loop over episodes explicitly. Algorithm 6.8 shows the algorithm without looping over episodes explicitly. Although Algo. 6.8 has a different controlling structure, the results are equivalent.

Algorithm 6.8 DQN policy optimization with experience replay (without looping over episodes explicitly).

Parameters: parameters for experience replay (for example, the capacity of the storage, number of replayed samples in each batch), and other parameters (such as the optimizer, and the discount factor).

1. Initialize:

    1.1. (Initialize network parameter) Initialize $\mathbf{w}$.

    1.2. (Initialize experience storage) $\mathcal{D} \leftarrow \varnothing$.

    1.3. (Initialize state) Choose the initial state S.

2. Loop:

2.1. (Collect experiences) Do the following once or multiple times:

    2.1.1. (Decide) Use the policy derived from $q(S, \cdot; \mathbf{w})$ (for example, ε-greedy policy) to determine the action A.

    2.1.2. (Sample) Execute the action A, and then observe the reward R, the next state S', and the indicator of episode end D'.

    2.1.3. Save the experience (S, A, R, S', D') in the storage $\mathcal{D}$.

    2.1.4. If the episode does not end, $S \leftarrow S'$. Otherwise, choose an initial state S for the next episode.

2.2. (Use experiences) Do the following once or multiple times:

    2.2.1. (Replay) Sample a batch of experience $\mathcal{B}$ from the storage $\mathcal{D}$. Each entry is in the form of (S, A, R, S', D').

    2.2.2. (Calculate TD return) $U \leftarrow R + \gamma(1 - D')\max_a q(S', a; \mathbf{w})$ $((S, A, R, S', D') \in \mathcal{B})$.

    2.2.3. (Update value parameter) Update $\mathbf{w}$ to reduce $\frac{1}{|\mathcal{B}|}\sum_{(S,A,R,S',D')\in\mathcal{B}}\left[U - q(S, A; \mathbf{w})\right]^2$. (For example, $\mathbf{w} \leftarrow \mathbf{w} + \alpha\frac{1}{|\mathcal{B}|}\sum_{(S,A,R,S',D')\in\mathcal{B}}\left[U - q(S, A; \mathbf{w})\right]\nabla q(S, A; \mathbf{w})$.)

Experience replayer can be further classified to centralized replayer or distributed replayer according to how the experiences are stored.

- **Centralized replay**: The agent interacts with one environment only, and all experiences are stored in a storage.
- **Distributed replay**: Multiple agent workers interact with their own environments, and then store all experiences in a storage. This way uses more resources to generate experiences more quickly, so the convergence can be faster in terms of time-consuming.

From the aspect of replay, experience replay can be uniformly replay or prioritized replay.

- **Uniformly experience replay**: When the replayer selects experiences from the storage, all experiences can be selected with equal probability.
- **Prioritized Experience Replay** (**PER**) (Schaul, 2016): Designate a priority value for each experience. When the replayer selects experiences from the storage, some experiences with higher priority can be selected with larger probability.

A common way of PER is to select an experience i with priority p_i with the probability

$$\frac{p_i^\alpha}{\sum_\iota p_\iota^\alpha},$$

where the parameter $\alpha \geq 0$. $\alpha = 0$ corresponds to the uniform replay.

The ways to set priority for an experience include proportional priority, rank-based priority, and so on.

- **Proportional priority**: The priority for experience i is

$$p_i = |\delta_i| + \varepsilon,$$

where δ_i is the TD error, defined as $\delta_i \stackrel{\text{def}}{=} U_i - q(S_i, A_i; \mathbf{w})$ or $\delta_i \stackrel{\text{def}}{=} U_i - v(S_i; \mathbf{w})$, and ε is a pre-designated small positive.

- **Rank-based priority**: The priority for experience i is

$$p_i = \frac{1}{\text{rank}_i},$$

where rank_i is the rank of experience i sorting descend according to $|\delta_i|$, starting from 1.

Shortcoming of PER: Selecting experience requires lots of computation, and this computation can not be accelerated using GPU. In order to efficiently select experiences, developers usually use trees (such as sum-tree and binary indexed tree) to maintain priorities, which need some codes.

Interdisciplinary Reference 6.3
Data Structure: Sum Tree and Binary Indexed Tree

Consider a full binary tree of depth $n + 1$. **Sum tree** assigns a value for each node, such that the value of every non-leaf node equals the summation of values of its two children nodes. Each time the value of the node is changed, the values of its n parent nodes need to be changed, too. Sum tree can be used to calculate the summation of values of the first i nodes ($0 \le i \le 2^n$).

Binary Indexed Tree (BIT) is a storage-efficient implementation of sum tree. We can notice that, in the sum tree, the value of the right child of a node can be calculated by subtracting the value of the left child from the value of the node itself. Therefore, we may choose not to save the value of right child explicitly. When we need to use the value of right child, we can calculate it on the fly. Therefore, a tree with 2^n leaf nodes only needs to store 2^n values, and these values can be saved in an array. However, this improvement can only save half of storage, and may introduce more computation. So it is not always worth implementing.

Combining the distributed experience replay with PER results in distributed PER (Horgan, 2018).

6.4.2 Deep Q Learning with Target Network

In Sect. 6.2.2, we have known that, since TD learning uses bootstrapping, both the TD return and value estimate depend on the parameter $\mathbf{w}$. The changes in the parameter $\mathbf{w}$ will change both TD return and value estimates. During training, it will be unstable if the action value estimates are chasing something moving. Therefore, we need to use semi-gradient descent, which avoids calculating the gradients of TD return U_t

with respect to the parameter. One way to avoid the gradient calculation is to copy the parameter as $\mathbf{w}_{\text{target}}$, and then use $\mathbf{w}_{\text{target}}$ to calculate the TD error U_t.

Based on this idea, (Mnih, 2015) proposed the **target network**. The target network is a neural network with the same network architecture as the original network, which is called **evaluation network**. During training, we use the target network to calculate TD return and use the TD return as the learning target. Updating will only update the parameters of the evaluation network, and will not update the parameters of the target network. In this way, the target of learning will be relatively constant. After some iterations, we can assign the parameters of the evaluation network to the target network, so the target network can be updated too. Since the target network does not change during some iterations, the learning is more stable. Therefore, using the target network has become a customary practice in RL.

Algorithm 6.9 shows the DQN algorithm with target network. Step 1.1 initializes the evaluation network and target network using the same parameters, while the initialization of the evaluation network should follow the practice in deep learning. Step 2.2.2.2 calculates the TD return using the target network only. Step 2.2.2.4 updates the target network.

Algorithm 6.9 DQN with experience replay and target network.

---

1. Initialize:

   1.1. (Initialize network parameter) Initialize the parameter of the evaluation network $\mathbf{w}$.
   Initialize the parameters of the target network $\mathbf{w}_{\text{target}} \leftarrow \mathbf{w}$.
   1.2. (Initialize experience storage) $\mathcal{D} \leftarrow \varnothing$.

2. For each episode:

   2.1. (Initialize state) Choose the initial state S.
   2.2. Loop until the episode ends:
      2.2.1. (Collect experiences) Do the following once or multiple times:
         2.2.1.1. (Decide) Use the policy derived from $q(S, \cdot; \mathbf{w})$ (for example, ε-greedy policy) to determine the action A.
         2.2.1.2. (Sample) Execute the action A, and then observe the reward R, the next state S', and the indicator of episode end D'.
         2.2.1.3. Save the experience (S, A, R, S', D') in the storage $\mathcal{D}$.
         2.2.1.4. $S \leftarrow S'$.
      2.2.2. (Use experiences) Do the following once or multiple times:
         2.2.2.1. (Replay) Sample a batch of experience $\mathcal{B}$ from the storage $\mathcal{D}$. Each entry is in the form of (S, A, R, S', D').

> **2.2.2.2.** (Calculate TD return) $U \leftarrow R + \gamma(1 - D')$ $\max_a q\left(S', a; \mathbf{w}_{\text{target}}\right) ((S, A, R, S', D') \in \mathcal{B})$.
>
> **2.2.2.3.** (Update action value parameter) Update $\mathbf{w}$ to reduce $\frac{1}{|\mathcal{B}|} \sum_{(S,A,R,S',D') \in \mathcal{B}} [U - q(S, A; \mathbf{w})]^2$. (For example, $\mathbf{w} \leftarrow \mathbf{w} + \alpha \frac{1}{|\mathcal{B}|} \sum_{(S,A,R,S',D') \in \mathcal{B}} [U - q(S, A; \mathbf{w})] \nabla q(S, A; \mathbf{w})$.)
>
> **2.2.2.4.** (Update target network) Under some condition (say, every several updates), update the parameters of target network $\mathbf{w}_{\text{target}} \leftarrow \left(1 - \alpha_{\text{target}}\right) \mathbf{w}_{\text{target}} + \alpha_{\text{target}} \mathbf{w}$.

Step 2.2.2.4 updates the target network using **Polyak average**. Polyak average introduces a learning rate α_{target}, and updates using $\mathbf{w}_{\text{target}} \leftarrow \left(1 - \alpha_{\text{target}}\right) \mathbf{w}_{\text{target}} + \alpha_{\text{target}} \mathbf{w}$. Specially, when $\alpha_{\text{target}} = 1$, Polyak average degrades to the simple assignment $\mathbf{w}_{\text{target}} \leftarrow \mathbf{w}$. For distributed learning, there are many workers who want to modify the target networks at the same time, so they will pick the learning rate $\alpha_{\text{target}} \in (0, 1)$.

Section 5.2.1 told us that some implementations calculate the mask $(1 - D')$ or the discounted mask $\gamma(1 - D')$ as intermediate results, since the indicator of episode end D' is usually used in the form of $\gamma(1 - D')$. If so, the experience in the storage can also be of the form $(S, A, R, S', 1 - D')$ or $(S, A, R, S', \gamma(1 - D'))$.

6.4.3 Double DQN

Section 5.3.3 mentioned that Q learning may introduce maximization bias, and double learning can help reduce such bias. The tabular version of double Q learning maintains two copies of action value estimates $q^{(0)}$ and $q^{(1)}$, and updates one of them in every update.

Applying double learning into DQN leads to the **Double Deep Q Network (Double DQN)** algorithm (Hasselt, 2015). Since DQN algorithm already has two networks, i.e. the evaluation network and the target network, Double DQN can still use these two networks. Each update chooses one network as the evaluation network to select an action, and uses another network as the target network to calculate TD return.

The implementation of Double DQN can be obtained by modifying Algo. 6.9: Just changing

$$U \leftarrow R + \gamma\left(1 - D'\right) \max_a q\left(S', a; \mathbf{w}_{\text{target}}\right)$$

to

$$U \leftarrow R + \gamma\left(1 - D'\right) q\left(S', \arg\max_a q(S', a; \mathbf{w}); \mathbf{w}_{\text{target}}\right)$$

completes the Double DQN algorithm.

6.4.4 Dueling DQN

First, let us define an important concept called "advantage". **Advantage** is the difference between action values and state values, i.e.

$$a(s, a) \stackrel{\text{def}}{=} q(s, a) - v(s), \quad s \in \mathcal{S}, a \in \mathcal{A}(s).$$

In some RL tasks, the difference among multiple advantages of different actions with the same state is much smaller than the difference among different state values. For those tasks, it is beneficial to learn state values. Accordingly, a network architecture called **dueling network** was proposed (Wang, 2015). Dueling network is still used to approximate the action values $q(\cdot, \cdot; \mathbf{w})$, but the action value network $q(\cdot, \cdot; \mathbf{w})$ is implemented as the summation of a state value network and an advantage network, i.e.

$$q(s, a; \mathbf{w}) = v(s; \mathbf{w}) + a(s, a; \mathbf{w}), \quad s \in \mathcal{S}, a \in \mathcal{A}(s),$$

where both $v(\cdot; \mathbf{w})$ and $a(\cdot, \cdot; \mathbf{w})$ may use part of elements of $\mathbf{w}$. During training, $v(\cdot; \mathbf{w})$ and $a(\cdot, \cdot; \mathbf{w})$ are trained jointly, and the training process has no difference compared to the training of a normal Q network.

Remarkably, there are an infinite number of ways to partition a set of action value estimates $q(s, a; \mathbf{w})$ ($s \in \mathcal{S}, a \in \mathcal{A}(s)$) into a set of state values $v(s; \mathbf{w})$ ($s \in \mathcal{S}$) and a set of advantages $a(s, a; \mathbf{w})$ ($s \in \mathcal{S}, a \in \mathcal{A}(s)$). Specifically, if $q(s, a; \mathbf{w})$ can be partitioned as the summation of $v(s; \mathbf{w})$ and $a(s, a; \mathbf{w})$, $q(s, a; \mathbf{w})$ can also be partitioned as the summation of $v(s; \mathbf{w}) + c(s)$ and $a(s, a; \mathbf{w}) - c(s)$, where $c(s)$ is a function only depends on the state s. In order to avoid unnecessary troubles in training, we usually design special network architecture on the advantage network so that the advantage is unique. Common ways include:

- Limit the advantage after partition (denoted as a_{duel}) so that its simple average over different actions is 0. That is, the partition result should satisfy

$$\sum_a a_{\text{duel}}(s, a; \mathbf{w}) = 0, \quad s \in \mathcal{S}.$$

This can be achieved by the following network structure:

$$a_{\text{duel}}(s, a; \mathbf{w}) = a(s, a; \mathbf{w}) - \frac{1}{|\mathcal{A}|} \sum_{a'} a(s, a'; \mathbf{w}), \quad s \in \mathcal{S}, a \in \mathcal{A}(s).$$

- Limit the advantage after partition a_{duel} so that its maximum value over different actions is 0. That is, the partition result should satisfy

$$\max_a a_{\text{duel}}(s, a; \mathbf{w}) = 0, \quad s \in \mathcal{S}.$$

This can be achieved by the following network structure:

$$a_{\text{duel}}(s, a; \mathbf{w}) = a(s, a; \mathbf{w}) - \frac{1}{|\mathcal{A}|}\max_{a'} a(s, a'; \mathbf{w}), \quad s \in \mathcal{S}, a \in \mathcal{A}(s).$$

At this point, we have learned four tricks of DQN: experience replay, target network, double learning, and dueling network. We always use experience replay and target network for DQN, but double learning and dueling network are selected for suitable tasks. Table 6.3 summarizes the tricks used by different algorithms.

Table 6.3　Tricks used by different algorithms.

algorithm	experience replay	target network	double learning	dueling network
DQN	✓	✓		
Double DQN	✓	✓	✓	
Dueling DQN	✓	✓		✓
Dueling Double DQN (D3QN)	✓	✓	✓	✓

6.5 Case Study: MountainCar

This section considers a classical control task MountainCar-v0 (Moore, 1990). As Fig. 6.3, a car wants to climb up to the top of the right mountain. On the horizontal axis, its position is within $[-1.2, 0.6]$, and its velocity is within $[-0.07, 0.07]$. The agent can enforce one action among three choices: push left, no push, and push right. The position and the velocity of the car, as well as the action, together determine the position and the velocity of the car in the next step. The episode ends successfully when the position of the car exceeds 0.5. If we reach the maximum number of steps of an episode before the car reaches the goal, the episode ends with failure. We want to reach the goal in the fewest possible steps. The task is solved when the average step to reach the goal in 100 successive episodes is ≤ 110.

In this task, simply pushing right can not make the car reaches the goal.

This section assumes that the agent does not know the dynamics of the environment. The dynamics of the environment is in fact as follows: Let X_t ($X_t \in [-1.2, 0.6]$) and V_t ($V_t \in [-0.07, 0.07]$) denote the position and the velocity of the car at step t ($t = 0, 1, 2, \ldots$) respectively. Let A_t ($A_t \in \{0, 1, 2\}$) denote the action at step t. In the beginning, $X_0 \in [-0.6, -0.4)$ and $V_0 = 0$. The dynamics from step t to step $t + 1$ is

$$X_{t+1} = \text{clip}(X_t + V_t, -1.2, 0.6)$$

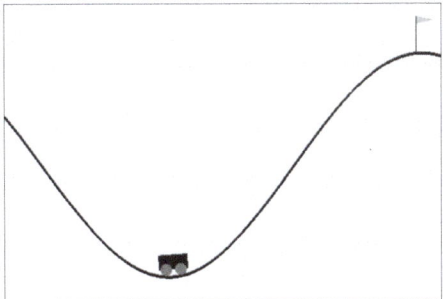

Fig. 6.3   The task MountainCar-v0.
This figure is adapted from (Moore, 1990).

$$V_{t+1} = \mathrm{clip}\big(V_t + 0.001(A_t - 1) - 0.0025\cos(3X_t), -0.07, 0.07\big),$$

where the function clip() limits the range of position and velocity:

$$\mathrm{clip}(x, x_{\min}, x_{\max}) = \begin{cases} x_{\min}, & x \le x_{\min} \\ x, & x_{\min} < x < x_{\max} \\ x_{\max}, & x \ge x_{\max}. \end{cases}$$

6.5.1 Use Environment

In Gym's task MountainCar-v0, the reward of each step is -1, and the episode reward is the negative of step number. Code 6.1 imports this environment, and checks its observation space, action space, the range of position and velocity. The maximum step number of an episode is 200. The threshold of episode reward for solving this task is -110. Code 6.2 tries to use this environment. The policy in Code 6.2 always pushes right. Result shows that always pushing right can not make the car reach the goal (Fig. 6.4).

Code 6.1   Import the environment of MountainCar-v0.
MountainCar-v0_SARSA_demo.ipynb

```
1  import gym
2  env = gym.make('MountainCar-v0')
3  logging.info('observation space = %s', env.observation_space)
4  logging.info('action space = %s', env.action_space)
5  logging.info('range of positions = %s ~ %s',
6          env.min_position, env.max_position)
7  logging.info('range of speeds = %s ~ %s', -env.max_speed, env.max_speed)
8  logging.info('goal position = %s', env.goal_position)
9  logging.info('reward threshold = %s', env.spec.reward_threshold)
10 logging.info('max episode steps = %s', env.spec.max_episode_steps)
```

Code 6.2   The agent that always pushes right.

`MountainCar-v0_SARSA_demo.ipynb`

```
1  positions, velocities = [], []
2  observation, _ = env.reset()
3  while True:
4      positions.append(observation[0])
5      velocities.append(observation[1])
6      next_observation, reward, terminated, truncated, _ = env.step(2)
7      if terminated or truncated:
8          break
9      observation = next_observation
10
11 if next_observation[0] > 0.5:
12     logging.info('succeed')
13 else:
14     logging.info('fail')
15
16 # plot positions and velocities
17 fig, ax = plt.subplots()
18 ax.plot(positions, label='position')
19 ax.plot(velocities, label='velocity')
20 ax.legend()
```

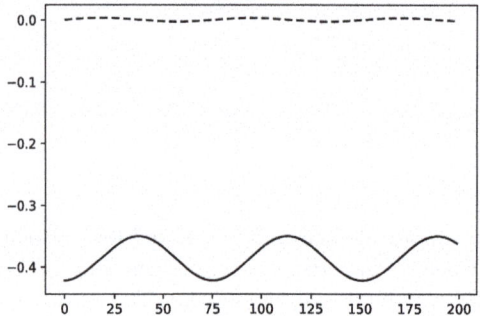

Fig. 6.4   Position and velocity of the car when it is always pushed right.
Solid curve shows position, and dash curve shows velocity.

6.5.2 Linear Approximation

This section considers policy optimization with linear model in the form of $q(s, a; \mathbf{w}) = \left[\mathbf{x}(s, a) \right]^{\top} \mathbf{w}$.

Interdisciplinary Reference 6.4
Feature Engineering: One-Hot Coding and Tile Coding

One-hot coding and tile-coding are two feature engineering methods that can discretize continuous inputs to finite features.

For example, in the task `MountainCar-v0`, the observation has two continuous elements: position and velocity. The simplest way to construct finite features from this continuous observation space is one-hot coding. As Fig. 6.5(a), we can partition the 2-dimension "position–velocity" space into grids. The total length of position axis is l_x, and the position length of each cell is δ_x, so there are $b_x = \lceil l_x \div \delta_x \rceil$ cells in the position axis. Similarly, the total length of velocity axis is l_v, the length of velocity length of each cell is δ_v, so there are $b_v = \lceil l_v \div \delta_v \rceil$ cells in the velocity axis. Therefore, there are $b_x b_v$ features in total. One-hot coding approximates the values such that all state–action pairs in the same cell will have the same feature values. We can reduce δ_x and δ_v to make the approximation more accurate, but it will also increase the number of features.

Tile coding tries to reduce the number of features without scarifying the precision. As Fig. 6.5(b), tile coding introduces multiple level of large grids. For tile coding with m layer ($m > 1$), the large grid in each layer is m times wide and m times high as the small grid in one-hot coding. For every two adjacent layers of larger grids, the difference of positions between these two layers in either dimension equals the length of a small cell. Given an arbitrary position-velocity pair, it will fall into one large cell in every layer. Therefore, we can conduct one-hot coding on the large grid, and each layer has $\approx b_x/m \times b_v/m$ features. All m layers together have $\approx b_x b_v/m$ features, which is much less than the feature numbers of native one-hot coding.

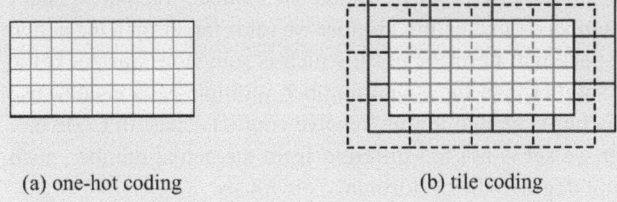

(a) one-hot coding   (b) tile coding

Fig. 6.5   One-hot coding and tile coding.
State spaces are in shadow.

The class `TileCoder` in Code 6.3 implements the tile coding. The constructor of the class `TileCoder` has two parameters: The parameter `layer_count` indicates the number of large-grid layers, the parameter `feature_count` indicates the number of features, i.e. the number of $x(s, a)$, which is also the number of elements in **w**. After an object of the class `TileCoder` is constructed, we can call this object to determine which features are activated for each input. The features that are activated are with the value 1, while the features that are not activated are with the value 0. When we call this object, the parameter floats needs to be a tuple whose elements are within the range $[0, 1]$, and the parameter ints needs to be a tuple of int values. The values of ints are not fed into the core of the tile coder. The return value will be a list of ints, which indicates the indices of activated features.

Code 6.3   Tile coding.
`MountainCar-v0_SARSA_demo.ipynb`

```python
class TileCoder:
    def __init__(self, layer_count, feature_count):
        self.layer_count = layer_count
        self.feature_count = feature_count
        self.codebook = {}

    def get_feature(self, codeword):
        if codeword in self.codebook:
            return self.codebook[codeword]
        count = len(self.codebook)
        if count >= self.feature_count:  # resolve conflicts
            return hash(codeword) % self.feature_count
        self.codebook[codeword] = count
        return count

    def __call__(self, floats=(), ints=()):
        dim = len(floats)
        scaled_floats = tuple(f * (self.layer_count ** 2) for f in floats)
        features = []
        for layer in range(self.layer_count):
            codeword = (layer,) + tuple(
                    int((f + (1 + dim * i) * layer) / self.layer_count)
                    for i, f in enumerate(scaled_floats)) + ints
            feature = self.get_feature(codeword)
            features.append(feature)
        return features
```

When using tile coding, we can just roughly estimate the number of features, rather than calculate the exact number. If the feature number we set is larger than the actual number, some allocated space will never be used, which is somehow wasted. If the feature number we set is smaller than the actual number, multiple cells need to use the same feature, which is implemented in the "resolve conflicts" part in Code 6.3. When the feature number we set is not too different from the actual number, such waste and conflicts will not degrade the performance obviously.

Current academia tends to use neural networks, rather than tile coding, to construct features. The tile coding is rarely used now.

When we use Code 6.3 for the task `MountainCar-v0`, we may choose 8-layer tile coding. If so, the layer 0 will have $8 \times 8 = 64$ large cells, since the length of each large cell is exactly 1/8 of total length, so each axis has 8 large cells. Each of the remaining $8 - 1 = 7$ layers has $(8 + 1) \times (8 + 1) = 81$ large cells, since offsets in those layers require $8 + 1 = 9$ large cells to cover the range of each axis. In total, all 8 layers have $64 + 7 \times 81 = 631$ cells. Additionally, the action space is of size 3, so there are $631 \times 3 = 1893$ features in total.

The class `SARSAAgent` in Code 6.4 and the class `SARSALambdaAgent` in Code 6.5 are the agent classes of function approximation SARSA agent and SARSA(λ) agent respectively. Both of them use linear approximation and tile coding. We once again use Code 5.3 to train (with modification of terminal condition when necessary) and Code 1.4 to test.

Code 6.4   SARSA agent with function approximation.

MountainCar-v0_SARSA_demo.ipynb

```
class SARSAAgent:
    def __init__(self, env):
        self.action_n = env.action_space.n
        self.obs_low = env.observation_space.low
        self.obs_scale = env.observation_space.high - \
                env.observation_space.low
        self.encoder = TileCoder(8, 1896)
        self.w = np.zeros(self.encoder.feature_count)
        self.gamma = 1.
        self.learning_rate = 0.03

    def encode(self, observation, action):
        states = tuple((observation - self.obs_low) / self.obs_scale)
        actions = (action,)
        return self.encoder(states, actions)

    def get_q(self, observation, action):  # action value
        features = self.encode(observation, action)
        return self.w[features].sum()

    def reset(self, mode=None):
        self.mode = mode
        if self.mode == 'train':
            self.trajectory = []

    def step(self, observation, reward, terminated):
        if self.mode == 'train' and np.random.rand() < 0.001:
            action = np.random.randint(self.action_n)
        else:
            qs = [self.get_q(observation, action) for action in
                    range(self.action_n)]
            action = np.argmax(qs)
        if self.mode == 'train':
            self.trajectory += [observation, reward, terminated, action]
            if len(self.trajectory) >= 8:
                self.learn()
        return action

    def close(self):
        pass

    def learn(self):
        observation, _, _, action, next_observation, reward, terminated, \
                next_action = self.trajectory[-8:]
        target = reward + (1. - terminated) * self.gamma * \
                self.get_q(next_observation, next_action)
        td_error = target - self.get_q(observation, action)
        features = self.encode(observation, action)
        self.w[features] += (self.learning_rate * td_error)


agent = SARSAAgent(env)
```

Code 6.5   SARSA(λ) agent with function approximation.

MountainCar-v0_SARSAlambda.ipynb

```
 1  class SARSALambdaAgent:
 2      def __init__(self, env):
 3          self.action_n = env.action_space.n
 4          self.obs_low = env.observation_space.low
 5          self.obs_scale = env.observation_space.high - \
 6                  env.observation_space.low
 7          self.encoder = TileCoder(8, 1896)
 8          self.w = np.zeros(self.encoder.feature_count)
 9          self.gamma = 1.
10          self.learning_rate = 0.03
11
12      def encode(self, observation, action):
13          states = tuple((observation - self.obs_low) / self.obs_scale)
14          actions = (action,)
15          return self.encoder(states, actions)
16
17      def get_q(self, observation, action):  # action value
18          features = self.encode(observation, action)
19          return self.w[features].sum()
20
21      def reset(self, mode=None):
22          self.mode = mode
23          if self.mode == 'train':
24              self.trajectory = []
25              self.z = np.zeros(self.encoder.feature_count)  # eligibility trace
26
27      def step(self, observation, reward, terminated):
28          if self.mode == 'train' and np.random.rand() < 0.001:
29              action = np.random.randint(self.action_n)
30          else:
31              qs = [self.get_q(observation, action) for action in
32                      range(self.action_n)]
33              action = np.argmax(qs)
34          if self.mode == 'train':
35              self.trajectory += [observation, reward, terminated, action]
36              if len(self.trajectory) >= 8:
37                  self.learn()
38          return action
39
40      def close(self):
41          pass
42
43      def learn(self):
44          observation, _, _, action, next_observation, reward, terminated, \
45                  next_action = self.trajectory[-8:]
46          target = reward + (1. - terminated) * self.gamma * \
47                  self.get_q(next_observation, next_action)
48          td_error = target - self.get_q(observation, action)
49
50          # update replace trace
51          self.z *= (self.gamma * 0.9)  # 0.9 is the lambda value
52          features = self.encode(observation, action)
53          self.z[features] = 1.
54
55          self.w += (self.learning_rate * td_error * self.z)
56
57
58  agent = SARSALambdaAgent(env)
```

For the task `MountainCar-v0`, SARSA(λ) algorithm outperforms SARSA algorithm. SARSA(λ) can solve the task in around 100 episodes. In fact, SARSA(λ) is one of the best algorithms to solve this task.

6.5.3 DQN and its Variants

Starting from this subsection, we will implement DRL algorithms. We will use TensorFlow and/or PyTorch in the implementation. Their CPU versions suffice.

The GitHub of this book provides the method to install TensorFlow and/or PyTorch. To put it simply, for Windows 10/11 users, they need to first install the latest version of Visual Studio, and execute the following commands in Anaconda Prompt as an administrator:

```
pip install --upgrade tensorflow tensorflow_probability
conda install pytorch cpuonly -c pytorch
```

Next, we will use DQN and its variants to find the optimal policy.

First, let us implement experience replay. The class `DQNReplayer` in Code 6.6 implements the experience replay. Its constructor has a parameter `capacity`, which is an `int` value to show how many entries can be saved in the storage. When the number of entries in the storage reaches `capacity`, upcoming experience entries will overwrite existing experience entries.

Code 6.6   Experience replayer.
`MountainCar-v0_DQN_tf.ipynb`

```
class DQNReplayer:
    def __init__(self, capacity):
        self.memory = pd.DataFrame(index=range(capacity), columns=
                ['state', 'action', 'reward', 'next_state', 'terminated'])
        self.i = 0
        self.count = 0
        self.capacity = capacity

    def store(self, *args):
        self.memory.loc[self.i] = np.asarray(args, dtype=object)
        self.i = (self.i + 1) % self.capacity
        self.count = min(self.count + 1, self.capacity)

    def sample(self, size):
        indices = np.random.choice(self.count, size=size)
        return (np.stack(self.memory.loc[indices, field]) for field in
                self.memory.columns)
```

Next, let us consider the algorithms with function approximation. We use the vector version of action values $q(\cdot; \mathbf{w})$, and the form of approximation function is a fully connected network. Both Codes 6.7 and 6.8 implement the DQN agent with target network. Code 6.7 is based on TensorFlow 2, and Code 6.8 is based on PyTorch. They have the same functionality, so we can use either of them. It is also okay if you want to use other deep learning packages such as Keras. But if you want to use packages other than TensorFlow 2 or PyTorch, you need to implement the agent by yourself. The codes for training and testing are still Codes 5.3 and 1.4 respectively.

Code 6.7   DQN agent with target network (TensorFlow version).

MountainCar-v0_DQN_tf.ipynb

```python
class DQNAgent:
    def __init__(self, env):
        self.action_n = env.action_space.n
        self.gamma = 0.99

        self.replayer = DQNReplayer(10000)

        self.evaluate_net = self.build_net(
                input_size=env.observation_space.shape[0],
                hidden_sizes=[64, 64], output_size=self.action_n)
        self.target_net = models.clone_model(self.evaluate_net)

    def build_net(self, input_size, hidden_sizes, output_size):
        model = keras.Sequential()
        for layer, hidden_size in enumerate(hidden_sizes):
            kwargs = dict(input_shape=(input_size,)) if not layer else {}
            model.add(layers.Dense(units=hidden_size,
                    activation=nn.relu, **kwargs))
        model.add(layers.Dense(units=output_size))
        optimizer = optimizers.Adam(0.001)
        model.compile(loss=losses.mse, optimizer=optimizer)
        return model

    def reset(self, mode=None):
        self.mode = mode
        if self.mode == 'train':
            self.trajectory = []
            self.target_net.set_weights(self.evaluate_net.get_weights())

    def step(self, observation, reward, terminated):
        if self.mode == 'train' and np.random.rand() < 0.001:
            # epsilon-greedy policy in train mode
            action = np.random.randint(self.action_n)
        else:
            qs = self.evaluate_net.predict(observation[np.newaxis], verbose=0)
            action = np.argmax(qs)
        if self.mode == 'train':
            self.trajectory += [observation, reward, terminated, action]
            if len(self.trajectory) >= 8:
                state, _, _, act, next_state, reward, terminated, _ = \
                        self.trajectory[-8:]
                self.replayer.store(state, act, reward, next_state, terminated)
            if self.replayer.count >= self.replayer.capacity * 0.95:
                    # skip first few episodes for speed
                self.learn()
        return action

    def close(self):
        pass

    def learn(self):
        # replay
        states, actions, rewards, next_states, terminateds = \
                self.replayer.sample(1024)

        # update value net
        next_qs = self.target_net.predict(next_states, verbose=0)
        next_max_qs = next_qs.max(axis=-1)
        us = rewards + self.gamma * (1. - terminateds) * next_max_qs
        targets = self.evaluate_net.predict(states, verbose=0)
        targets[np.arange(us.shape[0]), actions] = us
        self.evaluate_net.fit(states, targets, verbose=0)
```

```
63
64
65  agent = DQNAgent(env)
```

Code 6.8   DQN agent with target network (PyTorch version).
MountainCar-v0_DQN_torch.ipynb

```
1   class DQNAgent:
2       def __init__(self, env):
3           self.action_n = env.action_space.n
4           self.gamma = 0.99
5
6           self.replayer = DQNReplayer(10000)
7
8           self.evaluate_net = self.build_net(
9                   input_size=env.observation_space.shape[0],
10                  hidden_sizes=[64, 64], output_size=self.action_n)
11          self.optimizer = optim.Adam(self.evaluate_net.parameters(), lr=0.001)
12          self.loss = nn.MSELoss()
13
14      def build_net(self, input_size, hidden_sizes, output_size):
15          layers = []
16          for input_size, output_size in zip(
17                  [input_size,] + hidden_sizes, hidden_sizes + [output_size,]):
18              layers.append(nn.Linear(input_size, output_size))
19              layers.append(nn.ReLU())
20          layers = layers[:-1]
21          model = nn.Sequential(*layers)
22          return model
23
24      def reset(self, mode=None):
25          self.mode = mode
26          if self.mode == 'train':
27              self.trajectory = []
28              self.target_net = copy.deepcopy(self.evaluate_net)
29
30      def step(self, observation, reward, terminated):
31          if self.mode == 'train' and np.random.rand() < 0.001:
32              # epsilon-greedy policy in train mode
33              action = np.random.randint(self.action_n)
34          else:
35              state_tensor = torch.as_tensor(observation,
36                      dtype=torch.float).squeeze(0)
37              q_tensor = self.evaluate_net(state_tensor)
38              action_tensor = torch.argmax(q_tensor)
39              action = action_tensor.item()
40          if self.mode == 'train':
41              self.trajectory += [observation, reward, terminated, action]
42              if len(self.trajectory) >= 8:
43                  state, _, _, act, next_state, reward, terminated, _ = \
44                          self.trajectory[-8:]
45                  self.replayer.store(state, act, reward, next_state, terminated)
46              if self.replayer.count >= self.replayer.capacity * 0.95:
47                      # skip first few episodes for speed
48                  self.learn()
49          return action
50
51      def close(self):
52          pass
53
54      def learn(self):
55          # replay
56          states, actions, rewards, next_states, terminateds = \
57                  self.replayer.sample(1024)
58          state_tensor = torch.as_tensor(states, dtype=torch.float)
```

```
59    action_tensor = torch.as_tensor(actions, dtype=torch.long)
60    reward_tensor = torch.as_tensor(rewards, dtype=torch.float)
61    next_state_tensor = torch.as_tensor(next_states, dtype=torch.float)
62    terminated_tensor = torch.as_tensor(terminateds, dtype=torch.float)
63
64    # update value net
65    next_q_tensor = self.target_net(next_state_tensor)
66    next_max_q_tensor, _ = next_q_tensor.max(axis=-1)
67    target_tensor = reward_tensor + self.gamma * \
68            (1. - terminated_tensor) * next_max_q_tensor
69    pred_tensor = self.evaluate_net(state_tensor)
70    q_tensor = pred_tensor.gather(1, action_tensor.unsqueeze(1)).squeeze(1)
71    loss_tensor = self.loss(target_tensor, q_tensor)
72    self.optimizer.zero_grad()
73    loss_tensor.backward()
74    self.optimizer.step()
75
76
77 agent = DQNAgent(env)
```

The running results of DRL implementation can be found on GitHub. Without special notice, the results in GitHub are generated by CPU version of TensorFlow or PyTorch. Using GPU version of TensorFlow or PyTorch will probably lead to different running results.

The implementation of double DQN can be found in Code 6.9 (TensorFlow version) or Code 6.10 (PyTorch version).

Code 6.9   **Double DQN agent (TensorFlow version).**
MountainCar-v0_DoubleDQN_tf.ipynb

```
1  class DoubleDQNAgent:
2      def __init__(self, env):
3          self.action_n = env.action_space.n
4          self.gamma = 0.99
5
6          self.replayer = DQNReplayer(10000)
7
8          self.evaluate_net = self.build_net(
9                  input_size=env.observation_space.shape[0],
10                 hidden_sizes=[64, 64], output_size=self.action_n)
11         self.target_net = models.clone_model(self.evaluate_net)
12
13     def build_net(self, input_size, hidden_sizes, output_size):
14         model = keras.Sequential()
15         for layer, hidden_size in enumerate(hidden_sizes):
16             kwargs = dict(input_shape=(input_size,)) if not layer else {}
17             model.add(layers.Dense(units=hidden_size,
18                     activation=nn.relu, **kwargs))
19         model.add(layers.Dense(units=output_size))
20         optimizer = optimizers.Adam(0.001)
21         model.compile(loss=losses.mse, optimizer=optimizer)
22         return model
23
24     def reset(self, mode=None):
25         self.mode = mode
26         if self.mode == 'train':
27             self.trajectory = []
28             self.target_net.set_weights(self.evaluate_net.get_weights())
29
30     def step(self, observation, reward, terminated):
31         if self.mode == 'train' and np.random.rand() < 0.001:
32             # epsilon-greedy policy in train mode
33             action = np.random.randint(self.action_n)
```

```
34        else:
35            qs = self.evaluate_net.predict(observation[np.newaxis], verbose=0)
36            action = np.argmax(qs)
37        if self.mode == 'train':
38            self.trajectory += [observation, reward, terminated, action]
39            if len(self.trajectory) >= 8:
40                state, _, _, act, next_state, reward, terminated, _ = \
41                        self.trajectory[-8:]
42                self.replayer.store(state, act, reward, next_state, terminated)
43                if self.replayer.count >= self.replayer.capacity * 0.95:
44                        # skip first few episodes for speed
45                    self.learn()
46        return action
47
48    def close(self):
49        pass
50
51    def learn(self):
52        # replay
53        states, actions, rewards, next_states, terminateds = \
54                self.replayer.sample(1024)
55
56        # update value net
57        next_eval_qs = self.evaluate_net.predict(next_states, verbose=0)
58        next_actions = next_eval_qs.argmax(axis=-1)
59        next_qs = self.target_net.predict(next_states, verbose=0)
60        next_max_qs = next_qs[np.arange(next_qs.shape[0]), next_actions]
61        us = rewards + self.gamma * next_max_qs * (1. - terminateds)
62        targets = self.evaluate_net.predict(states, verbose=0)
63        targets[np.arange(us.shape[0]), actions] = us
64        self.evaluate_net.fit(states, targets, verbose=0)
65
66
67 agent = DoubleDQNAgent(env)
```

Code 6.10   Double DQN agent (PyTorch version).

MountainCar-v0_DoubleDQN_torch.ipynb

```
1  class DoubleDQNAgent:
2      def __init__(self, env):
3          self.action_n = env.action_space.n
4          self.gamma = 0.99
5
6          self.replayer = DQNReplayer(10000)
7
8          self.evaluate_net = self.build_net(
9                  input_size=env.observation_space.shape[0],
10                 hidden_sizes=[64, 64], output_size=self.action_n)
11         self.optimizer = optim.Adam(self.evaluate_net.parameters(), lr=0.001)
12         self.loss = nn.MSELoss()
13
14     def build_net(self, input_size, hidden_sizes, output_size):
15         layers = []
16         for input_size, output_size in zip(
17                 [input_size,] + hidden_sizes, hidden_sizes + [output_size,]):
18             layers.append(nn.Linear(input_size, output_size))
19             layers.append(nn.ReLU())
20         layers = layers[:-1]
21         model = nn.Sequential(*layers)
22         return model
23
24     def reset(self, mode=None):
25         self.mode = mode
26         if self.mode == 'train':
27             self.trajectory = []
```

```python
28          self.target_net = copy.deepcopy(self.evaluate_net)
29
30      def step(self, observation, reward, terminated):
31          if self.mode == 'train' and np.random.rand() < 0.001:
32              # epsilon-greedy policy in train mode
33              action = np.random.randint(self.action_n)
34          else:
35              state_tensor = torch.as_tensor(observation,
36                      dtype=torch.float).reshape(1, -1)
37              q_tensor = self.evaluate_net(state_tensor)
38              action_tensor = torch.argmax(q_tensor)
39              action = action_tensor.item()
40          if self.mode == 'train':
41              self.trajectory += [observation, reward, terminated, action]
42              if len(self.trajectory) >= 8:
43                  state, _, _, act, next_state, reward, terminated, _ = \
44                          self.trajectory[-8:]
45                  self.replayer.store(state, act, reward, next_state, terminated)
46                  if self.replayer.count >= self.replayer.capacity * 0.95:
47                      # skip first few episodes for speed
48                      self.learn()
49          return action
50
51      def close(self):
52          pass
53
54      def learn(self):
55          # replay
56          states, actions, rewards, next_states, terminateds = \
57                  self.replayer.sample(1024)
58          state_tensor = torch.as_tensor(states, dtype=torch.float)
59          action_tensor = torch.as_tensor(actions, dtype=torch.long)
60          reward_tensor = torch.as_tensor(rewards, dtype=torch.float)
61          next_state_tensor = torch.as_tensor(next_states, dtype=torch.float)
62          terminated_tensor = torch.as_tensor(terminateds, dtype=torch.float)
63
64          # update value net
65          next_eval_q_tensor = self.evaluate_net(next_state_tensor)
66          next_action_tensor = next_eval_q_tensor.argmax(axis=-1)
67          next_q_tensor = self.target_net(next_state_tensor)
68          next_max_q_tensor = torch.gather(next_q_tensor, 1,
69                  next_action_tensor.unsqueeze(1)).squeeze(1)
70          target_tensor = reward_tensor + self.gamma * \
71                  (1. - terminated_tensor) * next_max_q_tensor
72          pred_tensor = self.evaluate_net(state_tensor)
73          q_tensor = pred_tensor.gather(1, action_tensor.unsqueeze(1)).squeeze(1)
74          loss_tensor = self.loss(target_tensor, q_tensor)
75          self.optimizer.zero_grad()
76          loss_tensor.backward()
77          self.optimizer.step()
78
79
80  agent = DoubleDQNAgent(env)
```

At last, let us implement Dueling DQN. Either Code 6.11 or Code 6.12 implements the dueling network, which is constrained such that the simple average is 0. Code 6.13 or Code 6.14 implements the agent.

Code 6.11 Dueling network (TensorFlow version).

`MountainCar-v0_DuelDQN_tf.ipynb`

```
class DuelNet(keras.Model):
    def __init__(self, input_size, output_size):
        super().__init__()
        self.common_net = keras.Sequential([
                layers.Dense(64, input_shape=(input_size,),
                activation=nn.relu)])
        self.advantage_net = keras.Sequential([
                layers.Dense(32, input_shape=(64,), activation=nn.relu),
                layers.Dense(output_size)])
        self.v_net = keras.Sequential([
                layers.Dense(32, input_shape=(64,), activation=nn.relu),
                layers.Dense(1)])

    def call(self, s):
        h = self.common_net(s)
        adv = self.advantage_net(h)
        adv = adv - tf.math.reduce_mean(adv, axis=1, keepdims=True)
        v = self.v_net(h)
        q = v + adv
        return q
```

Code 6.12 Dueling network (PyTorch version).

`MountainCar-v0_DuelDQN_torch.ipynb`

```
class DuelNet(nn.Module):
    def __init__(self, input_size, output_size):
        super().__init__()
        self.common_net = nn.Sequential(nn.Linear(input_size, 64), nn.ReLU())
        self.advantage_net = nn.Sequential(nn.Linear(64, 32), nn.ReLU(),
                nn.Linear(32, output_size))
        self.v_net = nn.Sequential(nn.Linear(64, 32), nn.ReLU(),
                nn.Linear(32, 1))

    def forward(self, s):
        h = self.common_net(s)
        adv = self.advantage_net(h)
        adv = adv - adv.mean(1).unsqueeze(1)
        v = self.v_net(h)
        q = v + adv
        return q
```

Code 6.13 Dueling DQN agent (TensorFlow version).

`MountainCar-v0_DuelDQN_tf.ipynb`

```
class DuelDQNAgent:
    def __init__(self, env):
        self.action_n = env.action_space.n
        self.gamma = 0.99

        self.replayer = DQNReplayer(10000)

        self.evaluate_net = self.build_net(
                input_size=env.observation_space.shape[0],
                output_size=self.action_n)
        self.target_net = self.build_net(
                input_size=env.observation_space.shape[0],
                output_size=self.action_n)
```

```
14
15    def build_net(self, input_size, output_size):
16        net = DuelNet(input_size=input_size, output_size=output_size)
17        optimizer = optimizers.Adam(0.001)
18        net.compile(loss=losses.mse, optimizer=optimizer)
19        return net
20
21    def reset(self, mode=None):
22        self.mode = mode
23        if self.mode == 'train':
24            self.trajectory = []
25            self.target_net.set_weights(self.evaluate_net.get_weights())
26
27    def step(self, observation, reward, terminated):
28        if self.mode == 'train' and np.random.rand() < 0.001:
29            # epsilon-greedy policy in train mode
30            action = np.random.randint(self.action_n)
31        else:
32            qs = self.evaluate_net.predict(observation[np.newaxis], verbose=0)
33            action = np.argmax(qs)
34        if self.mode == 'train':
35            self.trajectory += [observation, reward, terminated, action]
36            if len(self.trajectory) >= 8:
37                state, _, _, act, next_state, reward, terminated, _ = \
38                        self.trajectory[-8:]
39                self.replayer.store(state, act, reward, next_state, terminated)
40                if self.replayer.count >= self.replayer.capacity * 0.95:
41                        # skip first few episodes for speed
42                    self.learn()
43        return action
44
45    def close(self):
46        pass
47
48    def learn(self):
49        # replay
50        states, actions, rewards, next_states, terminateds = \
51                self.replayer.sample(1024)
52
53        # update value net
54        next_eval_qs = self.evaluate_net.predict(next_states, verbose=0)
55        next_actions = next_eval_qs.argmax(axis=-1)
56        next_qs = self.target_net.predict(next_states, verbose=0)
57        next_max_qs = next_qs[np.arange(next_qs.shape[0]), next_actions]
58        us = rewards + self.gamma * next_max_qs * (1. - terminateds)
59        targets = self.evaluate_net.predict(states, verbose=0)
60        targets[np.arange(us.shape[0]), actions] = us
61        self.evaluate_net.fit(states, targets, verbose=0)
62
63
64  agent = DuelDQNAgent(env)
```

Code 6.14   Dueling DQN agent (PyTorch version).
MountainCar-v0_DuelDQN_torch.ipynb

```
1  class DuelDQNAgent:
2      def __init__(self, env):
3          self.action_n = env.action_space.n
4          self.gamma = 0.99
5
6          self.replayer = DQNReplayer(10000)
7
8          self.evaluate_net = DuelNet(input_size=env.observation_space.shape[0],
9                  output_size=self.action_n)
10         self.optimizer = optim.Adam(self.evaluate_net.parameters(), lr=0.001)
```

```
11          self.loss = nn.MSELoss()
12
13      def reset(self, mode=None):
14          self.mode = mode
15          if self.mode == 'train':
16              self.trajectory = []
17              self.target_net = copy.deepcopy(self.evaluate_net)
18
19      def step(self, observation, reward, terminated):
20          if self.mode == 'train' and np.random.rand() < 0.001:
21              # epsilon-greedy policy in train mode
22              action = np.random.randint(self.action_n)
23          else:
24              state_tensor = torch.as_tensor(observation,
25                      dtype=torch.float).reshape(1, -1)
26              q_tensor = self.evaluate_net(state_tensor)
27              action_tensor = torch.argmax(q_tensor)
28              action = action_tensor.item()
29          if self.mode == 'train':
30              self.trajectory += [observation, reward, terminated, action]
31              if len(self.trajectory) >= 8:
32                  state, _, _, act, next_state, reward, terminated, _ = \
33                          self.trajectory[-8:]
34                  self.replayer.store(state, act, reward, next_state, terminated)
35              if self.replayer.count >= self.replayer.capacity * 0.95:
36                      # skip first few episodes for speed
37                  self.learn()
38          return action
39
40      def close(self):
41          pass
42
43      def learn(self):
44          # replay
45          states, actions, rewards, next_states, terminateds = \
46                  self.replayer.sample(1024)
47          state_tensor = torch.as_tensor(states, dtype=torch.float)
48          action_tensor = torch.as_tensor(actions, dtype=torch.long)
49          reward_tensor = torch.as_tensor(rewards, dtype=torch.float)
50          next_state_tensor = torch.as_tensor(next_states, dtype=torch.float)
51          terminated_tensor = torch.as_tensor(terminateds, dtype=torch.float)
52
53          # update value net
54          next_eval_q_tensor = self.evaluate_net(next_state_tensor)
55          next_action_tensor = next_eval_q_tensor.argmax(axis=-1)
56          next_q_tensor = self.target_net(next_state_tensor)
57          next_max_q_tensor = torch.gather(next_q_tensor, 1,
58                  next_action_tensor.unsqueeze(1)).squeeze(1)
59          target_tensor = reward_tensor + self.gamma * \
60                  (1. - terminated_tensor) * next_max_q_tensor
61          pred_tensor = self.evaluate_net(state_tensor)
62          unsqueeze_tensor = action_tensor.unsqueeze(1)
63          q_tensor = pred_tensor.gather(1, action_tensor.unsqueeze(1)).squeeze(1)
64          loss_tensor = self.loss(target_tensor, q_tensor)
65          self.optimizer.zero_grad()
66          loss_tensor.backward()
67          self.optimizer.step()
68
69
70  agent = DuelDQNAgent(env)
```

We can still use Code 5.3 to train (with optional parameter changes) and Code 1.4 to test. The results of DQN algorithm and its variants should underperform the SARSA(λ) algorithm.

6.6 Summary

- Function approximation method uses function with parameters to approximate the value estimates (including state value estimates and action value estimates).
- Common forms of function approximation include linear approximation and neural networks.
- Tabular algorithms can be viewed as special cases of function approximation.
- Stochastic Gradient Descent (SGD) is used for the case without bootstrapping, while semi-gradient descent is used for the case with bootstrapping. Semi-gradient descent only calculates the gradient of value estimates with respect to the parameters, but does not calculate the gradient of TD return with respect to the parameters.
- Function approximation can be used together with eligibility trace. There is one-on-one mapping between the trace parameters and function parameters.
- The off-policy algorithm with bootstrapping and function approximation may not converge.
- Deep Q Network (DQN) is an off-policy DRL algorithm. DQN uses experience replay and target network.
- Experience replay first saves the experience, and then selects the experience randomly during learning. Experience replay makes the data distribution stable. Uniform replay selects experience with equal probability, while prioritized replay selects the experience with large absolute values of TD error with larger probability.
- Target network implements the semi-gradient descent.
- Double DQN randomly picks one network as target network to calculate TD return, and another network is used as evaluation network.
- Advantage is defined as:

$$a(s, a) \stackrel{\text{def}}{=} q(s, a) - v(s), \quad s \in \mathcal{S}, a \in \mathcal{A}.$$

- Dueling DQN partitions the action value network as the summation of state value network and advantage network.

6.7 Exercises

6.7.1 Multiple Choices

6.1 On bootstrapping, choose the correct one: (   )

A. Stochastic gradient descent does NOT use bootstrapping.
B. Semi-gradient descent does NOT use bootstrapping.
C. Semi-gradient descent with eligibility trace does NOT use bootstrapping.

6.2 On the convergence, choose the correct one: (   )

A. Q learning algorithm with linear function approximation and suitable learning rate sequence will certainly converge.
B. Q learning algorithm with non-linear function approximation and suitable learning rate sequence will certainly converge.
C. SARSA algorithm with non-linear function approximation and suitable learning rate sequence will certainly converge.

6.3 On DQN, choose the correct one: (   )

A. DQN uses neural network to improve the sample efficiency.
B. DQN uses experience replay to improve the sample efficiency.
C. DQN uses target network to improve the sample efficiency.

6.4 On the definition of advantage, choose the correct one: (   )

A. $a(s, a) = v(s) \div q(s, a)$.
B. $a(s, a) = q(s, a) \div v(s)$.
C. $a(s, a) = q(s, a) - v(s)$.

6.7.2 Programming

6.5 Solve the task `CartPole-v1` using DQN algorithm.

6.7.3 Mock Interview

6.6 What are the advantages and disadvantages of experience replay?

6.7 Why does DQN use target network?

Chapter 7
PG: Policy Gradient

This chapter covers

- function approximation for policy
- Policy Gradient (PG) theorem
- relationship between PG and maximum likelihood estimate
- Vanilla Policy Gradient (VPG) algorithm
- baseline
- VPG with importance sampling

The policy optimization algorithms in Chaps. 2–6 use the optimal value estimates to find the optimal policy, so those algorithms are called **optimal value algorithm**. However, estimating optimal values are not necessary for policy optimization. This chapter introduces a policy optimization method that does not need to estimate optimal values. This method uses a parametric function to approximate the optimal policy, and then adjusts the policy parameters to maximize the episode return. Since updating policy parameters relies on the calculation of the gradient of return with respect to the policy parameters, such method is called **Policy Gradient (PG)** method.

7.1 Theory of PG

The intuition of policy gradient method includes:

- Use parametric function to approximate the optimal policy.
- Calculate the policy gradient and use it to update the policy parameter.

© The Author(s), under exclusive license to Springer Nature Singapore Pte Ltd. 2024
Z. Xiao, *Reinforcement Learning*, https://doi.org/10.1007/978-981-19-4933-3_7

We will learn these intuitions in this section.

7.1.1 Function Approximation for Policy

Intuitively, we can use a parametric function $\pi(a|s; \boldsymbol{\theta})$ to approximate the optimal policy $\pi_*(a|s)$.

In the previous chapter, we use parametric functions to approximate values. Compared to the function approximation for values, the function approximation for policy has additional considerations. Specifically, every policy π should satisfy:

- $\pi(a|s) \geq 0$ holds for all $s \in \mathcal{S}, a \in \mathcal{A}(s)$.
- $\sum_a \pi(a|s) = 1$ holds for all $s \in \mathcal{S}$.

We also want the approximated policy $\pi(a|s; \boldsymbol{\theta})$ to satisfy these two constraints.

For a task with discrete action space, it usually introduces an **action preference function** $h(s, a; \boldsymbol{\theta})$ $(s \in \mathcal{S}, a \in \mathcal{A}(s))$, whose softmax is $\pi(a|s; \boldsymbol{\theta})$, i.e.

$$\pi(a|s; \boldsymbol{\theta}) = \frac{\exp h(s, a; \boldsymbol{\theta})}{\sum_{a'} \exp h(s, a'; \boldsymbol{\theta})}, \quad s \in \mathcal{S}, a \in \mathcal{A}(s).$$

The resulting approximated policy will satisfy the two constraints.

Remarkably, different action preferences may map to the same policy. Especially, action preference $h(s, a; \boldsymbol{\theta}) + f(s; \boldsymbol{\theta})$ $(s \in \mathcal{S}, a \in \mathcal{A}(s))$ and $h(s, a; \boldsymbol{\theta})$ $(s \in \mathcal{S}, a \in \mathcal{A}(s))$ map to the same policy, where $f(\cdot; \boldsymbol{\theta})$ is an arbitrary deterministic function.

In Chaps. 2–6, the optimal policy estimation derived from optimal action values has some certain forms (for example, greedy policy, or ε-greedy policy). Compared to those methods, the optimal policy derived from action preference is more flexible, since every action can be selected using different probabilities. If we iteratively update the policy $\boldsymbol{\theta}$, $\pi(a|s; \boldsymbol{\theta})$ can approximate the deterministic policy gradually without adjusting some exploration parameters such as ε.

Action preference can be linear models, neural networks, or other forms. Once the form of action preference is designated, determining the parameter $\boldsymbol{\theta}$ will determine the whole optimal policy estimates. Since the parameter $\boldsymbol{\theta}$ is usually learned using a gradient-based method, the action preference function needs to be differentiable with respect to the parameter $\boldsymbol{\theta}$.

It is easy to calculate softmax when the action space is finite. But it is somehow difficult if the action space is continuous. Therefore, for continuous action space, we may limit the policy as a parametric distribution, and use parametric function for the parameters of the distributions. For example, when the action space is $\mathcal{A} = \mathbb{R}^n$, we usually use the Gaussian distribution as the policy distribution: We use parametric function $\boldsymbol{\mu}(\boldsymbol{\theta})$ and $\boldsymbol{\sigma}(\boldsymbol{\theta})$ as the mean vector and the standard deviation vector of Gaussian distribution, and let the form of the action be $A = \boldsymbol{\mu}(\boldsymbol{\theta}) + N \circ \boldsymbol{\sigma}(\boldsymbol{\theta})$, where $N \sim \text{normal}(\mathbf{0}, \mathbf{I})$ is the n-dimension Gaussian random variables, and the operator "$\circ$" means element-wise multiplication. For example, when the action space is $\mathcal{A} = [-1, +1]^n$, we can use the tanh transformation of Gaussian distribution as the policy distribution, and let the form of the action be $A = \tanh(\boldsymbol{\mu}(\boldsymbol{\theta}) + N \circ \boldsymbol{\sigma}(\boldsymbol{\theta}))$, where the function tanh() is element-wise. There are even more complex form of approximation functions, such as Gaussian mixture model.

7.1.2 PG Theorem

Policy gradient theorem is the foundation of PG method. It provides a way to calculate the gradient of the return expectation $g_{\pi(\boldsymbol{\theta})}$ with respect to the policy parameter $\boldsymbol{\theta}$. Upon we get the gradient, changing the parameter $\boldsymbol{\theta}$ in the gradient direction may increase the return expectation.

Policy gradient theorem has many forms. Two of the forms are as follows: The gradient of $g_{\pi(\boldsymbol{\theta})} \stackrel{\text{def}}{=} \mathrm{E}_{\pi(\boldsymbol{\theta})}[G_0]$ with respect to $\boldsymbol{\theta}$ can be expressed as

$$\nabla g_{\pi(\boldsymbol{\theta})} = \mathrm{E}_{\pi(\boldsymbol{\theta})}\left[\sum_{t=0}^{T-1} G_0 \nabla \ln\pi(A_t|S_t; \boldsymbol{\theta})\right]$$

$$= \mathrm{E}_{\pi(\boldsymbol{\theta})}\left[\sum_{t=0}^{T-1} \gamma^t G_t \nabla \ln\pi(A_t|S_t; \boldsymbol{\theta})\right].$$

The right-hand side of the above equation is the expectation of summation, where summation is over $G_0 \nabla \ln\pi(A_t|S_t; \boldsymbol{\theta})$ or $\gamma^t G_t \nabla \ln\pi(A_t|S_t; \boldsymbol{\theta})$, which contains the parameter $\boldsymbol{\theta}$ in $\nabla \ln\pi(A_t|S_t; \boldsymbol{\theta})$.

The policy gradient algorithm tells us, we can get the policy gradient if we know $\nabla \ln\pi(A_t|S_t; \boldsymbol{\theta})$ and other easily obtained values such as G_0.

Next, we will use two different methods to prove these two forms. Both proof methods are commonly used in RL, and each of them can be extended to some cases more conveniently than the other.

The first proof method is the trajectory method. Given policy $\pi(\boldsymbol{\theta})$, let $\pi(T; \boldsymbol{\theta})$ denote the probability of trajectory $T = (S_0, A_0, R_1, \ldots, S_T)$. We can express the return expectation $g_{\pi(\boldsymbol{\theta})} \stackrel{\text{def}}{=} \mathrm{E}_{\pi(\boldsymbol{\theta})}[G_0]$ as

$$g_{\pi(\boldsymbol{\theta})} = \mathrm{E}_{\pi(\boldsymbol{\theta})}[G_0] = \sum_t G_0 \pi(t; \boldsymbol{\theta}).$$

Therefore,

$$\nabla g_{\pi(\boldsymbol{\theta})} = \sum_t G_0 \nabla \pi(t; \boldsymbol{\theta}).$$

Since $\nabla \ln\pi(t; \boldsymbol{\theta}) = \frac{\nabla \pi(t;\boldsymbol{\theta})}{\pi(t;\boldsymbol{\theta})}$, we have $\nabla \pi(t; \boldsymbol{\theta}) = \pi(t; \boldsymbol{\theta}) \nabla \ln\pi(t; \boldsymbol{\theta})$. Therefore,

$$\nabla g_{\pi(\boldsymbol{\theta})} = \sum_t \pi(t; \boldsymbol{\theta}) G_0 \nabla \ln\pi(t; \boldsymbol{\theta}) = \mathrm{E}_{\pi(\boldsymbol{\theta})}[G_0 \nabla \ln\pi(T; \boldsymbol{\theta})].$$

Considering

$$\pi(T; \boldsymbol{\theta}) = p_{S_0}(S_0) \prod_{t=0}^{T-1} \pi(A_t|S_t; \boldsymbol{\theta}) p(S_{t+1}|S_t, A_t),$$

we have

$$\ln \pi(T; \boldsymbol{\theta}) = \ln p_{S_0}(S_0) + \sum_{t=0}^{T-1} \left[\ln \pi(A_t|S_t; \boldsymbol{\theta}) + \ln p(S_{t+1}|S_t, A_t) \right],$$

$$\nabla \ln\pi(T; \boldsymbol{\theta}) = \sum_{t=0}^{T-1} \left[\nabla \ln\pi(A_t|S_t; \boldsymbol{\theta}) \right].$$

Therefore,

$$\nabla g_{\pi(\boldsymbol{\theta})} = \mathrm{E}_{\pi(\boldsymbol{\theta})} \left[\sum_{t=0}^{T-1} G_0 \nabla \ln\pi(A_t|S_t; \boldsymbol{\theta}) \right].$$

The second proof method is recursively calculating value functions. Bellman expectation equations tell us, the values of policy $\pi(\boldsymbol{\theta})$ satisfy

$$v_{\pi(\boldsymbol{\theta})}(s) = \sum_a \pi(a|s; \boldsymbol{\theta}) q_{\pi(\boldsymbol{\theta})}(s, a), \qquad s \in S$$

$$q_{\pi(\boldsymbol{\theta})}(s, a) = r(s, a) + \gamma \sum_{s'} p(s'|s, a) v_{\pi(\boldsymbol{\theta})}(s'), \quad s \in S, a \in \mathcal{A}(s).$$

Calculating the gradient of the above two equations with respect to the parameter $\boldsymbol{\theta}$, we have

$$\nabla v_{\pi(\boldsymbol{\theta})}(s) = \sum_a q_{\pi(\boldsymbol{\theta})}(s, a) \nabla \pi(a|s; \boldsymbol{\theta}) + \sum_a \pi(a|s; \boldsymbol{\theta}) \nabla q_{\pi(\boldsymbol{\theta})}(s, a), \quad s \in S$$

$$\nabla q_{\pi(\boldsymbol{\theta})}(s, a) = \gamma \sum_{s'} p(s'|s, a) \nabla v_{\pi(\boldsymbol{\theta})}(s'), \qquad s \in S, a \in \mathcal{A}(s).$$

Plugging the expression of $\nabla q_{\pi(\boldsymbol{\theta})}(s, a)$ into the expression of $\nabla v_{\pi(\boldsymbol{\theta})}(s)$, we have

$$\nabla v_{\pi(\boldsymbol{\theta})}(s) = \sum_a q_{\pi(\boldsymbol{\theta})}(s, a) \nabla \pi(a|s; \boldsymbol{\theta}) + \sum_a \pi(a|s; \boldsymbol{\theta}) \gamma \sum_{s'} p(s'|s, a) \nabla v_{\pi(\boldsymbol{\theta})}(s')$$

$$= \sum_a q_{\pi(\boldsymbol{\theta})}(s, a) \nabla \pi(a|s; \boldsymbol{\theta}) + \sum_{s'} \mathrm{Pr}_{\pi(\boldsymbol{\theta})} \left[S_{t+1} = s'|S_t = s \right] \gamma \nabla v_{\pi(\boldsymbol{\theta})}(s'),$$

$$s \in S.$$

Calculating the expectation of the aforementioned equation over S_t generated by the policy $\pi(\boldsymbol{\theta})$ leads to

$$\mathrm{E}_{\pi(\boldsymbol{\theta})} \left[\nabla v_{\pi(\boldsymbol{\theta})}(S_t) \right]$$

$$= \sum_s \Pr_{\pi(\boldsymbol{\theta})}[S_t = s] \nabla v_{\pi(\boldsymbol{\theta})}(s)$$

$$= \sum_s \Pr_{\pi(\boldsymbol{\theta})}[S_t = s] \left[\sum_a q_{\pi(\boldsymbol{\theta})}(s, a) \nabla \pi(a|s; \boldsymbol{\theta}) \right.$$

$$\left. + \sum_{s'} \Pr_{\pi(\boldsymbol{\theta})}[S_{t+1} = s'|S_t = s] \gamma \nabla v_{\pi(\boldsymbol{\theta})}(s') \right]$$

$$= \sum_s \Pr_{\pi(\boldsymbol{\theta})}[S_t = s] \sum_a q_{\pi(\boldsymbol{\theta})}(s, a) \nabla \pi(a|s; \boldsymbol{\theta})$$

$$+ \sum_s \Pr_{\pi(\boldsymbol{\theta})}[S_t = s] \sum_{s'} \Pr_{\pi(\boldsymbol{\theta})}[S_{t+1} = s'|S_t = s] \gamma \nabla v_{\pi(\boldsymbol{\theta})}(s')$$

$$= \sum_s \Pr_{\pi(\boldsymbol{\theta})}[S_t = s] \sum_a q_{\pi(\boldsymbol{\theta})}(s, a) \nabla \pi(a|s; \boldsymbol{\theta}) + \gamma \sum_{s'} \Pr_{\pi(\boldsymbol{\theta})}[S_{t+1} = s'] \nabla v_{\pi(\boldsymbol{\theta})}(s')$$

$$= \mathrm{E}_{\pi(\boldsymbol{\theta})} \left[\sum_a q_{\pi(\boldsymbol{\theta})}(S_t, a) \nabla \pi(a|S_t; \boldsymbol{\theta}) \right] + \gamma \mathrm{E}_{\pi(\boldsymbol{\theta})} \left[\nabla v_{\pi(\boldsymbol{\theta})}(S_{t+1}) \right].$$

Therefore, we get the recursive expression from $\mathrm{E}_{\pi(\boldsymbol{\theta})}\left[\nabla v_{\pi(\boldsymbol{\theta})}(S_t)\right]$ to $\mathrm{E}_{\pi(\boldsymbol{\theta})}\left[\nabla v_{\pi(\boldsymbol{\theta})}(S_{t+1})\right]$. Note that the gradient we finally care about is

$$g_{\pi(\boldsymbol{\theta})} = \nabla \mathrm{E}\left[v_{\pi(\boldsymbol{\theta})}(S_0)\right] = \mathrm{E}\left[\nabla v_{\pi(\boldsymbol{\theta})}(S_0)\right].$$

Therefore,

$$\nabla g_{\pi(\boldsymbol{\theta})} = \mathrm{E}_{\pi(\boldsymbol{\theta})}\left[\nabla v_{\pi(\boldsymbol{\theta})}(S_0)\right]$$

$$= \mathrm{E}_{\pi(\boldsymbol{\theta})} \left[\sum_a q_{\pi(\boldsymbol{\theta})}(S_0, a) \nabla \pi(a|S_0; \boldsymbol{\theta}) \right] + \mathrm{E}_{\pi(\boldsymbol{\theta})} \left[\nabla v_{\pi(\boldsymbol{\theta})}(S_1) \right]$$

$$= \mathrm{E}_{\pi(\boldsymbol{\theta})} \left[\sum_a q_{\pi(\boldsymbol{\theta})}(S_0, a) \nabla \pi(a|S_0; \boldsymbol{\theta}) \right]$$

$$+ \mathrm{E}_{\pi(\boldsymbol{\theta})} \left[\sum_a q_{\pi(\boldsymbol{\theta})}(S_1, a) \nabla \pi(a|S_1; \boldsymbol{\theta}) \right] + \mathrm{E}_{\pi(\boldsymbol{\theta})} \left[\nabla v_{\pi(\boldsymbol{\theta})}(S_2) \right]$$

$$= \cdots$$

$$= \sum_{t=0}^{+\infty} \mathrm{E}_{\pi(\boldsymbol{\theta})} \left[\sum_a \gamma^t q_{\pi(\boldsymbol{\theta})}(S_t, a) \nabla \pi(a|S_t; \boldsymbol{\theta}) \right].$$

Since

$$\nabla \pi(a|S_t; \boldsymbol{\theta}) = \pi(a|S_t; \boldsymbol{\theta}) \nabla \ln \pi(a|S_t; \boldsymbol{\theta}),$$

we have

$$E_{\pi(\theta)}\left[\sum_a \gamma^t q_{\pi(\theta)}(S_t, a)\nabla \pi(a|S_t; \theta)\right]$$

$$= E_{\pi(\theta)}\left[\sum_a \pi(a|S_t; \theta)\gamma^t q_{\pi(\theta)}(S_t, a)\nabla \ln\pi(a|S_t; \theta)\right]$$

$$= E_{\pi(\theta)}\left[\gamma^t q_{\pi(\theta)}(S_t, A_t)\nabla \ln\pi(A_t|S_t; \theta)\right].$$

Furthermore,

$$\nabla g_{\pi(\theta)} = \sum_{t=0}^{+\infty} E_{\pi(\theta)}\left[\gamma^t q_{\pi(\theta)}(S_t, A_t)\nabla \ln\pi(A_t|S_t; \theta)\right]$$

$$= E_{\pi(\theta)}\left[\sum_{t=0}^{+\infty} \gamma^t q_{\pi(\theta)}(S_t, A_t)\nabla \ln\pi(A_t|S_t; \theta)\right].$$

Plugging $q_{\pi(\theta)}(S_t, A_t) = E_{\pi(\theta)}[G_t|S_t, A_t]$ into the above formula and applying the total expectation law will lead to

$$\nabla g_{\pi(\theta)} = E_{\pi(\theta)}\left[\sum_{t=0}^{+\infty} \gamma^t G_t \nabla \ln\pi(A_t|S_t; \theta)\right].$$

We can also directly prove the equivalency of these two forms. That is,

$$E_{\pi(\theta)}\left[\sum_{t=0}^{T-1} G_0 \nabla \ln\pi(A_t|S_t; \theta)\right] = E_{\pi(\theta)}\left[\sum_{t=0}^{T-1} \gamma^t G_t \nabla \ln\pi(A_t|S_t; \theta)\right].$$

In order to prove this equation, we introduce the concept of baseline. A **baseline** function $B(s)$ ($s \in S$) can be an arbitrary deterministic function or stochastic function such that its value can depend on the state s, but can not depend on the action a. After satisfying this condition, the baseline function will satisfy

$$E_{\pi(\theta)}\left[\gamma^t (G_t - B(S_t))\nabla \ln\pi(A_t|S_t; \theta)\right] = E_{\pi(\theta)}\left[\gamma^t G_t \nabla \ln\pi(A_t|S_t; \theta)\right].$$

(Proof: Since B does not depend on the action a,

$$\sum_a B(S_t)\nabla \pi(a|S_t; \theta) = B(S_t)\nabla \sum_a \pi(a|S_t; \theta) = B(S_t)\nabla 1 = 0.$$

Therefore,

$$E_{\pi(\theta)}\left[\gamma^t (G_t - B(S_t))\nabla \ln\pi(a|S_t; \theta)\right]$$

$$= \sum_a \gamma^t (G_t - B(S_t))\nabla \pi(a|S_t; \theta)$$

$$= \sum_{a} \gamma^{t} G_{t} \nabla \pi(a|S_{t}; \boldsymbol{\theta})$$

$$= E_{\pi(\boldsymbol{\theta})} \left[\gamma^{t} G_{t} \nabla \ln \pi(a|S_{t}; \boldsymbol{\theta}) \right]$$

The proof completes.) Selecting the baseline function as the stochastic function $B(S_t) = -\sum_{\tau=0}^{t-1} \gamma^{t-\tau} R_{\tau+1}$, we will have $\gamma^t(G_t - B(S_t)) = G_0$. Therefore, we prove the equivalence of these two forms.

Let us consider another form of PG theorem at the end of this section: The second proof in this section has proved that

$$\nabla g_{\pi(\boldsymbol{\theta})} = E_{\pi(\boldsymbol{\theta})} \left[\sum_{t=0}^{+\infty} \gamma^{t} q_{\pi(\boldsymbol{\theta})}(S_{t}, A_{t}) \nabla \ln \pi(A_{t}|S_{t}; \boldsymbol{\theta}) \right].$$

So we can use discounted expectation to express this as

$$\nabla g_{\pi(\boldsymbol{\theta})} = E_{(S,A) \sim \rho_{\pi(\boldsymbol{\theta})}} \left[q_{\pi(\boldsymbol{\theta})}(S, A) \nabla \ln \pi(A|S; \boldsymbol{\theta}) \right]$$

You may try to prove this by yourself.

7.1.3 Relationship between PG and Maximum Likelihood Estimate

The update in PG method tries to maximize $E_{\pi(\boldsymbol{\theta})} \left[\Psi_{t} \ln \pi(A_{t}|S_{t}; \boldsymbol{\theta}) \right]$, where Ψ_{t} can be G_0, $\gamma^t G_t$, or others. We can compare this procedure with the following Maximum Likelihood Estimation (MLE) in supervised learning: Suppose we want to use parametric function $\pi(\boldsymbol{\theta})$ to approximate a policy π. We can use MLE to estimate $\boldsymbol{\theta}$. Specifically, we use the policy π to generate lots of samples, and the log-likelihood of these samples with respect to the policy $\pi(\boldsymbol{\theta})$ is proportion to $E_{\pi(\boldsymbol{\theta})} \left[\ln \pi(A_{t}|S_{t}; \boldsymbol{\theta}) \right]$. MLE tries to adjust the parameter $\boldsymbol{\theta}$ to increase $E_{\pi(\boldsymbol{\theta})} \left[\ln \pi(A_{t}|S_{t}; \boldsymbol{\theta}) \right]$. Here, we notice that $E_{\pi(\boldsymbol{\theta})} \left[\ln \pi(A_{t}|S_{t}; \boldsymbol{\theta}) \right]$ exactly equals $E_{\pi(\boldsymbol{\theta})} \left[\Psi_{t} \ln \pi(A_{t}|S_{t}; \boldsymbol{\theta}) \right]$ by picking $\Psi_{t} = 1$. Therefore, there are some similarities between PG and MLE.

PG can be viewed as the weighted average of log-likelihood sample $\ln \pi(A_{t}|S_{t}; \boldsymbol{\theta})$, where the weight Ψ_{t} can be positive, negative, or 0. Especially, if Ψ_{t} is constant throughout the episode (say $\Psi_{t} = g_0$), the policy gradient in this episode, i.e. $E_{\pi(\boldsymbol{\theta})} \left[g_0 \ln \pi(A_{t}|S_{t}; \boldsymbol{\theta}) \right] = g_0 E_{\pi(\boldsymbol{\theta})} \left[\ln \pi(A_{t}|S_{t}; \boldsymbol{\theta}) \right]$ is the weighted values of episode log-likelihood. If an episode has good return, i.e. g_0 is a large positive value, the likelihood value of $E_{\pi(\boldsymbol{\theta})} \left[\ln \pi(A_{t}|S_{t}; \boldsymbol{\theta}) \right]$ of this episode will be weighted by a large positive weight, and similar episodes tend to reoccur in the future. If an episode has bad return, i.e. g_0 is a negative return with large absolute value, the likelihood will be weighted by a negative weight, so similar episodes will be unlikely to appear again. This explains why policy gradient can make the policy better.

7.2 On-Policy PG

Policy gradient theorem tells us, changing the policy parameter θ in the direction of $\nabla g_{\pi(\theta)} = \mathrm{E}_{\pi(\theta)} \left[\sum_{t=0}^{+\infty} \gamma^t G_t \nabla \ln\pi(A_t|S_t; \theta) \right]$ may increase return expectation. Accordingly, we can design PG algorithms. This section considers on-policy PG algorithms.

7.2.1 VPG: Vanilla Policy Gradient

The simplest form of policy gradient is the **Vanilla Policy Gradient (VPG)** algorithm. VPG updates the parameter θ using the following form

$$\theta \leftarrow \theta + \alpha\gamma^t G_t \nabla \ln\pi(A_t|S_t; \theta)$$

for every state–action pair (S_t, A_t) in the episode. (Williams, 1992) named this algorithm as "REward Increment = Nonnegative Factor × Offset Reinforcement × Characteristic Eligibility" (REINFORCE), meaning that the increment $\alpha\gamma^t G_t \nabla \ln\pi(A_t|S_t; \theta)$ is the product of three elements. Updating using all state–action pairs in the episode is equivalent to

$$\theta \leftarrow \theta + \alpha \sum_{t=0}^{+\infty} \gamma^t G_t \nabla \ln\pi(A_t|S_t; \theta).$$

When we use software packages such as TensorFlow or PyTorch, we can define one-step loss as $-\gamma^t G_t \ln\pi(A_t|S_t; \theta)$, and try to minimize the average loss over the episode. In total, we change θ in the direction of $\sum_{t=0}^{+\infty} \gamma^t G_t \nabla\pi(A_t|S_t; \theta)$.

VPG algorithm is shown in Algo. 7.1. On the initialization in Step 1, if the policy is in the form of a neural network, the initialization should follow the practice in deep learning. All initialization in this book should be understood in this way.

Algorithm 7.1   VPG policy optimization.

Input: environment (without mathematical model).
Output: optimal policy estimate $\pi(\theta)$.
Parameters: optimizer (containing learning rate α), discount factor γ, and parameters to control the number of episodes and maximum steps in an episode.

1. (Initialize) Initialize the policy parameter θ.
2. (MC update) For each episode:

    2.1. (Decide and sample) Use the policy $\pi(\theta)$ to generate the trajectory
        $S_0, A_0, R_1, S_1, \cdots, S_{T-1}, A_{T-1}, R_T, S_T$.

> **2.2.** (Initialize return) $G \leftarrow 0$.
> **2.3.** (Update) For $t \leftarrow T - 1, T - 2, \ldots, 0$:
>   **2.3.1.** (Calculate return) $G \leftarrow \gamma G + R_{t+1}$.
>   **2.3.2.** (Update policy parameter) Update θ to reduce $-\gamma^t G \ln \pi(A_t | S_t; \theta)$. (For example, $\theta \leftarrow \theta + \alpha \gamma^t G \nabla \ln \pi(A_t | S_t; \theta)$.)

The most outstanding shortcoming of VPG algorithm is that its variance is usually very large.

7.2.2 PG with Baseline

VPG algorithm usually has very large variance. This section uses baseline function to reduce this variance.

Section 7.1.2 has introduced the concept of baseline function $B(s)$ $(s \in S)$, which is a stochastic or deterministic function that depends on the state s but does not depend on the action a. The baseline function has the following property:

$$\mathrm{E}_{\pi(\theta)} \left[\gamma^t (G_t - B(S_t)) \nabla \ln \pi(A_t | S_t; \theta) \right] = \mathrm{E}_{\pi(\theta)} \left[\gamma^t G_t \nabla \ln \pi(A_t | S_t; \theta) \right].$$

We can arbitrarily choose the baseline function. Some examples include:

- Choose the baseline function as a stochastic function determined by part of trajectory $B(S_t) = -\sum_{\tau=0}^{t-1} \gamma^{\tau-t} R_{\tau+1}$, and then $\gamma^t (G_t - B(S_t)) = G_0$. In this case, the policy gradient is $\mathrm{E}_{\pi(\theta)} \left[G_0 \nabla \ln \pi(A_t | S_t; \theta) \right]$.
- Choose the baseline function as $B(S_t) = \gamma^t v_*(S_t)$. In this case, the policy gradient is $\mathrm{E}_{\pi(\theta)} \left[\gamma^t (G_t - v_*(S_t)) \nabla \ln \pi(A_t | S_t; \theta) \right]$.

We need to follow the following guideline when we select the baseline function:

- The baseline function should reduce variance. It is usually difficult to judge whether a baseline function can reduce variance in theory. We usually know that from practice.
- The baseline function needs to be attainable. For example, the actual optimal values are something we will never know, so we can not use them as baseline. However, we can estimate them, so the estimates can be used as baseline. The optimal value estimates can be updated with iteration, too.

State value estimates are a baseline that can efficiently reduce variance. Algorithm 7.2 shows the algorithm that uses state value estimates as baseline. This algorithm has two suites of parameters θ and $\mathbf{w}$, which are the parameters of optimal policy estimates and optimal state value estimates respectively. Steps 2.3.2 and 2.3.3 update these two parameters. These two parameters can be trained using different optimizers at different paces. Usually, we assign a smaller learning rate for optimizer of policy parameters, and assign a larger learning rate for optimizer of value parameters. Additionally, both parameter update uses $G - v(S_t; \mathbf{w})$, so we can combine these two computations to one.

Algorithm 7.2   VPG policy optimization with baseline.

Input: environment (without mathematical model).
Output: optimal policy estimate $\pi(\boldsymbol{\theta})$.
Parameters: optimizer (containing learning rate $\alpha^{(\boldsymbol{\theta})}$ and $\alpha^{(\mathbf{w})}$), discount factor γ, and parameters to control the number of episodes and maximum steps in an episode.

1. (Initialize) Initialize the policy parameter $\boldsymbol{\theta}$ and value parameter $\mathbf{w}$.
2. (MC update) For each episode:

    2.1. (Decide and sample) Use the policy $\pi(\boldsymbol{\theta})$ to generate the trajectory $S_0, A_0, R_1, S_1, \cdots, S_{T-1}, A_{T-1}, R_T, S_T$.
    2.2. (Initialize return) $G \leftarrow 0$.
    2.3. (Update) For $t \leftarrow T-1, T-2, \ldots, 0$:
        2.3.1. (Calculate return) $G \leftarrow \gamma G + R_{t+1}$.
        2.3.2. (Update value parameter) Update $\mathbf{w}$ to reduce $[G - v(S_t; \mathbf{w})]^2$. (For example, $\mathbf{w} \leftarrow \mathbf{w} + \alpha^{(\mathbf{w})}[G - v(S_t; \mathbf{w})]\nabla v(S_t; \mathbf{w})$.)
        2.3.3. (Update policy parameter) Update $\boldsymbol{\theta}$ to reduce $-\gamma^t G \ln \pi(A_t|S_t; \boldsymbol{\theta})$. (For example, $\boldsymbol{\theta} \leftarrow \boldsymbol{\theta} + \alpha^{(\boldsymbol{\theta})}\gamma^t[G - v(S_t; \mathbf{w})]\nabla \ln\pi(A_t|S_t; \boldsymbol{\theta})$.)

In PG with baseline, both policy parameters and value parameters need to be updated. For such cases, the convergence needs to be analyzed using two timescale Robbins–Monro algorithm.

At the end of this section, let us analyze why baseline function can minimize the variance. The variance of $\mathrm{E}\left[\gamma^t(G_t - B(S_t))\nabla \ln\pi(A_t|S_t; \boldsymbol{\theta})\right]$ is

$$\mathrm{E}\left[\gamma^t(G_t - B(S_t))\nabla \ln\pi(A_t|S_t; \boldsymbol{\theta})\right]^2 - \left[\mathrm{E}\left[\gamma^t(G_t - B(S_t))\nabla \ln\pi(A_t|S_t; \boldsymbol{\theta})\right]\right]^2.$$

Its partial derivative with respect to $B(S_t)$ is

$$\mathrm{E}_{\pi(\boldsymbol{\theta})}\left[-2\gamma^{2t}(G_t - B(S_t))\left[\nabla \ln\pi(A_t|S_t; \boldsymbol{\theta})\right]^2\right],$$

since $\frac{\partial}{\partial B(S_t)}\mathrm{E}_{\pi(\boldsymbol{\theta})}\left[\gamma^t(G_t - B(S_t))\left[\nabla \ln\pi(A_t|S_t; \boldsymbol{\theta})\right]^2\right] = 0$. Setting this partial derivative to 0, and assuming

$$\mathrm{E}_{\pi(\boldsymbol{\theta})}\left[B(S_t)\left[\nabla \ln\pi(A_t|S_t; \boldsymbol{\theta})\right]^2\right] = \mathrm{E}_{\pi(\boldsymbol{\theta})}\left[B(S_t)\right]\mathrm{E}_{\pi(\boldsymbol{\theta})}\left[\left[\nabla \ln\pi(A_t|S_t; \boldsymbol{\theta})\right]^2\right],$$

we have

$$E_{\pi(\theta)}\big[B(S_t)\big] = \frac{E_{\pi(\theta)}\Big[B(S_t)\big[\nabla \ln\pi(A_t|S_t;\theta)\big]^2\Big]}{E_{\pi(\theta)}\Big[\big[\nabla \ln\pi(A_t|S_t;\theta)\big]^2\Big]},$$

This means that the optimal baseline function should be close to the weighted average of the return G_t with the weight $\big[\nabla \ln\pi(A_t|S_t;\theta)\big]^2$. However, this average can not be known, so we can not use it as baseline function in real world.

7.3 Off-Policy PG

This section applies importance sampling on VPG to get the off-policy PG algorithm.
    Let $b(a|s)$ ($s \in \mathcal{S}, a \in \mathcal{A}(s)$) be the behavior policy. We have

$$\sum_a \pi(a|s;\theta)\gamma^t G_t \nabla \ln\pi(a|s;\theta)$$

$$= \sum_a b(a|s)\frac{\pi(a|s;\theta)}{b(a|s)}\gamma^t G_t \nabla \ln\pi(a|s;\theta)$$

$$= \sum_a b(a|s)\frac{1}{b(a|s)}\gamma^t G_t \nabla \pi(a|s;\theta).$$

That is,

$$E_{\pi(\theta)}\big[\gamma^t G_t \nabla \ln\pi(A_t|S_t;\theta)\big] = E_b\left[\frac{1}{b(A_t|S_t)}\gamma^t G_t \nabla \pi(A_t|S_t;\theta)\right].$$

Therefore, the importance sampling off-policy algorithm changes the policy gradient from $\gamma^t G_t \nabla \pi(A_t|S_t;\theta)$ to $\frac{1}{b(A_t|S_t)}\gamma^t G_t \nabla \pi(A_t|S_t;\theta)$. When updating the parameter θ, it tries to increase $\frac{1}{b(A_t|S_t)}\gamma^t G_t \nabla \pi(A_t|S_t;\theta)$. Algorithm 7.3 shows this algorithm.

Algorithm 7.3   Importance sampling PG policy optimization.

---

1. (Initialize) Initialize the policy parameter θ.
2. (MC update) For each episode:

    2.1. (Behavior policy) Designate behavior policy b such that $\pi(\theta) \ll b$.
    2.2. (Decide and sample) Use the behavior policy b to generate the trajectory $S_0, A_0, R_1, S_1, \cdots, S_{T-1}, A_{T-1}, R_T, S_T$.
    2.3. (Initialize return) $G \leftarrow 0$.
    2.4. (Update) For $t \leftarrow T-1, T-2, \ldots, 0$:
        2.4.1. (Calculate return) $G \leftarrow \gamma G + R_{t+1}$.

2.4.2. (Update policy parameter) Update $\boldsymbol{\theta}$ to reduce $-\frac{1}{b(A_t|S_t)}\gamma^t G\nabla\pi(A_t|S_t;\boldsymbol{\theta})$. (For example, $\boldsymbol{\theta} \leftarrow \boldsymbol{\theta} + \alpha\frac{1}{b(A_t|S_t)}\gamma^t G\nabla\pi(A_t|S_t;\boldsymbol{\theta})$.)

The baseline can be applied to off-policy PG algorithm to reduce variance.

7.4 Case Study: CartPole

This section considers a cart-pole task in Gym (Barto, 1983). As Fig. 7.1, a cart moves along a track. A pole connects to a cart. The initial position and angle of the cart are randomly generated. The agent can push the cart left or right. An episode ends when one of the following conditions is met:

- The tilt angle of the pole exceeds 12°;
- The position of the cart exceeds ±2.4;
- The number of steps reaches the upper limit.

We can get 1 unit of reward for every step. We want to make an episode longer.

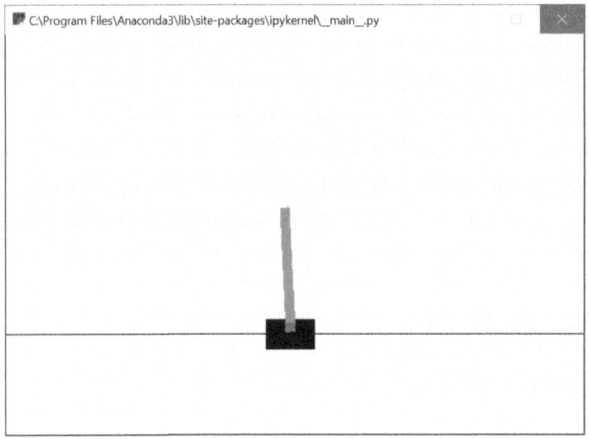

Fig. 7.1   The cart-pole problem.
This figure is adapted from (Barto, 1983).

The observation has 4 elements, representing the position of the cart, the velocity of the cart, the angle of the pole, and the angular velocity of the pole. The range of each element is shown in Table 7.1. The action can be either 0 or 1, representing pushing left and pushing right respectively.

There are two versions of cart-pole problem in Gym, i.e. CartPole-v0 and CartPole-v1. These two tasks only differ in the maximum numbers of steps and

Table 7.1   Observations in the cart-pole problem.

observation component index	meaning	minimum	maximum
0	position of cart	-4.8	$+4.8$
1	velocity of cart	$-\infty$	$+\infty$
2	angle of pole	$\approx -41.8°$	$\approx +41.8°$
3	angular velocity of pole	$-\infty$	$+\infty$

the reward threshold for solving the task. The task `CartPole-v0` has at most 200 steps, and its reward threshold is 195; while the task `CartPole-v1` has at most 500 steps, and its reward threshold is 475. These two tasks have similar difficulties.

Section 1.8.2 introduces a closed-form solution for this task. Codes are available in GitHub.

This section focuses on `CartPole-v0`. The episode reward of a random policy is around 9 to 10.

7.4.1 On-Policy PG

This section uses on-policy PG algorithm to find an optimal policy. VPG without baseline is implemented by Code 7.1 (TensorFlow version) or Code 7.2 (PyTorch version). The optimal policy estimate is approximated by a one-layer neural network. The training uses the form $\nabla g_{\pi(\boldsymbol{\theta})} = \mathrm{E}_{\pi(\boldsymbol{\theta})}\left[\sum_{t=0}^{+\infty} \gamma^t G_t \nabla \ln\pi(A_t|S_t; \boldsymbol{\theta})\right]$ to calculate gradient descent. Code 7.1 (the TensorFlow version) uses weighted cross entropy loss with return as weights, while Code 7.2 (the PyTorch version) implements the calculation directly. In Code 7.2, function `torch.clamp()` is used to clip the input of `torch.log()` to improve the numerical stability.

Code 7.1   On-policy VPG agent (TensorFlow version).
`CartPole-v0_VPG_tf.ipynb`

```
1   class VPGAgent:
2       def __init__(self, env):
3           self.action_n = env.action_space.n
4           self.gamma = 0.99
5
6           self.policy_net = self.build_net(hidden_sizes=[],
7                   output_size=self.action_n, output_activation=nn.softmax,
8                   loss=losses.categorical_crossentropy)
9
10      def build_net(self, hidden_sizes, output_size,
11              activation=nn.relu, output_activation=None,
12              use_bias=False, loss=losses.mse, learning_rate=0.005):
13          model = keras.Sequential()
14          for hidden_size in hidden_sizes:
15              model.add(layers.Dense(units=hidden_size,
16                  activation=activation, use_bias=use_bias))
17          model.add(layers.Dense(units=output_size,
18                  activation=output_activation, use_bias=use_bias))
19          optimizer = optimizers.Adam(learning_rate)
```

```
20        model.compile(optimizer=optimizer, loss=loss)
21        return model
22
23    def reset(self, mode=None):
24        self.mode = mode
25        if self.mode == 'train':
26            self.trajectory = []
27
28    def step(self, observation, reward, terminated):
29        probs = self.policy_net.predict(observation[np.newaxis], verbose=0)[0]
30        action = np.random.choice(self.action_n, p=probs)
31        if self.mode == 'train':
32            self.trajectory += [observation, reward, terminated, action]
33        return action
34
35    def close(self):
36        if self.mode == 'train':
37            self.learn()
38
39    def learn(self):
40        df = pd.DataFrame(
41                np.array(self.trajectory, dtype=object).reshape(-1, 4),
42                columns=['state', 'reward', 'terminated', 'action'])
43        df['discount'] = self.gamma ** df.index.to_series()
44        df['discounted_reward'] = df['discount'] * df['reward']
45        df['discounted_return'] = df['discounted_reward'][::-1].cumsum()
46        states = np.stack(df['state'])
47        actions = np.eye(self.action_n)[df['action'].astype(int)]
48        sample_weight = df[['discounted_return',]].values.astype(float)
49        self.policy_net.fit(states, actions, sample_weight=sample_weight,
50                verbose=0)
51
52
53 agent = VPGAgent(env)
```

Code 7.2   On-policy VPG agent (PyTorch version).
CartPole-v0_VPG_torch.ipynb

```
1  class VPGAgent:
2      def __init__(self, env):
3          self.action_n = env.action_space.n
4          self.gamma = 0.99
5
6          self.policy_net = self.build_net(
7                  input_size=env.observation_space.shape[0], hidden_sizes=[],
8                  output_size=self.action_n, output_activator=nn.Softmax(1))
9          self.optimizer = optim.Adam(self.policy_net.parameters(), lr=0.005)
10
11     def build_net(self, input_size, hidden_sizes, output_size,
12             output_activator=None, use_bias=False):
13         layers = []
14         for input_size, output_size in zip(
15                 [input_size,] + hidden_sizes, hidden_sizes + [output_size,]):
16             layers.append(nn.Linear(input_size, output_size, bias=use_bias))
17             layers.append(nn.ReLU())
18         layers = layers[:-1]
19         if output_activator:
20             layers.append(output_activator)
21         model = nn.Sequential(*layers)
22         return model
23
24     def reset(self, mode=None):
25         self.mode = mode
26         if self.mode == 'train':
27             self.trajectory = []
```

```
28
29    def step(self, observation, reward, terminated):
30        state_tensor = torch.as_tensor(observation,
31                dtype=torch.float).unsqueeze(0)
32        prob_tensor = self.policy_net(state_tensor)
33        action_tensor = distributions.Categorical(prob_tensor).sample()
34        action = action_tensor.numpy()[0]
35        if self.mode == 'train':
36            self.trajectory += [observation, reward, terminated, action]
37        return action
38
39    def close(self):
40        if self.mode == 'train':
41            self.learn()
42
43    def learn(self):
44        state_tensor = torch.as_tensor(self.trajectory[0::4],
45                dtype=torch.float)
46        reward_tensor = torch.as_tensor(self.trajectory[1::4],
47                dtype=torch.float)
48        action_tensor = torch.as_tensor(self.trajectory[3::4],
49                dtype=torch.long)
50        arange_tensor = torch.arange(state_tensor.shape[0], dtype=torch.float)
51        discount_tensor = self.gamma ** arange_tensor
52        discounted_reward_tensor = discount_tensor * reward_tensor
53        discounted_return_tensor = \
54                discounted_reward_tensor.flip(0).cumsum(0).flip(0)
55        all_pi_tensor = self.policy_net(state_tensor)
56        pi_tensor = torch.gather(all_pi_tensor, 1,
57                action_tensor.unsqueeze(1)).squeeze(1)
58        log_pi_tensor = torch.log(torch.clamp(pi_tensor, 1e-6, 1.))
59        loss_tensor = -(discounted_return_tensor * log_pi_tensor).mean()
60        self.optimizer.zero_grad()
61        loss_tensor.backward()
62        self.optimizer.step()
63
64
65 agent = VPGAgent(env)
```

VPG with baseline is implemented by Code 7.3 (TensorFlow version) and Code 7.4 (PyTorch version). The baseline function they use is the state value estimates approximated by neural networks.

Code 7.3 On-policy VPG agent with baseline (TensorFlow version).
CartPole-v0_VPGwBaseline_tf.ipynb

```
1  class VPGwBaselineAgent:
2      def __init__(self, env):
3          self.action_n = env.action_space.n
4          self.gamma = 0.99
5
6          self.trajectory = []
7
8          self.policy_net = self.build_net(hidden_sizes=[],
9                  output_size=self.action_n,
10                 output_activation=nn.softmax,
11                 loss=losses.categorical_crossentropy,
12                 learning_rate=0.005)
13         self.baseline_net = self.build_net(hidden_sizes=[],
14                 learning_rate=0.01)
15
16     def build_net(self, hidden_sizes, output_size=1,
17             activation=nn.relu, output_activation=None,
18             use_bias=False, loss=losses.mse, learning_rate=0.005):
19         model = keras.Sequential()
```

```
20      for hidden_size in hidden_sizes:
21          model.add(layers.Dense(units=hidden_size,
22                  activation=activation, use_bias=use_bias))
23      model.add(layers.Dense(units=output_size,
24              activation=output_activation, use_bias=use_bias))
25      optimizer = optimizers.Adam(learning_rate)
26      model.compile(optimizer=optimizer, loss=loss)
27      return model
28
29  def reset(self, mode=None):
30      self.mode = mode
31      if self.mode == 'train':
32          self.trajectory = []
33
34  def step(self, observation, reward, terminated):
35      probs = self.policy_net.predict(observation[np.newaxis], verbose=0)[0]
36      action = np.random.choice(self.action_n, p=probs)
37      if self.mode == 'train':
38          self.trajectory += [observation, reward, terminated, action]
39      return action
40
41  def close(self):
42      if self.mode == 'train':
43          self.learn()
44
45  def learn(self):
46      df = pd.DataFrame(np.array(self.trajectory, dtype=object).reshape(-1,
47              4), columns=['state', 'reward', 'terminated', 'action'])
48
49      # update baseline
50      df['discount'] = self.gamma ** df.index.to_series()
51      df['discounted_reward'] = df['discount'] * df['reward'].astype(float)
52      df['discounted_return'] = df['discounted_reward'][::-1].cumsum()
53      df['return'] = df['discounted_return'] / df['discount']
54      states = np.stack(df['state'])
55      returns = df[['return',]].values
56      self.baseline_net.fit(states, returns, verbose=0)
57
58      # update policy
59      df['baseline'] = self.baseline_net.predict(states, verbose=0)
60      df['psi'] = df['discounted_return'] - df['baseline'] * df['discount']
61      actions = np.eye(self.action_n)[df['action'].astype(int)]
62      sample_weight = df[['discounted_return',]].values
63      self.policy_net.fit(states, actions, sample_weight=sample_weight,
64              verbose=0)
65
66
67  agent = VPGwBaselineAgent(env)
```

Code 7.4   On-policy VPG agent with baseline (PyTorch version).
CartPole-v0_VPGwBaseline_torch.ipynb

```
1   class VPGwBaselineAgent:
2       def __init__(self, env,):
3           self.action_n = env.action_space.n
4           self.gamma = 0.99
5
6           self.policy_net = self.build_net(
7                   input_size=env.observation_space.shape[0],
8                   hidden_sizes=[],
9                   output_size=self.action_n, output_activator=nn.Softmax(1))
10          self.policy_optimizer = optim.Adam(self.policy_net.parameters(),
11                  lr=0.005)
12          self.baseline_net = self.build_net(
13                  input_size=env.observation_space.shape[0],
```

```
14              hidden_sizes=[])
15      self.baseline_optimizer = optim.Adam(self.policy_net.parameters(),
16              lr=0.01)
17      self.baseline_loss = nn.MSELoss()
18
19  def build_net(self, input_size, hidden_sizes, output_size=1,
20          output_activator=None, use_bias=False):
21      layers = []
22      for input_size, output_size in zip(
23              [input_size,] + hidden_sizes, hidden_sizes + [output_size,]):
24          layers.append(nn.Linear(input_size, output_size, bias=use_bias))
25          layers.append(nn.ReLU())
26      layers = layers[:-1]
27      if output_activator:
28          layers.append(output_activator)
29      model = nn.Sequential(*layers)
30      return model
31
32  def reset(self, mode=None):
33      self.mode = mode
34      if self.mode == 'train':
35          self.trajectory = []
36
37  def step(self, observation, reward, terminated):
38      state_tensor = torch.as_tensor(observation,
39              dtype=torch.float).unsqueeze(0)
40      prob_tensor = self.policy_net(state_tensor)
41      action_tensor = distributions.Categorical(prob_tensor).sample()
42      action = action_tensor.numpy()[0]
43      if self.mode == 'train':
44          self.trajectory += [observation, reward, terminated, action]
45      return action
46
47  def close(self):
48      if self.mode == 'train':
49          self.learn()
50
51  def learn(self):
52      state_tensor = torch.as_tensor(self.trajectory[0::4],
53              dtype=torch.float)
54      reward_tensor = torch.as_tensor(self.trajectory[1::4],
55              dtype=torch.float)
56      action_tensor = torch.as_tensor(self.trajectory[3::4],
57              dtype=torch.long)
58      arange_tensor = torch.arange(state_tensor.shape[0], dtype=torch.float)
59
60      # update baseline
61      discount_tensor = self.gamma ** arange_tensor
62      discounted_reward_tensor = discount_tensor * reward_tensor
63      discounted_return_tensor = \
64              discounted_reward_tensor.flip(0).cumsum(0).flip(0)
65      return_tensor = discounted_return_tensor / discount_tensor
66      pred_tensor = self.baseline_net(state_tensor)
67      psi_tensor = (discounted_return_tensor - discount_tensor *
68              pred_tensor).detach()
69      baseline_loss_tensor = self.baseline_loss(pred_tensor,
70              return_tensor.unsqueeze(1))
71      self.baseline_optimizer.zero_grad()
72      baseline_loss_tensor.backward()
73      self.baseline_optimizer.step()
74
75      # update policy
76      all_pi_tensor = self.policy_net(state_tensor)
77      pi_tensor = torch.gather(all_pi_tensor, 1,
78              action_tensor.unsqueeze(1)).squeeze(1)
79      log_pi_tensor = torch.log(torch.clamp(pi_tensor, 1e-6, 1.))
80      policy_loss_tensor = -(psi_tensor * log_pi_tensor).mean()
```

```
81          self.policy_optimizer.zero_grad()
82          policy_loss_tensor.backward()
83          self.policy_optimizer.step()
84
85
86  agent = VPGwBaselineAgent(env)
```

We still use Code 5.3 to train and Code 1.4 to test.

7.4.2 Off-Policy PG

This section implements off-policy PG algorithm with importance sampling. The off-policy algorithm without baseline is shown in Code 7.5 (TensorFlow version) and Code 7.6 (PyTorch version). The off-policy algorithm with baseline is shown in Code 7.7 (TensorFlow version) and Code 7.8 (PyTorch version). The behavior policy is the random policy $\pi(a|s) = 0.5$ $(s \in \mathcal{S}, a \in \mathcal{A}(s))$.

Code 7.5   Off-policy PG agent (TensorFlow version).
CartPole-v0_OffPolicyVPG_tf.ipynb

```
1   class OffPolicyVPGAgent:
2       def __init__(self, env):
3           self.action_n = env.action_space.n
4           self.gamma = 0.99
5
6           def dot(y_true, y_pred):
7               return -tf.reduce_sum(y_true * y_pred, axis=-1)
8
9           self.policy_net = self.build_net(hidden_sizes=[],
10                  output_size=self.action_n, output_activation=nn.softmax,
11                  loss=dot, learning_rate=0.06)
12
13      def build_net(self, hidden_sizes, output_size,
14              activation=nn.relu, output_activation=None,
15              use_bias=False, loss=losses.mse, learning_rate=0.001):
16          model = keras.Sequential()
17          for hidden_size in hidden_sizes:
18              model.add(layers.Dense(units=hidden_size,
19                      activation=activation, use_bias=use_bias))
20          model.add(layers.Dense(units=output_size,
21                  activation=output_activation, use_bias=use_bias))
22          optimizer = optimizers.Adam(learning_rate)
23          model.compile(optimizer=optimizer, loss=loss)
24          return model
25
26      def reset(self, mode=None):
27          self.mode = mode
28          if self.mode == 'train':
29              self.trajectory = []
30
31      def step(self, observation, reward, terminated):
32          if self.mode == 'train':
33              action = np.random.choice(self.action_n)   # use random policy
34              self.trajectory += [observation, reward, terminated, action]
35          else:
36              probs = self.policy_net.predict(observation[np.newaxis],
37                      verbose=0)[0]
38              action = np.random.choice(self.action_n, p=probs)
39          return action
```

```
40
41      def close(self):
42          if self.mode == 'train':
43              self.learn()
44
45      def learn(self):
46          df = pd.DataFrame(np.array(self.trajectory, dtype=object).reshape(-1,
47                  4), columns=['state', 'reward', 'terminated', 'action'])
48          df['discount'] = self.gamma ** df.index.to_series()
49          df['discounted_reward'] = df['discount'] * df['reward'].astype(float)
50          df['discounted_return'] = df['discounted_reward'][::-1].cumsum()
51          states = np.stack(df['state'])
52          actions = np.eye(self.action_n)[df['action'].astype(int)]
53          df['behavior_prob'] = 1. / self.action_n
54          df['sample_weight'] = df['discounted_return'] / df['behavior_prob']
55          sample_weight = df[['sample_weight',]].values
56          self.policy_net.fit(states, actions, sample_weight=sample_weight,
57                  verbose=0)
58
59
60  agent = OffPolicyVPGAgent(env)
```

Code 7.6   Off-policy PG agent (PyTorch version).
CartPole-v0_OffPolicyVPG_torch.ipynb

```
 1  class OffPolicyVPGAgent:
 2      def __init__(self, env):
 3          self.action_n = env.action_space.n
 4          self.gamma = 0.99
 5
 6          self.policy_net = self.build_net(
 7                  input_size=env.observation_space.shape[0], hidden_sizes=[],
 8                  output_size=self.action_n, output_activator=nn.Softmax(1))
 9          self.optimizer = optim.Adam(self.policy_net.parameters(), lr=0.06)
10
11      def build_net(self, input_size, hidden_sizes, output_size,
12              output_activator=None, use_bias=False):
13          layers = []
14          for input_size, output_size in zip(
15                  [input_size,] + hidden_sizes, hidden_sizes + [output_size,]):
16              layers.append(nn.Linear(input_size, output_size, bias=use_bias))
17              layers.append(nn.ReLU())
18          layers = layers[:-1]
19          if output_activator:
20              layers.append(output_activator)
21          model = nn.Sequential(*layers)
22          return model
23
24      def reset(self, mode=None):
25          self.mode = mode
26          if self.mode == 'train':
27              self.trajectory = []
28
29      def step(self, observation, reward, terminated):
30          if self.mode == 'train':
31              action = np.random.choice(self.action_n)  # use random policy
32              self.trajectory += [observation, reward, terminated, action]
33          else:
34              state_tensor = torch.as_tensor(observation,
35                      dtype=torch.float).unsqueeze(0)
36              prob_tensor = self.policy_net(state_tensor)
37              action_tensor = distributions.Categorical(prob_tensor).sample()
38              action = action_tensor.numpy()[0]
39          return action
40
```

```
41      def close(self):
42          if self.mode == 'train':
43              self.learn()
44
45      def learn(self):
46          state_tensor = torch.as_tensor(self.trajectory[0::4],
47                  dtype=torch.float)
48          reward_tensor = torch.as_tensor(self.trajectory[1::4],
49                  dtype=torch.float)
50          action_tensor = torch.as_tensor(self.trajectory[3::4],
51                  dtype=torch.long)
52          arange_tensor = torch.arange(state_tensor.shape[0], dtype=torch.float)
53          discount_tensor = self.gamma ** arange_tensor
54          discounted_reward_tensor = discount_tensor * reward_tensor
55          discounted_return_tensor = \
56                  discounted_reward_tensor.flip(0).cumsum(0).flip(0)
57          all_pi_tensor = self.policy_net(state_tensor)
58          pi_tensor = torch.gather(all_pi_tensor, 1,
59                  action_tensor.unsqueeze(1)).squeeze(1)
60          behavior_prob = 1. / self.action_n
61          loss_tensor = -(discounted_return_tensor / behavior_prob *
62                  pi_tensor).mean()
63          self.optimizer.zero_grad()
64          loss_tensor.backward()
65          self.optimizer.step()
66
67
68  agent = OffPolicyVPGAgent(env)
```

Code 7.7   Off-policy PG agent with baseline (TensorFlow version).
CartPole-v0_OffPolicyVPGwBaseline_tf.ipynb

```
1   class OffPolicyVPGwBaselineAgent:
2       def __init__(self, env):
3           self.action_n = env.action_space.n
4           self.gamma = 0.99
5
6           def dot(y_true, y_pred):
7               return -tf.reduce_sum(y_true * y_pred, axis=-1)
8
9           self.policy_net = self.build_net(hidden_sizes=[],
10                  output_size=self.action_n, output_activation=nn.softmax,
11                  loss=dot, learning_rate=0.06)
12          self.baseline_net = self.build_net(hidden_sizes=[],
13                  learning_rate=0.1)
14
15      def build_net(self, hidden_sizes, output_size=1,
16              activation=nn.relu, output_activation=None,
17              use_bias=False, loss=losses.mse, learning_rate=0.001):
18          model = keras.Sequential()
19          for hidden_size in hidden_sizes:
20              model.add(layers.Dense(units=hidden_size,
21                      activation=activation, use_bias=use_bias))
22          model.add(layers.Dense(units=output_size,
23                  activation=output_activation, use_bias=use_bias))
24          optimizer = optimizers.Adam(learning_rate)
25          model.compile(optimizer=optimizer, loss=loss)
26          return model
27
28      def reset(self, mode=None):
29          self.mode = mode
30          if self.mode == 'train':
31              self.trajectory = []
32
33      def step(self, observation, reward, terminated):
```

```
34        if self.mode == 'train':
35            action = np.random.choice(self.action_n)  # use random policy
36            self.trajectory += [observation, reward, terminated, action]
37        else:
38            probs = self.policy_net.predict(observation[np.newaxis],
39                    verbose=0)[0]
40            action = np.random.choice(self.action_n, p=probs)
41        return action
42
43    def close(self):
44        if self.mode == 'train':
45            self.learn()
46
47    def learn(self):
48        df = pd.DataFrame(np.array(self.trajectory, dtype=object).reshape(-1,
49                4), columns=['state', 'reward', 'terminated', 'action'])
50
51        # update baseline
52        df['discount'] = self.gamma ** df.index.to_series()
53        df['discounted_reward'] = df['discount'] * df['reward'].astype(float)
54        df['discounted_return'] = df['discounted_reward'][::-1].cumsum()
55        df['return'] = df['discounted_return'] / df['discount']
56        states = np.stack(df['state'])
57        returns = df[['return',]].values
58        self.baseline_net.fit(states, returns, verbose=0)
59
60        # update policy
61        states = np.stack(df['state'])
62        df['baseline'] = self.baseline_net.predict(states, verbose=0)
63        df['psi'] = df['discounted_return'] - df['baseline'] * df['discount']
64        df['behavior_prob'] = 1. / self.action_n
65        df['sample_weight'] = df['psi'] / df['behavior_prob']
66        actions = np.eye(self.action_n)[df['action'].astype(int)]
67        sample_weight = df[['sample_weight',]].values
68        self.policy_net.fit(states, actions, sample_weight=sample_weight,
69                verbose=0)
70
71
72 agent = OffPolicyVPGwBaselineAgent(env)
```

Code 7.8   Off-policy PG agent with baseline (PyTorch version).
CartPole-v0_OffPolicyVPGwBaseline_torch.ipynb

```
1  class OffPolicyVPGwBaselineAgent:
2      def __init__(self, env):
3          self.action_n = env.action_space.n
4          self.gamma = 0.99
5
6          self.policy_net = self.build_net(
7                  input_size=env.observation_space.shape[0], hidden_sizes=[],
8                  output_size=self.action_n, output_activator=nn.Softmax(1))
9          self.policy_optimizer = optim.Adam(self.policy_net.parameters(),
10                 lr=0.06)
11         self.baseline_net = self.build_net(
12                 input_size=env.observation_space.shape[0], hidden_sizes=[])
13         self.baseline_optimizer = optim.Adam(self.policy_net.parameters(),
14                 lr=0.1)
15         self.baseline_loss = nn.MSELoss()
16
17     def build_net(self, input_size, hidden_sizes, output_size=1,
18             output_activator=None, use_bias=False):
19         layers = []
20         for input_size, output_size in zip([input_size,] + hidden_sizes,
21                 hidden_sizes + [output_size,]):
22             layers.append(nn.Linear(input_size, output_size, bias=use_bias))
```

```
23          layers.append(nn.ReLU())
24      layers = layers[:-1]
25      if output_activator:
26          layers.append(output_activator)
27      model = nn.Sequential(*layers)
28      return model
29
30  def reset(self, mode=None):
31      self.mode = mode
32      if self.mode == 'train':
33          self.trajectory = []
34
35  def step(self, observation, reward, terminated):
36      if self.mode == 'train':
37          action = np.random.choice(self.action_n)  # use random policy
38          self.trajectory += [observation, reward, terminated, action]
39      else:
40          state_tensor = torch.as_tensor(observation,
41                  dtype=torch.float).unsqueeze(0)
42          prob_tensor = self.policy_net(state_tensor)
43          action_tensor = distributions.Categorical(prob_tensor).sample()
44          action = action_tensor.numpy()[0]
45      return action
46
47  def close(self):
48      if self.mode == 'train':
49          self.learn()
50
51  def learn(self):
52      state_tensor = torch.as_tensor(self.trajectory[0::4],
53              dtype=torch.float)
54      reward_tensor = torch.as_tensor(self.trajectory[1::4],
55              dtype=torch.float)
56      action_tensor = torch.as_tensor(self.trajectory[3::4],
57              dtype=torch.long)
58      arange_tensor = torch.arange(state_tensor.shape[0], dtype=torch.float)
59
60      # update baseline
61      discount_tensor = self.gamma ** arange_tensor
62      discounted_reward_tensor = discount_tensor * reward_tensor
63      discounted_return_tensor = discounted_reward_tensor.flip(
64              0).cumsum(0).flip(0)
65      return_tensor = discounted_return_tensor / discount_tensor
66      pred_tensor = self.baseline_net(state_tensor)
67      psi_tensor = (discounted_return_tensor - discount_tensor
68              * pred_tensor).detach()
69      baseline_loss_tensor = self.baseline_loss(pred_tensor,
70              return_tensor.unsqueeze(1))
71      self.baseline_optimizer.zero_grad()
72      baseline_loss_tensor.backward()
73      self.baseline_optimizer.step()
74
75      # update policy
76      all_pi_tensor = self.policy_net(state_tensor)
77      pi_tensor = torch.gather(all_pi_tensor, 1,
78              action_tensor.unsqueeze(1)).squeeze(1)
79      behavior_prob = 1. / self.action_n
80      policy_loss_tensor = -(psi_tensor / behavior_prob * pi_tensor).mean()
81      self.policy_optimizer.zero_grad()
82      policy_loss_tensor.backward()
83      self.policy_optimizer.step()
84
85
86  agent = OffPolicyVPGwBaselineAgent(env)
```

7.5 Summary

- Policy Gradient (PG) method uses the parametric function $\pi(a|s;\boldsymbol{\theta})$ to approximate the optimal policy. For discrete action space, the policy approximation can be of the form

$$\pi(a|s;\boldsymbol{\theta}) = \frac{\exp h(s,a;\boldsymbol{\theta})}{\sum_{a'}\exp h(s,a';\boldsymbol{\theta})}, \quad s \in \mathcal{S}, a \in \mathcal{A}(s),$$

where $h(s,a;\boldsymbol{\theta})$ is the action preference function. For continuous action space, the policy approximation can be of the form

$$A = \boldsymbol{\mu}(\boldsymbol{\theta}) + N \circ \boldsymbol{\sigma}(\boldsymbol{\theta}),$$

where $\boldsymbol{\mu}(\boldsymbol{\theta})$ and $\boldsymbol{\sigma}(\boldsymbol{\theta})$ as the mean vector and the standard deviation vector,
- PG theorem shows that the gradient of return with respect to the policy parameter is in the direction of $E_{\pi(\boldsymbol{\theta})}\left[\Psi_t \nabla \ln\pi(A_t|S_t;\boldsymbol{\theta})\right]$, where Ψ_t can be G_0, $\gamma^t G_t$, and so on.
- PG algorithm updates the parameter $\boldsymbol{\theta}$ to increase $\Psi_t \ln \pi(A_t|S_t;\boldsymbol{\theta})$.
- Baseline function $B(S_t)$ can be a stochastic function or a deterministic function. The baseline function should not depend on the action A_t. The baseline function should be able to reduce the variance.
- PG with importance sampling is the off-policy PG algorithm.

7.6 Exercises

7.6.1 Multiple Choices

7.1 On the update $\boldsymbol{\theta} \leftarrow \boldsymbol{\theta}+\alpha\Psi_t\nabla\ln\pi(A_t|S_t;\boldsymbol{\theta})$ in VPG algorithm, choose the correct one: (   )

A. Ψ_t can be G_0, but shall not be $\gamma^t G_t$.
B. Ψ_t can be $\gamma^t G_t$, but shall not be G_0.
C. Ψ_t can be either G_0 or $\gamma^t G_t$.

7.2 On the baseline in the policy gradient algorithm, choose the correct one: (   )

A. The baseline in PG algorithm is primarily used to reduce bias.
B. The baseline in PG algorithm is primarily used to reduce variance.
C. The baseline in PG algorithm is primarily used to reduce both bias and variance.

7.6.2 Programming

7.3 Use a PG algorithm to solve the task `Blackjack-v1`.

7.6.3 Mock Interview

7.4 Prove the policy gradient theorem.

7.5 What is the greatest weakness of policy gradient algorithms?

7.6 What is baseline? Why do PG algorithms need baseline?

Chapter 8
AC: Actor–Critic

This chapter covers

- Actor–Critic (AC) method
- action-value AC algorithm
- advantage AC algorithm
- performance difference lemma
- surrogate advantage
- Proximal Policy Optimization (PPO) algorithm
- Natural Policy Gradient (NPG) algorithm
- Trust Region Policy Optimization (TRPO) algorithm
- Off-Policy AC (OffPAC) algorithm

8.1 Intuition of AC

Actor–critic method combines the policy gradient method and bootstrapping. On the one hand, it uses policy gradient theorem to calculate policy gradient and update parameters. This part is called **actor**. On the other hand, it estimates values, and uses the value estimate to bootstrap. This part is called **critic**. Using policy gradient and value bootstrapping together is called **Actor–Critic (AC)**.

The previous chapter uses a parametric function $\pi(a|s; \boldsymbol{\theta})$ ($s \in \mathcal{S}$, $a \in \mathcal{A}(s)$), which satisfies $\sum_a \pi(a|s; \boldsymbol{\theta}) = 1$ ($s \in \mathcal{S}$) to approximate optimal policy, and updates the policy parameter $\boldsymbol{\theta}$ in the direction of $\mathrm{E}_{\pi(\boldsymbol{\theta})}\left[\Psi_t \nabla \ln\pi(A_t|S_t; \boldsymbol{\theta})\right]$ according to the policy gradient theorem, where $\Psi_t = \gamma^t\left(G_t - B(s)\right)$. (Schulman, 2016) pointed out that Ψ_t can also be in the following forms:

- (action value) $\Psi_t = \gamma^t q_\pi(S_t, A_t)$. (This can be proved using $\mathrm{E}_{\pi(\boldsymbol{\theta})}\left[\gamma^t G_t \nabla \ln\pi(A_t|S_t; \boldsymbol{\theta})\right] = \mathrm{E}_{\pi(\boldsymbol{\theta})}\left[\gamma^t q_{\pi(\boldsymbol{\theta})}(S_t, A_t) \nabla \ln\pi(A_t|S_t; \boldsymbol{\theta})\right]$.)

© The Author(s), under exclusive license to Springer Nature Singapore Pte Ltd. 2024
Z. Xiao, *Reinforcement Learning*, https://doi.org/10.1007/978-981-19-4933-3_8

- (advantage) $\Psi_t = \gamma^t \left[R_t + \gamma v_\pi(S_{t+1}) - v_\pi(S_t) \right]$. (Add a baseline in the previous form.)
- (TD error) $\Psi_t = \gamma^t \left[R_t + \gamma v_\pi(S_{t+1}) - v_\pi(S_t) \right]$.

All aforementioned three forms use bootstrapping. For $\Psi_t = \gamma^t q_\pi(S_t, A_t)$, it uses $q_\pi(S_t, A_t)$ to estimate return, so it uses bootstrapping. For $\Psi_t = \gamma^t \left[q_\pi(S_t, A_t) - v_\pi(S_t) \right]$, it still uses $q_\pi(S_t, A_t)$ to estimate return, but with an additional baseline function $B(s) = v_\pi(s)$. For $\Psi_t = \gamma^t \left[R_t + \gamma v_\pi(S_{t+1}) - v_\pi(S_t) \right]$, it uses one-step TD return $R_t + \gamma v_\pi(S_{t+1})$ to estimate return, with additional baseline $V(s) = v_\pi(s)$. So it uses bootstrapping, too.

During actual training, we do not know the actual values. We can only estimate those values. We can use the function approximation, which uses parametric function $v(s; \mathbf{w})$ $(s \in S)$ to approximate v_π, or use $q(s, a; \mathbf{w})$ $(s \in S, a \in \mathcal{A}(s))$ to estimate q_π. In the previous chapter, PG with baseline used a parametric function $v(s; \mathbf{w})$ $(s \in S)$ as baseline. We can take a step further by replacing the return part in Ψ_t by U_t, the return obtained by bootstrapping. For example, we can use one-step TD return $U_t = R_{t+1} + \gamma v(S_{t+1}; \mathbf{w})$ so that $\Psi_t = \gamma^t \left[R_{t+1} + v(S_{t+1}; \mathbf{w}) - v(S_t; \mathbf{w}) \right]$. Here, state value estimates $v(s; \mathbf{w})$ $(s \in S)$ are critic, and such algorithm is an AC algorithm.

! Note

Only algorithms that use bootstrapping to estimate return, which introduces bias, are AC algorithms. Using value estimates as the baseline function does not introduce bias, since the baseline function can be selected arbitrarily anyway. Therefore, VPG with baseline is not an AC algorithm.

8.2 On-Policy AC

This section covers the basic on-policy algorithm, including action-value AC algorithm and advantage AC algorithm. We will learn eligibility-trace AC algorithm, too.

8.2.1 Action-Value AC

Action-value AC algorithm updates the policy parameter θ to increase $\Psi_t \ln \pi(A_t | S_t; \theta)$, where $\Psi_t = \gamma^t q(S_t, A_t; \mathbf{w})$. The return estimates are bootstrapped from the action value estimate $q(S_t, A_t; \mathbf{w})$.

Algorithm 8.1 shows the action-value AC algorithm. It uses a variable Γ to store the accumulative discount factor. This value is 1 at the beginning of an episode, and it is multiplied by γ every step. Therefore, it will be γ^t at step t.

Algorithm 8.1 Action-value on-policy AC.

Input: environment (without mathematical model).
Output: optimal policy estimate $\pi(\theta)$.
Parameters: optimizers (containing learning rate $\alpha^{(\theta)}$ and $\alpha^{(\mathbf{w})}$), discount factor γ, and parameters to control the number of episodes and maximum steps in an episode.

1. (Initialize) Initialize the policy parameter θ and value parameter $\mathbf{w}$.
2. (AC update) For each episode:

   2.1. (Initialize accumulative discount factor) $\Gamma \leftarrow 1$.
   2.2. (Initialize state–action pair) Select the initial state S, and use $\pi(\cdot|S;\theta)$ to select the action A.
   2.3. Loop until the episode ends:
   2.3.1. (Sample) Execute the action A, and then observe the reward R, the next state S', and the indicator of episode end D'.
   2.3.2. (Decide) Use $\pi(\cdot|S';\theta)$ to determine the action A'. (The action can be arbitrary if $D' = 1$.)
   2.3.3. (Calculate TD return) $U \leftarrow R + \gamma(1 - D')q(S', A'; \mathbf{w})$.
   2.3.4. (Update policy parameter) Update θ to reduce $-\Gamma q(S, A; \mathbf{w}) \ln \pi(A|S; \theta)$. (For example, $\theta \leftarrow \theta + \alpha^{(\theta)}\Gamma q(S, A; \mathbf{w})\nabla \ln\pi(A|S; \theta)$).
   2.3.5. (Update value parameter) Update $\mathbf{w}$ to reduce $[U - q(S, A; \mathbf{w})]^2$. (For example, $\mathbf{w} \leftarrow \mathbf{w} + \alpha^{(\mathbf{w})}[U - q(S, A; \mathbf{w})]\nabla q(S, A; \mathbf{w})$.)
   2.3.6. (Update accumulative discount factor) $\Gamma \leftarrow \gamma\Gamma$.
   2.3.7. $S \leftarrow S', A \leftarrow A'$.

8.2.2 Advantage AC

The action-value AC algorithm in the previous section uses $\Psi_t = \gamma^t q(S_t, A_t; \mathbf{w})$. Introducing an additional baseline function $B(S_t) = v(S_t; \mathbf{w})$ will lead to $\Psi_t = \gamma^t[q(S_t, A_t; \mathbf{w}) - v(S_t; \mathbf{w})]$. Here, $q(S_t, A_t; \mathbf{w}) - v(S_t; \mathbf{w})$ can be viewed as an estimate of advantage. The algorithm is named as **Advantage AC** algorithm accordingly. However, if we use the form $q(S_t, A_t; \mathbf{w}) - v(S_t; \mathbf{w})$ to approximate the advantage function, we need to maintain two networks for both $q(\cdot, \cdot; \mathbf{w})$ and $v(\cdot; \mathbf{w})$. In order to avoid this troublesome, we use $U_t = R_{t+1} + \gamma v(S_{t+1}; \mathbf{w})$ as the target. Therefore, the advantage function estimates are in the form of $R_{t+1} + \gamma v(S_{t+1}; \mathbf{w}) - v(S_t; \mathbf{w})$. Algorithm 8.2 shows the advantage AC algorithm.

Algorithm 8.2   Advantage AC.

Input: environment (without mathematical model).
Output: optimal policy estimate $\pi(\boldsymbol{\theta})$.
Parameters: optimizers (containing learning rate $\alpha^{(\boldsymbol{\theta})}$ and $\alpha^{(\mathbf{w})}$), discount factor γ, and parameters to control the number of episodes and maximum steps in an episode.

1. (Initialize) Initialize the policy parameter $\boldsymbol{\theta}$ and value parameter $\mathbf{w}$.
2. (AC update) For each episode:
   - **2.1.** (Initialize accumulative discount factor) $\Gamma \leftarrow 1$.
   - **2.2.** (Initialize state) Select the initial state S.
   - **2.3.** Loop until the episode ends:
     - **2.3.1.** (Decide) Use $\pi(\cdot|S; \boldsymbol{\theta})$ to determine the action A.
     - **2.3.2.** (Sample) Execute the action A, and then observe the reward R, the next state S', and the indicator of episode end D'.
     - **2.3.3.** (Calculate TD return) $U \leftarrow R + \gamma(1 - D')q(S', A'; \mathbf{w})$.
     - **2.3.4.** (Update policy parameter) Update $\boldsymbol{\theta}$ to reduce $-\Gamma q(S, A; \mathbf{w}) \ln \pi(A|S; \boldsymbol{\theta})$. (For example, $\boldsymbol{\theta} \leftarrow \boldsymbol{\theta} + \alpha^{(\boldsymbol{\theta})} \Gamma q(S, A; \mathbf{w}) \nabla \ln \pi(A|S; \boldsymbol{\theta})$).
     - **2.3.5.** (Update value parameter) Update $\mathbf{w}$ to reduce $\left[U - q(S, A; \mathbf{w})\right]^2$. (For example, $\mathbf{w} \leftarrow \mathbf{w} + \alpha^{(\mathbf{w})}\left[U - q(S, A; \mathbf{w})\right] \nabla q(S, A; \mathbf{w})$.)
     - **2.3.6.** (Update accumulative discount factor) $\Gamma \leftarrow \gamma\Gamma$.
     - **2.3.7.** $S \leftarrow S'$.

Asynchronous Advantage Actor–Critic (**A3C**) is a famous RL algorithm in history. It is a distributed version of Advantage AC algorithm. Algorithm 8.3 shows the main idea of this algorithm. A3C has multiple workers. Therefore, besides the global value parameter $\mathbf{w}$ and policy parameter $\boldsymbol{\theta}$, each thread maintains its own value parameter $\mathbf{w}'$ and policy parameter $\boldsymbol{\theta}'$. When each worker learns, it first fetches parameters from global parameters. Then it learns and updates its own parameters. At last, it synchronizes what it had learned to the global parameters. The bootstrapping of A3C can use one-step TD or multiple-step TD.

Algorithm 8.3   A3C (one-step TD version, showing the behavior of one worker).

Input: environment (without mathematical model).
Output: optimal policy estimate $\pi(\boldsymbol{\theta})$.

Parameters: optimizers (containing learning rate $\alpha^{(\boldsymbol{\theta})}$ and $\alpha^{(\mathbf{w})}$), discount factor γ, and parameters to control the number of episodes and maximum steps in an episode.

1. (Initialize) Fetch the policy parameter $\boldsymbol{\theta}' \leftarrow \boldsymbol{\theta}$ and value parameter $\mathbf{w}' \leftarrow \mathbf{w}$.

2. (AC update) For each episode:

   2.1. Use the policy $\pi(\boldsymbol{\theta}')$ to generate the trajectory $S_0, A_0, R_1, S_1, A_1, R_2, \ldots, S_{T-1}, R_T, S_T$, until the episode ends or reaches the maximum number of steps T.

   2.2. Initialize for gradient calculation:

      2.2.1. (Initialize episode) If S_T is the terminal states, set $U \leftarrow 0$. Otherwise, set $U \leftarrow v(S_T; \mathbf{w}')$.

      2.2.2. (Initialize gradients) $\mathbf{g}^{(\boldsymbol{\theta})} \leftarrow \mathbf{0}$, and $\mathbf{g}^{(\mathbf{w})} \leftarrow \mathbf{0}$.

   2.3. (Calculate gradients asynchronically) For $t = T - 1, T - 2, \ldots, 0$, execute the following:

      2.3.1. (Calculate TD return) $U \leftarrow \gamma U + R_{t+1}$.

      2.3.2. (Update policy parameter) $\mathbf{g}^{(\boldsymbol{\theta})} \leftarrow \mathbf{g}^{(\boldsymbol{\theta})} + \gamma^t \left[U - v(S_t; \mathbf{w}') \right] \nabla \ln \pi(A_t | S_t; \boldsymbol{\theta}')$.

      2.3.3. (Update value gradient) $\mathbf{g}^{(\mathbf{w})} \leftarrow \mathbf{g}^{(\mathbf{w})} + \gamma^t \left[U - v(S_t; \mathbf{w}') \right] \nabla v(S_t; \mathbf{w}')$.

3. Update global parameters:

   3.1. (Update global policy parameter) Update the policy parameter $\boldsymbol{\theta}$ in the direction of $\mathbf{g}^{(\boldsymbol{\theta})}$ (For example, $\boldsymbol{\theta} \leftarrow \boldsymbol{\theta} + \alpha^{(\boldsymbol{\theta})} \mathbf{g}^{(\boldsymbol{\theta})}$).

   3.2. (Update global value parameter) Update the policy parameter $\mathbf{w}$ in the direction of $\mathbf{g}^{(\mathbf{w})}$ (For example, $\mathbf{w} \leftarrow \mathbf{w} + \alpha^{(\mathbf{w})} \mathbf{g}^{(\mathbf{w})}$).

8.2.3 Eligibility Trace AC

AC method uses bootstrapping, so it can use eligibility trace, too. Algorithm 8.4 shows the eligibility trace advantage AC algorithm. This algorithm has two eligibility trace $\mathbf{z}^{(\boldsymbol{\theta})}$ and $\mathbf{z}^{(\mathbf{w})}$, which are for the policy parameter $\boldsymbol{\theta}$ and value parameter $\mathbf{w}$ respectively. They can have their own eligibility trace parameters. Specifically, the eligibility trace $\mathbf{z}^{(\boldsymbol{\theta})}$ corresponds to the policy parameter $\boldsymbol{\theta}$, and it uses the accumulative trace with gradient $\nabla \ln \pi(A | S; \boldsymbol{\theta})$ and decay parameter $\lambda^{(\boldsymbol{\theta})}$. The accumulative discount factor γ^t can be integrated into the eligibility trace, too. The eligibility trace $\mathbf{z}^{(\mathbf{w})}$ corresponds to the value parameter $\mathbf{w}$. It uses the accumulative trace with gradient $\nabla v(S; \mathbf{w})$ and decay parameter $\lambda^{(\mathbf{w})}$.

Algorithm 8.4   Advantage AC with eligibility trace.

Input: environment (without mathematical model).
Output: optimal policy estimate $\pi(\boldsymbol{\theta})$.
Parameters: eligibility trace parameters $\lambda^{(\boldsymbol{\theta})}$ and $\lambda^{(\mathbf{w})}$, optimizers (containing learning rate $\alpha^{(\boldsymbol{\theta})}$ and $\alpha^{(\mathbf{w})}$), discount factor γ, and parameters to control the number of episodes and maximum steps in an episode.

1. (Initialize) Initialize the policy parameter $\boldsymbol{\theta}$ and value parameter $\mathbf{w}$.
2. (AC update) For each episode:

    2.1. (Initialize accumulative discount factor) $\Gamma \leftarrow 1$.
    2.2. (Initialize state) Select the initial state S.
    2.3. (Initialize eligibility trace) $\mathbf{z}^{(\boldsymbol{\theta})} \leftarrow \mathbf{0}$ and $\mathbf{z}^{(\mathbf{w})} \leftarrow \mathbf{0}$.
    2.4. Loop until the episode ends:
        2.4.1. (Decide) Use $\pi(\cdot|S;\boldsymbol{\theta})$ to determine the action A.
        2.4.2. (Sample) Execute the action A, and then observe the reward R, the next state S', and the indicator of episode end D'.
        2.4.3. (Calculate TD return) $U \leftarrow R + \gamma(1 - D')q(S', A'; \mathbf{w})$.
        2.4.4. (Update policy eligibility trace) $\mathbf{z}^{(\boldsymbol{\theta})} \leftarrow \gamma\lambda^{(\boldsymbol{\theta})}\mathbf{z}^{(\boldsymbol{\theta})} + \Gamma\nabla\ln\pi(A|S;\boldsymbol{\theta})$.
        2.4.5. (Update policy parameter) $\boldsymbol{\theta} \leftarrow \boldsymbol{\theta} + \alpha^{(\boldsymbol{\theta})}\big[U - v(S;\mathbf{w})\big]\mathbf{z}^{(\boldsymbol{\theta})}$.
        2.4.6. (Update value eligibility trace) $\mathbf{z}^{(\mathbf{w})} \leftarrow \gamma\lambda^{(\mathbf{w})}\mathbf{z}^{(\mathbf{w})} + \nabla v(S;\mathbf{w})$.
        2.4.7. (Update value parameter) $\mathbf{w} \leftarrow \mathbf{w} + \alpha^{(\mathbf{w})}\big[U - v(S;\mathbf{w})\big]\mathbf{z}^{(\mathbf{w})}$.
        2.4.8. (Update accumulative discount factor) $\Gamma \leftarrow \gamma\Gamma$.
        2.4.9. $S \leftarrow S'$.

8.3 On-Policy AC with Surrogate Objective

This section introduces AC algorithms with surrogate objectives. These algorithms do not try to maximize the return expectation directly during training. They try to maximize the surrogate advantage to obtain better performance.

8.3.1 Performance Difference Lemma

Performance difference lemma uses advantages to present the return expectation difference between two policies. It is the foundation of algorithms in this section.

    Performance Difference Lemma (Kakade, 2002: PDL): Consider two policies π' and π'' on the same environment. The difference of the return expectation between these two policies can be presented as

$$g_{\pi'} - g_{\pi''} = \mathrm{E}_{\pi'}\left[\sum_{t=0}^{+\infty} \gamma^t a_{\pi''}(S_t, A_t)\right].$$

It can also be expressed as the following using discounted expectation:

$$g_{\pi'} - g_{\pi''} = \mathrm{E}_{(S,A)\sim\rho_{\pi'}}\left[a_{\pi''}(S, A)\right].$$

(Proof: Since $a_{\pi''}(S_t, A_t) = q_{\pi''}(S_t, A_t) - v_{\pi''}(S_t) = \mathrm{E}\left[R_{t+1} + \gamma v_{\pi''}(S_{t+1})\right] - v_{\pi''}(S_t)$, we have

$$\mathrm{E}_{\pi'}\left[\sum_{t=0}^{+\infty} \gamma^t a_{\pi''}(S_t, A_t)\right]$$

$$= \mathrm{E}_{\pi'}\left[\sum_{t=0}^{+\infty} \gamma^t \left(R_{t+1} + \gamma v_{\pi''}(S_{t+1}) - v_{\pi''}(S_t)\right)\right]$$

$$= -\mathrm{E}\left[v_{\pi''}(S_0)\right] + \mathrm{E}_{\pi'}\left[\sum_{t=0}^{+\infty} \gamma^t R_{t+1}\right]$$

$$= -g_{\pi''} + g_{\pi'}.$$

The proof completes.)

Performance difference lemma provides a way to update policy parameters. If we view π'' as the policy estimate before an iteration, and we want to find a new policy π' in the iteration such that its return expectation $g_{\pi'}$ can be as large as possible. Then that iteration can maximize $\mathrm{E}_{(S,A)\sim\rho_{\pi'}}\left[a_{\pi''}(S, A)\right]$ to maximize $g_{\pi'}$. We will further discuss this in subsequent subsections.

8.3.2 Surrogate Advantage

Performance difference lemma tells us, we can try to maximize an advantage expectation in each iteration to maximize the return expectation. For policy that is approximated by a parametric function, we can use $\pi(\theta_k)$ to denote the old policy before the k-th iteration, where θ_k is the old policy parameter. We can maximize $\mathrm{E}_{(S,A)\sim\rho_{\pi(\theta)}}\left[a_{\pi(\theta_k)}(S, A)\right]$ to find a better policy $\pi(\theta)$. However, the expectation $\mathrm{E}_{(S,A)\sim\rho_{\pi(\theta)}}\left[a_{\pi(\theta_k)}(S, A)\right]$ is over $\rho_{\pi(\theta)}$, which make it impractical to generate trajectories using this distribution. It will be more convenient for us to use the old policy $\pi(\theta_k)$ to generate trajectories. Can we covert the expectation over $\rho_{\pi(\theta)}$ to the expectation over $\pi(\theta_k)$?

We can first use importance sampling to resolve actions. The importance sampling ratio from $\rho_{\pi(\theta)}$ to $\pi(\theta_k)$ is $\frac{\pi(A|S;\theta)}{\pi(A|S;\theta_k)}$, since $\rho_{\pi(\theta)}(S, A) = \eta_{\pi(\theta)}(S)\pi(A|S; \theta)$ and $\mathrm{Pr}_{\pi(\theta_k)}\left[S_t = s, A_t = a\right] = \mathrm{Pr}_{\pi(\theta_k)}\left[S_t = s\right]\pi(a|s; \theta)$. Accordingly,

$$E_{(S,A)\sim\rho_{\pi(\theta)}}\left[a_{\pi(\theta_k)}(S,A)\right] = E_{S\sim\eta_{\pi(\theta)},A\sim\pi(\cdot|S;\theta_k)}\left[\frac{\pi(A|S;\theta)}{\pi(A|S;\theta_k)}a_{\pi(\theta_k)}(S,A)\right].$$

However, the expectation over $S \sim \eta_{\pi(\theta)}$ is still intractable.

For some optimization problems, it may be difficult to optimize on the original objective. Therefore, we may consider optimizing another parametric objective, that is, adjusting the new parametric function in each iteration. The new objective is called surrogate objective.

MM algorithm is an example to use surrogate objective.

Interdisciplinary Reference 8.1
Optimization: MM Algorithm

MM algorithm, which is short for either Majorize–Minimize algorithm or Minorize–Maximize algorithm, is an optimization method that leverages a surrogate objective during optimization.

The idea of Minorize–Maximize algorithm is as follows: When we want to maximize a function $f(\theta)$ over θ, the Minorize–Maximize algorithm first proposes a surrogate function $l(\theta|\theta_k)$ such that $l(\theta|\theta_k)$ minorizes $f(\theta)$, i.e.

$$l(\theta|\theta_k) \geq f(\theta), \quad \text{all } \theta$$
$$l(\theta_k|\theta_k) = f(\theta_k).$$

Then we maximize $l(\theta|\theta_k)$ (rather than directly maximizing $f(\theta)$). $l(\theta|\theta_k)$ is called surrogate objective (Fig. 8.1).

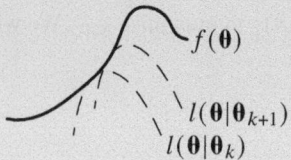

Fig. 8.1   Illustration of MM algorithm.
This figure is adapted from https://en.wikipedia.org/wiki/File:Mmalgorithm.jpg.

Ascent property: If $g(\theta_{k+1}|\theta_k) \geq g(\theta_k|\theta_k)$, $f(\theta_{k+1}) \geq g(\theta_{k+1}|\theta_k) \geq g(\theta_k|\theta_k)$ = $f(\theta_k)$. (Proof: $f(\theta_{k+1}) \geq g(\theta_{k+1}|\theta_k) \geq g(\theta_k|\theta_k) = f(\theta_k)$.) This property guarantees that the MM algorithm can optimize successfully.

Surrogate advantage approximates the expectation over $S \sim \eta_{\pi(\theta)}$ to the expectation over $S_t \sim \pi(\theta_k)$, besides applying the importance sampling over actions. That is,

$$E_{(S,A)\sim\rho_{\pi(\theta)}}\left[a_{\pi(\theta_k)}(S,A)\right] \approx E_{(S_t,A_t)\sim\pi(\theta_k)}\left[\frac{\pi(A_t|S_t;\theta)}{\pi(A_t|S_t;\theta_k)}a_{\pi(\theta_k)}(S_t,A_t)\right].$$

Accordingly, we can get an approximated expression of $g_{\pi(\theta)}$, denoted as $l(\theta|\theta_k)$, where

$$l(\theta|\theta_k) \stackrel{\text{def}}{=} g_{\pi(\theta_k)} + \mathrm{E}_{(S_t, A_t) \sim \pi(\theta_k)} \left[\frac{\pi(A_t|S_t; \theta)}{\pi(A_t|S_t; \theta_k)} a_{\pi(\theta_k)}(S_t, A_t) \right].$$

It is easy to use performance difference lemma to prove that $g_{\pi(\theta)}$ and $l(\theta|\theta_k)$ share the same value (i.e. $g_{\pi(\theta_k)}$) and gradient at $\theta = \theta_k$. Therefore, we approximate the expectation over $g_{\pi(\theta)}$ as the expectation over the old policy $\pi(\theta_k)$.

Since $g_{\pi(\theta)}$ and $l(\theta|\theta_k)$ share the same value and gradient at $\theta = \theta_k$, we can update the policy parameter θ in the direction of

$$\mathrm{E}_{(S_t, A_t) \sim \pi(\theta_k)} \left[\sum_{t=0}^{+\infty} \frac{\pi(A_t|S_t; \theta)}{\pi(A_t|S_t; \theta_k)} a_{\pi(\theta_k)}(S_t, A_t) \right]$$

to increase $g_{\pi(\theta)}$. This is the fundamental idea of AC algorithms with surrogate advantages.

8.3.3 PPO: Proximal Policy Optimization

We have known that the surrogate advantage and the return expectation share the same value and gradient at $\theta = \theta_k$. However, if θ and θ_k are very different, the approximation will not hold. Therefore, the optimized policy in each iteration should not go too far away from the policy before iteration. Based on this idea, (Schulman, 2017: PPO) proposed **Proximal Policy Optimization (PPO)** algorithm, which sets the optimization objective as

$$\mathrm{E}_{\pi(\theta_k)} \left[\min \left\{ \frac{\pi(A_t|S_t; \theta)}{\pi(A_t|S_t; \theta_k)} a_{\pi(\theta_k)}(S_t, A_t), a_{\pi(\theta_k)}(S_t, A_t) + \varepsilon |a_{\pi(\theta_k)}(S_t, A_t)| \right\} \right],$$

where $\varepsilon \in (0, 1)$ is a designated hyperparameter. When we optimize using this objective, the objective is at most $\varepsilon |a_{\pi(\theta_k)}(S_t, A_t)|$ larger than $a_{\pi(\theta_k)}(S_t, A_t)$. Consequently, the optimization problem does not need to make the surrogate advantage $\frac{\pi(A_t|S_t; \theta)}{\pi(A_t|S_t; \theta_k)} a_{\pi(\theta_k)}(S_t, A_t)$ very large, so the policy after an update will not be too different from the policy before the update. Since the objective is a clipped version of $\frac{\pi(A_t|S_t; \theta)}{\pi(A_t|S_t; \theta_k)} a_{\pi(\theta_k)}(S_t, A_t)$, so this algorithm is also called clipped PPO.

Algorithm 8.5 shows a simplified version of clipped PPO algorithm. In Step 2.2 of this algorithm, the update of advantage estimates uses a **Generalized Advantage Estimate (GAE)**: It introduces an eligibility-trace parameter $\lambda \in [0, 1]$, which can control the tradeoff between bias and variance. If set $\lambda = 1$, the GAE degrades to the basic form of advantage estimate.

Algorithm 8.5   Clipped PPO (simplified version).

Inputs: environment (without mathematical model).
Outputs: optimal policy estimate $\pi(\theta)$.
Parameters: parameter ε ($\varepsilon > 0$) that limits the objective, parameters used for advantage estimation (such as $\lambda \in [0, 1]$), optimizers, the discount factor γ, and parameters to control the number of episodes and maximum steps in an episode.

1. (Initialize) Initialize the policy parameter θ and value parameter $\mathbf{w}$.
2. (AC update) For each episode:

    2.1. (Decide and sample) Use the policy $\pi(\theta)$ to generate a trajectory.
    2.2. (Calculate old advantage) Use the trajectory to calculate the old advantage, bootstrapping from the value estimates parameterized by $\mathbf{w}$. (For example, $a(S_t, A_t) \leftarrow \sum_{\tau=t}^{T-1} (\gamma\lambda)^{\tau-t} \left[U_{\tau:\tau+1}^{(v)} - v(S_\tau; \mathbf{w}) \right]$.)
    2.3. (Update policy parameter) Update θ to increase $\min\left\{ \frac{\pi(A_t|S_t;\theta)}{\pi(A_t|S_t;\theta_k)} a_{\pi(\theta_k)}(S_t, A_t), a_{\pi(\theta_k)}(S_t, A_t) + \varepsilon \left| a_{\pi(\theta_k)}(S_t, A_t) \right| \right\}$.
    2.4. (Update value parameter) Update $\mathbf{w}$ to reduce value estimate errors. (For example, minimize $\left[G_t - v(S_t; \mathbf{w}) \right]^2$.)

Practically, PPO is used to experience replay. The method is as follows: Every time it uses a policy to generate a trajectory, it calculates the probability $\pi(A_t|S_t; \theta)$, advantage estimate $a(S_t, A_t; \mathbf{w})$, and return G_t for each step, and saves the experience in the form of $(S_t, A_t, \pi(A_t|S_t; \theta), a(S_t, A_t; \mathbf{w}), G_t)$ in the storage. $\pi(A_t|S_t; \theta)$ can also be saved in the form of $\ln \pi(A_t|S_t; \theta)$. During the replaying, select a batch of experiences $\mathcal{B}$ from the storage, and learn from this batch. $S_t, A_t, \pi(A_t|S_t; \theta), a(S_t, A_t; \mathbf{w})$ are used to update policy parameters, while S_t and G_t are used to update value parameters. We can repeat the replaying multiple times to re-use the experiences. The storage is emptied after the policy parameters are updated.

Remarkably, PPO algorithm only uses experiences that are generated by the recent policy. So the PPO algorithm is an on-policy algorithm. Every time PPO updates its policy parameters, the policy becomes meaningless. Therefore, we need to empty the experience storage.

Algorithm 8.6 shows the PPO algorithm with on-policy replayer.

Algorithm 8.6   Clipped PPO (with on-policy experience replay).

Inputs: environment (without mathematical model).
Outputs: optimal policy estimate $\pi(\theta)$.

Parameters: parameter ε ($\varepsilon > 0$) that limits the objective, parameters used for advantage estimation (such as $\lambda \in [0, 1]$), optimizers, the discount factor γ, and parameters to control the number of episodes and maximum steps in an episode.

1. (Initialize) Initialize the policy parameter $\boldsymbol{\theta}$ and value parameter $\mathbf{w}$.
2. Loop:

    2.1. (Initialize experience storage) $\mathcal{D} \leftarrow \varnothing$.

    2.2. (Collect experiences) Do the following one or multiple times:

        2.2.1. (Decide and sample) Use the policy $\pi(\boldsymbol{\theta})$ to generate a trajectory.

        2.2.2. (Calculate old advantage) Use the trajectory to calculate the old advantage, bootstrapping from the value estimates parameterized by $\mathbf{w}$. (For example,
$$a(S_t, A_t) \leftarrow \sum_{\tau=t}^{T-1} (\gamma\lambda)^{\tau-t} \left[U_{\tau:\tau+1}^{(v)} - v(S_\tau; \mathbf{w}) \right].)$$

        2.2.3. (Store) Save the experience $(S_t, A_t, \pi(A_t|S_t; \boldsymbol{\theta}), a(S_t, A_t; \mathbf{w}), G_t)$ in the storage $\mathcal{D}$.

    2.3. (Use experiences) Do the following once or multiple times:

        2.3.1. (Replay) Sample a batch of experiences $\mathcal{B}$ from the storage $\mathcal{D}$. Each entry in the form of (S, A, Π, A, G).

        2.3.2. (Update policy parameter) Update $\boldsymbol{\theta}$ to increase $\frac{1}{|\mathcal{B}|} \sum_{(S,A,\Pi,A,G)\in\mathcal{B}} \min\left\{ \frac{\pi(A|S;\boldsymbol{\theta})}{\Pi} A, A + \varepsilon|A| \right\}$.

        2.3.3. (Update value parameter) Update $\mathbf{w}$ to reduce value estimate errors. (For example, minimize $\frac{1}{|\mathcal{B}|} \sum_{(S,A,\Pi,A,G)\in\mathcal{B}} \left[G - v(S; \mathbf{w}) \right]^2$.)

8.4 Natural PG and Trust Region Algorithm

Combining the trust region method with the surrogate objective, we can get Natural PG algorithm and Trust Region Policy Optimization algorithm. This selection will first introduce the trust region method, and then introduce the trust-region based algorithms.

Interdisciplinary Reference 8.2
Optimization: Trust Region Method

Trust Region Method (TRM) is a method for solving constrained optimization. Consider the following constrained optimization problem:

$$\underset{\boldsymbol{\theta}}{\text{maximize}} \quad f(\boldsymbol{\theta})$$

s.t. θ satisfies constraints

where $f(\theta)$ is a second-order differentiable function on $\mathbb{R}^n$. Trust region method tries to solve it iteratively. Let θ_k denote the value before k-th iteration. We can define a neighborhood $\mathcal{U}_k$ as trust region. In the trust region $\mathcal{U}_k$, we assume the objective $f(\theta)$ can be approximated as a second-order function in the form of

$$f(\theta) \approx f(\theta_k) + \left[\mathbf{g}(\theta_k)\right]^\top (\theta - \theta_k) + \frac{1}{2}(\theta - \theta_k)^\top \mathbf{F}(\theta_k)(\theta - \theta_k),$$

where $\mathbf{g}(\theta) = \nabla f(\theta)$ and $\mathbf{F}(\theta) = \nabla^2 f(\theta)$. In this way, we obtain the following trust-region sub-problem:

$$\underset{\theta}{\text{maximize}} \quad f(\theta_k) + \left[\mathbf{g}(\theta_k)\right]^\top (\theta - \theta_k) + \frac{1}{2}(\theta - \theta_k)^\top \mathbf{F}(\theta_k)(\theta - \theta_k)$$

$$\text{s.t.} \qquad \|\theta - \theta_k\| \le \delta_k^{(\text{TR})}.$$

8.4.1 Kullback–Leibler Divergence and Fisher Information Matrix

This section recaps the definition of Kullback–Leibler divergence and its property. The algorithms in this section need this information.

Interdisciplinary Reference 8.3
Information Theory: Kullback–Leibler Divergence

When we talked about the concept of importance sampling, we know that if two distributions $p(x)\,(x \in X)$ and $q(x)\,(x \in X)$ satisfy the property that $q(x) > 0$ holds if $q(x) > 0$, we say that the distribution p is absolutely continuous with respect to the distribution q (denoted as $p \ll q$). Under this condition, we can define the Kullback–Leibler divergence (KLD) from q to p as

$$d_{\text{KL}}(p\|q) \stackrel{\text{def}}{=} \mathrm{E}_{X \sim p}\left[\ln \frac{p(X)}{q(X)}\right].$$

The KLD from q to p is 0, if and only if p equal to q almost everywhere.

This section will consider $d_{\text{KL}}\big(p(\theta_k)\|p(\theta)\big)$, the KLD from $p(\theta)$ and $p(\theta_k)$, which are two distributions with the same form but different parameters. We will consider its second-order approximation at $\theta = \theta_k$. This second-order approximation directly relates to the Fisher Information Matrix.

Interdisciplinary Reference 8.4
Information Geometry: Fisher Information Matrix

Consider the parametric distribution $p(x; \boldsymbol{\theta})$ $(x \in \mathcal{X})$, where $\boldsymbol{\theta}$ is the parameter. Define the score vector $\nabla \ln p(X; \boldsymbol{\theta})$ as the gradient of log-likelihood function $\ln p(X; \boldsymbol{\theta})$ with respect to the parameter $\boldsymbol{\theta}$. We can prove that the expectation of the score vector is $\mathbf{0}$. (Proof: $\mathrm{E}_{X \sim p(\boldsymbol{\theta})}\big[\nabla \ln p(X; \boldsymbol{\theta})\big] = \sum_x p(x; \boldsymbol{\theta}) \nabla \ln p(x; \boldsymbol{\theta}) = \sum_x \nabla p(x; \boldsymbol{\theta}) = \nabla \sum_x p(x; \boldsymbol{\theta}) = \nabla 1 = \mathbf{0}$.) Therefore, the covariance of $\nabla \ln p(X; \boldsymbol{\theta})$ can be expressed as

$$\mathbf{F} \overset{\text{def}}{=} \mathrm{E}_{X \sim p(\boldsymbol{\theta})}\Big[\big[\nabla \ln p(X; \boldsymbol{\theta})\big]\big[\nabla \ln p(X; \boldsymbol{\theta})\big]^{\top}\Big].$$

The matrix $\mathbf{F}$ is called **Fisher Information Matrix** (**FIM**).

FIM has the following property: The negative expected Hessian matrix of log-likelihood equals to the Fisher matrix, that is,

$$\mathbf{F} = -\mathrm{E}_X\Big[\nabla^2 \ln p(X; \boldsymbol{\theta})\Big].$$

(Proof: Take the expectation of the following equation

$$\begin{aligned}
\nabla^2 &\ln p(X; \boldsymbol{\theta}) \\
&= \nabla\left(\frac{\nabla p(X; \boldsymbol{\theta})}{p(X; \boldsymbol{\theta})}\right) \\
&= \frac{\big[\nabla^2 p(X; \boldsymbol{\theta})\big] p(X; \boldsymbol{\theta}) - \big[\nabla p(X; \boldsymbol{\theta})\big]\big[p(X; \boldsymbol{\theta})\big]^{\top}}{p(X; \boldsymbol{\theta}) p(X; \boldsymbol{\theta})} \\
&= \frac{\nabla^2 p(X; \boldsymbol{\theta})}{p(X; \boldsymbol{\theta})} - \big[\nabla \ln p(X; \boldsymbol{\theta})\big]\big[\nabla \ln p(X; \boldsymbol{\theta})\big]^{\top}.
\end{aligned}$$

Then plug in

$$\begin{aligned}
\mathrm{E}_X\left[\frac{\nabla^2 p(X; \boldsymbol{\theta})}{p(X; \boldsymbol{\theta})}\right] &= \sum_x p(x; \boldsymbol{\theta}) \frac{\nabla^2 p(x; \boldsymbol{\theta})}{p(x; \boldsymbol{\theta})} \\
&= \sum_x \nabla^2 p(x; \boldsymbol{\theta}) \\
&= \nabla^2 \sum_x p(x; \boldsymbol{\theta}) \\
&= \nabla^2 1 \\
&= \mathbf{O}.
\end{aligned}$$

The proof completes.)

Interdisciplinary Reference 8.5
Information Geometry: Second-order Approximation of KL Divergence

The second-order approximation of $d_{\mathrm{KL}}(p(\boldsymbol{\theta}_k)\|p(\boldsymbol{\theta}))$ at $\boldsymbol{\theta} = \boldsymbol{\theta}_k$ is

$$d_{\mathrm{KL}}(p(\boldsymbol{\theta}_k)\|p(\boldsymbol{\theta})) \approx \frac{1}{2}(\boldsymbol{\theta} - \boldsymbol{\theta}_k)^{\top}\mathbf{F}(\boldsymbol{\theta}_k)(\boldsymbol{\theta} - \boldsymbol{\theta}_k)$$

where $\mathbf{F}(\boldsymbol{\theta}_k) = \mathrm{E}_{X \sim p(\boldsymbol{\theta}_k)}\left[\left[\nabla \ln p(X; \boldsymbol{\theta}_k)\right]\left[\nabla \ln p(X; \boldsymbol{\theta}_k)\right]^{\top}\right]$ is the FIM. (Proof: To calculate the second-order approximation of $d_{\mathrm{KL}}(p(\boldsymbol{\theta}_k)\|p(\boldsymbol{\theta}))$ at $\boldsymbol{\theta} = \boldsymbol{\theta}_k$, we need to calculate the values of $d_{\mathrm{KL}}(p(\boldsymbol{\theta}_k)\|p(\boldsymbol{\theta}))$, $\nabla d_{\mathrm{KL}}(p(\boldsymbol{\theta}_k)\|p(\boldsymbol{\theta}))$, and $\nabla^2 d_{\mathrm{KL}}(p(\boldsymbol{\theta}_k)\|p(\boldsymbol{\theta}))$ at $\boldsymbol{\theta} = \boldsymbol{\theta}_k$. The computation is as follows:

- The value of $d_{\mathrm{KL}}(p(\boldsymbol{\theta}_k)\|p(\boldsymbol{\theta}))$ at $\boldsymbol{\theta} = \boldsymbol{\theta}_k$:

$$\left[d_{\mathrm{KL}}(p(\boldsymbol{\theta}_k)\|p(\boldsymbol{\theta}))\right]_{\boldsymbol{\theta}=\boldsymbol{\theta}_k} = \mathrm{E}_{p(\boldsymbol{\theta}_k)}\left[\ln p(\boldsymbol{\theta}_k) - \ln p(\boldsymbol{\theta}_k)\right] = 0.$$

- The value of $\nabla d_{\mathrm{KL}}(p(\boldsymbol{\theta}_k)\|p(\boldsymbol{\theta}))$ at $\boldsymbol{\theta} = \boldsymbol{\theta}_k$: Since

$$d_{\mathrm{KL}}(p(\boldsymbol{\theta}_k)\|p(\boldsymbol{\theta})) = \mathrm{E}_{X \sim p(\boldsymbol{\theta}_k)}\left[\ln p(X; \boldsymbol{\theta}_k) - \ln p(X; \boldsymbol{\theta})\right],$$

we have

$$\nabla d_{\mathrm{KL}}(p(\boldsymbol{\theta}_k)\|p(\boldsymbol{\theta}))$$
$$= \mathrm{E}_{X \sim p(\boldsymbol{\theta}_k)}\left[-\nabla \ln p(X; \boldsymbol{\theta})\right]$$
$$= \mathrm{E}_{X \sim p(\boldsymbol{\theta}_k)}\left[-\frac{\nabla p(X; \boldsymbol{\theta})}{p(X; \boldsymbol{\theta})}\right]$$
$$= -\sum_x p(x; \boldsymbol{\theta}_k)\frac{\nabla p(x; \boldsymbol{\theta})}{p(x; \boldsymbol{\theta})}.$$

Therefore,

$$\left[\nabla d_{\mathrm{KL}}(p(\boldsymbol{\theta}_k)\|p(\boldsymbol{\theta}))\right]_{\boldsymbol{\theta}=\boldsymbol{\theta}_k}$$
$$= -\sum_x p(x; \boldsymbol{\theta}_k)\frac{\nabla p(x; \boldsymbol{\theta}_k)}{p(x; \boldsymbol{\theta}_k)}$$
$$= -\nabla \sum_x p(x; \boldsymbol{\theta}_k)$$
$$= -\nabla 1$$
$$= 0.$$

- The value of $\nabla^2 d_{\mathrm{KL}}(p(\boldsymbol{\theta}_k)\|p(\boldsymbol{\theta}))$ at $\boldsymbol{\theta} = \boldsymbol{\theta}_k$: Obviously, we have

$$\nabla^2 d_{\mathrm{KL}}\big(p(\boldsymbol{\theta}_k)\big\|p(\boldsymbol{\theta})\big) = \mathrm{E}_{X\sim p(\boldsymbol{\theta}_k)}\Big[-\nabla^2 \ln p(X;\boldsymbol{\theta})\Big].$$

It equals $-\mathrm{E}_{X\sim p(\boldsymbol{\theta}_k)}\big[-\nabla^2 \ln p(X;\boldsymbol{\theta}_k)\big]$ at $\boldsymbol{\theta} = \boldsymbol{\theta}_k$, which is the expectation of the Hessian matrix of the log-likelihood. Since FIM has the following property:

$$\mathbf{F}(\boldsymbol{\theta}_k) = \mathrm{E}_{X\sim p(\boldsymbol{\theta}_k)}\Big[\big[\nabla \ln p(X;\boldsymbol{\theta}_k)\big]\big[\nabla \ln p(X;\boldsymbol{\theta}_k)\big]^{\top}\Big]$$

$$= -\mathrm{E}_{X\sim p(\boldsymbol{\theta}_k)}\Big[\nabla^2 \ln p(X;\boldsymbol{\theta}_k)\Big].$$

Therefore, $\big[\nabla^2 d_{\mathrm{KL}}\big(p(\boldsymbol{\theta}_k)\big\|p(\boldsymbol{\theta})\big)\big]_{\boldsymbol{\theta}=\boldsymbol{\theta}_k} = \mathbf{F}(\boldsymbol{\theta}_k)$.

Therefore, the second-order approximation of $d_{\mathrm{KL}}\big(p(\boldsymbol{\theta}_k)\big\|p(\boldsymbol{\theta})\big)$ at $\boldsymbol{\theta} = \boldsymbol{\theta}_k$ is

$$d_{\mathrm{KL}}\big(p(\boldsymbol{\theta}_k)\big\|p(\boldsymbol{\theta})\big) \approx 0 + \mathbf{0}^{\top}(\boldsymbol{\theta} - \boldsymbol{\theta}_k) + \frac{1}{2}(\boldsymbol{\theta} - \boldsymbol{\theta}_k)^{\top}\mathbf{F}(\boldsymbol{\theta}_k)(\boldsymbol{\theta} - \boldsymbol{\theta}_k)$$

which completes the proof.)

8.4.2 Trust Region of Surrogate Objective

The performance difference lemma tells us, we can use $l(\boldsymbol{\theta}|\boldsymbol{\theta}_k)$, which relates to the surrogate advantage, to approximate the return expectation $g_{\pi(\boldsymbol{\theta})}$. Although this approximation is accurate around $\boldsymbol{\theta} = \boldsymbol{\theta}_k$, the approximation may be inaccurate when $\boldsymbol{\theta}$ is far away from $\boldsymbol{\theta}_k$. (Schulman, 2017: TRPO) proved that

$$g_{\pi(\boldsymbol{\theta})} \geq l(\boldsymbol{\theta}|\boldsymbol{\theta}_k) - c \max_s d_{\mathrm{KL}}\big(\pi(\cdot|s;\boldsymbol{\theta}_k)\big\|\pi(\cdot|s;\boldsymbol{\theta})\big),$$

where $c \stackrel{\mathrm{def}}{=} \frac{4\gamma}{(1-\gamma)^2} \max_{s,a}|a_{\pi(\boldsymbol{\theta})}(s,a)|$. This result tells us, the error between $l(\boldsymbol{\theta}|\boldsymbol{\theta}_k)$ and $g_{\pi(\boldsymbol{\theta})}$ is limited. We can control the error if we limit the KLD between $\pi(\cdot|s;\boldsymbol{\theta}_k)$ and $\pi(\cdot|s;\boldsymbol{\theta})$. On the other hand,

$$l_c(\boldsymbol{\theta}|\boldsymbol{\theta}_k) \stackrel{\mathrm{def}}{=} l(\boldsymbol{\theta}|\boldsymbol{\theta}_k) - c \max_s d_{\mathrm{KL}}\big(\pi(\cdot|s;\boldsymbol{\theta}_k)\big\|\pi(\cdot|s;\boldsymbol{\theta})\big)$$

can be viewed as a lower bound on $g_{\pi(\boldsymbol{\theta})}$. Since the values and gradients of $d_{\mathrm{KL}}\big(\pi(\cdot|s;\boldsymbol{\theta}_k)\big\|\pi(\cdot|s;\boldsymbol{\theta})\big)$ are zero at $\boldsymbol{\theta} = \boldsymbol{\theta}_k$, this lower bound is still an approximation of $g_{\pi(\boldsymbol{\theta})}$, but it is smaller than $g_{\pi(\boldsymbol{\theta})}$. The relationship among $g_{\pi(\boldsymbol{\theta})}$, $l(\boldsymbol{\theta}|\boldsymbol{\theta}_k)$, and $l_c(\boldsymbol{\theta}|\boldsymbol{\theta}_k)$ is depicted in Fig. 8.2.

It is difficult to estimate $\max_s d_{\mathrm{KL}}\big(\pi(\cdot|s;\boldsymbol{\theta}_k)\big\|\pi(\cdot|s;\boldsymbol{\theta})\big)$ accurately in practice. Therefore, we can use the expectation

$$\bar{d}_{\mathrm{KL}}\big(\pi(\cdot|s;\boldsymbol{\theta}_k)\big\|\pi(\cdot|s;\boldsymbol{\theta})\big) \stackrel{\mathrm{def}}{=} \mathrm{E}_{S\sim\pi(\boldsymbol{\theta}_k)}\big[d_{\mathrm{KL}}\big(\pi(\cdot|S;\boldsymbol{\theta}_k)\big\|\pi(\cdot|S;\boldsymbol{\theta})\big)\big]$$

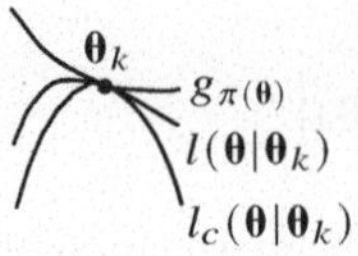

Fig. 8.2   Relationship among $g_{\pi(\theta)}, l(\theta|\theta_k)$, **and** $l_c(\theta|\theta_k)$.

to replace the maximum $\max_s d_{\mathrm{KL}}\big(\pi(\cdot|s; \theta_k)\|\pi(\cdot|s; \theta)\big)$.

The algorithms in this section use the trust region method to control the difference between $l(\theta|\theta_k)$ and $g_{\pi(\theta)}$. We can designate a threshold δ, and do not allow $\bar{d}_{\mathrm{KL}}$ to exceed this threshold. Using this method, we can get the trust region $\big\{\theta : \bar{d}_{\mathrm{KL}}(\theta_k\|\theta) \leq \delta\big\}$.

The previous section told us that the KL divergence can be approximated by a quadratic function consisting of FIM matrix. Using the second-order approximation, it is easy to know that $\bar{d}_{\mathrm{KL}}(\theta_k\|\theta)$ has the following second-order approximation at $\theta = \theta_k$:

$$\bar{d}_{\mathrm{KL}}(\theta_k\|\theta) \approx \frac{1}{2}(\theta - \theta_k)^{\top}\mathbf{F}(\theta_k)(\theta - \theta_k).$$

Therefore, the second-order approximation of the trust region is

$$\left\{\theta : \frac{1}{2}(\theta - \theta_k)^{\top}\mathbf{F}(\theta_k)(\theta - \theta_k) \leq \delta\right\}.$$

8.4.3 NPG: Natural Policy Gradient

Natural Policy Gradient (NPG) algorithm, proposed in (Kakade, 2002: NPG), is an iterative algorithm that uses surrogate advantage and trust region. The intuition of this algorithm is to constrain the policy parameter at the trust region while maximizing the surrogate advantage. Mathematically, it considers the following optimization problem:

$$\underset{\theta}{\text{maximize}} \quad \mathrm{E}_{\pi(\theta_k)}\left[\frac{\pi(A|S; \theta)}{\pi(A|S; \theta_k)}a_{\pi(\theta_k)}(S, A)\right]$$

$$\text{s.t.} \qquad \mathrm{E}_{S\sim\pi(\theta_k)}\big[d_{\mathrm{KL}}\big(\pi(\cdot|S; \theta_k)\|\pi(\cdot|S; \theta)\big)\big] \leq \delta,$$

where δ is a hyperparameter. Both the objective and the constraint of this optimization are complex, and they need to be further simplified.

One simplification is to Tayler expand the objective at $\theta = \theta_k$ and use the first two terms:

$$\mathrm{E}_{\pi(\theta_k)}\left[\frac{\pi(A|S; \theta)}{\pi(A|S; \theta_k)}a_{\pi(\theta_k)}(S, A)\right] \approx 0 + \big[\mathbf{g}(\theta_k)\big]^{\top}(\theta - \theta_k).$$

Further, we use the second-order approximation for the constraint. In this way, we get a simplified optimization problem:

$$\underset{\boldsymbol{\theta}}{\text{maximize}} \quad \left[\mathbf{g}(\boldsymbol{\theta}_k)\right]^{\top}(\boldsymbol{\theta} - \boldsymbol{\theta}_k)$$

$$\text{s.t.} \quad \frac{1}{2}(\boldsymbol{\theta} - \boldsymbol{\theta}_k)^{\top}\mathbf{F}^{-1}(\boldsymbol{\theta}_k)\mathbf{g}(\boldsymbol{\theta}_k) \leq \delta.$$

This simplified optimization problem has a closed-form solution

$$\boldsymbol{\theta}_{k+1} = \boldsymbol{\theta}_k + \sqrt{\frac{2\delta}{\left[\mathbf{g}(\boldsymbol{\theta}_k)\right]^{\top}\mathbf{F}^{-1}(\boldsymbol{\theta}_k)\mathbf{g}(\boldsymbol{\theta}_k)}}\mathbf{F}^{-1}(\boldsymbol{\theta}_k)\mathbf{g}(\boldsymbol{\theta}_k).$$

Here, $\sqrt{\frac{2\delta}{\left[\mathbf{g}(\boldsymbol{\theta}_k)\right]^{\top}\mathbf{F}^{-1}(\boldsymbol{\theta}_k)\mathbf{g}(\boldsymbol{\theta}_k)}}\mathbf{F}^{-1}(\boldsymbol{\theta}_k)\mathbf{g}(\boldsymbol{\theta}_k)$ is called **natural gradient** (Amari, 1998). The above formula shows how natural policy gradient descent updates.

We can control the learning rate of this update by controlling the parameter δ. The learning rate is in approximately proportion to $\sqrt{\delta}$.

Algorithm 8.7 shows the NPG algorithm.

Algorithm 8.7 Vanilla NPG.

Inputs: environment (without mathematical model).
Outputs: optimal policy estimate $\pi(\boldsymbol{\theta})$.
Parameters: upper bound on KL divergence δ, parameters that control the trajectory generation, and parameters for estimating advantage (such as the discount factor γ).

1. (Initialize) Initialize the policy parameter $\boldsymbol{\theta}$ and value parameter $\mathbf{w}$.
2. (AC update) For each episode:

   2.1. (Decide and sample) Use the policy $\pi(\boldsymbol{\theta})$ to generate a trajectory.
   2.2. (Calculate natural gradient) Use the generated trajectory to calculate the policy gradient $\mathbf{g}$ and Fisher information matrix $\mathbf{F}$ at the parameter $\boldsymbol{\theta}$. Calculate the natural gradient $\sqrt{\frac{2\delta}{\mathbf{g}^{\top}\mathbf{F}^{-1}\mathbf{g}}}\mathbf{F}^{-1}\mathbf{g}$.
   2.3. (Update policy parameter) $\boldsymbol{\theta} \leftarrow \boldsymbol{\theta} + \sqrt{\frac{2\delta}{\mathbf{g}^{\top}\mathbf{F}^{-1}\mathbf{g}}}\mathbf{F}^{-1}\mathbf{g}$.
   2.4. (Update value parameter) Update $\mathbf{w}$ to reduce the error of value estimates.

The greatest weakness of NPG algorithm is that it is very compute-intensive to calculate FIM and its inverse. Specifically, NPG needs to calculate $\mathbf{F}^{-1}\mathbf{g}$, which includes the inverse of FIM $\mathbf{F}^{-1}$. The complexity of calculating matrix inverse is the cube of row number (or column number). The row number equals the number of elements in the parameter $\boldsymbol{\theta}$. The number of elements in $\boldsymbol{\theta}$ can be very big, especially when neural networks are used.

One method to somehow relieve the computation burden of calculating the inverse of FIM is the Conjugate Gradient (CG) method. CG can calculate $\mathbf{F}^{-1}\mathbf{g}$ directly without calculating $\mathbf{F}^{-1}$.

Interdisciplinary Reference 8.6
Numerical Linear Algebra: Conjugate Gradient

Conjugate Gradient (CG) algorithm is an algorithm to solve the linear equation set $\mathbf{Fx} = \mathbf{g}$, where the efficient matrix $\mathbf{F}$ is a real symmetric positive-definite matrix.

The definition of **conjugate** is as follows: For two vectors $\mathbf{p}_i$ and $\mathbf{p}_j$, they conjugate with each other with respect to the matrix $\mathbf{F}$ if and only if $\mathbf{p}_i^\top \mathbf{F} \mathbf{p}_j = 0$.

The intuition of CG algorithm is as follows: The solution of the linear equation set $\mathbf{Fx} = \mathbf{g}$ is the minimal point of the quadratic function $\frac{1}{2}\mathbf{x}^\top \mathbf{F}\mathbf{x} - \mathbf{g}^\top\mathbf{x}$, so we can convert the problem of finding the solution of $\mathbf{Fx} = \mathbf{g}$ to an optimization problem. This problem can be solved iteratively. That is, starting from $\mathbf{x}_0$, we change the value of $\mathbf{x}$ in some direction with some step in each iteration. Previously we always used the gradient descent method to find the minimal point, where the direction always opposes the gradient, and the learning rate is designated beforehand. This procedure only uses one-order gradient. In fact, we can do something smarter in minimizing $\frac{1}{2}\mathbf{x}^\top \mathbf{F}\mathbf{x} - \mathbf{g}^\top\mathbf{x}$, since we completely know the matrix $\mathbf{F}$. CG algorithm does it smarter: It picks the gradient that is conjugate to all previously used gradients, and directly uses the most appropriate step. In this way, CG algorithm can reach the optimal point in fewer iterations. Specifically, in k-th iteration ($k = 0, 1, 2, \ldots$), let $\mathbf{x}_k$ denote the value of $\mathbf{x}$ before the iteration. At this time, the negative gradient of $\frac{1}{2}\mathbf{x}^\top \mathbf{F}\mathbf{x} - \mathbf{g}^\top\mathbf{x}$ is exactly the residual error of the linear equation set we need to solve, i.e. $\mathbf{r}_k \stackrel{\text{def}}{=} \mathbf{g} - \mathbf{F}\mathbf{x}_k$. Next, we need to find a direction that is conjugate to all of $\mathbf{p}_0, \mathbf{p}_1, \ldots, \mathbf{p}_{k-1}$. In order to find such direction, we first assume the direction is

$$\mathbf{p}_k = \mathbf{r}_k - \sum_{\kappa=0}^{k-1} \beta_{k,\kappa}\mathbf{p}_\kappa,$$

Using $\mathbf{p}_k^\top \mathbf{F}\mathbf{p}_\kappa = 0 \ (0 \le \kappa < k)$ leads to

$$\beta_{k,\kappa} = \frac{\mathbf{p}_\kappa^\top \mathbf{F}\mathbf{r}_\kappa}{\mathbf{p}_\kappa^\top \mathbf{F}\mathbf{p}_\kappa}, \quad 0 \le \kappa < k.$$

In this way, we have determined the direction of the k-th iteration. Then we try to determine the learning rate α_k. The choice of learning rate needs to make the objective $\frac{1}{2}\mathbf{x}^\top \mathbf{F}\mathbf{x} - \mathbf{g}^T\mathbf{x}$ at the updated value $\mathbf{x}_{k+1} = \mathbf{x}_k + \alpha_k\mathbf{p}_k$ as small as possible. Since

$$\frac{\partial}{\partial \alpha_k}\left(\frac{1}{2}(\mathbf{x}_k + \alpha_k\mathbf{p}_k)^\top \mathbf{F}(\mathbf{x}_k + \alpha_k\mathbf{p}_k) - \mathbf{g}^\top(\mathbf{x}_k + \alpha_k\mathbf{p}_k)\right) = \alpha_k\mathbf{p}_k^\top \mathbf{F}\mathbf{p}_k + \mathbf{p}_k^\top(\mathbf{F}\mathbf{x}_k - \mathbf{g}).$$

Setting this to 0 leads to

$$\alpha_k = \frac{\mathbf{p}_k^\top (\mathbf{g} - \mathbf{F}\mathbf{x}_k)}{\mathbf{p}_k^\top \mathbf{F}\mathbf{p}_k}.$$

In this way, we determine how CG updates.

In practice, the computation can be further simplified. Define $\rho_k \overset{\text{def}}{=} \mathbf{r}_k^\top \mathbf{r}_k$, $\mathbf{z}_k \overset{\text{def}}{=} \mathbf{F}\mathbf{p}_k$ ($k = 0, 1, 2 \ldots$), we have

$$\alpha_k = \frac{\rho_k}{\mathbf{p}_k^\top \mathbf{z}_k}$$

$$\mathbf{r}_{k+1} = \mathbf{r}_k - \alpha_k \mathbf{z}_k$$

$$\mathbf{p}_{k+1} = \mathbf{r}_{k+1} + \frac{\rho_{k+1}}{\rho_k} \mathbf{p}_k.$$

(Proof is omitted due to its complexity.)

Accordingly, the CG algorithm is shown in Algo. 8.8. Step 2.2 also introduces a small positive number ε_{CG} (say, $\varepsilon_{\text{CG}} = 10^{-8}$) to increase the numerical stability.

Algorithm 8.8   CG.

Inputs: matrix $\mathbf{F}$ and vector $\mathbf{g}$.

Outputs: $\mathbf{x}$ (the solution of the linear equation set $\mathbf{F}\mathbf{x} = \mathbf{g}$).

Parameters: parameters that control the iterations (for example, the maximum iteration number n_{CG} or tolerance ρ_{tol}), and parameter for numerical stability $\varepsilon_{\text{CG}} > 0$.

1. (Initialize) Set the start point $\mathbf{x} \leftarrow$ arbitrary values, the residual $\mathbf{r} \leftarrow \mathbf{g} - \mathbf{F}\mathbf{x}$, base $\mathbf{p} \leftarrow \mathbf{r}$, and $\rho \leftarrow \mathbf{r}^\top \mathbf{r}$.
2. (Iteration) For $k = 1, \ldots, n_{\text{CG}}$:

    2.1. $\mathbf{z} \leftarrow \mathbf{F}\mathbf{p}$.
    2.2. (Calculate learning rate) $\alpha \leftarrow \frac{\rho}{\mathbf{p}^\top \mathbf{z} + \varepsilon_{\text{CG}}}$.
    2.3. (Update value) $\mathbf{x} \leftarrow \mathbf{x} + \alpha \mathbf{p}$.
    2.4. (Update residual) $\mathbf{r} \leftarrow \mathbf{r} - \alpha \mathbf{z}$.
    2.5. (Update base) $\rho_{\text{new}} \leftarrow \mathbf{r}^\top \mathbf{r}$, $\mathbf{p} \leftarrow \mathbf{r} + \frac{\rho_{\text{new}}}{\rho} \mathbf{p}$.
    2.6. $\rho \leftarrow \rho_{\text{new}}$. (If the tolerance ρ_{tol} is set and $\rho < \rho_{\text{tol}}$, break the loop.)

Algorithm 8.9 shows the NPG algorithm with conjugate gradient.

Algorithm 8.9   NPG with CG.

Inputs: environment (without mathematical model).

Outputs: optimal policy estimate $\pi(\boldsymbol{\theta})$.

Parameters: parameters for CG (say, n_{CG}, ρ_{tol}, and ε_{CG}), upper bound on KL divergence δ, parameters that control the trajectory generation, and parameters for estimating advantage (such as discount factor γ).

1. (Initialize) Initialize the policy parameter θ and value parameter $\mathbf{w}$.
2. (AC update) For each episode:

    2.1. (Decide and sample) Use the policy $\pi(\theta)$ to generate a trajectory.

    2.2. (Calculate natural gradient) Use the generated trajectory to calculate the policy gradient $\mathbf{g}$ and Fisher information matrix $\mathbf{F}$ at the parameter θ. Use CG algorithm to calculate. Calculate the natural gradient $\sqrt{\frac{2\delta}{\mathbf{g}^\top \mathbf{F}^{-1}\mathbf{g}}}\mathbf{F}^{-1}\mathbf{g}$.

    2.3. (Update policy parameter) $\theta \leftarrow \theta + \sqrt{\frac{2\delta}{\mathbf{g}^\top \mathbf{F}^{-1}\mathbf{g}}}\mathbf{F}^{-1}\mathbf{g}$.

    2.4. (Update value parameter) Update $\mathbf{w}$ to reduce the error of value estimates.

8.4.4 TRPO: Trust Region Policy Optimization

Trust Region Policy Optimization (**TRPO**) algorithm is modified from NPG algorithm (Schulman, 2017: TRPO). Recap that NPG considered the following optimization problem:

$$\underset{\theta}{\text{maximize}} \quad \mathrm{E}_{\pi(\theta_k)}\left[\frac{\pi(A|S;\theta)}{\pi(A|S;\theta_k)}a_{\pi(\theta_k)}(S,A)\right]$$
$$\text{s.t.} \qquad \mathrm{E}_{S\sim\pi(\theta_k)}\big[d_{\mathrm{KL}}\big(\pi(\cdot|S;\theta_k)\|\pi(\cdot|S;\theta)\big)\big] \leq \delta.$$

However, NPG does not solve it directly. It considers an approximated version, and finds the optimal solution for that programming, which is not necessarily the optimal point in the original programming. In the extreme case, the solution in the approximated programming may even make the original programming worse. In order to solve this issue, TRPO changes the updating process to

$$\theta_{k+1} = \theta_k + \alpha^j\sqrt{\frac{2\delta}{[\mathbf{x}(\theta_k)]^\top \mathbf{F}(\theta_k)\mathbf{x}(\theta_k)}}\mathbf{x}(\theta_k),$$

where $\alpha \in (0,1)$ is a learning parameter, j is a natural number. NPG can be viewed as the case that $j = 0$. TRPO uses the following method to determine the value of j: We check $j = 0, 1, 2, \ldots$ one-by-one, and find the smallest integer that both satisfies the KL divergence constraints and improves the surrogate advantage. Since the approximation is good generally, most cases will observe $j = 0$. Some cases will observe $j = 1$. Other values are much less observed. Although most of j is 0, this change can avoid the occasional devastating degradation. Due to the introduction of α^j, we usually pick larger δ compared to the δ in NPG.

Algorithm 8.10 shows the TRPO algorithm.

Algorithm 8.10   TRPO.

Inputs: environment (without mathematical model).

Outputs: optimal policy estimate $\pi(\boldsymbol{\theta})$.

Parameters: trust region parameter α, parameters for CG (say, n_{CG}, ρ_{tol}, and ε_{CG}), upper bound on KL divergence δ, parameters that control the trajectory generation, and parameters for estimating advantage (such as discount factor γ).

1. (Initialize) Initialize the policy parameter $\boldsymbol{\theta}$ and value parameter $\mathbf{w}$.
2. (AC update) For each episode:

   2.1. (Decide and sample) Use the policy $\pi(\boldsymbol{\theta})$ to generate a trajectory.
   2.2. (Calculate natural gradient) Use the generated trajectory to calculate the policy gradient $\mathbf{g}$ and Fisher information matrix $\mathbf{F}$ at the parameter $\boldsymbol{\theta}$. Use CG algorithm to calculate. Calculate the natural gradient $\sqrt{\frac{2\delta}{\mathbf{g}^\top \mathbf{F}^{-1}\mathbf{g}}}\mathbf{F}^{-1}\mathbf{g}$.
   2.3. (Update policy parameter) Select a natural number j such that the new policy is in the trust region, and surrogate advantage is improved. Set $\boldsymbol{\theta} \leftarrow \boldsymbol{\theta} + \alpha^j \sqrt{\frac{2\delta}{\mathbf{g}^\top \mathbf{F}^{-1}\mathbf{g}}}\mathbf{F}^{-1}\mathbf{g}$.
   2.4. (Update value parameter) Update $\mathbf{w}$ to reduce the error of value estimates.

The implementation of NPG or TRPO is much more complex than PPO, so they are not as widely used as PPO.

8.5 Importance Sampling Off-Policy AC

This section introduces **Off-Policy Actor–Critic (OffPAC)**, an off-policy AC algorithm based on importance sampling. This algorithm introduces a behavior policy $b(\cdot|\cdot)$, and then the gradient changes from $\mathrm{E}_{\pi(\boldsymbol{\theta})}\left[\Psi_t \nabla \ln\pi(A_t|S_t; \boldsymbol{\theta})\right]$ to $\mathrm{E}_b\left[\frac{\pi(A_t|S_t;\boldsymbol{\theta})}{b(A_t|S_t)}\Psi_t \nabla \ln\pi(A_t|S_t; \boldsymbol{\theta})\right] = \mathrm{E}_b\left[\frac{1}{b(A_t|S_t)}\Psi_t \nabla \pi(A_t|S_t; \boldsymbol{\theta})\right]$. Therefore, it updates the policy parameter $\boldsymbol{\theta}$ by reducing $-\frac{1}{b(A_t|S_t)}\Psi_t \nabla \pi(A_t|S_t; \boldsymbol{\theta})$. Algorithm 8.11 shows the OffPAC algorithm.

Algorithm 8.11   OffPAC.

Inputs: environment (without mathematical model).
Outputs: optimal policy estimate $\pi(\boldsymbol{\theta})$.
Parameters: optimizers (containing learning rate $\alpha^{(\boldsymbol{\theta})}$ and $\alpha^{(\mathbf{w})}$), discount factor γ, and parameters to control the number of episodes and maximum steps in an episode.

1. (Initialize) Initialize the policy parameter $\boldsymbol{\theta}$ and value parameter $\mathbf{w}$.
2. (AC update) For each episode:

    2.1. (Behavior policy) Designate a behavior policy b.
    2.2. (Initialize accumulative discount factor) $\Gamma \leftarrow 1$.
    2.3. (Initialize state–action pair) Select the initial state S, and use the behavior policy $b(\cdot|S)$ to get the action A.
    2.4. Loop until the episode ends:
        2.4.1. (Sample) Execute the action A, and then observe the reward R, the next state S', and the indicator of episode end D'.
        2.4.2. (Decide) Use $b(\cdot|S')$ to determine the action A'. (The action can be arbitrary if $D' = 1$.)
        2.4.3. (Calculate TD return) $U \leftarrow R + \gamma(1 - D')q(S', A'; \mathbf{w})$.
        2.4.4. (Update policy parameter) Update $\boldsymbol{\theta}$ to reduce $-\frac{1}{b(A|S)}\Gamma q(S, A; \mathbf{w})\pi(A|S; \boldsymbol{\theta})$. (For example, $\boldsymbol{\theta} \leftarrow \boldsymbol{\theta} + \alpha^{(\boldsymbol{\theta})}\frac{1}{b(A|S)}\Gamma q(S, A; \mathbf{w})\pi(A|S; \boldsymbol{\theta})$).
        2.4.5. (Update value parameter) Update $\mathbf{w}$ to reduce $\frac{\pi(A|S; \boldsymbol{\theta})}{b(A|S)}\big[U - q(S, A; \mathbf{w})\big]^2$. (For example, $\mathbf{w} \leftarrow \mathbf{w} + \alpha^{(\mathbf{w})}\frac{\pi(A|S; \boldsymbol{\theta})}{b(A|S)}\big[U - q(S, A; \mathbf{w})\big]\nabla q(S, A; \mathbf{w})$.)
        2.4.6. (Update accumulative discount factor) $\Gamma \leftarrow \gamma\Gamma$.
        2.4.7. $S \leftarrow S', A \leftarrow A'$.

8.6 Case Study: Acrobot

This section considers the task `Acrobot-v1` in Gym (DeJong, 1994). As Fig. 8.3, there are two sticks in a 2-D vertical plane. One stick connects to the original point at one hand, and connects to the other stick at the other hand. The second stick has a free end. We can construct an absolute coordinate system $X'Y'$ in the vertical plane, where X'-axis points downward, and Y'-axis points in the right direction. We can build another coordination system $X''Y''$ according to the position of the stick that connects to the original: X''-axis outward, and Y''-axis is perpendicular to the X''-axis. At any time t ($t = 0, 1, 2, \ldots$), we can observe the position of the connection point in the absolute coordinate system $(X'_t, Y'_t) = (\cos \Theta'_t, \sin \Theta'_t)$ and the position of the free end at the relative coordination system $(X_t'', Y_t'') = (\cos \Theta_t'', \sin \Theta_t'')$. Besides, there are two angle velocity $\dot{\Theta}_t'$ and $\dot{\Theta}_t''$ (Note that the dots above the letters represent velocity). We can apply an action to the connection point between these two sticks. The action space is $\mathcal{A} = \{0, 1, 2\}$. Every step will be punished by the reward -1. The episode ends when the X' position of free end at the absolute coordinate system is ≤ -1 (i.e. $\cos \Theta' + \cos (\Theta' + \Theta'') < -1$), or the number of steps in the episode reaches 500. We want to finish the episode in the fewest steps.

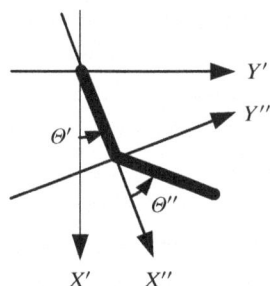

Fig. 8.3   The task Acrobot-v1.
This figure is adapted from (Dejong, 1994).

In fact, at time t, the environment is fully determined by the state $S_t = (\Theta_t', \Theta_t'', \dot{\Theta}_t', \dot{\Theta}_t'')$. The state $S_t = (\Theta_t', \Theta_t'', \dot{\Theta}_t', \dot{\Theta}_t'')$ can be calculated from the observation $O_t = (\cos \Theta_t', \sin \Theta_t', \cos \Theta_t'', \sin \Theta_t'', \dot{\Theta}_t', \dot{\Theta}_t'')$, so this environment is fully observable. Executing the action A_t at state S_t will change the angular acceleration $\ddot{\Theta}_t'$ and $\ddot{\Theta}_t''$ as

$$\ddot{\Theta}_t'' = \left(A_t - 1 + \frac{D_t''}{D_t'} \Phi_t' - \frac{1}{2}\left(\dot{\Theta}_t''\right)^2 \sin \Theta_t'' - \Phi_t'' \right)\left(\frac{5}{4} - \frac{(D_t'')^2}{D_t'} \right)$$

$$\ddot{\Theta}_t' = -\frac{1}{D_t'}\left(D_t'' \ddot{\Theta}_t'' + \Phi_t'\right),$$

where

$$D_t' \overset{\text{def}}{=} \cos \Theta_t'' + \frac{7}{2}$$

$$D_t'' \overset{\text{def}}{=} \frac{1}{2}\cos \Theta_t'' + \frac{5}{4}$$

$$\Phi_t'' \overset{\text{def}}{=} \frac{1}{2}g \sin \left(\Theta_t' + \Theta_t''\right)$$

$$\Phi_t'' \overset{\text{def}}{=} -\frac{1}{2}\left(\dot{\Theta}_t''\right)^2 \sin \Theta_t'' - \dot{\Theta}_t' \dot{\Theta}_t'' \sin \Theta_t'' + \frac{3}{2}g \sin \Theta_t' + \Phi_t''.$$

And the acceleration of gravity $g = 9.8$. After obtaining the angular acceleration $\ddot{\Theta}'_t$ and $\ddot{\Theta}''_t$, we can calculate integral over 0.2 continuous-time unit to get the state of the next discrete time. During the calculation, the function clip() is applied to bound the angle velocity such that $\dot{\Theta}'_t \in [-4\pi, 4\pi]$ and $\dot{\Theta}''_t \in [-9\pi, 9\pi]$.

! Note

In this environment, the interval between two discrete time steps is 0.2 continuous-time units, rather than 1 continuous-time unit. It once again demonstrates that the discrete-time index does not need to match the exact values in the continuous-time index.

This dynamics is very complex. We can not find the closed-form of the optimal policy even we know the dynamics.

8.6.1 On-Policy AC

This section considers on-policy AC policy optimizations.

Codes 8.1 and 8.2 show the agent of action value AC. Inside the agent class QActorCriticAgent, the actor is approximated by actor_net, while the critic is approximated by critic_net. The member function learn() trains these two networks. Training of actor network calculates $\ln \pi(A|S; \boldsymbol{\theta})$. To increase numerical stability, $\pi(A|S; \boldsymbol{\theta})$ is clipped within $\left[10^{-6}, 1\right]$ before fed to the logarithm. The interaction between agents and the environment is implemented as Code 1.3. The agents are trained and tested using Codes 5.3 and 1.4 respectively.

Code 8.1 Action-value AC agent (TensorFlow version).
Acrobot-v1_QActorCritic_tf.ipynb

```
class QActorCriticAgent:
    def __init__(self, env):
        self.action_n = env.action_space.n
        self.gamma = 0.99

        self.actor_net = self.build_net(hidden_sizes=[100,],
                output_size=self.action_n, output_activation=nn.softmax,
                loss=losses.categorical_crossentropy, learning_rate=0.0001)
        self.critic_net = self.build_net(hidden_sizes=[100,],
                output_size=self.action_n, learning_rate=0.0002)

    def build_net(self, hidden_sizes, output_size, input_size=None,
                activation=nn.relu, output_activation=None,
                loss=losses.mse, learning_rate=0.01):
        model = keras.Sequential()
        for hidden_size in hidden_sizes:
            model.add(layers.Dense(units=hidden_size, activation=activation))
        model.add(layers.Dense(units=output_size,
                activation=output_activation))
        optimizer = optimizers.Adam(learning_rate)
        model.compile(optimizer=optimizer, loss=loss)
```

```
22              return model
23
24      def reset(self, mode=None):
25          self.mode = mode
26          if self.mode == 'train':
27              self.trajectory = []
28              self.discount = 1.
29
30      def step(self, observation, reward, terminated):
31          probs = self.actor_net.predict(observation[np.newaxis], verbose=0)[0]
32          action = np.random.choice(self.action_n, p=probs)
33          if self.mode == 'train':
34              self.trajectory += [observation, reward, terminated, action]
35              if len(self.trajectory) >= 8:
36                  self.learn()
37              self.discount *= self.gamma
38          return action
39
40      def close(self):
41          pass
42
43      def learn(self):
44          state, _, _, action, next_state, reward, terminated, next_action = \
45                  self.trajectory[-8:]
46
47          # update actor
48          states = state[np.newaxis]
49          preds = self.critic_net.predict(states, verbose=0)
50          q = preds[0, action]
51          state_tensor = tf.convert_to_tensor(states, dtype=tf.float32)
52          with tf.GradientTape() as tape:
53              pi_tensor = self.actor_net(state_tensor)[0, action]
54              log_pi_tensor = tf.math.log(tf.clip_by_value(pi_tensor, 1e-6, 1.))
55              loss_tensor = -self.discount * q * log_pi_tensor
56          grad_tensors = tape.gradient(loss_tensor, self.actor_net.variables)
57          self.actor_net.optimizer.apply_gradients(zip(grad_tensors,
58                  self.actor_net.variables))
59
60          # update critic
61          next_q = self.critic_net.predict(next_state[np.newaxis],
62                  verbose=0)[0, next_action]
63          preds[0, action] = reward + (1. - terminated) * self.gamma * next_q
64          self.critic_net.fit(states, preds, verbose=0)
65
66
67  agent = QActorCriticAgent(env)
```

Code 8.2   Action-value AC agent (PyTorch version).
Acrobot-v1_QActorCritic_torch.ipynb

```
1   class QActorCriticAgent:
2       def __init__(self, env):
3           self.gamma = 0.99
4
5           self.actor_net = self.build_net(
6                   input_size=env.observation_space.shape[0], hidden_sizes=[100,],
7                   output_size=env.action_space.n, output_activator=nn.Softmax(1))
8           self.actor_optimizer = optim.Adam(self.actor_net.parameters(), 0.001)
9           self.critic_net = self.build_net(
10                  input_size=env.observation_space.shape[0], hidden_sizes=[100,],
11                  output_size=env.action_space.n)
12          self.critic_optimizer = optim.Adam(self.critic_net.parameters(), 0.002)
13          self.critic_loss = nn.MSELoss()
14
15      def build_net(self, input_size, hidden_sizes, output_size=1,
```

```
16          output_activator=None):
17      layers = []
18      for input_size, output_size in zip([input_size,] + hidden_sizes,
19              hidden_sizes + [output_size,]):
20          layers.append(nn.Linear(input_size, output_size))
21          layers.append(nn.ReLU())
22      layers = layers[:-1]
23      if output_activator:
24          layers.append(output_activator)
25      net = nn.Sequential(*layers)
26      return net
27
28  def reset(self, mode=None):
29      self.mode = mode
30      if self.mode == 'train':
31          self.trajectory = []
32          self.discount = 1.
33
34  def step(self, observation, reward, terminated):
35      state_tensor = torch.as_tensor(observation,
36              dtype=torch.float).reshape(1, -1)
37      prob_tensor = self.actor_net(state_tensor)
38      action_tensor = distributions.Categorical(prob_tensor).sample()
39      action = action_tensor.numpy()[0]
40      if self.mode == 'train':
41          self.trajectory += [observation, reward, terminated, action]
42          if len(self.trajectory) >= 8:
43              self.learn()
44          self.discount *= self.gamma
45      return action
46
47  def close(self):
48      pass
49
50  def learn(self):
51      state, _, _, action, next_state, reward, terminated, next_action = \
52              self.trajectory[-8:]
53      state_tensor = torch.as_tensor(state, dtype=torch.float).unsqueeze(0)
54      next_state_tensor = torch.as_tensor(next_state,
55              dtype=torch.float).unsqueeze(0)
56
57      # update actor
58      q_tensor = self.critic_net(state_tensor)[0, action]
59      pi_tensor = self.actor_net(state_tensor)[0, action]
60      logpi_tensor = torch.log(pi_tensor.clamp(1e-6, 1.))
61      actor_loss_tensor = -self.discount * q_tensor * logpi_tensor
62      self.actor_optimizer.zero_grad()
63      actor_loss_tensor.backward()
64      self.actor_optimizer.step()
65
66      # update critic
67      next_q_tensor = self.critic_net(next_state_tensor)[:, next_action]
68      target_tensor = reward + (1. - terminated) * self.gamma * next_q_tensor
69      pred_tensor = self.critic_net(state_tensor)[:, action]
70      critic_loss_tensor = self.critic_loss(pred_tensor, target_tensor)
71      self.critic_optimizer.zero_grad()
72      critic_loss_tensor.backward()
73      self.critic_optimizer.step()
74
75
76  agent = QActorCriticAgent(env)
```

Codes 8.3 and 8.4 show the agents of advantage AC. They are trained and tested using Codes 5.3 and 1.4 respectively.

Code 8.3　Advantage AC agent (TensorFlow version).

Acrobot-v1_AdvantageActorCritic_tf.ipynb

```python
class AdvantageActorCriticAgent:
    def __init__(self, env):
        self.action_n = env.action_space.n
        self.gamma = 0.99

        self.actor_net = self.build_net(hidden_sizes=[100,],
                output_size=self.action_n, output_activation=nn.softmax,
                loss=losses.categorical_crossentropy,
                learning_rate=0.0001)
        self.critic_net = self.build_net(hidden_sizes=[100,],
                learning_rate=0.0002)

    def build_net(self, hidden_sizes, output_size=1,
                activation=nn.relu, output_activation=None,
                loss=losses.mse, learning_rate=0.001):
        model = keras.Sequential()
        for hidden_size in hidden_sizes:
            model.add(layers.Dense(units=hidden_size,
                    activation=activation))
        model.add(layers.Dense(units=output_size,
                activation=output_activation))
        optimizer = optimizers.Adam(learning_rate)
        model.compile(optimizer=optimizer, loss=loss)
        return model

    def reset(self, mode=None):
        self.mode = mode
        if self.mode == 'train':
            self.trajectory = []
            self.discount = 1.

    def step(self, observation, reward, terminated):
        probs = self.actor_net.predict(observation[np.newaxis], verbose=0)[0]
        action = np.random.choice(self.action_n, p=probs)
        if self.mode == 'train':
            self.trajectory += [observation, reward, terminated, action]
            if len(self.trajectory) >= 8:
                self.learn()
            self.discount *= self.gamma
        return action

    def close(self):
        pass

    def learn(self):
        state, _, _, action, next_state, reward, terminated, _ = \
                self.trajectory[-8:]
        states = state[np.newaxis]
        v = self.critic_net.predict(states, verbose=0)
        next_v = self.critic_net.predict(next_state[np.newaxis], verbose=0)
        target = reward + (1. - terminated) * self.gamma * next_v
        td_error = target - v

        # update actor
        state_tensor = tf.convert_to_tensor(states, dtype=tf.float32)
        with tf.GradientTape() as tape:
            pi_tensor = self.actor_net(state_tensor)[0, action]
            logpi_tensor = tf.math.log(tf.clip_by_value(pi_tensor, 1e-6, 1.))
            loss_tensor = -self.discount * td_error * logpi_tensor
        grad_tensors = tape.gradient(loss_tensor, self.actor_net.variables)
        self.actor_net.optimizer.apply_gradients(zip(grad_tensors,
                self.actor_net.variables))
```

```
63
64          # update critic
65          self.critic_net.fit(states, np.array([[target,],]), verbose=0)
66
67
68   agent = AdvantageActorCriticAgent(env)
```

Code 8.4   Advantage AC agent (PyTorch version).
Acrobot-v1_AdvantageActorCritic_torch.ipynb

```
1    class AdvantageActorCriticAgent:
2        def __init__(self, env):
3            self.gamma = 0.99
4
5            self.actor_net = self.build_net(
6                    input_size=env.observation_space.shape[0], hidden_sizes=[100,],
7                    output_size=env.action_space.n, output_activator=nn.Softmax(1))
8            self.actor_optimizer = optim.Adam(self.actor_net.parameters(), 0.0001)
9            self.critic_net = self.build_net(
10                   input_size=env.observation_space.shape[0], hidden_sizes=[100,])
11           self.critic_optimizer = optim.Adam(self.critic_net.parameters(),
12                   0.0002)
13           self.critic_loss = nn.MSELoss()
14
15       def build_net(self, input_size, hidden_sizes, output_size=1,
16               output_activator=None):
17           layers = []
18           for input_size, output_size in zip([input_size,] + hidden_sizes,
19                   hidden_sizes + [output_size,]):
20               layers.append(nn.Linear(input_size, output_size))
21               layers.append(nn.ReLU())
22           layers = layers[:-1]
23           if output_activator:
24               layers.append(output_activator)
25           net = nn.Sequential(*layers)
26           return net
27
28       def reset(self, mode=None):
29           self.mode = mode
30           if self.mode == 'train':
31               self.trajectory = []
32               self.discount = 1.
33
34       def step(self, observation, reward, terminated):
35           state_tensor = torch.as_tensor(observation,
36                   dtype=torch.float).reshape(1, -1)
37           prob_tensor = self.actor_net(state_tensor)
38           action_tensor = distributions.Categorical(prob_tensor).sample()
39           action = action_tensor.numpy()[0]
40           if self.mode == 'train':
41               self.trajectory += [observation, reward, terminated, action]
42               if len(self.trajectory) >= 8:
43                   self.learn()
44               self.discount *= self.gamma
45           return action
46
47       def close(self):
48           pass
49
50       def learn(self):
51           state, _, _, action, next_state, reward, terminated, next_action = \
52                   self.trajectory[-8:]
53           state_tensor = torch.as_tensor(state, dtype=torch.float).unsqueeze(0)
54           next_state_tensor = torch.as_tensor(next_state,
55                   dtype=torch.float).unsqueeze(0)
```

```
56
57    # calculate TD error
58    next_v_tensor = self.critic_net(next_state_tensor)
59    target_tensor = reward + (1. - terminated) * self.gamma * next_v_tensor
60    v_tensor = self.critic_net(state_tensor)
61    td_error_tensor = target_tensor - v_tensor
62
63    # update actor
64    pi_tensor = self.actor_net(state_tensor)[0, action]
65    logpi_tensor = torch.log(pi_tensor.clamp(1e-6, 1.))
66    actor_loss_tensor = -(self.discount * td_error_tensor *
67            logpi_tensor).squeeze()
68    self.actor_optimizer.zero_grad()
69    actor_loss_tensor.backward(retain_graph=True)
70    self.actor_optimizer.step()
71
72    # update critic
73    pred_tensor = self.critic_net(state_tensor)
74    critic_loss_tensor = self.critic_loss(pred_tensor, target_tensor)
75    self.critic_optimizer.zero_grad()
76    critic_loss_tensor.backward()
77    self.critic_optimizer.step()
78
79
80  agent = AdvantageActorCriticAgent(env)
```

Codes 8.5 and 8.6 show the eligibility trace AC agent. The trace is accumulating trace. At the beginning of a training episode, the member function `reset()` of the agent class resets the accumulative discount factor as 1, and eligibility trace parameters as 0.

Code 8.5 Eligibility-trace AC agent (TensorFlow version).
Acrobot-v1_EligibilityTraceAC_tf.ipynb

```
1   class ElibilityTraceActorCriticAgent:
2       def __init__(self, env):
3           self.action_n = env.action_space.n
4           self.gamma = 0.99
5           self.actor_lambda = 0.9
6           self.critic_lambda = 0.9
7
8           self.actor_net = self.build_net(
9                   input_size=env.observation_space.shape[0], hidden_sizes=[100,],
10                  output_size=self.action_n, output_activation=nn.softmax,
11                  loss=losses.categorical_crossentropy, learning_rate=0.0001)
12          self.critic_net = self.build_net(
13                  input_size=env.observation_space.shape[0],
14                  hidden_sizes=[100,], learning_rate=0.0002)
15
16      def build_net(self, input_size, hidden_sizes, output_size=1,
17              activation=nn.relu, output_activation=None,
18              loss=losses.mse, learning_rate=0.001):
19          model = keras.Sequential()
20          for layer, hidden_size in enumerate(hidden_sizes):
21              kwargs = {'input_shape': (input_size,)} if layer == 0 else {}
22              model.add(layers.Dense(units=hidden_size, activation=activation,
23                      **kwargs))
24          model.add(layers.Dense(units=output_size,
25                  activation=output_activation))
26          optimizer = optimizers.Adam(learning_rate)
27          model.compile(optimizer=optimizer, loss=loss)
28          return model
29
30      def reset(self, mode=None):
31          self.mode = mode
```

```
32      if self.mode == 'train':
33          self.trajectory = []
34          self.discount = 1.
35          self.actor_trace_tensors = [0. * weight for weight in
36                  self.actor_net.get_weights()]
37          self.critic_trace_tensors = [0. * weight for weight in
38                  self.critic_net.get_weights()]
39
40  def step(self, observation, reward, terminated):
41      probs = self.actor_net.predict(observation[np.newaxis], verbose=0)[0]
42      action = np.random.choice(self.action_n, p=probs)
43      if self.mode == 'train':
44          self.trajectory += [observation, reward, terminated, action]
45          if len(self.trajectory) >= 8:
46              self.learn()
47          self.discount *= self.gamma
48      return action
49
50  def close(self):
51      pass
52
53  def learn(self):
54      state, _, _, action, next_state, reward, terminated, _ = \
55              self.trajectory[-8:]
56      states = state[np.newaxis]
57      q = self.critic_net.predict(states, verbose=0)[0, 0]
58      next_v = self.critic_net.predict(next_state[np.newaxis],
59              verbose=0)[0, 0]
60      target = reward + (1. - terminated) * self.gamma * next_v
61      td_error = target - q
62
63      # update actor
64      state_tensor = tf.convert_to_tensor(states, dtype=tf.float32)
65      with tf.GradientTape() as tape:
66          pi_tensor = self.actor_net(state_tensor)[0, action]
67          logpi_tensor = tf.math.log(tf.clip_by_value(pi_tensor, 1e-6, 1.))
68      grad_tensors = tape.gradient(logpi_tensor, self.actor_net.variables)
69      self.actor_trace_tensors = [self.gamma * self.actor_lambda * trace +
70              self.discount * grad for trace, grad in
71              zip(self.actor_trace_tensors, grad_tensors)]
72      actor_grads = [-td_error * trace for trace in self.actor_trace_tensors]
73      actor_grads_and_vars = tuple(zip(actor_grads,
74              self.actor_net.variables))
75      self.actor_net.optimizer.apply_gradients(actor_grads_and_vars)
76
77      # update critic
78      with tf.GradientTape() as tape:
79          v_tensor = self.critic_net(state_tensor)[0, 0]
80      grad_tensors = tape.gradient(v_tensor, self.critic_net.variables)
81      self.critic_trace_tensors = [self.gamma * self.critic_lambda * trace +
82              grad for trace, grad in
83              zip(self.critic_trace_tensors, grad_tensors)]
84      critic_grads = [-td_error * trace for trace in
85              self.critic_trace_tensors]
86      critic_grads_and_vars = tuple(zip(critic_grads,
87              self.critic_net.variables))
88      self.critic_net.optimizer.apply_gradients(critic_grads_and_vars)
89
90
91  agent = ElibilityTraceActorCriticAgent(env)
```

Code 8.6   Eligibility-trace AC agent (PyTorch version).

Acrobot-v1_EligibilityTraceAC_torch.ipynb

```
class ElibilityTraceActorCriticAgent:
    def __init__(self, env):
        self.action_n = env.action_space.n
        self.gamma = 0.99
        self.actor_lambda = 0.9
        self.critic_lambda = 0.9

        self.actor_net = self.build_net(
                input_size=env.observation_space.shape[0],
                hidden_sizes=[100,],
                output_size=env.action_space.n, output_activator=nn.Softmax(1))
        self.actor_optimizer = optim.Adam(self.actor_net.parameters(), 0.0001)
        self.actor_trace = copy.deepcopy(self.actor_net)

        self.critic_net = self.build_net(
                input_size=env.observation_space.shape[0],
                hidden_sizes=[100,], output_size=self.action_n)
        self.critic_optimizer = optim.Adam(self.critic_net.parameters(),
                0.0002)
        self.critic_loss = nn.MSELoss()
        self.critic_trace = copy.deepcopy(self.critic_net)

    def build_net(self, input_size, hidden_sizes, output_size,
            output_activator=None):
        layers = []
        for input_size, output_size in zip(
                [input_size,] + hidden_sizes, hidden_sizes + [output_size,]):
            layers.append(nn.Linear(input_size, output_size))
            layers.append(nn.ReLU())
        layers = layers[:-1]
        if output_activator:
            layers.append(output_activator)
        net = nn.Sequential(*layers)
        return net

    def reset(self, mode=None):
        self.mode = mode
        if self.mode == 'train':
            self.trajectory = []
            self.discount = 1.

            def weights_init(m):
                if isinstance(m, nn.Linear):
                    init.zeros_(m.weight)
                    init.zeros_(m.bias)
            self.actor_trace.apply(weights_init)
            self.critic_trace.apply(weights_init)

    def step(self, observation, reward, terminated):
        state_tensor = torch.as_tensor(observation,
                dtype=torch.float).unsqueeze(0)
        prob_tensor = self.actor_net(state_tensor)
        action_tensor = distributions.Categorical(prob_tensor).sample()
        action = action_tensor.numpy()[0]
        if self.mode == 'train':
            self.trajectory += [observation, reward, terminated, action]
            if len(self.trajectory) >= 8:
                self.learn()
            self.discount *= self.gamma
        return action

    def close(self):
```

```
63          pass
64
65      def update_net(self, target_net, evaluate_net, target_weight,
66              evaluate_weight):
67          for target_param, evaluate_param in zip(
68                  target_net.parameters(), evaluate_net.parameters()):
69              target_param.data.copy_(evaluate_weight * evaluate_param.data
70                      + target_weight * target_param.data)
71
72      def learn(self):
73          state, _, _, action, next_state, reward, terminated, next_action = \
74                  self.trajectory[-8:]
75          state_tensor = torch.as_tensor(state,
76                  dtype=torch.float).unsqueeze(0)
77          next_state_tensor = torch.as_tensor(state,
78                  dtype=torch.float).unsqueeze(0)
79
80          pred_tensor = self.critic_net(state_tensor)
81          pred = pred_tensor.detach().numpy()[0, 0]
82          next_v_tesnor = self.critic_net(next_state_tensor)
83          next_v = next_v_tesnor.detach().numpy()[0, 0]
84          target = reward + (1. - terminated) * self.gamma * next_v
85          td_error = target - pred
86
87          # update actor
88          pi_tensor = self.actor_net(state_tensor)[0, action]
89          logpi_tensor = torch.log(torch.clamp(pi_tensor, 1e-6, 1.))
90          self.actor_optimizer.zero_grad()
91          logpi_tensor.backward(retain_graph=True)
92          for param, trace in zip(self.actor_net.parameters(),
93                  self.actor_trace.parameters()):
94              trace.data.copy_(self.gamma * self.actor_lambda * trace.data +
95                      self.discount * param.grad)
96              param.grad.copy_(-td_error * trace)
97          self.actor_optimizer.step()
98
99          # update critic
100         v_tensor = self.critic_net(state_tensor)[0, 0]
101         self.critic_optimizer.zero_grad()
102         v_tensor.backward()
103         for param, trace in zip(self.critic_net.parameters(),
104                 self.critic_trace.parameters()):
105             trace.data.copy_(self.gamma * self.critic_lambda * trace.data +
106                     param.grad)
107             param.grad.copy_(-td_error * trace)
108         self.critic_optimizer.step()
109
110
111 agent = ElibilityTraceActorCriticAgent(env)
```

8.6.2 On-Policy AC with Surrogate Objective

This section considers PPO algorithm. First, let us consider clipped PPO with on-policy experience replay. Code 8.7 is the replayer for PPO. Since PPO is an on-policy algorithm, all experiences in an object of PPOReplayer are generated by the same policy. If the policy is improved afterward, we need to construct a new PPOPlayer object.

Code 8.7   Replayer for PPO.

`Acrobot-v1_PPO_tf.ipynb`

```
class PPOReplayer:
    def __init__(self):
        self.fields = ['state', 'action', 'prob', 'advantage', 'return']
        self.memory = pd.DataFrame(columns=self.fields)

    def store(self, df):
        self.memory = pd.concat([self.memory, df[self.fields]],
                ignore_index=True)

    def sample(self, size):
        indices = np.random.choice(self.memory.shape[0], size=size)
        return (np.stack(self.memory.loc[indices, field]) for field in
                self.fields)
```

Codes 8.8 and 8.9 show the PPO agent. The member function `save_trajectory_to_replay()` estimates advantages and returns, and saves experiences. It chooses the parameter $\lambda = 1$ when it estimates advantages. The member function `close()` calls the member function `learn()` mutiple times, so the same experiences can be replayed and reused multiple times, which can make fully use of collected experience. The member function `learn()` samples a batch of experiences, and uses the batch to update the actor network and the critic network. When updating the actor network, it uses the parameter $\varepsilon = 0.1$ to clip the training objective. Code 5.3 trains the agent, and Code 1.4 tests the agent.

Code 8.8   PPO agent (TensorFlow version).

`Acrobot-v1_PPO_tf.ipynb`

```
class PPOAgent:
    def __init__(self, env):
        self.action_n = env.action_space.n
        self.gamma = 0.99

        self.replayer = PPOReplayer()

        self.actor_net = self.build_net(hidden_sizes=[100,],
                output_size=self.action_n, output_activation=nn.softmax,
                learning_rate=0.001)
        self.critic_net = self.build_net(hidden_sizes=[100,],
                learning_rate=0.002)

    def build_net(self, input_size=None, hidden_sizes=None, output_size=1,
                activation=nn.relu, output_activation=None,
                loss=losses.mse, learning_rate=0.001):
        model = keras.Sequential()
        for hidden_size in hidden_sizes:
            model.add(layers.Dense(units=hidden_size,
                    activation=activation))
        model.add(layers.Dense(units=output_size,
                activation=output_activation))
        optimizer = optimizers.Adam(learning_rate)
        model.compile(optimizer=optimizer, loss=loss)
        return model

    def reset(self, mode=None):
        self.mode = mode
        if self.mode == 'train':
```

```
30              self.trajectory = []
31
32      def step(self, observation, reward, terminated):
33          probs = self.actor_net.predict(observation[np.newaxis], verbose=0)[0]
34          action = np.random.choice(self.action_n, p=probs)
35          if self.mode == 'train':
36              self.trajectory += [observation, reward, terminated, action]
37          return action
38
39      def close(self):
40          if self.mode == 'train':
41              self.save_trajectory_to_replayer()
42              if len(self.replayer.memory) >= 1000:
43                  for batch in range(5):  # learn multiple times
44                      self.learn()
45                  self.replayer = PPOReplayer()
46                          # reset replayer after the agent changes itself
47
48      def save_trajectory_to_replayer(self):
49          df = pd.DataFrame(
50                  np.array(self.trajectory, dtype=object).reshape(-1, 4),
51                  columns=['state', 'reward', 'terminated', 'action'],
52                  dtype=object)
53          states = np.stack(df['state'])
54          df['v'] = self.critic_net.predict(states, verbose=0)
55          pis = self.actor_net.predict(states, verbose=0)
56          df['prob'] = [pi[action] for pi, action in zip(pis, df['action'])]
57          df['next_v'] = df['v'].shift(-1).fillna(0.)
58          df['u'] = df['reward'] + self.gamma * df['next_v']
59          df['delta'] = df['u'] - df['v']
60          df['advantage'] = signal.lfilter([1.,], [1., -self.gamma],
61                  df['delta'][::-1])[::-1]
62          df['return'] = signal.lfilter([1.,], [1., -self.gamma],
63                  df['reward'][::-1])[::-1]
64          self.replayer.store(df)
65
66      def learn(self):
67          states, actions, old_pis, advantages, returns = \
68                  self.replayer.sample(size=64)
69          state_tensor = tf.convert_to_tensor(states, dtype=tf.float32)
70          action_tensor = tf.convert_to_tensor(actions, dtype=tf.int32)
71          old_pi_tensor = tf.convert_to_tensor(old_pis, dtype=tf.float32)
72          advantage_tensor = tf.convert_to_tensor(advantages, dtype=tf.float32)
73
74          # update actor
75          with tf.GradientTape() as tape:
76              all_pi_tensor = self.actor_net(state_tensor)
77              pi_tensor = tf.gather(all_pi_tensor, action_tensor, batch_dims=1)
78              surrogate_advantage_tensor = (pi_tensor / old_pi_tensor) * \
79                      advantage_tensor
80              clip_times_advantage_tensor = 0.1 * surrogate_advantage_tensor
81              max_surrogate_advantage_tensor = advantage_tensor + \
82                      tf.where(advantage_tensor > 0.,
83                      clip_times_advantage_tensor, -clip_times_advantage_tensor)
84              clipped_surrogate_advantage_tensor = tf.minimum(
85                      surrogate_advantage_tensor, max_surrogate_advantage_tensor)
86              loss_tensor = -tf.reduce_mean(clipped_surrogate_advantage_tensor)
87          actor_grads = tape.gradient(loss_tensor, self.actor_net.variables)
88          self.actor_net.optimizer.apply_gradients(
89                  zip(actor_grads, self.actor_net.variables))
90
91          # update critic
92          self.critic_net.fit(states, returns, verbose=0)
93
94
95  agent = PPOAgent(env)
```

Code 8.9   PPO agent (PyTorch version).

Acrobot-v1_PPO_torch.ipynb

```python
class PPOAgent:
    def __init__(self, env):
        self.gamma = 0.99

        self.replayer = PPOReplayer()

        self.actor_net = self.build_net(
                input_size=env.observation_space.shape[0], hidden_sizes=[100,],
                output_size=env.action_space.n, output_activator=nn.Softmax(1))
        self.actor_optimizer = optim.Adam(self.actor_net.parameters(), 0.001)
        self.critic_net = self.build_net(
                input_size=env.observation_space.shape[0], hidden_sizes=[100,])
        self.critic_optimizer = optim.Adam(self.critic_net.parameters(), 0.002)
        self.critic_loss = nn.MSELoss()

    def build_net(self, input_size, hidden_sizes, output_size=1,
            output_activator=None):
        layers = []
        for input_size, output_size in zip(
                [input_size,] + hidden_sizes, hidden_sizes + [output_size,]):
            layers.append(nn.Linear(input_size, output_size))
            layers.append(nn.ReLU())
        layers = layers[:-1]
        if output_activator:
            layers.append(output_activator)
        net = nn.Sequential(*layers)
        return net

    def reset(self, mode=None):
        self.mode = mode
        if self.mode == 'train':
            self.trajectory = []

    def step(self, observation, reward, terminated):
        state_tensor = torch.as_tensor(observation,
                dtype=torch.float).unsqueeze(0)
        prob_tensor = self.actor_net(state_tensor)
        action_tensor = distributions.Categorical(prob_tensor).sample()
        action = action_tensor.numpy()[0]
        if self.mode == 'train':
            self.trajectory += [observation, reward, terminated, action]
        return action

    def close(self):
        if self.mode == 'train':
            self.save_trajectory_to_replayer()
            if len(self.replayer.memory) >= 1000:
                for batch in range(5):  # learn multiple times
                    self.learn()
                self.replayer = PPOReplayer()
                        # reset replayer after the agent changes itself

    def save_trajectory_to_replayer(self):
        df = pd.DataFrame(
                np.array(self.trajectory, dtype=object).reshape(-1, 4),
                columns=['state', 'reward', 'terminated', 'action'])
        state_tensor = torch.as_tensor(np.stack(df['state']),
                dtype=torch.float)
        action_tensor = torch.as_tensor(df['action'], dtype=torch.long)
        v_tensor = self.critic_net(state_tensor)
        df['v'] = v_tensor.detach().numpy()
        prob_tensor = self.actor_net(state_tensor)
```

```
63        pi_tensor = prob_tensor.gather(-1, action_tensor.unsqueeze(1)
64                ).squeeze(1)
65        df['prob'] = pi_tensor.detach().numpy()
66        df['next_v'] = df['v'].shift(-1).fillna(0.)
67        df['u'] = df['reward'] + self.gamma * df['next_v']
68        df['delta'] = df['u'] - df['v']
69        df['advantage'] = signal.lfilter([1.,], [1., -self.gamma],
70                df['delta'][::-1])[::-1]
71        df['return'] = signal.lfilter([1.,], [1., -self.gamma],
72                df['reward'][::-1])[::-1]
73        self.replayer.store(df)
74
75    def learn(self):
76        states, actions, old_pis, advantages, returns = \
77                self.replayer.sample(size=64)
78        state_tensor = torch.as_tensor(states, dtype=torch.float)
79        action_tensor = torch.as_tensor(actions, dtype=torch.long)
80        old_pi_tensor = torch.as_tensor(old_pis, dtype=torch.float)
81        advantage_tensor = torch.as_tensor(advantages, dtype=torch.float)
82        return_tensor = torch.as_tensor(returns,
83                dtype=torch.float).unsqueeze(1)
84
85        # update actor
86        all_pi_tensor = self.actor_net(state_tensor)
87        pi_tensor = all_pi_tensor.gather(1,
88                action_tensor.unsqueeze(1)).squeeze(1)
89        surrogate_advantage_tensor = (pi_tensor / old_pi_tensor) * \
90                advantage_tensor
91        clip_times_advantage_tensor = 0.1 * surrogate_advantage_tensor
92        max_surrogate_advantage_tensor = advantage_tensor + \
93                torch.where(advantage_tensor > 0.,
94                clip_times_advantage_tensor, -clip_times_advantage_tensor)
95        clipped_surrogate_advantage_tensor = torch.min(
96                surrogate_advantage_tensor, max_surrogate_advantage_tensor)
97        actor_loss_tensor = -clipped_surrogate_advantage_tensor.mean()
98        self.actor_optimizer.zero_grad()
99        actor_loss_tensor.backward()
100       self.actor_optimizer.step()
101
102       # update critic
103       pred_tensor = self.critic_net(state_tensor)
104       critic_loss_tensor = self.critic_loss(pred_tensor, return_tensor)
105       self.critic_optimizer.zero_grad()
106       critic_loss_tensor.backward()
107       self.critic_optimizer.step()
108
109
110  agent = PPOAgent(env)
```

8.6.3 NPG and TRPO

This section implements NPG algorithm and TRPO algorithm.

We use CG algorithm to solve the linear equation set $\mathbf{Fx} = \mathbf{g}$. Codes 8.10 and 8.11 shows how to use CG algorithm to solve the equations: it first initializes $\mathbf{x}_0 \leftarrow \mathbf{0}$, and have at most 10 iterations. The iteration can break early when rho is smaller than the threshold tol. The return values include both the solution $\mathbf{x}$ and the function value $\mathbf{Fx}$.

Code 8.10   Calculate CG (TensorFlow version).

`Acrobot-v1_NPG_tf.ipynb`

```
def conjugate_gradient(f, b, iter_count=10, epsilon=1e-12, tol=1e-6):
    x = b * 0.
    r = tf.identity(b)
    p = tf.identity(b)
    rho = tf.reduce_sum(r * r)
    for i in range(iter_count):
        z = f(p)
        alpha = rho / (tf.reduce_sum(p * z) + epsilon)
        x += alpha * p
        r -= alpha * z
        rho_new = tf.reduce_sum(r * r)
        p = r + (rho_new / rho) * p
        rho = rho_new
        if rho < tol:
            break
    return x, f(x)
```

Code 8.11   Calculate CG (PyTorch version).

`Acrobot-v1_NPG_torch.ipynb`

```
def conjugate_gradient(f, b, iter_count=10, epsilon=1e-12, tol=1e-6):
    x = b * 0.
    r = b.clone()
    p = b.clone()
    rho = torch.dot(r, r)
    for i in range(iter_count):
        z = f(p)
        alpha = rho / (torch.dot(p, z) + epsilon)
        x += alpha * p
        r -= alpha * z
        rho_new = torch.dot(r, r)
        p = r + (rho_new / rho) * p
        rho = rho_new
        if rho < tol:
            break
    return x, f(x)
```

Codes 8.12 and 8.13 show the agents of NPG. The member function `learn()` trains the agent. The training of actor network is somehow complex. The codes of actor training are organized into the following four blocks: (1) estimate the gradient of KL divergence; (2) use CG to calculate **x** and **Fx**. Especially, there is an auxiliary function `f()`, whose input is **x** and output is **Fx**. The function `f()` calculate gradients twice to estimate FIM. Use the function `f()` together with the CG algorithm, we can get **x** and **Fx**. (3) use **x** and **Fx** to get natural gradient. These three code blocks correspond to Step 2.2 in Algo. 8.9. (4) Use natural gradient to update actor parameters, which corresponds to Step 2.3 in Algo. 8.9.

Code 8.12   NPG agent (TensorFlow version).

`Acrobot-v1_NPG_tf.ipynb`

```python
class NPGAgent:
    def __init__(self, env):
        self.action_n = env.action_space.n
        self.gamma = 0.99

        self.replayer = PPOReplayer()
        self.trajectory = []

        self.max_kl = 0.0005
        self.actor_net = self.build_net(hidden_sizes=[100,],
                output_size=self.action_n, output_activation=nn.softmax)
        self.critic_net = self.build_net(hidden_sizes=[100,],
                learning_rate=0.002)

    def build_net(self, input_size=None, hidden_sizes=None, output_size=1,
                activation=nn.relu, output_activation=None,
                loss=losses.mse, learning_rate=0.001):
        model = keras.Sequential()
        for hidden_size in hidden_sizes:
            model.add(layers.Dense(units=hidden_size,
                    activation=activation))
        model.add(layers.Dense(units=output_size,
                activation=output_activation))
        optimizer = optimizers.Adam(learning_rate)
        model.compile(optimizer=optimizer, loss=loss)
        return model

    def reset(self, mode=None):
        self.mode = mode
        if self.mode == 'train':
            self.trajectory = []

    def step(self, observation, reward, terminated):
        probs = self.actor_net.predict(observation[np.newaxis], verbose=0)[0]
        action = np.random.choice(self.action_n, p=probs)
        if self.mode == 'train':
            self.trajectory += [observation, reward, terminated, action]
        return action

    def close(self):
        if self.mode == 'train':
            self.save_trajectory_to_replayer()
            if len(self.replayer.memory) >= 1000:
                for batch in range(5):  # learn multiple times
                    self.learn()
                self.replayer = PPOReplayer()
                        # reset replayer after the agent changes itself

    def save_trajectory_to_replayer(self):
        df = pd.DataFrame(
                np.array(self.trajectory, dtype=object).reshape(-1, 4),
                columns=['state', 'reward', 'terminated', 'action'],
                dtype=object)
        states = np.stack(df['state'])
        df['v'] = self.critic_net.predict(states, verbose=0)
        pis = self.actor_net.predict(states, verbose=0)
        df['prob'] = [pi[action] for pi, action in zip(pis, df['action'])]
        df['next_v'] = df['v'].shift(-1).fillna(0.)
        df['u'] = df['reward'] + self.gamma * df['next_v']
        df['delta'] = df['u'] - df['v']
        df['advantage'] = signal.lfilter([1.,], [1., -self.gamma],
                df['delta'][::-1])[::-1]
```

```
63          df['return'] = signal.lfilter([1.,], [1., -self.gamma],
64                  df['reward'][::-1])[::-1]
65          self.replayer.store(df)
66
67      def learn(self):
68          states, actions, old_pis, advantages, returns = \
69                  self.replayer.sample(size=64)
70          state_tensor = tf.convert_to_tensor(states, dtype=tf.float32)
71          action_tensor = tf.convert_to_tensor(actions, dtype=tf.int32)
72          old_pi_tensor = tf.convert_to_tensor(old_pis, dtype=tf.float32)
73          advantage_tensor = tf.convert_to_tensor(advantages, dtype=tf.float32)
74
75          # update actor
76          # ... calculate first order gradient of KL divergence
77          with tf.GradientTape() as tape:
78              all_pi_tensor = self.actor_net(state_tensor)
79              pi_tensor = tf.gather(all_pi_tensor, action_tensor, batch_dims=1)
80              surrogate_tensor = (pi_tensor / old_pi_tensor) * advantage_tensor
81          actor_grads = tape.gradient(surrogate_tensor, self.actor_net.variables)
82          loss_grad = tf.concat([tf.reshape(grad, (-1,)) for grad in actor_grads
83                  ], axis=0)
84
85          # ... calculate conjugate gradient: Fx = g
86          def f(x):  # calculate Fx
87              with tf.GradientTape() as tape2:  # tape for 2nd-order gradient
88                  with tf.GradientTape() as tape1:  # tape for 1st-order gradient
89                      prob_tensor = self.actor_net(state_tensor)
90                      prob_old_tensor = tf.stop_gradient(prob_tensor)
91                      kld_tensor = tf.reduce_sum(prob_old_tensor * (tf.math.log(
92                              prob_old_tensor) - tf.math.log(prob_tensor)),
93                              axis=1)
94                      kld_loss_tensor = tf.reduce_mean(kld_tensor)
95                  grads = tape1.gradient(kld_loss_tensor,
96                          self.actor_net.variables)
97                  flatten_grad_tensor = tf.concat(
98                          [tf.reshape(grad, (-1,)) for grad in grads], axis=-1)
99                  grad_matmul_x = tf.tensordot(flatten_grad_tensor, x,
100                         axes=[[-1], [-1]])
101             grad_grads = tape2.gradient(grad_matmul_x,
102                     self.actor_net.variables)
103             flatten_grad_grad = tf.stop_gradient(tf.concat(
104                     [tf.reshape(grad_grad, (-1,)) for grad_grad in grad_grads],
105                     axis=-1))
106             fx = flatten_grad_grad + x * 1e-2
107             return fx
108         x, fx = conjugate_gradient(f, loss_grad)
109
110         # ... calculate natural gradient
111         natural_gradient_tensor = tf.sqrt(2 * self.max_kl /
112                 tf.reduce_sum(fx * x)) * x
113         # ....... refactor the flatten gradient into un-flatten version
114         flatten_natural_gradient = natural_gradient_tensor.numpy()
115         weights = []
116         begin = 0
117         for weight in self.actor_net.get_weights():
118             end = begin + weight.size
119             weight += flatten_natural_gradient[begin:end].reshape(weight.shape)
120             weights.append(weight)
121             begin = end
122         self.actor_net.set_weights(weights)
123
124         # update critic
125         self.critic_net.fit(states, returns, verbose=0)
126
127
128 agent = NPGAgent(env)
```

Code 8.13   NPG agent (PyTorch version).

Acrobot-v1_NPG_torch.ipynb

```python
class NPGAgent:
    def __init__(self, env):
        self.gamma = 0.99

        self.replayer = PPOReplayer()
        self.trajectory = []

        self.actor_net = self.build_net(
                input_size=env.observation_space.shape[0],
                hidden_sizes=[100,],
                output_size=env.action_space.n, output_activator=nn.Softmax(1))
        self.max_kl = 0.001
        self.critic_net = self.build_net(
                input_size=env.observation_space.shape[0],
                hidden_sizes=[100,])
        self.critic_optimizer = optim.Adam(self.critic_net.parameters(), 0.002)
        self.critic_loss = nn.MSELoss()

    def build_net(self, input_size, hidden_sizes, output_size=1,
                output_activator=None):
        layers = []
        for input_size, output_size in zip(
                [input_size,] + hidden_sizes, hidden_sizes + [output_size,]):
            layers.append(nn.Linear(input_size, output_size))
            layers.append(nn.ReLU())
        layers = layers[:-1]
        if output_activator:
            layers.append(output_activator)
        net = nn.Sequential(*layers)
        return net

    def reset(self, mode=None):
        self.mode = mode
        if self.mode == 'train':
            self.trajectory = []

    def step(self, observation, reward, terminated):
        state_tensor = torch.as_tensor(observation,
                dtype=torch.float).unsqueeze(0)
        prob_tensor = self.actor_net(state_tensor)
        action_tensor = distributions.Categorical(prob_tensor).sample()
        action = action_tensor.numpy()[0]
        if self.mode == 'train':
            self.trajectory += [observation, reward, terminated, action]
        return action

    def close(self):
        if self.mode == 'train':
            self.save_trajectory_to_replayer()
            if len(self.replayer.memory) >= 1000:
                for batch in range(5):  # learn multiple times
                    self.learn()
                self.replayer = PPOReplayer()
                        # reset replayer after the agent changes itself

    def save_trajectory_to_replayer(self):
        df = pd.DataFrame(
                np.array(self.trajectory, dtype=object).reshape(-1, 4),
                columns=['state', 'reward', 'terminated', 'action'])
        state_tensor = torch.as_tensor(np.stack(df['state']),
                dtype=torch.float)
        action_tensor = torch.as_tensor(df['action'], dtype=torch.long)
```

```
63        v_tensor = self.critic_net(state_tensor)
64        df['v'] = v_tensor.detach().numpy()
65        prob_tensor = self.actor_net(state_tensor)
66        pi_tensor = prob_tensor.gather(-1,
67                action_tensor.unsqueeze(1)).squeeze(1)
68        df['prob'] = pi_tensor.detach().numpy()
69        df['next_v'] = df['v'].shift(-1).fillna(0.)
70        df['u'] = df['reward'] + self.gamma * df['next_v']
71        df['delta'] = df['u'] - df['v']
72        df['advantage'] = signal.lfilter([1.,], [1., -self.gamma],
73                df['delta'][::-1])[::-1]
74        df['return'] = signal.lfilter([1.,], [1., -self.gamma],
75                df['reward'][::-1])[::-1]
76        self.replayer.store(df)
77
78    def learn(self):
79        states, actions, old_pis, advantages, returns = \
80                self.replayer.sample(size=64)
81        state_tensor = torch.as_tensor(states, dtype=torch.float)
82        action_tensor = torch.as_tensor(actions, dtype=torch.long)
83        old_pi_tensor = torch.as_tensor(old_pis, dtype=torch.float)
84        advantage_tensor = torch.as_tensor(advantages, dtype=torch.float)
85        return_tensor = torch.as_tensor(returns,
86                dtype=torch.float).unsqueeze(1)
87
88        # update actor
89        # ... calculate first order gradient: g
90        all_pi_tensor = self.actor_net(state_tensor)
91        pi_tensor = all_pi_tensor.gather(1,
92                action_tensor.unsqueeze(1)).squeeze(1)
93        surrogate_tensor = (pi_tensor / old_pi_tensor) * advantage_tensor
94        loss_tensor = surrogate_tensor.mean()
95        loss_grads = autograd.grad(loss_tensor, self.actor_net.parameters())
96        loss_grad = torch.cat([grad.view(-1) for grad in loss_grads]).detach()
97                # flatten for calculating conjugate gradient
98
99        # ... calculate conjugate gradient: Fx = g
100        def f(x):  # calculate Fx
101            prob_tensor = self.actor_net(state_tensor)
102            prob_old_tensor = prob_tensor.detach()
103            kld_tensor = (prob_old_tensor * (torch.log((prob_old_tensor /
104                    prob_tensor).clamp(1e-6, 1e6)))).sum(axis=1)
105            kld_loss_tensor = kld_tensor.mean()
106            grads = autograd.grad(kld_loss_tensor, self.actor_net.parameters(),
107                    create_graph=True)
108            flatten_grad_tensor = torch.cat([grad.view(-1) for grad in grads])
109            grad_matmul_x = torch.dot(flatten_grad_tensor, x)
110            grad_grads = autograd.grad(grad_matmul_x,
111                    self.actor_net.parameters())
112            flatten_grad_grad = torch.cat([grad.contiguous().view(-1) for grad
113                    in grad_grads]).detach()
114            fx = flatten_grad_grad + x * 1e-2
115            return fx
116        x, fx = conjugate_gradient(f, loss_grad)
117
118        # ... calculate natural gradient: sqrt(...) g
119        natural_gradient = torch.sqrt(2 * self.max_kl / torch.dot(fx, x)) * x
120
121        # ... update actor net
122        begin = 0
123        for param in self.actor_net.parameters():
124            end = begin + param.numel()
125            param.data.copy_(natural_gradient[begin:end].view(param.size()) +
126                    param.data)
127            begin = end
128
129        # update critic
```

```
130        pred_tensor = self.critic_net(state_tensor)
131        critic_loss_tensor = self.critic_loss(pred_tensor, return_tensor)
132        self.critic_optimizer.zero_grad()
133        critic_loss_tensor.backward()
134        self.critic_optimizer.step()
135
136
137 agent = NPGAgent(env)
```

Codes 8.14 and 8.15 show TRPO agents. They differ from Codes 8.12 and 8.13 only in how to use natural gradient to update actor network parameters.

Code 8.14   TRPO agent (TensorFlow version).
Acrobot-v1_TRPO_tf.ipynb

```
1  class TRPOAgent:
2      def __init__(self, env):
3          self.action_n = env.action_space.n
4          self.gamma = 0.99
5
6          self.replayer = PPOReplayer()
7          self.trajectory = []
8
9          self.max_kl = 0.01
10         self.actor_net = self.build_net(hidden_sizes=[100,],
11                 output_size=self.action_n, output_activation=nn.softmax)
12         self.critic_net = self.build_net(hidden_sizes=[100,],
13                 learning_rate=0.002)
14
15     def build_net(self, input_size=None, hidden_sizes=None, output_size=1,
16                 activation=nn.relu, output_activation=None,
17                 loss=losses.mse, learning_rate=0.001):
18         model = keras.Sequential()
19         for hidden_size in hidden_sizes:
20             model.add(layers.Dense(units=hidden_size,
21                 activation=activation))
22         model.add(layers.Dense(units=output_size,
23                 activation=output_activation))
24         optimizer = optimizers.Adam(learning_rate)
25         model.compile(optimizer=optimizer, loss=loss)
26         return model
27
28     def reset(self, mode=None):
29         self.mode = mode
30         if self.mode == 'train':
31             self.trajectory = []
32
33     def step(self, observation, reward, terminated):
34         probs = self.actor_net.predict(observation[np.newaxis], verbose=0)[0]
35         action = np.random.choice(self.action_n, p=probs)
36         if self.mode == 'train':
37             self.trajectory += [observation, reward, terminated, action]
38         return action
39
40     def close(self):
41         if self.mode == 'train':
42             self.save_trajectory_to_replayer()
43             if len(self.replayer.memory) >= 1000:
44                 for batch in range(5):  # learn multiple times
45                     self.learn()
46                 self.replayer = PPOReplayer()
47                     # reset replayer after the agent changes itself
48
49     def save_trajectory_to_replayer(self):
50         df = pd.DataFrame(
51                 np.array(self.trajectory, dtype=object).reshape(-1, 4),
```

```python
52          columns=['state', 'reward', 'terminated', 'action'],
53          dtype=object)
54      states = np.stack(df['state'])
55      df['v'] = self.critic_net.predict(states, verbose=0)
56      pis = self.actor_net.predict(states, verbose=0)
57      df['prob'] = [pi[action] for pi, action in zip(pis, df['action'])]
58      df['next_v'] = df['v'].shift(-1).fillna(0.)
59      df['u'] = df['reward'] + self.gamma * df['next_v']
60      df['delta'] = df['u'] - df['v']
61      df['advantage'] = signal.lfilter([1.,], [1., -self.gamma],
62              df['delta'][::-1])[::-1]
63      df['return'] = signal.lfilter([1.,], [1., -self.gamma],
64              df['reward'][::-1])[::-1]
65      self.replayer.store(df)
66
67  def learn(self):
68      states, actions, old_pis, advantages, returns = \
69              self.replayer.sample(size=64)
70      state_tensor = tf.convert_to_tensor(states, dtype=tf.float32)
71      action_tensor = tf.convert_to_tensor(actions, dtype=tf.int32)
72      old_pi_tensor = tf.convert_to_tensor(old_pis, dtype=tf.float32)
73      advantage_tensor = tf.convert_to_tensor(advantages, dtype=tf.float32)
74
75      # update actor
76      # ... calculate first order gradient of KL divergence
77      with tf.GradientTape() as tape:
78          all_pi_tensor = self.actor_net(state_tensor)
79          pi_tensor = tf.gather(all_pi_tensor, action_tensor, batch_dims=1)
80          surrogate_tensor = (pi_tensor / old_pi_tensor) * advantage_tensor
81      actor_grads = tape.gradient(surrogate_tensor, self.actor_net.variables)
82      loss_grad = tf.concat([tf.reshape(grad, (-1,)) for grad in actor_grads
83              ], axis=0)
84
85      # ... calculate conjugate gradient: Fx = g
86      def f(x):  # calculate Fx
87          with tf.GradientTape() as tape2:  # tape for 2nd-order gradient
88              with tf.GradientTape() as tape1:  # tape for 1st-order gradient
89                  prob_tensor = self.actor_net(state_tensor)
90                  prob_old_tensor = tf.stop_gradient(prob_tensor)
91                  kld_tensor = tf.reduce_sum(prob_old_tensor * (tf.math.log(
92                          prob_old_tensor) - tf.math.log(prob_tensor)),
93                          axis=1)
94                  kld_loss_tensor = tf.reduce_mean(kld_tensor)
95              grads = tape1.gradient(kld_loss_tensor,
96                      self.actor_net.variables)
97              flatten_grad_tensor = tf.concat(
98                      [tf.reshape(grad, (-1,)) for grad in grads], axis=-1)
99              grad_matmul_x = tf.tensordot(flatten_grad_tensor, x,
100                         axes=[[-1], [-1]])
101         grad_grads = tape2.gradient(grad_matmul_x,
102                 self.actor_net.variables)
103         flatten_grad_grad = tf.stop_gradient(tf.concat([tf.reshape(
104                 grad_grad, (-1,)) for grad_grad in grad_grads], axis=-1))
105         fx = flatten_grad_grad + x * 1e-2
106         return fx
107     x, fx = conjugate_gradient(f, loss_grad)
108
109     # ... calculate natural gradient
110     natural_gradient_tensor = tf.sqrt(2 * self.max_kl /
111             tf.reduce_sum(fx * x)) * x
112     # ....... refactor the flatten gradient into un-flatten version
113     flatten_natural_gradient = natural_gradient_tensor.numpy()
114     natural_grads = []
115     begin = 0
116     for weight in self.actor_net.get_weights():
117         end = begin + weight.size
118         natural_grad = flatten_natural_gradient[begin:end].reshape(
```

```
119                         weight.shape)
120                     natural_grads.append(natural_grad)
121                     begin = end
122
123             # ... line search
124             old_weights = self.actor_net.get_weights()
125             expected_improve = tf.reduce_sum(loss_grad *
126                     natural_gradient_tensor).numpy()
127             for learning_step in [0.,] + [.5 ** j for j in range(10)]:
128                 self.actor_net.set_weights([weight + learning_step * grad
129                         for weight, grad in zip(old_weights, natural_grads)])
130                 all_pi_tensor = self.actor_net(state_tensor)
131                 new_pi_tensor = tf.gather(all_pi_tensor,
132                     action_tensor[:, np.newaxis], axis=1)[:, 0]
133                 new_pi_tensor = tf.stop_gradient(new_pi_tensor)
134                 surrogate_tensor = (new_pi_tensor / pi_tensor) * advantage_tensor
135                 objective = tf.reduce_sum(surrogate_tensor).numpy()
136                 if np.isclose(learning_step, 0.):
137                     old_objective = objective
138                 else:
139                     if objective - old_objective > 0.1 * expected_improve * \
140                             learning_step:
141                         break # success, keep the weight
142             else:
143                 self.actor_net.set_weights(old_weights)
144
145             # update critic
146             self.critic_net.fit(states, returns, verbose=0)
147
148
149 agent = TRPOAgent(env)
```

Code 8.15   TRPO agent (PyTorch version).
Acrobot-v1_TRPO_torch.ipynb

```
1  class TRPOAgent:
2      def __init__(self, env):
3          self.gamma = 0.99
4
5          self.replayer = PPOReplayer()
6          self.trajectory = []
7
8          self.actor_net = self.build_net(
9                  input_size=env.observation_space.shape[0], hidden_sizes=[100,],
10                 output_size=env.action_space.n, output_activator=nn.Softmax(1))
11         self.max_kl = 0.01
12         self.critic_net = self.build_net(
13                 input_size=env.observation_space.shape[0], hidden_sizes=[100,])
14         self.critic_optimizer = optim.Adam(self.critic_net.parameters(), 0.002)
15         self.critic_loss = nn.MSELoss()
16
17     def build_net(self, input_size, hidden_sizes, output_size=1,
18             output_activator=None):
19         layers = []
20         for input_size, output_size in zip([input_size,] + hidden_sizes,
21                 hidden_sizes + [output_size,]):
22             layers.append(nn.Linear(input_size, output_size))
23             layers.append(nn.ReLU())
24         layers = layers[:-1]
25         if output_activator:
26             layers.append(output_activator)
27         net = nn.Sequential(*layers)
28         return net
29
30     def reset(self, mode=None):
```

```
31          self.mode = mode
32          if self.mode == 'train':
33              self.trajectory = []
34
35      def step(self, observation, reward, terminated):
36          state_tensor = torch.as_tensor(observation,
37                  dtype=torch.float).unsqueeze(0)
38          prob_tensor = self.actor_net(state_tensor)
39          action_tensor = distributions.Categorical(prob_tensor).sample()
40          action = action_tensor.numpy()[0]
41          if self.mode == 'train':
42              self.trajectory += [observation, reward, terminated, action]
43          return action
44
45      def close(self):
46          if self.mode == 'train':
47              self.save_trajectory_to_replayer()
48              if len(self.replayer.memory) >= 1000:
49                  for batch in range(5):  # learn multiple times
50                      self.learn()
51                  self.replayer = PPOReplayer()
52                          # reset replayer after the agent changes itself
53
54      def save_trajectory_to_replayer(self):
55          df = pd.DataFrame(
56                  np.array(self.trajectory, dtype=object).reshape(-1, 4),
57                  columns=['state', 'reward', 'terminated', 'action'])
58          state_tensor = torch.as_tensor(np.stack(df['state']),
59                  dtype=torch.float)
60          action_tensor = torch.as_tensor(df['action'], dtype=torch.long)
61          v_tensor = self.critic_net(state_tensor)
62          df['v'] = v_tensor.detach().numpy()
63          prob_tensor = self.actor_net(state_tensor)
64          pi_tensor = prob_tensor.gather(-1,
65                  action_tensor.unsqueeze(1)).squeeze(1)
66          df['prob'] = pi_tensor.detach().numpy()
67          df['next_v'] = df['v'].shift(-1).fillna(0.)
68          df['u'] = df['reward'] + self.gamma * df['next_v']
69          df['delta'] = df['u'] - df['v']
70          df['advantage'] = signal.lfilter([1.,], [1., -self.gamma],
71                  df['delta'][::-1])[::-1]
72          df['return'] = signal.lfilter([1.,], [1., -self.gamma],
73                  df['reward'][::-1])[::-1]
74          self.replayer.store(df)
75
76      def learn(self):
77          states, actions, old_pis, advantages, returns = \
78                  self.replayer.sample(size=64)
79          state_tensor = torch.as_tensor(states, dtype=torch.float)
80          action_tensor = torch.as_tensor(actions, dtype=torch.long)
81          old_pi_tensor = torch.as_tensor(old_pis, dtype=torch.float)
82          advantage_tensor = torch.as_tensor(advantages, dtype=torch.float)
83          return_tensor = torch.as_tensor(returns,
84                  dtype=torch.float).unsqueeze(1)
85
86          # update actor
87          # ... calculate first order gradient: g
88          all_pi_tensor = self.actor_net(state_tensor)
89          pi_tensor = all_pi_tensor.gather(1,
90                  action_tensor.unsqueeze(1)).squeeze(1)
91          surrogate_tensor = (pi_tensor / old_pi_tensor) * advantage_tensor
92          loss_tensor = surrogate_tensor.mean()
93          loss_grads = autograd.grad(loss_tensor, self.actor_net.parameters())
94          loss_grad = torch.cat([grad.view(-1) for grad in loss_grads]).detach()
95                  # flatten for calculating conjugate gradient
96
97          # ... calculate conjugate gradient: Fx = g
```

```
 98      def f(x):  # calculate Fx
 99          prob_tensor = self.actor_net(state_tensor)
100          prob_old_tensor = prob_tensor.detach()
101          kld_tensor = (prob_old_tensor * torch.log(
102                  (prob_old_tensor / prob_tensor).clamp(1e-6, 1e6)
103                  )).sum(axis=1)
104          kld_loss_tensor = kld_tensor.mean()
105          grads = autograd.grad(kld_loss_tensor, self.actor_net.parameters(),
106                  create_graph=True)
107          flatten_grad_tensor = torch.cat([grad.view(-1) for grad in grads])
108          grad_matmul_x = torch.dot(flatten_grad_tensor, x)
109          grad_grads = autograd.grad(grad_matmul_x,
110                  self.actor_net.parameters())
111          flatten_grad_grad = torch.cat([grad.contiguous().view(-1) for grad
112                  in grad_grads]).detach()
113          fx = flatten_grad_grad + x * 0.01
114          return fx
115      x, fx = conjugate_gradient(f, loss_grad)
116
117      # ... calculate natural gradient: sqrt(...) g
118      natural_gradient_tensor = torch.sqrt(2 * self.max_kl /
119              torch.dot(fx, x)) * x
120
121      # ... line search
122      def set_actor_net_params(flatten_params):
123          # auxiliary function to overwrite actor_net
124          begin = 0
125          for param in self.actor_net.parameters():
126              end = begin + param.numel()
127              param.data.copy_(flatten_params[begin:end].view(param.size()))
128              begin = end
129
130      old_param = torch.cat([param.view(-1) for param in
131              self.actor_net.parameters()])
132      expected_improve = torch.dot(loss_grad, natural_gradient_tensor)
133      for learning_step in [0.,] + [.5 ** j for j in range(10)]:
134          new_param = old_param + learning_step * natural_gradient_tensor
135          set_actor_net_params(new_param)
136          all_pi_tensor = self.actor_net(state_tensor)
137          new_pi_tensor = all_pi_tensor.gather(1,
138                  action_tensor.unsqueeze(1)).squeeze(1)
139          new_pi_tensor = new_pi_tensor.detach()
140          surrogate_tensor = (new_pi_tensor / pi_tensor) * advantage_tensor
141          objective = surrogate_tensor.mean().item()
142          if np.isclose(learning_step, 0.):
143              old_objective = objective
144          else:
145              if objective - old_objective > 0.1 * expected_improve * \
146                      learning_step:
147                  break  # success, keep the weight
148      else:
149          set_actor_net_params(old_param)
150
151      # update critic
152      pred_tensor = self.critic_net(state_tensor)
153      critic_loss_tensor = self.critic_loss(pred_tensor, return_tensor)
154      self.critic_optimizer.zero_grad()
155      critic_loss_tensor.backward()
156      self.critic_optimizer.step()
157
158
159  agent = TRPOAgent(env)
```

The interaction between agents and the environment is still Code 1.3. The agents are trained and tested using Codes 5.3 and 1.4 respectively.

8.6.4 Importance Sampling Off-Policy AC

This section considers off-policy AC algorithms. Codes 8.16 and 8.17 implement the
OffPAC agent.

Code 8.16   OffPAC agent (TensorFlow version).
Acrobot-v1_OffPAC_tf.ipynb

```
class OffPACAgent:
    def __init__(self, env):
        self.action_n = env.action_space.n
        self.gamma = 0.99

        self.actor_net = self.build_net(hidden_sizes=[100,],
                output_size=self.action_n,
                output_activation=nn.softmax, learning_rate=0.0001)
        self.critic_net = self.build_net(hidden_sizes=[100,],
                output_size=self.action_n, learning_rate=0.0002)

    def build_net(self, hidden_sizes, output_size,
                activation=nn.relu, output_activation=None,
                loss=losses.mse, learning_rate=0.001):
        model = keras.Sequential()
        for hidden_size in hidden_sizes:
            model.add(layers.Dense(units=hidden_size,
                    activation=activation))
        model.add(layers.Dense(units=output_size,
                activation=output_activation))
        optimizer = optimizers.SGD(learning_rate)
        model.compile(optimizer=optimizer, loss=loss)
        return model

    def reset(self, mode=None):
        self.mode = mode
        if self.mode == 'train':
            self.trajectory = []
            self.discount = 1.

    def step(self, observation, reward, terminated):
        if self.mode == 'train':
            action = np.random.choice(self.action_n)
            self.trajectory += [observation, reward, terminated, action]
            if len(self.trajectory) >= 8:
                self.learn()
            self.discount *= self.gamma
        else:
            probs = self.actor_net.predict(observation[np.newaxis],
                    verbose=0)[0]
            action = np.random.choice(self.action_n, p=probs)
        return action

    def close(self):
        pass

    def learn(self):
        state, _, _, action, next_state, reward, terminated, next_action = \
                self.trajectory[-8:]
        behavior_prob = 1. / self.action_n
        pi = self.actor_net.predict(state[np.newaxis], verbose=0)[0, action]
        ratio = pi / behavior_prob # importance sampling ratio

        # update actor
        q = self.critic_net.predict(state[np.newaxis], verbose=0)[0, action]
        state_tensor = tf.convert_to_tensor(state[np.newaxis],
```

```
57          dtype=tf.float32)
58      with tf.GradientTape() as tape:
59          pi_tensor = self.actor_net(state_tensor)[0, action]
60          actor_loss_tensor = -self.discount * q / behavior_prob * pi_tensor
61      grad_tensors = tape.gradient(actor_loss_tensor,
62              self.actor_net.variables)
63      self.actor_net.optimizer.apply_gradients(zip(grad_tensors,
64              self.actor_net.variables))
65
66      # update critic
67      next_q = self.critic_net.predict(next_state[np.newaxis], verbose=0)[0,
68              next_action]
69      target = reward + (1. - terminated) * self.gamma * next_q
70      target_tensor = tf.convert_to_tensor(target, dtype=tf.float32)
71      with tf.GradientTape() as tape:
72          q_tensor = self.critic_net(state_tensor)[:, action]
73          mse_tensor = losses.MSE(target_tensor, q_tensor)
74          critic_loss_tensor = ratio * mse_tensor
75      grad_tensors = tape.gradient(critic_loss_tensor,
76              self.critic_net.variables)
77      self.critic_net.optimizer.apply_gradients(zip(grad_tensors,
78              self.critic_net.variables))
79
80
81  agent = OffPACAgent(env)
```

Code 8.17   OffPAC agent (PyTorch version).
Acrobot-v1_OffPAC_torch.ipynb

```
1   class OffPACAgent:
2       def __init__(self, env):
3           self.action_n = env.action_space.n
4           self.gamma = 0.99
5
6           self.actor_net = self.build_net(
7                   input_size=env.observation_space.shape[0],
8                   hidden_sizes=[100,],
9                   output_size=env.action_space.n, output_activator=nn.Softmax(1))
10          self.actor_optimizer = optim.Adam(self.actor_net.parameters(), 0.0002)
11          self.critic_net = self.build_net(
12                  input_size=env.observation_space.shape[0],
13                  hidden_sizes=[100,], output_size=self.action_n)
14          self.critic_optimizer = optim.Adam(self.critic_net.parameters(),
15                  0.0004)
16          self.critic_loss = nn.MSELoss()
17
18      def build_net(self, input_size, hidden_sizes, output_size,
19              output_activator=None):
20          layers = []
21          for input_size, output_size in zip(
22                  [input_size,] + hidden_sizes, hidden_sizes + [output_size,]):
23              layers.append(nn.Linear(input_size, output_size))
24              layers.append(nn.ReLU())
25          layers = layers[:-1]
26          if output_activator:
27              layers.append(output_activator)
28          net = nn.Sequential(*layers)
29          return net
30
31      def reset(self, mode=None):
32          self.mode = mode
33          if self.mode == 'train':
34              self.trajectory = []
35              self.discount = 1.
36
```

```
37    def step(self, observation, reward, terminated):
38        if self.mode == 'train':
39            action = np.random.choice(self.action_n)
40            self.trajectory += [observation, reward, terminated, action]
41            if len(self.trajectory) >= 8:
42                self.learn()
43            self.discount *= self.gamma
44        else:
45            state_tensor = torch.as_tensor(observation,
46                    dtype=torch.float).unsqueeze(0)
47            prob_tensor = self.actor_net(state_tensor)
48            action_tensor = distributions.Categorical(prob_tensor).sample()
49            action = action_tensor.numpy()[0]
50        return action
51
52    def close(self):
53        pass
54
55    def learn(self):
56        state, _, _, action, next_state, reward, terminated, next_action = \
57                self.trajectory[-8:]
58        state_tensor = torch.as_tensor(state, dtype=torch.float).unsqueeze(0)
59        next_state_tensor = torch.as_tensor(state,
60                dtype=torch.float).unsqueeze(0)
61
62        # update actor
63        q_tensor = self.critic_net(state_tensor)[0, action]
64        pi_tensor = self.actor_net(state_tensor)[0, action]
65        behavior_prob = 1. / self.action_n
66        actor_loss_tensor = -self.discount * q_tensor / behavior_prob * \
67                pi_tensor
68        self.actor_optimizer.zero_grad()
69        actor_loss_tensor.backward()
70        self.actor_optimizer.step()
71
72        # update critic
73        next_q_tensor = self.critic_net(next_state_tensor)[:, next_action]
74        target_tensor = reward + (1. - terminated) * self.gamma * next_q_tensor
75        pred_tensor = self.critic_net(state_tensor)[:, action]
76        critic_loss_tensor = self.critic_loss(pred_tensor, target_tensor)
77        pi_tensor = self.actor_net(state_tensor)[0, action]
78        ratio_tensor = pi_tensor / behavior_prob  # importance sampling ratio
79        critic_loss_tensor *= ratio_tensor
80        self.critic_optimizer.zero_grad()
81        critic_loss_tensor.backward()
82        self.critic_optimizer.step()
83
84
85 agent = OffPACAgent(env)
```

The interaction between agents and the environment is still Code 1.3. The agents are trained and tested using Codes 5.3 and 1.4 respectively.

8.7 Summary

- Actor–Critic (AC) method combines the policy gradient method and bootstrapping.
- On-policy AC updates the policy parameter $\boldsymbol{\theta}$ to increase $\Psi_t \ln \pi(A_t|S_t; \boldsymbol{\theta})$, where Ψ_t uses bootstrapping.

- For eligibility trace AC algorithm, both policy parameter and value parameter have their own eligibility trace.
- Performance difference lemma:

$$g_{\pi'} - g_{\pi''} = \mathrm{E}_{(S,A)\sim\rho_{\pi'}} \left[a_{\pi''}(S, A) \right]$$

- Surrogate advantage is:

$$\mathrm{E}_{(S_t,A_t)\sim\pi(\theta_k)} \left[\frac{\pi(A_t|S_t; \theta)}{\pi(A_t|S_t; \theta_k)} a_{\pi(\theta_k)}(S_t, A_t) \right].$$

- Proximal Policy Optimization (PPO) is an on-policy algorithm that maximizes the upper bounded surrogate advantage. It is often used with experience replay.
- The Kullback–Leibler Divergence (KLD) $d_{\mathrm{KL}}\big(\pi(\cdot|S; \theta_k)\|\pi(\cdot|S; \theta)\big)$ can be expanded at $\theta = \theta_k$ as

$$d_{\mathrm{KL}}\big(\pi(\cdot|S; \theta_k)\|\pi(\cdot|S; \theta)\big) \approx \frac{1}{2}(\theta - \theta_k)^\top \mathbf{F}(\theta_k)(\theta - \theta_k)$$

where $\mathbf{F}(\theta)$ is the FIM.
- The trust region of NPG or TRPO can be expressed as

$$\mathrm{E}_{S\sim\pi(\theta_k)} \left[d_{\mathrm{KL}}\big(\pi(\cdot|S; \theta_k)\|\pi(\cdot|S; \theta))\big) \right] \leq \delta.$$

- OffPAC is an AC algorithm that uses importance sampling.

8.8 Exercises

8.8.1 Multiple Choices

8.1 On AC algorithms, choose the correct one: (   )

A. The policy gradient algorithm with state value estimates as baseline is an AC algorithm.
B. AC algorithm can not use baseline functions.
C. AC algorithm can use state value estimates as baselines.

8.2 On performance difference lemma, choose the correct one: (   )

A. The difference between the return expectation of the policy $\pi(\theta)$ and the return expectation of the policy $\pi(\theta_k)$ is $\mathrm{E}_{(S,A)\sim\rho_{\pi(\theta)}} \left[a_{\pi(\theta_k)}(S, A) \right]$.
B. The difference between the return expectation of the policy $\pi(\theta)$ and the return expectation of the policy $\pi(\theta_k)$ is $\mathrm{E}_{(S,A)\sim\rho_{\pi(\theta_k)}} \left[a_{\pi(\theta_k)}(S, A) \right]$.
C. The difference between the return expectation of the policy $\pi(\theta)$ and the return expectation of the policy $\pi(\theta_k)$ is $\mathrm{E}_{(S,A)\sim\rho_{\pi(\theta_k)}} \left[a_{\pi(\theta)}(S, A) \right]$.

8.3 On PPO algorithm, choose the correct one: (   )

  A. PPO algorithm is an on-policy algorithm. It can not use experience replay.
  B. PPO algorithm is an on-policy algorithm. It can use experience replay.
  C. PPO algorithm is an off-policy algorithm. It can use experience replay.

8.4 On Fisher information matrix, choose the correct one: (   )

  A. Advantage AC algorithm uses Fisher information matrix.
  B. PPO algorithm uses Fisher information matrix.
  C. NPG algorithm uses Fisher information matrix.

8.8.2 Programming

8.5 Use PPO algorithm to solve the task `CartPole-v0`.

8.8.3 Mock Interview

8.6 What is the actor and what is the critic in the actor–critic method?

8.7 Prove the performance difference lemma.

8.8 What is surrogate advantage? Why do we need the surrogate advantage?

Chapter 9
DPG: Deterministic Policy Gradient

This chapter covers

- Deterministic Policy Gradient (DPG) theorem
- DPG algorithm
- Off-Policy Deterministic Actor–Critic (OPDAC) algorithm
- Deep Deterministic Policy Gradient (DDPG) algorithm
- Twin Delay Deep Deterministic Policy Gradient (TD3) algorithm
- exploration noises

When calculating policy gradient using the vanilla PG algorithm, we need to take the expectation over states as well as actions. It is not difficult if the action space is finite, but the sampling will be very inefficient if the action space is continuous, especially when the dimension of action space is high. Therefore, (Silver, 2014) proposed the **Deterministic Policy Gradient** (**DPG**) theorem and DPG method to deal with the task with continuous action spaces.

This chapter only considers the cases where the action space $\mathcal{A}$ is continuous.

9.1 DPG Theorem

This section will introduce the DPG theorem.

For a deterministic policy $\pi(\theta)$ with continuous action space, $\pi(\cdot|s; \theta)$ $(s \in \mathcal{S})$ is not a conventional function, so its gradient with respect to the parameter θ, i.e. $\nabla\pi(\cdot|s; \theta)$, does not exist either. Therefore, the vanilla AC algorithms can not be used anymore. Luckily, Sect. 2.1.3 has mentioned that a deterministic policy can be

© The Author(s), under exclusive license to Springer Nature Singapore Pte Ltd. 2024
Z. Xiao, *Reinforcement Learning*, https://doi.org/10.1007/978-981-19-4933-3_9

presented as $\pi(s; \boldsymbol{\theta})$ ($s \in S$). Such presentation can help bypass the problem caused by the fact that $\pi(\cdot|s; \boldsymbol{\theta})$ ($s \in S$) is not a conventional function.

DPG theorem shows the gradient of return expectation $g_{\pi(\boldsymbol{\theta})}$ with respect to the policy parameter $\boldsymbol{\theta}$ when the policy $\pi(s; \boldsymbol{\theta})$ ($s \in S$) is a deterministic policy and the action space is continuous. DPG has many forms, too. One of the forms is:

$$\nabla g_{\pi(\boldsymbol{\theta})} = \mathrm{E}_{\pi(\boldsymbol{\theta})}\left[\sum_{t=0}^{\infty} \gamma^t \nabla \pi(S_t; \boldsymbol{\theta})\left[\nabla_a q_{\pi(\boldsymbol{\theta})}(S_t, a)\right]_{a=\pi(S_t;\boldsymbol{\theta})}\right].$$

(Proof: According to Bellman expectation equations,

$$v_{\pi(\boldsymbol{\theta})}(s) \qquad\qquad = q_{\pi(\boldsymbol{\theta})}(s, \pi(s; \boldsymbol{\theta})), \qquad\qquad\qquad s \in S$$
$$q_{\pi(\boldsymbol{\theta})}(s, \pi(s; \boldsymbol{\theta})) = r(s, \pi(s; \boldsymbol{\theta})) + \gamma \sum_{s'} p(s'|s, \pi(s; \boldsymbol{\theta})) v_{\pi(\boldsymbol{\theta})}(s'), \quad s \in S.$$

Take the gradients on the above two equations with respect to $\boldsymbol{\theta}$, and we have

$$\nabla v_{\pi(\boldsymbol{\theta})}(s) = \nabla q_{\pi(\boldsymbol{\theta})}(s, \pi(s; \boldsymbol{\theta})), \qquad s \in S,$$
$$\nabla q_{\pi(\boldsymbol{\theta})}(s, \pi(s; \boldsymbol{\theta})) = \left[\nabla_a r(s, a)\right]_{a=\pi(s;\boldsymbol{\theta})} \nabla \pi(s; \boldsymbol{\theta})$$
$$+ \gamma \sum_{s'}\left(\left[\nabla_a p(s'|s, a)\right]_{a=\pi(s;\boldsymbol{\theta})}\left[\nabla \pi(s; \boldsymbol{\theta})\right] v_{\pi(\boldsymbol{\theta})}(s')\right.$$
$$\left. + p(s'|s, \pi(s; \boldsymbol{\theta})) \nabla v_{\pi(\boldsymbol{\theta})}(s')\right)$$
$$= \nabla \pi(s; \boldsymbol{\theta})\left[\nabla_a r(s, a) + \gamma \sum_{a'} \nabla_a p(s'|s, a) v_{\pi(\boldsymbol{\theta})}(s')\right]_{a=\pi(s;\boldsymbol{\theta})}$$
$$+ \gamma \sum_{s'} p(s'|s, \pi(s; \boldsymbol{\theta})) \nabla v_{\pi(\boldsymbol{\theta})}(s')$$
$$= \nabla \pi(s; \boldsymbol{\theta})\left[\nabla_a q_{\pi(\boldsymbol{\theta})}(s, a)\right]_{a=\pi(s;\boldsymbol{\theta})} + \gamma \sum_{s'} p(s'|s, \pi(s; \boldsymbol{\theta})) \nabla v_{\pi(\boldsymbol{\theta})}(s'), \quad s \in S.$$

Plugging the expression of $\nabla q_{\pi(\boldsymbol{\theta})}(s, \pi(s; \boldsymbol{\theta}))$ into the expression of $\nabla v_{\pi(\boldsymbol{\theta})}(s)$ leads to

$$\nabla v_{\pi(\boldsymbol{\theta})}(s) = \nabla \pi(s; \boldsymbol{\theta})\left[\nabla_a q_{\pi(\boldsymbol{\theta})}(s, a)\right]_{a=\pi(s;\boldsymbol{\theta})} + \gamma \sum_{s'} p(s'|s, \pi(s; \boldsymbol{\theta})) \nabla v_{\pi(\boldsymbol{\theta})}(s'),$$

$$s \in S.$$

Calculating the expectation of the aforementioned equation over the state $S_t = s$, we have

$$\mathrm{E}_{\pi(\boldsymbol{\theta})}\left[\nabla v_{\pi(\boldsymbol{\theta})}(S_t)\right]$$
$$= \sum_{s} \mathrm{Pr}_{\pi(\boldsymbol{\theta})}[S_t = s] \nabla v_{\pi(\boldsymbol{\theta})}(s)$$

$$= \sum_s \Pr_{\pi(\boldsymbol{\theta})}[S_t = s] \left[\nabla \pi(s; \boldsymbol{\theta}) \left[\nabla_a q_{\pi(\boldsymbol{\theta})}(s, a) \right]_{a = \pi(s; \boldsymbol{\theta})} \right.$$

$$\left. + \gamma \sum_{s'} p(s' | s, \pi(s; \boldsymbol{\theta})) \nabla v_{\pi(\boldsymbol{\theta})}(s') \right]$$

$$= \sum_s \Pr_{\pi(\boldsymbol{\theta})}[S_t = s] \left[\nabla \pi(s; \boldsymbol{\theta}) \left[\nabla_a q_{\pi(\boldsymbol{\theta})}(s, a) \right]_{a = \pi(s; \boldsymbol{\theta})} \right.$$

$$\left. + \gamma \sum_{s'} \Pr_{\pi(\boldsymbol{\theta})}\left[S_{t+1} = s' | S_t = s\right] \nabla v_{\pi(\boldsymbol{\theta})}(s') \right]$$

$$= \sum_a \Pr_{\pi(\boldsymbol{\theta})}[S_t = s] \left[\nabla \pi(s; \boldsymbol{\theta}) \left[\nabla_a q_{\pi(\boldsymbol{\theta})}(s, a) \right]_{a = \pi(s; \boldsymbol{\theta})} \right]$$

$$+ \gamma \sum_s \Pr_{\pi(\boldsymbol{\theta})}[S_t = s] \sum_{s'} \Pr_{\pi(\boldsymbol{\theta})}\left[S_{t+1} = s' | S_t = s\right] \nabla v_{\pi(\boldsymbol{\theta})}(s')$$

$$= \sum_s \Pr_{\pi(\boldsymbol{\theta})}[S_t = s] \left[\nabla \pi(s; \boldsymbol{\theta}) \left[\nabla_a q_{\pi(\boldsymbol{\theta})}(s, a) \right]_{a = \pi(s; \boldsymbol{\theta})} \right]$$

$$+ \gamma \sum_{s'} \Pr_{\pi(\boldsymbol{\theta})}\left[S_{t+1} = s'\right] \nabla v_{\pi(\boldsymbol{\theta})}(s')$$

$$= \mathrm{E}_{\pi(\boldsymbol{\theta})}\left[\nabla \pi(S; \boldsymbol{\theta}) \left[\nabla_a q_{\pi(\boldsymbol{\theta})}(S, a) \right]_{a = \pi(S; \boldsymbol{\theta})} \right] + \gamma \mathrm{E}_{\pi(\boldsymbol{\theta})}\left[\nabla v_{\pi(\boldsymbol{\theta})}(S_{t+1}) \right]$$

Now we get the recursive formula from $\mathrm{E}_{\pi(\boldsymbol{\theta})}\left[\nabla v_{\pi(\boldsymbol{\theta})}(S_t) \right]$ to $\mathrm{E}_{\pi(\boldsymbol{\theta})}\left[\nabla v_{\pi(\boldsymbol{\theta})}(S_{t+1}) \right]$. Note that the gradient we want to calculate is

$$g_{\pi(\boldsymbol{\theta})} = \nabla \mathrm{E}_{\pi(\boldsymbol{\theta})}\left[v_{\pi(\boldsymbol{\theta})}(S_0) \right] = \mathrm{E}_{\pi(\boldsymbol{\theta})}\left[\nabla v_{\pi(\boldsymbol{\theta})}(S_0) \right],$$

so

$$\nabla g_{\pi(\boldsymbol{\theta})} = \mathrm{E}_{\pi(\boldsymbol{\theta})}\left[\nabla v_{\pi(\boldsymbol{\theta})}(S_0) \right]$$

$$= \mathrm{E}_{\pi(\boldsymbol{\theta})}\left[\nabla \pi(S_0; \boldsymbol{\theta}) \left[\nabla_a q_{\pi(\boldsymbol{\theta})}(S_0, a) \right]_{a = \pi(S_0, \boldsymbol{\theta})} \right] + \gamma \mathrm{E}_{\pi(\boldsymbol{\theta})}\left[\nabla v_{\pi(\boldsymbol{\theta})}(S_1) \right]$$

$$= \mathrm{E}_{\pi(\boldsymbol{\theta})}\left[\nabla \pi(S_0; \boldsymbol{\theta}) \left[\nabla_a q_{\pi(\boldsymbol{\theta})}(S_0, a) \right]_{a = \pi(S_0, \boldsymbol{\theta})} \right]$$

$$+ \mathrm{E}_{\pi(\boldsymbol{\theta})}\left[\nabla \pi(S_0; \boldsymbol{\theta}) \left[\nabla_a q_{\pi(\boldsymbol{\theta})}(S_1, a) \right]_{a = \pi(S_1, \boldsymbol{\theta})} \right] + \gamma^2 \mathrm{E}_{\pi(\boldsymbol{\theta})}\left[\nabla v_{\pi(\boldsymbol{\theta})}(S_2) \right]$$

$$= \dots$$

$$= \sum_{t=0}^{+\infty} \mathrm{E}_{\pi(\boldsymbol{\theta})}\left[\gamma^t \nabla \pi(S_t; \boldsymbol{\theta}) \left[\nabla_a q_{\pi(\boldsymbol{\theta})}(S_t, a) \right]_{a = \pi(S_t; \boldsymbol{\theta})} \right].$$

The proof completes.)

Using the concept of discounted expectation, DPG theorem can also be expressed as

$$\nabla g_{\pi(\boldsymbol{\theta})} = \mathrm{E}_{S \sim \eta_{\pi(\boldsymbol{\theta})}}\left[\nabla \pi(S; \boldsymbol{\theta}) \left[\nabla_a q_{\pi(\boldsymbol{\theta})}(S, a) \right]_{a = \pi(S; \boldsymbol{\theta})} \right],$$

where the expectation is over discounted state visitation frequency.

Compared to vanilla PG algorithms, DPG algorithms have the advantage that they can calculate policy gradient more sample-efficiently.

The shortcomings of DPG algorithms include that the exploration needs to rely on the noise process on the action. The choice of noise process is brittle and not easy to tune. This will be discussed in Sect. 9.4.

9.2 On-Policy DPG

This section considers vanilla on-policy deterministic AC algorithm.

Deterministic on-policy AC algorithm also uses $q(s, a; \mathbf{w})$ to approximate $q_{\pi(\boldsymbol{\theta})}(s, a)$. After using such approximation, the deterministic policy gradient can be approximated as

$$
\mathrm{E}\left[\sum_{t=0}^{+\infty} \gamma^t \nabla \pi(S_t; \boldsymbol{\theta}) \left[\nabla_a q(S_t, a; \mathbf{w})\right]_{a=\pi(S_t;\boldsymbol{\theta})}\right] = \mathrm{E}\left[\sum_{t=0}^{+\infty} \nabla\left[\gamma^t q(S_t, \pi(S_t; \boldsymbol{\theta}); \mathbf{w})\right]\right].
$$

Therefore, we can update $\boldsymbol{\theta}$ to reduce $-\gamma^t q(S_t, \pi(S_t; \boldsymbol{\theta}); \mathbf{w})$. The corresponding update can be in the form of

$$
\boldsymbol{\theta} \leftarrow \boldsymbol{\theta} + \alpha^{(\boldsymbol{\theta})} \gamma^t \nabla \pi(S_t; \boldsymbol{\theta}) \left[\nabla_a q(S_t, a; \mathbf{w})\right]_{a=\pi(S_t;\boldsymbol{\theta})}.
$$

Algorithm 9.1 shows the vanilla on-policy deterministic AC algorithm. Steps 2.2 and 2.3.2 add additional noise on the action determined by the policy estimates to explore. Especially, for the state S_t, the action determined by the deterministic policy $\pi(\boldsymbol{\theta})$ is $\pi(S_t; \boldsymbol{\theta})$. In order to explore, we introduce the noises N_t, which can be Gaussian Process (GP) or other processes. The action with the addictive noise is $\pi(S_t; \boldsymbol{\theta}) + N_t$. This action has the exploration functionality. Section 9.4 discusses more about the noises.

Algorithm 9.1   Vanilla on-policy deterministic AC.

Input: environment (without mathematical model).
Output: optimal policy estimate $\pi(\boldsymbol{\theta})$.
Parameters: optimizers (containing learning rate $\alpha^{(\boldsymbol{\theta})}$ and $\alpha^{(\mathbf{w})}$), discount factor γ, and parameters to control the number of episodes and maximum steps in an episode.

1. (Initialize) Initialize the policy parameter $\boldsymbol{\theta}$ and value parameter $\mathbf{w}$.
2. (AC update) For each episode:

    2.1. (Initialize accumulative discount factor) $\Gamma \leftarrow 1$.

2.2. (Initialize state–action pair) Select the initial state S, and use $\pi(S;\boldsymbol{\theta})$ to select the action A.

2.3. Loop until the episode ends:

   2.3.1. (Sample) Execute the action A, and then observe the reward R, the next state S', and the indicator of episode end D'.

   2.3.2. (Decide) Use $\pi(S';\boldsymbol{\theta})$ to determine the action A'. (The action can be arbitrary if $D' = 1$.)

   2.3.3. (Calculate TD return) $U \leftarrow R + \gamma(1 - D')q(S', A';\mathbf{w})$.

   2.3.4. (Update policy parameter) Update $\boldsymbol{\theta}$ to reduce $-\Gamma q(S, A;\mathbf{w})\ln\pi(A|S;\boldsymbol{\theta})$. (For example, $\boldsymbol{\theta} \leftarrow \boldsymbol{\theta} + \alpha^{(\boldsymbol{\theta})}\Gamma q(S, A;\mathbf{w})\nabla\ln\pi(A|S;\boldsymbol{\theta})$).

   2.3.5. (Update value parameter) Update $\mathbf{w}$ to reduce $\left[U - q(S, A;\mathbf{w})\right]^2$. (For example, $\mathbf{w} \leftarrow \mathbf{w} + \alpha^{(\mathbf{w})}\left[U - q(S, A;\mathbf{w})\right]\nabla q(S, A;\mathbf{w})$.)

   2.3.6. (Update accumulative discount factor) $\Gamma \leftarrow \gamma\Gamma$.

   2.3.7. $S \leftarrow S', A \leftarrow A'$.

9.3 Off-Policy DPG

9.3.1 OPDAC: Off-Policy Deterministic Actor–Critic

In the previous section, on-policy deterministic AC algorithm uses $\pi(\boldsymbol{\theta})$ to generate trajectories, calculate the return expectation, and try to maximize the return on each trajectory by maximizing the return expectation. In fact, if we want to maximize the return of every trajectory, we want to maximize the return expectation no matter what policy is used to generate those trajectories. Therefore, off-policy deterministic AC algorithm introduces a deterministic behavior policy b, and uses this policy to generate trajectories. Trying to maximize the return expectation over those trajectories leads to the off-policy policy gradient:

$$\nabla E_{\rho_b}\left[q_{\pi(\boldsymbol{\theta})}(S, \pi(S;\boldsymbol{\theta}))\right] = E_{\rho_b}\left[\nabla\pi(S;\boldsymbol{\theta})\left[\nabla_a q_{\pi(\boldsymbol{\theta})}(S, a)\right]_{a=\pi(S;\boldsymbol{\theta})}\right].$$

It shares the same expression inside the expectation with on-policy algorithm, but over different distributions. Therefore, off-policy deterministic algorithm shares the same update formula with on-policy deterministic algorithm.

The reason why off-policy deterministic AC algorithms can sometimes outperform on-policy deterministic AC algorithms is that the behavior policy may help exploration. The behavior policy may generate trajectories that can not be generated by adding simplified noises, so that we can consider more diverse trajectories. Therefore, maximizing the return expectation over those trajectories may better maximize the returns of all possible trajectories.

In previous chapters, the off-policy algorithms based on importance sampling usually use the important sampling ratio $\frac{\pi(A_t|S_t;\boldsymbol{\theta})}{b(A_t|S_t)}$. However, the off-policy algorithm in this section does not use this ratio. This is because, the deterministic behavior policy b is not absolutely continuous with the target policy $\pi(\boldsymbol{\theta})$, so we can not define the importance sampling ratio. Therefore, the importance sampling ratio is not used here.

Algorithm 9.2 shows the **Off-Policy Deterministic Actor–Critic (OPDAC)** algorithm. Remarkably, in the expression of the off-policy deterministic policy gradient $\mathrm{E}_{\rho_b}\Big[\nabla\pi(S;\boldsymbol{\theta})\big[\nabla q_{\pi(\boldsymbol{\theta})}(S,a)\big]_{a=\pi(S;\boldsymbol{\theta})}\Big]$, the behavior policy is only used for calculation expectation, while the action a is still generated by $a = \pi(S;\boldsymbol{\theta})$. Therefore, when Step 2.4.4 calculates the policy gradient, the action used for action value estimates should be generated by the policy $\pi(\boldsymbol{\theta})$ rather than the behavior policy b.

Algorithm 9.2   OPDAC.

Input: environment (without mathematical model).
Output: optimal policy estimate $\pi(\boldsymbol{\theta})$.
Parameters: optimizers (containing learning rate $\alpha^{(\boldsymbol{\theta})}$ and $\alpha^{(\mathbf{w})}$), discount factor γ, and parameters to control the number of episodes and maximum steps in an episode.

1. (Initialize) Initialize the policy parameter $\boldsymbol{\theta}$ and value parameter $\mathbf{w}$.
2. (AC update) For each episode:

    2.1. (Designate behavior policy) Designate a behavior policy b.
    2.2. (Initialize accumulative discount factor) $\Gamma \leftarrow 1$.
    2.3. (Initialize state–action pair) Select the initial state S, and use $\pi(S;\boldsymbol{\theta})$ to select the action A.
    2.4. Loop until the episode ends:
        2.4.1. (Decide) Use the policy $b(S)$ to determine the action A. (The action can be arbitrary if $D' = 1$.)
        2.4.2. (Sample) Execute the action A, and then observe the reward R, the next state S', and the indicator of episode end D'.
        2.4.3. (Calculate TD return) $U \leftarrow R + \gamma(1 - D')q(S', A'; \mathbf{w})$.
        2.4.4. (Update policy parameter) Update $\boldsymbol{\theta}$ to reduce $-\Gamma q\big(S, \pi(S;\boldsymbol{\theta});\mathbf{w}\big)$. (For example, $\boldsymbol{\theta} \leftarrow \boldsymbol{\theta} + \alpha^{(\boldsymbol{\theta})}\Gamma\nabla\pi(S;\boldsymbol{\theta})\big[\nabla q(S,a;\mathbf{w})\big]_{a=\pi(S;\boldsymbol{\theta})}$).
        2.4.5. (Update value parameter) Update $\mathbf{w}$ to reduce $\big[U - q(S,A;\mathbf{w})\big]^2$. (For example, $\mathbf{w} \leftarrow \mathbf{w} + \alpha^{(\mathbf{w})}\big[U - q(S,A;\mathbf{w})\big]\nabla q(S,A;\mathbf{w})$.)
        2.4.6. (Update accumulative discount factor) $\Gamma \leftarrow \gamma\Gamma$.
        2.4.7. $S \leftarrow S'$.

9.3.2 DDPG: Deep Deterministic Policy Gradient

Deep Deterministic Policy Gradient (**DDPG**) algorithm applies the tricks of DQN to deterministic AC algorithm (Lillicrap, 2016). Specially speaking, DDPG uses the following tricks:

- Experience replay: Save the experience in the form of (S, A, R, S', D'), and then replay them in batch to update parameters.
- Target network: Besides the conventional value parameter $\mathbf{w}$ and policy parameter $\mathbf{\theta}$, allocate a suite of target value parameter $\mathbf{w}_{\text{target}}$ and a suite of target policy parameter $\mathbf{\theta}_{\text{target}}$. It also introduces the learning rate for the target network $\alpha_{\text{target}} \in (0, 1)$ when updating the target networks.

Algorithm 9.3 shows the DDPG algorithm. Most of steps are exactly the same as those in the DQN algorithm (Algo. 6.7), since DDPG algorithm uses the same tricks as DQN algorithm. We can also use the looping structure in Algo. 6.8.

Algorithm 9.3   DDPG.

Input: environment (without mathematical model).
Output: optimal policy estimate $\pi(\mathbf{\theta})$.
Parameters: optimizers (containing learning rate $\alpha^{(\mathbf{\theta})}$ and $\alpha^{(\mathbf{w})}$), discount factor γ, parameters to control the number of episodes and maximum steps in an episode, parameters for experience replay, and the learning rate for target networks α_{target}.

1. Initialize:

   1.1. (Initialize network parameters) Initialize the policy parameter $\mathbf{\theta}$ and value parameter $\mathbf{w}$. $\mathbf{\theta}_{\text{target}} \leftarrow \mathbf{\theta}$, $\mathbf{w}_{\text{target}} \leftarrow \mathbf{w}$.
   1.2. (Initialize experience storage) $\mathcal{D} \leftarrow \varnothing$.

2. For each episode:

   2.1. (Initialize state) Choose the initial state S.
   2.2. Loop until the episode ends:
       2.2.1. (Collect experiences) Do the following once or multiple times:
           2.2.1.1. (Decide) Use the policy derived from $q(S, \cdot; \mathbf{w})$ (for example, ε-greedy policy) to determine the action A.
           2.2.1.2. (Sample) Execute the action A, and then observe the reward R, the next state S', and the indicator of episode end D'.
           2.2.1.3. Save the experience (S, A, R, S', D') in the storage $\mathcal{D}$.
           2.2.1.4. $S \leftarrow S'$.
       2.2.2. (Use experiences) Do the following once or multiple times:

2.2.2.1. (Replay) Sample a batch of experience $\mathcal{B}$ from the storage $\mathcal{D}$. Each entry is in the form of (S, A, R, S', D').

2.2.2.2. (Calculate TD return) $U \leftarrow R + \gamma(1 - D')$ $\max_a q\left(S', a; \mathbf{w}_{\text{target}}\right) ((S, A, R, S', D') \in \mathcal{B})$.

2.2.2.3. (Update policy parameter) Update $\boldsymbol{\theta}$ to reduce $-\frac{1}{|\mathcal{B}|} \sum_{(S,A,R,S',D') \in \mathcal{B}} q(S, \pi(S; \boldsymbol{\theta}); \mathbf{w})$. (For example, $\boldsymbol{\theta} \leftarrow \boldsymbol{\theta} + \alpha^{(\boldsymbol{\theta})} \frac{1}{|\mathcal{B}|} \sum_{(S,A,R,S',D') \in \mathcal{B}} \nabla \pi(S; \boldsymbol{\theta}) \left[\nabla_a q(S, a; \mathbf{w})\right]_{a=\pi(S;\boldsymbol{\theta})} .)$

2.2.2.4. (Update value parameter) Update $\mathbf{w}$ to reduce $\frac{1}{|\mathcal{B}|} \sum_{(S,A,R,S',D') \in \mathcal{B}} \left[U - q(S, A; \mathbf{w})\right]^2$. (For example, $\mathbf{w} \leftarrow \mathbf{w} + \alpha^{(\mathbf{w})} \frac{1}{|\mathcal{B}|} \sum_{(S,A,R,S',D') \in \mathcal{B}} \left[U - q(S, A; \mathbf{w})\right] \nabla q(S, A; \mathbf{w}) .)$

2.2.2.5. (Update target networks) Update the target networks when appropriate: $\boldsymbol{\theta}_{\text{target}} \leftarrow \left(1 - \alpha_{\text{target}}\right) \boldsymbol{\theta}_{\text{target}} + \alpha_{\text{target}} \boldsymbol{\theta}$, and $\mathbf{w}_{\text{target}} \leftarrow \left(1 - \alpha_{\text{target}}\right) \mathbf{w}_{\text{target}} + \alpha_{\text{target}} \mathbf{w}$.

9.3.3 TD3: Twin Delay Deep Deterministic Policy Gradient

Twin Delay Deep Deterministic Policy Gradient (TD3) is an algorithm that combines DDPG and double learning (Fujimoto, 2018).

Section 5.3.3 introduced how to use double learning to reduce maximum bias in Q learning. Tabular double Q learning uses two suites of action value estimates $q^{(0)}(s, a)$ and $q^{(1)}(s, a)$ ($s \in \mathcal{S}, a \in \mathcal{A}$). In each update, one suite of action value estimates is used to calculate the action (i.e. $A' = \arg\max_a q^{(0)}(S', a)$), while another is used to calculate TD return (i.e., $q^{(1)}(S', A')$). Chapter 6 introduced Double DQN. In Double DQN, there are two suites of value networks: the evaluation network and the target network. It uses the evaluation network to calculate the optimal action (i.e., $A' = \arg\max_a q^{(0)}(S', a; \mathbf{w})$), and uses the target network to calculate TD return (i.e. $q\left(S', A'; \mathbf{w}_{\text{target}}\right)$). However, for the deterministic AC algorithm, the action $\pi(S'; \boldsymbol{\theta})$ has been designated by the parametric policy $\pi(\boldsymbol{\theta})$, how can we reduce the maximum bias? TD3 uses the following trick: Maintain two copies of value network $\mathbf{w}^{(i)}$ ($i = 0, 1$), each of which has its own target network $\mathbf{w}_{\text{target}}^{(i)}$ ($i = 0, 1$). When calculating the TD return, we calculate $\min_{i=0,1} q\left(\cdot, \cdot; \mathbf{w}_{\text{target}}^{(i)}\right)$, the minimum of two target networks.

Algorithm 9.4 shows TD3 algorithm. Remarkably, Step 2.2.2 adds noises for replayed actions.

Algorithm 9.4 TD3.

Input: environment (without mathematical model).
Output: optimal policy estimate $\pi(\boldsymbol{\theta})$.
Parameters: optimizers (containing learning rate $\alpha^{(\boldsymbol{\theta})}$ and $\alpha^{(\mathbf{w})}$), discount factor γ, parameters to control the number of episodes and maximum steps in an episode, parameters for experience replay, and the learning rate for target networks α_{target}.

1. Initialize:

   1.1. (Initialize network parameters) Initialize the policy parameter $\boldsymbol{\theta}$, and $\boldsymbol{\theta}_{\text{target}} \leftarrow \boldsymbol{\theta}$. Initialize the value parameters $\mathbf{w}^{(i)}$, $\mathbf{w}_{\text{target}}^{(i)} \leftarrow \mathbf{w}^{(i)}$ ($i \in \{0, 1\}$).

   1.2. (Initialize experience storage) $\mathcal{D} \leftarrow \varnothing$.

2. For each episode:

   2.1. (Initialize state) Choose the initial state S.

   2.2. Loop until the episode ends:

   2.2.1. (Collect experiences) Do the following once or multiple times:

   2.2.1.1. (Decide) Use the policy derived from $q(S, \cdot; \mathbf{w})$ (for example, ε-greedy policy) to determine the action A.

   2.2.1.2. (Sample) Execute the action A, and then observe the reward R, the next state S', and the indicator of episode end D'.

   2.2.1.3. Save the experience (S, A, R, S', D') in the storage $\mathcal{D}$.

   2.2.1.4. $S \leftarrow S'$.

   2.2.2. (Use experiences) Do the following once or multiple times:

   2.2.2.1. (Replay) Sample a batch of experience $\mathcal{B}$ from the storage $\mathcal{D}$. Each entry is in the form of (S, A, R, S', D').

   2.2.2.2. (Explore) Get the action A' by adding noises (such as Gaussian Process) on the action determined by the policy $\pi\left(S'; \boldsymbol{\theta}_{\text{target}}\right)$.

   2.2.2.3. (Calculate TD return) $U \leftarrow R + \gamma(1 - D')$ $\min_{i=0,1} q\left(S', A'; \mathbf{w}^{(i)}\right)$ $((S, A, R, S', D') \in \mathcal{B})$.

   2.2.2.4. (Update policy parameter) Update $\boldsymbol{\theta}$ to reduce $-\frac{1}{|\mathcal{B}|} \sum_{(S,A,R,S',D') \in \mathcal{B}} q\left(S, \pi(S; \boldsymbol{\theta}); \mathbf{w}^{(0)}\right)$.

(For example, $\boldsymbol{\theta} \leftarrow \boldsymbol{\theta} + \alpha^{(\boldsymbol{\theta})}\frac{1}{|\mathcal{B}|}$

$\sum_{(S,A,R,S',D')\in\mathcal{B}} \nabla\pi(S;\boldsymbol{\theta})\left[\nabla_a q\left(S,a;\mathbf{w}^{(0)}\right)\right]_{a=\pi(S;\boldsymbol{\theta})}$)

2.2.2.5. (Update value parameter) Update $\mathbf{w}^{(i)}$ to

reduce $\frac{1}{|\mathcal{B}|}\sum_{(S,A,R,S',D')\in\mathcal{B}}\left[U - q\left(S,A;\mathbf{w}^{(i)}\right)\right]^2$

($i = 0, 1$). (For example, $\mathbf{w}^{(i)} \leftarrow \mathbf{w}^{(i)} + \alpha^{(\mathbf{w})}\frac{1}{|\mathcal{B}|}$

$\sum_{(S,A,R,S',D')\in\mathcal{B}}\left[U - q\left(S,A;\mathbf{w}^{(i)}\right)\right]\nabla q\left(S,A;\mathbf{w}^{(i)}\right)$.)

2.2.2.6. (Update target networks) Update the target networks when appropriate: $\boldsymbol{\theta}_{\text{target}} \leftarrow \left(1 - \alpha_{\text{target}}\right)\boldsymbol{\theta}_{\text{target}} + \alpha_{\text{target}}\boldsymbol{\theta}$ and $\mathbf{w}_{\text{target}}^{(i)} \leftarrow \left(1 - \alpha_{\text{target}}\right)\mathbf{w}_{\text{target}}^{(i)} + \alpha_{\text{target}}\mathbf{w}^{(i)}$

($i = 0, 1$).

9.4 Exploration Process

Deterministic AC algorithm explores by adding noises on the action. The simplest form of noises is Gaussian Process (GP). In some tasks, the effect of action in the system is alike passing a low-brand filter, which requires multiple noises in the same direction to take some effects. In those cases, independent Gaussian noises can not explore effectively. For example, in a task, the direct impact of the action is the acceleration of a point. If we add independent Gaussian noises to this action, such noises may not provide consistent shifts in the position of the point such that the point can move to somewhere far away. For those tasks, Ornstein Uhlenbeck (OU) process may be better than Gaussian process, since OU noises at different time are positively correlated, so it may move the actions in the same direction.

Interdisciplinary Reference 9.1
Stochastic Process: Ornstein Uhlenbeck Process

Ornstein Uhlenbeck (OU) process is defined using the following stochastic difference equations (1-dimension case):

$$dN_t = \theta\left(\mu - N_t\right)dt + \sigma dB_t,$$

where θ, μ, σ are parameters ($\theta > 0, \sigma > 0$), and B_t is the standard Brownian motion. When the initial noise is the one-point distribution at the origin (i.e. N_0), and $\mu = 0$, the solution of the aforementioned equation is

$$N_t = \sigma \int_0^t e^{\theta(\tau - t)} dB_\tau, \quad t \geq 0.$$

(Proof: Plugging $dN_t = \theta(\mu - N_t)dt + \sigma dB_t$ into $d\left(N_t e^{\theta t}\right) = \theta N_t e^{\theta t} dt + e^{\theta t} dN_t$ leads to $d\left(N_t e^{\theta t}\right) = \mu \theta e^{\theta t} dt + \sigma e^{\theta t} dB_t$. Integrating this from 0 to t will get $N_t e^{\theta t} - N_0 = \mu\left(e^{\theta t} - 1\right) + \sigma \int_0^t e^{\theta \tau} dB_\tau$. Considering $N_0 = 0$ and $\mu = 0$, we will get the result.) The mean of this solution is 0, and the covariance of this solution is

$$\text{Cov}(N_t, N_s) = \frac{\sigma^2}{2\theta}\left(e^{-\theta |t-s|} - e^{-\theta(t+s)}\right).$$

(Proof: We can easily check the mean is 0. Therefore,

$$\text{Cov}(N_t, N_s) = \text{E}[N_t N_s] = \sigma^2 e^{-\theta(t+s)} \text{E}\left[\int_0^t e^{\theta \tau} dB_\tau \int_0^s e^{\theta \tau} dB_\tau\right].$$

Besides, Ito Isometry tells us $\text{E}\left[\int_0^t e^{\theta \tau} dB_\tau \int_0^s e^{\theta \tau} dB_\tau\right] = \text{E}\left[\int_0^{\min\{t,s\}} e^{2\theta \tau} d\tau\right]$, so $\text{Cov}(N_t, N_s) = \sigma^2 e^{-\theta(t+s)} \int_0^{\min\{t,s\}} d^{2\theta \tau} d\tau$. Simplifying this gets the result.) For $t, s > 0$, we always have $|t - s| < t + s$. Therefore, $\text{Cov}(N_t, N_s) \geq 0$.

For the tasks with bounded action space, we add to limit the range of actions after they were added by noises. For example, the action space in a task is lower-bounded and upper-bounded by a_{low} and a_{high} respectively. Let N_t denote the GP or OU noises, and let $\pi(S_t; \boldsymbol{\theta}) + N_t$ denote the action with noises. The action with noise may exceed the range of action space. In order to resolve this, we can use the following method to bound the action:

- Limit the range of actions using the clip() function, in the form of $\text{clip}\left(\pi(S_t; \boldsymbol{\theta}) + N_t, a_{\text{low}}, a_{\text{high}}\right)$.
- Limit the range of actions using the sigmoid() function, in the form of $a_{\text{low}} + \left(a_{\text{high}} - a_{\text{low}}\right)\text{sigmoid}\left(\pi(S_t; \boldsymbol{\theta}) + N_t\right)$. All operations are element-wise.

9.5 Case Study: Pendulum

This section considers the task Pendulum-v1 in Gym. As Fig. 9.1, there is a 2-dimension coordinate system in a vertical plane. Its Y-axis points downward, while its X-axis points in the left direction. There is a stick of length 1 in this vertical plane. One end of the stick connects to the origin, and another end is a free end. At time t ($t = 0, 1, 2, \ldots$), we can observe the position of the free end $(X_t, Y_t) = (\cos \Theta_t, \sin \Theta_t)$ ($\Theta_t \in [-\pi, +\pi)$) and the angular velocity is $\dot{\Theta}_t$ ($\dot{\Theta}_t \in [-8, +8]$). We can add a moment of force A_t ($A_t \in [-2, +2]$), and get the reward R_{t+1} and next observation $\left(\cos \Theta_{t+1}, \sin \Theta_{t+1}, \dot{\Theta}_{t+1}\right)$. Each episode has 200 time steps. We want to maximize the episode reward.

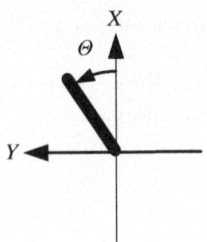

Fig. 9.1   The task `Pendulum-v1`.

In fact, at time t, the environment is fully specified by the state $\left(\Theta_t, \dot{\Theta}_t\right)$. The initial state $\left(\Theta_0, \dot{\Theta}_0\right)$ is uniformly selected from $[-\pi, \pi) \times [-1, 1]$. The dynamics from the state $S_t = \left(\Theta_t, \dot{\Theta}_t\right)$ and the action A_t to the reward R_{t+1} and the next state $S_{t+1} = \left(\Theta_{t+1}, \dot{\Theta}_{t+1}\right)$ is:

$$R_{t+1} \leftarrow -\left(\Theta_t^2 + 0.1\dot{\Theta}_t^2 + 0.001A_t^2\right)$$

$$\Theta_{t+1} \leftarrow \Theta_t + 0.05\left(\dot{\Theta}_t + 0.75 \sin \dot{\Theta}_t + 0.15A_t\right)$$

in the principal value interval $[-\pi, +\pi)$

$$\dot{\Theta}_{t+1} \leftarrow \text{clip}\left(\dot{\Theta}_t + 0.75 \sin \Theta_t + 0.15A_t, -8, +8\right).$$

Since the reward is larger when X_t is larger, and the reward is larger when the absolute angular velocity $\left|\dot{\Theta}_t\right|$ and absolute action $|A_t|$ is smaller, it is better to keep the strike straight and still. Therefore, this task is called "Pendulum".

This task has larger observation space and action space compared to the task `Acrobot-v1` in Sect. 8.6 (See Table 9.1 for comparison). Since the action space is continuous, we may use the deterministic algorithm in this chapter.

This task does not define a threshold for episode reward. Therefore, there are no such sayings that the task is solved if the average episode reward of 100 successive episodes exceeds a number.

Table 9.1 Compare the task `Acrobot-v1` and the task `Pendulum-v1`.

space	Acrobot-v1	Pendulum-v1
state space $\mathcal{S}$	$[-\pi, \pi)^2 \times [-4\pi, 4\pi] \times [-9\pi, 9\pi]$	$[-\pi, \pi) \times [-8, 8]$
observation space $\mathcal{O}$	$[-1, 1)^4 \times [-4\pi, 4\pi] \times [-9\pi, 9\pi]$	$[-1, 1)^2 \times [-8, 8]$
action space $\mathcal{A}$	$\{0, 1, 2\}$	$[-2, 2)$
reward space $\mathcal{R}$	$\{-1, 0\}$	$[-\pi^2 - 6.404, 0]$

9.5.1 DDPG

This section uses DDPG algorithm to solve the task.

Code 9.1 implements OU process. The class `OrnsteinUhlenbech` uses difference equations to approximate differential equations. In order to use this class, we need to create an object `noise` at the beginning of each episode, and call `noise()` to get a set of values.

Code 9.1 OU process.

`Pendulum-v1_DDPG_tf.ipynb`

```
1  class OrnsteinUhlenbeckProcess:
2      def __init__(self, x0):
3          self.x = x0
4
5      def __call__(self, mu=0., sigma=1., theta=.15, dt=.01):
6          n = np.random.normal(size=self.x.shape)
7          self.x += (theta * (mu - self.x) * dt + sigma * np.sqrt(dt) * n)
8          return self.x
```

Codes 9.2 and 9.3 implement DDPG. They use the class `DQNReplayer` in Code 6.6.

Code 9.2 DDPG agent (TensorFlow version).

`Pendulum-v1_DDPG_tf.ipynb`

```
1   class DDPGAgent:
2       def __init__(self, env):
3           state_dim = env.observation_space.shape[0]
4           self.action_dim = env.action_space.shape[0]
5           self.action_low = env.action_space.low
6           self.action_high = env.action_space.high
7           self.gamma = 0.99
8
9           self.replayer = DQNReplayer(20000)
10
11          self.actor_evaluate_net = self.build_net(
12                  input_size=state_dim, hidden_sizes=[32, 64],
13                  output_size=self.action_dim, output_activation=nn.tanh,
14                  learning_rate=0.0001)
15          self.actor_target_net = models.clone_model(self.actor_evaluate_net)
16          self.actor_target_net.set_weights(
17                  self.actor_evaluate_net.get_weights())
18
19          self.critic_evaluate_net = self.build_net(
20                  input_size=state_dim+self.action_dim, hidden_sizes=[64, 128],
21                  learning_rate=0.001)
```

```
22          self.critic_target_net = models.clone_model(self.critic_evaluate_net)
23          self.critic_target_net.set_weights(
24                  self.critic_evaluate_net.get_weights())
25
26      def build_net(self, input_size=None, hidden_sizes=None, output_size=1,
27                  activation=nn.relu, output_activation=None,
28                  loss=losses.mse, learning_rate=0.001):
29          model = keras.Sequential()
30          for layer, hidden_size in enumerate(hidden_sizes):
31              kwargs = {'input_shape' : (input_size,)} if layer == 0 else {}
32              model.add(layers.Dense(units=hidden_size,
33                      activation=activation, **kwargs))
34          model.add(layers.Dense(units=output_size,
35                  activation=output_activation))
36          optimizer = optimizers.Adam(learning_rate)
37          model.compile(optimizer=optimizer, loss=loss)
38          return model
39
40      def reset(self, mode=None):
41          self.mode = mode
42          if self.mode == 'train':
43              self.trajectory = []
44              self.noise = OrnsteinUhlenbeckProcess(np.zeros((self.action_dim,)))
45
46      def step(self, observation, reward, terminated):
47          if self.mode == 'train' and self.replayer.count < 3000:
48              action = np.random.uniform(self.action_low, self.action_high)
49          else:
50              action = self.actor_evaluate_net.predict(observation[np.newaxis],
51                      verbose=0)[0]
52          if self.mode == 'train':
53              # noisy action
54              noise = self.noise(sigma=0.1)
55              action = (action + noise).clip(self.action_low, self.action_high)
56
57              self.trajectory += [observation, reward, terminated, action]
58              if len(self.trajectory) >= 8:
59                  state, _, _, act, next_state, reward, terminated, _ = \
60                          self.trajectory[-8:]
61                  self.replayer.store(state, act, reward, next_state, terminated)
62
63                  if self.replayer.count >= 3000:
64                      self.learn()
65          return action
66
67      def close(self):
68          pass
69
70      def update_net(self, target_net, evaluate_net, learning_rate=0.005):
71          average_weights = [(1. - learning_rate) * t + learning_rate * e for
72                  t, e in zip(target_net.get_weights(),
73                  evaluate_net.get_weights())]
74          target_net.set_weights(average_weights)
75
76      def learn(self):
77          # replay
78          states, actions, rewards, next_states, terminateds = \
79                  self.replayer.sample(64)
80          state_tensor = tf.convert_to_tensor(states, dtype=tf.float32)
81
82          # update critic
83          next_actions = self.actor_target_net.predict(next_states, verbose=0)
84          next_noises = np.random.normal(0, 0.2, size=next_actions.shape)
85          next_actions = (next_actions + next_noises).clip(self.action_low,
86                  self.action_high)
87          state_actions = np.hstack([states, actions])
88          next_state_actions = np.hstack([next_states, next_actions])
```

```
89      next_qs = self.critic_target_net.predict(next_state_actions,
90          verbose=0)[:, 0]
91      targets = rewards + (1. - terminateds) * self.gamma * next_qs
92      self.critic_evaluate_net.fit(state_actions, targets[:, np.newaxis],
93          verbose=0)
94
95      # update actor
96      with tf.GradientTape() as tape:
97          action_tensor = self.actor_evaluate_net(state_tensor)
98          state_action_tensor = tf.concat([state_tensor, action_tensor],
99              axis=1)
100         q_tensor = self.critic_evaluate_net(state_action_tensor)
101         loss_tensor = -tf.reduce_mean(q_tensor)
102     grad_tensors = tape.gradient(loss_tensor,
103         self.actor_evaluate_net.variables)
104     self.actor_evaluate_net.optimizer.apply_gradients(zip(
105         grad_tensors, self.actor_evaluate_net.variables))
106
107     self.update_net(self.critic_target_net, self.critic_evaluate_net)
108     self.update_net(self.actor_target_net, self.actor_evaluate_net)
109
110
111 agent = DDPGAgent(env)
```

Code 9.3 DDPG agent (PyTorch version).

Pendulum-v1_DDPG_torch.ipynb

```
1   class DDPGAgent:
2       def __init__(self, env):
3           state_dim = env.observation_space.shape[0]
4           self.action_dim = env.action_space.shape[0]
5           self.action_low = env.action_space.low[0]
6           self.action_high = env.action_space.high[0]
7           self.gamma = 0.99
8
9           self.replayer = DQNReplayer(20000)
10
11          self.actor_evaluate_net = self.build_net(
12              input_size=state_dim, hidden_sizes=[32, 64],
13              output_size=self.action_dim)
14          self.actor_optimizer = optim.Adam(self.actor_evaluate_net.parameters(),
15              lr=0.0001)
16          self.actor_target_net = copy.deepcopy(self.actor_evaluate_net)
17
18          self.critic_evaluate_net = self.build_net(
19              input_size=state_dim+self.action_dim, hidden_sizes=[64, 128])
20          self.critic_optimizer = optim.Adam(
21              self.critic_evaluate_net.parameters(), lr=0.001)
22          self.critic_loss = nn.MSELoss()
23          self.critic_target_net = copy.deepcopy(self.critic_evaluate_net)
24
25      def build_net(self, input_size, hidden_sizes, output_size=1,
26              output_activator=None):
27          layers = []
28          for input_size, output_size in zip(
29                  [input_size,] + hidden_sizes, hidden_sizes + [output_size,]):
30              layers.append(nn.Linear(input_size, output_size))
31              layers.append(nn.ReLU())
32          layers = layers[:-1]
33          if output_activator:
34              layers.append(output_activator)
35          net = nn.Sequential(*layers)
36          return net
37
38      def reset(self, mode=None):
```

```
39      self.mode = mode
40      if self.mode == 'train':
41          self.trajectory = []
42          self.noise = OrnsteinUhlenbeckProcess(np.zeros((self.action_dim,)))
43
44  def step(self, observation, reward, terminated):
45      if self.mode == 'train' and self.replayer.count < 3000:
46          action = np.random.uniform(self.action_low, self.action_high)
47      else:
48          state_tensor = torch.as_tensor(observation,
49                  dtype=torch.float).reshape(1, -1)
50          action_tensor = self.actor_evaluate_net(state_tensor)
51          action = action_tensor.detach().numpy()[0]
52      if self.mode == 'train':
53          # noisy action
54          noise = self.noise(sigma=0.1)
55          action = (action + noise).clip(self.action_low, self.action_high)
56
57          self.trajectory += [observation, reward, terminated, action]
58          if len(self.trajectory) >= 8:
59              state, _, _, act, next_state, reward, terminated, _ = \
60                      self.trajectory[-8:]
61              self.replayer.store(state, act, reward, next_state, terminated)
62
63          if self.replayer.count >= 3000:
64              self.learn()
65      return action
66
67  def close(self):
68      pass
69
70  def update_net(self, target_net, evaluate_net, learning_rate=0.005):
71      for target_param, evaluate_param in zip(
72              target_net.parameters(), evaluate_net.parameters()):
73          target_param.data.copy_(learning_rate * evaluate_param.data
74                  + (1 - learning_rate) * target_param.data)
75
76  def learn(self):
77      # replay
78      states, actions, rewards, next_states, terminateds = \
79              self.replayer.sample(64)
80      state_tensor = torch.as_tensor(states, dtype=torch.float)
81      action_tensor = torch.as_tensor(actions, dtype=torch.long)
82      reward_tensor = torch.as_tensor(rewards, dtype=torch.float)
83      next_state_tensor = torch.as_tensor(next_states, dtype=torch.float)
84      terminated_tensor = torch.as_tensor(terminateds, dtype=torch.float)
85
86      # update critic
87      next_action_tensor = self.actor_target_net(next_state_tensor)
88      noise_tensor = (0.2 * torch.randn_like(action_tensor,
89              dtype=torch.float))
90      noisy_next_action_tensor = (next_action_tensor + noise_tensor).clamp(
91              self.action_low, self.action_high)
92      next_state_action_tensor = torch.cat([next_state_tensor,
93              noisy_next_action_tensor], 1)
94      next_q_tensor = self.critic_target_net(
95              next_state_action_tensor).squeeze(1)
96      critic_target_tensor = reward_tensor + (1. - terminated_tensor) * \
97              self.gamma * next_q_tensor
98      critic_target_tensor = critic_target_tensor.detach()
99
100     state_action_tensor = torch.cat([state_tensor, action_tensor], 1)
101     critic_pred_tensor = self.critic_evaluate_net(state_action_tensor
102             ).squeeze(1)
103     critic_loss_tensor = self.critic_loss(critic_pred_tensor,
104             critic_target_tensor)
105     self.critic_optimizer.zero_grad()
```

```
106    critic_loss_tensor.backward()
107    self.critic_optimizer.step()
108
109    # update actor
110    pred_action_tensor = self.actor_evaluate_net(state_tensor)
111    pred_action_tensor = pred_action_tensor.clamp(self.action_low,
112            self.action_high)
113    pred_state_action_tensor = torch.cat([state_tensor,
114            pred_action_tensor], 1)
115    critic_pred_tensor = self.critic_evaluate_net(pred_state_action_tensor)
116    actor_loss_tensor = -critic_pred_tensor.mean()
117    self.actor_optimizer.zero_grad()
118    actor_loss_tensor.backward()
119    self.actor_optimizer.step()
120
121    self.update_net(self.critic_target_net, self.critic_evaluate_net)
122    self.update_net(self.actor_target_net, self.actor_evaluate_net)
123
124
125 agent = DDPGAgent(env)
```

Especially, in the TensorFlow version (Code 9.2), the function `learn()` implements model parameters updating. Updating the value parameters **w** can use the Keras Sequential API as usual, but updating the policy parameter $\boldsymbol{\theta}$ can not since it needs to increase $q(S, \pi(S; \boldsymbol{\theta}); \mathbf{w})$. Therefore, the implementation uses the eager mode of TensorFlow.

Interaction between this agent and the environment uses Code 1.3.

9.5.2 TD3

This section uses TD3 algorithm to find the optimal policy. Codes 9.4 and 9.5 implement the TD3 algorithm. These codes differ from Codes 9.2 and 9.3 in the constructors and the `learn()` function.

Code 9.4   TD3 agent (TensorFlow version).
Pendulum-v1_TD3_tf.ipynb

```
1  class TD3Agent:
2      def __init__(self, env):
3          state_dim = env.observation_space.shape[0]
4          self.action_dim = env.action_space.shape[0]
5          self.action_low = env.action_space.low
6          self.action_high = env.action_space.high
7          self.gamma = 0.99
8
9          self.replayer = DQNReplayer(20000)
10
11         self.actor_evaluate_net = self.build_net(
12                 input_size=state_dim, hidden_sizes=[32, 64],
13                 output_size=self.action_dim, output_activation=nn.tanh)
14         self.actor_target_net = models.clone_model(self.actor_evaluate_net)
15         self.actor_target_net.set_weights(
16                 self.actor_evaluate_net.get_weights())
17
18         self.critic0_evaluate_net = self.build_net(
19                 input_size=state_dim+self.action_dim, hidden_sizes=[64, 128])
20         self.critic0_target_net = models.clone_model(self.critic0_evaluate_net)
21         self.critic0_target_net.set_weights(
```

```
22              self.critic0_evaluate_net.get_weights())
23
24          self.critic1_evaluate_net = self.build_net(
25                  input_size=state_dim+self.action_dim, hidden_sizes=[64, 128])
26          self.critic1_target_net = models.clone_model(self.critic1_evaluate_net)
27          self.critic1_target_net.set_weights(
28                  self.critic1_evaluate_net.get_weights())
29
30      def build_net(self, input_size=None, hidden_sizes=None, output_size=1,
31                  activation=nn.relu, output_activation=None,
32                  loss=losses.mse, learning_rate=0.001):
33          model = keras.Sequential()
34          for layer, hidden_size in enumerate(hidden_sizes):
35              kwargs = {'input_shape' : (input_size,)} if layer == 0 else {}
36              model.add(layers.Dense(units=hidden_size,
37                      activation=activation, **kwargs))
38          model.add(layers.Dense(units=output_size,
39                  activation=output_activation))
40          optimizer = optimizers.Adam(learning_rate)
41          model.compile(optimizer=optimizer, loss=loss)
42          return model
43
44      def reset(self, mode=None):
45          self.mode = mode
46          if self.mode == 'train':
47              self.trajectory = []
48              self.noise = OrnsteinUhlenbeckProcess(np.zeros((self.action_dim,)))
49
50      def step(self, observation, reward, terminated):
51          if self.mode == 'train' and self.replayer.count < 3000:
52              action = np.random.uniform(self.action_low, self.action_high)
53          else:
54              action = self.actor_evaluate_net.predict(observation[np.newaxis],
55                      verbose=0)[0]
56          if self.mode == 'train':
57              # noisy action
58              noise = self.noise(sigma=0.1)
59              action = (action + noise).clip(self.action_low, self.action_high)
60
61              self.trajectory += [observation, reward, terminated, action]
62              if len(self.trajectory) >= 8:
63                  state, _, _, act, next_state, reward, terminated, _ = \
64                          self.trajectory[-8:]
65                  self.replayer.store(state, act, reward, next_state, terminated)
66
67              if self.replayer.count >= 3000:
68                  self.learn()
69          return action
70
71      def close(self):
72          pass
73
74      def update_net(self, target_net, evaluate_net, learning_rate=0.005):
75          average_weights = [(1. - learning_rate) * t + learning_rate * e for
76                  t, e in zip(target_net.get_weights(),
77                  evaluate_net.get_weights())]
78          target_net.set_weights(average_weights)
79
80      def learn(self):
81          # replay
82          states, actions, rewards, next_states, terminateds = \
83                  self.replayer.sample(64)
84          state_tensor = tf.convert_to_tensor(states, dtype=tf.float32)
85
86          # update critic
87          next_actions = self.actor_target_net.predict(next_states, verbose=0)
88          next_noises = np.random.normal(0, 0.2, size=next_actions.shape)
```

```
 89        next_actions = (next_actions + next_noises).clip(self.action_low,
 90                self.action_high)
 91        state_actions = np.hstack([states, actions])
 92        next_state_actions = np.hstack([next_states, next_actions])
 93        next_q0s = self.critic0_target_net.predict(next_state_actions,
 94                verbose=0)[:, 0]
 95        next_q1s = self.critic1_target_net.predict(next_state_actions,
 96                verbose=0)[:, 0]
 97        next_qs = np.minimum(next_q0s, next_q1s)
 98        targets = rewards + (1. - terminateds) * self.gamma * next_qs
 99        self.critic0_evaluate_net.fit(state_actions, targets[:, np.newaxis],
100                verbose=0)
101        self.critic1_evaluate_net.fit(state_actions, targets[:, np.newaxis],
102                verbose=0)
103
104        # update actor
105        with tf.GradientTape() as tape:
106            action_tensor = self.actor_evaluate_net(state_tensor)
107            state_action_tensor = tf.concat([state_tensor, action_tensor],
108                    axis=1)
109            q_tensor = self.critic0_evaluate_net(state_action_tensor)
110            loss_tensor = -tf.reduce_mean(q_tensor)
111        grad_tensors = tape.gradient(loss_tensor,
112                self.actor_evaluate_net.variables)
113        self.actor_evaluate_net.optimizer.apply_gradients(zip(
114                grad_tensors, self.actor_evaluate_net.variables))
115
116        self.update_net(self.critic0_target_net, self.critic0_evaluate_net)
117        self.update_net(self.critic1_target_net, self.critic1_evaluate_net)
118        self.update_net(self.actor_target_net, self.actor_evaluate_net)
119
120
121 agent = TD3Agent(env)
```

Code 9.5   TD3 agent (PyTorch version).

Pendulum-v1_TD3_torch.ipynb

```
 1 class TD3Agent:
 2     def __init__(self, env):
 3         state_dim = env.observation_space.shape[0]
 4         self.action_dim = env.action_space.shape[0]
 5         self.action_low = env.action_space.low[0]
 6         self.action_high = env.action_space.high[0]
 7
 8         self.gamma = 0.99
 9
10         self.replayer = DQNReplayer(20000)
11
12         self.actor_evaluate_net = self.build_net(
13                 input_size=state_dim, hidden_sizes=[32, 64],
14                 output_size=self.action_dim)
15         self.actor_optimizer = optim.Adam(self.actor_evaluate_net.parameters(),
16                 lr=0.001)
17         self.actor_target_net = copy.deepcopy(self.actor_evaluate_net)
18
19         self.critic0_evaluate_net = self.build_net(
20                 input_size=state_dim+self.action_dim, hidden_sizes=[64, 128])
21         self.critic0_optimizer = optim.Adam(
22                 self.critic0_evaluate_net.parameters(), lr=0.001)
23         self.critic0_loss = nn.MSELoss()
24         self.critic0_target_net = copy.deepcopy(self.critic0_evaluate_net)
25
26         self.critic1_evaluate_net = self.build_net(
27                 input_size=state_dim+self.action_dim, hidden_sizes=[64, 128])
28         self.critic1_optimizer = optim.Adam(
```

```python
                       self.critic1_evaluate_net.parameters(), lr=0.001)
        self.critic1_loss = nn.MSELoss()
        self.critic1_target_net = copy.deepcopy(self.critic1_evaluate_net)

    def build_net(self, input_size, hidden_sizes, output_size=1,
            output_activator=None):
        layers = []
        for input_size, output_size in zip([input_size,] + hidden_sizes,
                hidden_sizes + [output_size,]):
            layers.append(nn.Linear(input_size, output_size))
            layers.append(nn.ReLU())
        layers = layers[:-1]
        if output_activator:
            layers.append(output_activator)
        net = nn.Sequential(*layers)
        return net

    def reset(self, mode=None):
        self.mode = mode
        if self.mode == 'train':
            self.trajectory = []
            self.noise = OrnsteinUhlenbeckProcess(np.zeros((self.action_dim,)))

    def step(self, observation, reward, terminated):
        state_tensor = torch.as_tensor(observation,
                dtype=torch.float).unsqueeze(0)
        action_tensor = self.actor_evaluate_net(state_tensor)
        action = action_tensor.detach().numpy()[0]

        if self.mode == 'train':
            # noisy action
            noise = self.noise(sigma=0.1)
            action = (action + noise).clip(self.action_low, self.action_high)

            self.trajectory += [observation, reward, terminated, action]
            if len(self.trajectory) >= 8:
                state, _, _, act, next_state, reward, terminated, _ = \
                        self.trajectory[-8:]
                self.replayer.store(state, act, reward, next_state, terminated)

            if self.replayer.count >= 3000:
                self.learn()
        return action

    def close(self):
        pass

    def update_net(self, target_net, evaluate_net, learning_rate=0.005):
        for target_param, evaluate_param in zip(
                target_net.parameters(), evaluate_net.parameters()):
            target_param.data.copy_(learning_rate * evaluate_param.data
                    + (1 - learning_rate) * target_param.data)

    def learn(self):
        # replay
        states, actions, rewards, next_states, terminateds = \
                self.replayer.sample(64)
        state_tensor = torch.as_tensor(states, dtype=torch.float)
        action_tensor = torch.as_tensor(actions, dtype=torch.long)
        reward_tensor = torch.as_tensor(rewards, dtype=torch.float)
        next_state_tensor = torch.as_tensor(next_states, dtype=torch.float)
        terminated_tensor = torch.as_tensor(terminateds, dtype=torch.float)

        # update critic
        next_action_tensor = self.actor_target_net(next_state_tensor)
        noise_tensor = (0.2 * torch.randn_like(action_tensor,
                dtype=torch.float))
```

```
 96        noisy_next_action_tensor = (next_action_tensor + noise_tensor
 97                ).clamp(self.action_low, self.action_high)
 98        next_state_action_tensor = torch.cat([next_state_tensor,
 99                noisy_next_action_tensor], 1)
100        next_q0_tensor = self.critic0_target_net(next_state_action_tensor
101                ).squeeze(1)
102        next_q1_tensor = self.critic1_target_net(next_state_action_tensor
103                ).squeeze(1)
104        next_q_tensor = torch.min(next_q0_tensor, next_q1_tensor)
105        critic_target_tensor = reward_tensor + (1. - terminated_tensor) * \
106                self.gamma * next_q_tensor
107        critic_target_tensor = critic_target_tensor.detach()
108
109        state_action_tensor = torch.cat([state_tensor, action_tensor], 1)
110        critic_pred0_tensor = self.critic0_evaluate_net(
111                state_action_tensor).squeeze(1)
112        critic0_loss_tensor = self.critic0_loss(critic_pred0_tensor,
113                critic_target_tensor)
114        self.critic0_optimizer.zero_grad()
115        critic0_loss_tensor.backward()
116        self.critic0_optimizer.step()
117
118        critic_pred1_tensor = self.critic1_evaluate_net(
119                state_action_tensor).squeeze(1)
120        critic1_loss_tensor = self.critic1_loss(critic_pred1_tensor,
121                critic_target_tensor)
122        self.critic1_optimizer.zero_grad()
123        critic1_loss_tensor.backward()
124        self.critic1_optimizer.step()
125
126        # update actor
127        pred_action_tensor = self.actor_evaluate_net(state_tensor)
128        pred_action_tensor = pred_action_tensor.clamp(self.action_low,
129                self.action_high)
130        pred_state_action_tensor = torch.cat([state_tensor,
131                pred_action_tensor], 1)
132        critic_pred_tensor = self.critic0_evaluate_net(
133                pred_state_action_tensor)
134        actor_loss_tensor = -critic_pred_tensor.mean()
135        self.actor_optimizer.zero_grad()
136        actor_loss_tensor.backward()
137        self.actor_optimizer.step()
138
139        self.update_net(self.critic0_target_net, self.critic0_evaluate_net)
140        self.update_net(self.critic1_target_net, self.critic1_evaluate_net)
141        self.update_net(self.actor_target_net, self.actor_evaluate_net)
142
143
144 agent = TD3Agent(env)
```

Interaction between this agent and the environment once again uses Code 1.3.

9.6 Summary

- The optimal deterministic policy in a task with continuous action space task can be approximated by $\pi(s; \boldsymbol{\theta})$ ($s \in \mathcal{S}$).
- DPG theorem with continuous action space is

$$\nabla g_{\pi(\boldsymbol{\theta})} = \mathrm{E}_{S \sim \eta_{\pi(\boldsymbol{\theta})}} \left[\nabla \pi(S; \boldsymbol{\theta}) \left[\nabla_a q_{\pi(\boldsymbol{\theta})}(S, a) \right]_{a=\pi(S;\boldsymbol{\theta})} \right].$$

- The vanilla on-policy and off-policy deterministic AC algorithm try to update policy parameter θ to maximize $\gamma^t q(S_t, \pi(S_t; \theta); \mathbf{w})$.
- Combined behavior policy and importance sampling to DPG can lead to off-policy DPG. DDPG algorithm further uses the techniques of experience replay and target network.
- DPG algorithms in continuous action space can use artificial noises to explore. The noises can be i.i.d Gaussian noises, OU process, and so on.

9.7 Exercises

9.7.1 Multiple Choices

9.1 On deterministic policy gradient theorem, choose the correct one: (   )

A. Deterministic policy gradient theorem considers the task whose state space is continuous.
B. Deterministic policy gradient theorem considers the task whose action space is continuous.
C. Deterministic policy gradient theorem considers the task whose reward space is continuous.

9.2 Introducing Ornstein Uhlenbeck process into action sequence can: (   )

A. Let the noises at different times be positively correlated.
B. Let the noises at different times be not correlated.
C. Let the noises at different times be negatively correlated.

9.3 On DDPG algorithm, choose the correct one: (   )

A. DDPG algorithm is a value-based algorithm.
B. DDPG algorithm is a policy-based algorithm.
C. DDPG algorithm is an actor–critic algorithm.

9.7.2 Programming

9.4 Use DDPG algorithm to solve the task `MountainCarContinuous-v0`.

9.7.3 Mock Interview

9.5 Prove the deterministic policy gradient theorem.

9.6 What are the advantages and disadvantages of deterministic policy gradients when they are used to solve tasks with continuous action space?

Chapter 10
Maximum-Entropy RL

This chapter covers

- maximum-entropy RL
- soft values, and their properties, including soft Bellman equations
- soft policy gradient theorem
- Soft Q Learning (SQL) algorithm
- Soft Actor–Critic (SAC) algorithm
- automatic entropy adjustment

This chapter introduces maximum-entropy RL, which uses the concept of entropy in information theory to encourage exploration.

10.1 Maximum-Entropy RL and Soft RL

10.1.1 Reward Engineering and Reward with Entropy

In RL, **reward engineering** is a trick that modifies the definition of reward in the original environment, and trains the modified environment so that the trained agent can work well in the original environment.

This chapter will consider maximum-entropy RL algorithms. These algorithms modify the reward in the original environment using the reward with entropy, and train using the RL task using rewards with entropy to solve the original RL tasks.

The tradeoff between exploitation and exploration is a key issue in RL. Given a state, we will let the action be more randomly selected if we want to encourage

© The Author(s), under exclusive license to Springer Nature Singapore Pte Ltd. 2024
Z. Xiao, *Reinforcement Learning*, https://doi.org/10.1007/978-981-19-4933-3_10

exploration more, and we will let the action more deterministic if we want to encourage exploitation more. The randomness of actions can be characterized by the entropy in information theory.

Interdisciplinary Reference 10.1
Information Theory: Entropy

Assume that a random variable X obeys the probability distribution p. The entropy of the random variable distribution (denoted as $\text{H}[X]$), or the entropy of the probability distribution (denoted as $\text{H}[p]$), is defined as

$$\text{H}[p] = \text{H}[X] \overset{\text{def}}{=} \text{E}_{X \sim p}\left[-\ln p(X)\right] = -\sum_x p(x) \ln p(x).$$

A random variable with larger randomness has a larger entropy. Therefore, the entropy can be viewed as a metric of stochasticity.

Example 10.1 For the uniform distributed random variable $X \sim \text{uniform}\,(a, b)$, its entropy is $\text{H}[X] = \ln(b - a)$.

Example 10.2 For an n-dimension Gaussian distributed random variable $\mathbf{X} \sim \text{normal}(\mathbf{\mu}, \mathbf{\Sigma})$, its entropy is $\text{H}[\mathbf{X}] = \frac{1}{2}(n \ln 2\pi\text{e} + \ln \det \mathbf{\Sigma})$.

Since larger entropy indicates more exploration, we can try to maximize the entropy to encourage exploration. At the same time, we still need to maximize the episode reward. Therefore, we can combine these two objectives, and define the reward with entropy as (Harrnoja, 2017)

$$R_{t+1}^{(\text{H})} \overset{\text{def}}{=} R_{t+1} + \alpha^{(\text{H})} \text{H}\left[\pi(\cdot | S_t)\right],$$

where $\alpha^{(\text{H})}$ is a parameter ($\alpha^{(\text{H})} > 0$) to trade-off between exploration and exploitation. Large $\alpha^{(\text{H})}$ means the entropy is important, and we care about exploration; while small $\alpha^{(\text{H})}$ means the entropy is not that important, and we care about exploitation.

Based on the definition of reward with entropy, we can further define the return with entropy as

$$G_t^{(\text{H})} \overset{\text{def}}{=} \sum_{\tau=0}^{+\infty} \gamma^\tau R_{t+\tau+1}^{(\text{H})}.$$

Maximum-entropy RL tries to find a policy that maximizes the expectation of return with entropy, i.e.

$$\pi_*^{(\text{H})} = \arg\max_\pi \text{E}_\pi\left[G_0^{(\text{H})}\right].$$

In the objective $E_\pi\left[G_0^{(H)}\right]$, the policy π not only impacts the distribution that expectation is over, but also impacts what is taken expectation.

Similarly, we can define the values with entropy as

$$v_\pi^{(H)}(s) \overset{\text{def}}{=} E_\pi\left[G_t^{(H)}\middle|S_t = s\right], \qquad s \in \mathcal{S}$$

$$q_\pi^{(H)}(s, a) \overset{\text{def}}{=} E_\pi\left[G_t^{(H)}\middle|S_t = s, A_t = a\right], \quad s \in \mathcal{S}, a \in \mathcal{A}(s).$$

The relationship between the values with entropy includes:

$$v_\pi^{(H)}(s) = \sum_a \pi(a|s)q_\pi^{(H)}(s, a), \qquad s \in \mathcal{S}$$

$$q_\pi^{(H)}(s, a) = r^{(H)}(s, a) + \gamma \sum_{s'} p(s'|s, a)v_\pi^{(H)}(s), \quad s \in \mathcal{S}, a \in \mathcal{A}(s),$$

where

$$r^{(H)}(s, a) \overset{\text{def}}{=} r(s, a) + H[\pi(\cdot|s)], \quad s \in \mathcal{S}, a \in \mathcal{A}(s).$$

The Bellman equation that uses action value with entropy to back up the action values with entropy is

$$q_\pi^{(H)}(s, a) = r^{(H)}(s, a) + \gamma \sum_{s',a'} p_\pi(s', a'|s, a)q_\pi^{(H)}(s', a'), \quad s \in \mathcal{S}, a \in \mathcal{A}(s).$$

It is easy to show that the maximum-entropy RL can also be formulated as

$$\pi_*^{(H)} = \arg\max_\pi E_{(S,A)\sim\rho_\pi}\left[q_\pi^{(H)}(S, A)\right].$$

Section 3.1 has proved that Bellman optimal operators in a contraction mapping in a complete metrics space, so the maximum-entropy RL exists a unique solution. This solution is of course different from the optimal solution of the original problem without entropy.

10.1.2 Soft Values

The action value with entropy $q_\pi^{(H)}(s, a)$ contains a term $\alpha^{(H)}H[\pi(\cdot|s)]$. However, the input of action value, the state–action pair (s, a), does not relate to the policy π. Therefore, we should exclude the term $\alpha^{(H)}H[\pi(\cdot|s)]$. Using this intuition, we can define soft values.

The **soft values** of a policy π are defined as:

- **Soft action values**

$$q_\pi^{(\text{soft})}(s, a) \overset{\text{def}}{=} q_\pi^{(H)}(s, a) - \alpha^{(H)}H[\pi(\cdot|s)], \quad s \in \mathcal{S}, a \in \mathcal{A}(s).$$

- **Soft state values**

$$v_\pi^{(\text{soft})}(s) \overset{\text{def}}{=} \alpha^{(\text{H})} \log \sum_a \exp\left(\frac{1}{\alpha^{(\text{H})}} q_\pi^{(\text{soft})}(s, a)\right), \quad s \in \mathcal{S}.$$

Using the following logsumexp() operator

$$\text{logsumexp}\, x \overset{\text{def}}{=} \log \sum_{x \in X} \exp x,$$
$$\quad\quad x \in X$$

the soft state values can also be written as

$$v_\pi^{(\text{soft})}(s) \overset{\text{def}}{=} \alpha^{(\text{H})} \operatorname*{logsumexp}_a\left(\frac{1}{\alpha^{(\text{H})}} q_\pi^{(\text{soft})}(s, a)\right), \quad s \in \mathcal{S}.$$

This formula uses the soft action value to back up soft state values. Remarkably, soft state values are merely the partition function of soft action values, and they are not actual state values. The reason why they are called soft state values is merely that, for the optimal maximum-entropy policy, the soft state values satisfy the form that uses action values to backup state values in the expectation. Details will be explained later. These values are called soft values, because the logsumexp() operator can be viewed as a "soft" version of max() operator.

Furthermore, we can define **soft advantages** as

$$a_\pi^{(\text{soft})}(s, a) \overset{\text{def}}{=} q_\pi^{(\text{soft})}(s, a) - v_\pi^{(\text{soft})}(s), \quad s \in \mathcal{S}, a \in \mathcal{A}(s).$$

Since

$$q_\pi^{(\text{H})}(s, a) = q_\pi^{(\text{soft})}(s, a) + \alpha^{(\text{H})} \text{H}\big[\pi(\cdot|s)\big], \quad s \in \mathcal{S}, a \in \mathcal{A}(s),$$

the maximum-entropy RL can also be formulated as

$$\pi_*^{(\text{H})} = \arg\max_\pi \text{E}_{(S,A)\sim\rho_\pi}\Big[q_\pi^{(\text{soft})}(S, A) + \alpha^{(\text{H})} \text{H}\big[\pi(\cdot|S)\big]\Big].$$

The Bellman equation that uses soft action values to backup soft action values is

$$q_\pi^{(\text{soft})}(s, a) = r(s, a) + \gamma \sum_{s',a'} p_\pi(s', a'|s, a)\Big[q_\pi^{(\text{soft})}(s', a') + \alpha^{(\text{H})} \text{H}\big[\pi(\cdot|s')\big]\Big],$$

$$s \in \mathcal{S}, a \in \mathcal{A}(s).$$

(Proof: Plugging the following equations

$$q_\pi^{(\text{H})}(s, a) = q_\pi^{(\text{soft})}(s, a) + \alpha^{(\text{H})} \text{H}\big[\pi(\cdot|s)\big], \quad s \in \mathcal{S}, a \in \mathcal{A}(s)$$
$$r^{(\text{H})}(s, a) = \quad r(s, a) + \alpha^{(\text{H})} \text{H}\big[\pi(\cdot|s)\big], \quad s \in \mathcal{S}, a \in \mathcal{A}(s)$$

into the Bellman equations of action values with entropy

$$q_{\pi}^{(H)}(s, a) = r^{(H)}(s, a) + \gamma \sum_{s', a'} p_{\pi}(s', a'|s, a) q_{\pi}^{(H)}(s', a'), \quad s \in \mathcal{S}, a \in \mathcal{A}(s)$$

will lead to

$$q_{\pi}^{(\text{soft})}(s, a) + \alpha^{(H)} H[\pi(\cdot|s)]$$
$$= r(s, a) + \alpha^{(H)} H[\pi(\cdot|s)] + \gamma \sum_{s', a'} p_{\pi}(s', a'|s, a) \left[q_{\pi}^{(\text{soft})}(s', a') + \alpha^{(H)} H[\pi(\cdot|s')] \right],$$
$$s \in \mathcal{S}, a \in \mathcal{A}(s).$$

Eliminating $\alpha^{(H)} H[\pi(\cdot|s)]$ from both sides of the above equation completes the proof.) The Bellman equation can be written as

$$q_{\pi}^{(\text{soft})}(S_t, A_t) = E_{\pi} \left[R_{t+1} + \gamma \left(E_{\pi} \left[q_{\pi}^{(\text{soft})}(S_{t+1}, A_{t+1}) \right] + \alpha^{(H)} H[\pi(\cdot|S_{t+1})] \right) \right].$$

10.1.3 Soft Policy Improvement Theorem and Numeric Iterative Algorithm

This section introduces a model-based numerical iterative algorithm for maximum-entropy RL. This algorithm uses the soft policy improvement theorem.

Soft policy improvement theorem: Given the policy π. Construct a new policy $\tilde{\pi}$ using the soft advantage of the policy $a_{\pi}^{(\text{soft})}(s, a) = q_{\pi}^{(\text{soft})}(s, a) - v_{\pi}^{(\text{soft})}(s)$ ($s \in \mathcal{S}, a \in \mathcal{A}(s)$) as follows:

$$\tilde{\pi}(a|s) = \exp \left(\frac{1}{\alpha^{(H)}} a_{\pi}^{(\text{soft})}(s, a) \right), \quad s \in \mathcal{S}, a \in \mathcal{A}(s).$$

If all action values and state values of both policies are bounded, the two policies satisfy

$$q_{\tilde{\pi}}^{(\text{soft})}(s, a) \geq q_{\pi}^{(\text{soft})}(s, a), \quad s \in \mathcal{S}, a \in \mathcal{A}(s).$$

Moreover, the inequality holds when the policy π does not satisfy

$$\pi(\cdot|s) \overset{\text{a.e.}}{=} \exp \left(\frac{1}{\alpha^{(H)}} a_{\pi}^{(\text{soft})}(s, \cdot) \right)$$

for some state s, where a.e. is short for almost everywhere. (Proof: The KL divergence of two arbitrary policies $\hat{\pi}$ and $\tilde{\pi}$ satisfies

$$d_{\text{KL}} \left(\hat{\pi}(\cdot|s) \| \tilde{\pi}(\cdot|s) \right)$$
$$= d_{\text{KL}} \left(\hat{\pi}(\cdot|s) \middle\| \exp \left(\frac{1}{\alpha^{(H)}} \left(q_{\pi}^{(\text{soft})}(s, \cdot) - v_{\pi}^{(\text{soft})}(s) \right) \right) \right)$$

$$= \mathrm{E}_{\hat{\pi}}\left[\ln \hat{\pi}(\cdot|S_t) - \frac{1}{\alpha^{(\mathrm{H})}}\left(q_{\pi}^{(\mathrm{soft})}(S_t, A_t) - v_{\pi}^{(\mathrm{soft})}(S_t)\right)\bigg| S_t = s\right]$$

$$= -\mathrm{H}\left[\hat{\pi}(\cdot|s)\right] - \frac{1}{\alpha^{(\mathrm{H})}}\left(\mathrm{E}_{\hat{\pi}}\left[q_{\pi}^{(\mathrm{soft})}(S_t, A_t)\big| S_t = s\right] - v_{\pi}^{(\mathrm{soft})}(s)\right).$$

Since

$$d_{\mathrm{KL}}\left(\pi(\cdot|s)\big\|\tilde{\pi}(\cdot|s)\right) \geq 0 = d_{\mathrm{KL}}\left(\tilde{\pi}(\cdot|s)\big\|\tilde{\pi}(\cdot|s)\right)$$

and the inequality holds when $d_{\mathrm{KL}}\left(\pi(\cdot|s)\big\|\tilde{\pi}(\cdot|s)\right) > 0$, we have

$$-\mathrm{H}\left[\pi(\cdot|s)\right] - \frac{1}{\alpha^{(\mathrm{H})}}\left(\mathrm{E}_{\pi}\left[q_{\pi}^{(\mathrm{soft})}(S_t, A_t)\big| S_t = s\right] - v_{\pi}^{(\mathrm{soft})}(s)\right) \geq$$

$$-\mathrm{H}\left[\tilde{\pi}(\cdot|s)\right] - \frac{1}{\alpha^{(\mathrm{H})}}\left(\mathrm{E}_{\tilde{\pi}}\left[q_{\pi}^{(\mathrm{soft})}(S_t, A_t)\big| S_t = s\right] - v_{\pi}^{(\mathrm{soft})}(s)\right).$$

Therefore,

$$\alpha^{(\mathrm{H})}\mathrm{H}\left[\pi(\cdot|s)\right] + \mathrm{E}_{\pi}\left[q_{\pi}^{(\mathrm{soft})}(S_t, A_t)\big| S_t = s\right]$$

$$\leq \alpha^{(\mathrm{H})}\mathrm{H}\left[\tilde{\pi}(\cdot|s)\right] + \mathrm{E}_{\tilde{\pi}}\left[q_{\pi}^{(\mathrm{soft})}(S_t, A_t)\big| S_t = s\right].$$

The inequality holds when $d_{\mathrm{KL}}\left(\pi(\cdot|s)\big\|\tilde{\pi}(\cdot|s)\right) > 0$, or equivalently when $\pi(\cdot|s) \overset{\mathrm{a.e.}}{=} \tilde{\pi}(\cdot|s)$ does not hold. Therefore,

$$q_{\pi}^{(\mathrm{soft})}(s, a)$$

$$= \mathrm{E}\left[R_1 + \gamma\left(\alpha^{(\mathrm{H})}\mathrm{H}\left[\pi(\cdot|S_1)\right]\right) + \mathrm{E}_{\pi}\left[q_{\pi}^{(\mathrm{soft})}(S_1, A_1)\right]\bigg| S_0 = s, A_0 = a\right]$$

$$\leq \mathrm{E}\left[R_1 + \gamma\left(\alpha^{(\mathrm{H})}\mathrm{H}\left[\tilde{\pi}(\cdot|S_1)\right]\right) + \mathrm{E}_{\tilde{\pi}}\left[q_{\pi}^{(\mathrm{soft})}(S_1, A_1)\right]\bigg| S_0 = s, A_0 = a\right]$$

$$= \mathrm{E}\left[R_1 + \gamma\alpha^{(\mathrm{H})}\mathrm{H}\left[\tilde{\pi}(\cdot|S_1)\right]\right.$$

$$\left. + \gamma^2\mathrm{E}_{\tilde{\pi}}\left[\alpha^{(\mathrm{H})}\mathrm{H}\left[\tilde{\pi}(\cdot|S_2)\right] + \mathrm{E}_{\pi}\left[q_{\pi}^{(\mathrm{soft})}(S_2, A_2)\right]\right]\bigg| S_0 = s, A_0 = a\right]$$

$$\leq \mathrm{E}\left[R_1 + \gamma\alpha^{(\mathrm{H})}\mathrm{H}\left[\tilde{\pi}(\cdot|S_1)\right]\right.$$

$$\left. + \gamma^2\mathrm{E}_{\tilde{\pi}}\left[\alpha^{(\mathrm{H})}\mathrm{H}\left[\tilde{\pi}(\cdot|S_2)\right] + \mathrm{E}_{\tilde{\pi}}\left[q_{\pi}^{(\mathrm{soft})}(S_2, A_2)\right]\right]\bigg| S_0 = s, A_0 = a\right]$$

$$\cdots$$

$$\leq \mathrm{E}\left[R_1 + \sum_{t=0}^{+\infty}\left(\alpha^{(\mathrm{H})}\left[\tilde{\pi}(\cdot|S_t)\right]\right) + R_{t+1}\bigg| S_0 = s, A_0 = a\right]$$

$$= q_{\tilde{\pi}}^{(\mathrm{soft})}(s, a), \qquad\qquad s \in \mathcal{S}, a \in \mathcal{A}(s).$$

Any states that make the KL divergence strictly positive will lead to inequality. The proof completes.)

Soft policy improvement theorem shows an iteration method to improve the objective. This iteration method can be represented by the soft Bellman optimal operator.

As usual, we use Q to denote the set of all possible action values. The soft Bellman operator $\mathfrak{b}_*^{(\text{soft})} : Q \to Q$ is defined as

$$\mathfrak{b}_*^{(\text{soft})}(q)(s, a) \stackrel{\text{def}}{=} r(s, a) + \gamma \sum_{s'} p(s'|s, a)\alpha^{(\text{H})} \operatorname*{logsumexp}_{a'} \frac{1}{\alpha^{(\text{H})}} q(s', a'),$$

$$q \in Q, s \in S, a \in \mathcal{A}(s).$$

The soft Bellman operator $\mathfrak{b}_*^{(\text{soft})}$ is a contraction mapping on the metric space (Q, d_∞). (Proof: For any $q', q'' \in Q$, since

$$d_\infty(q', q'') \stackrel{\text{def}}{=} \max_{s', a'} |q'(s', a') - q''(s', a')|,$$

we have

$$q'(s', a') \leq q''(s', a') + d_\infty(q', q''), \qquad s' \in S, a' \in \mathcal{A}(s')$$

$$\frac{1}{\alpha^{(\text{H})}} q'(s', a') \leq \frac{1}{\alpha^{(\text{H})}} q''(s', a') + \frac{1}{\alpha^{(\text{H})}} d_\infty(q', q''), \quad s' \in S, a' \in \mathcal{A}(s').$$

Furthermore,

$$\operatorname*{logsumexp}_{a'} \frac{1}{\alpha^{(\text{H})}} q'(s', a') \leq \operatorname*{logsumexp}_{a'} \frac{1}{\alpha^{(\text{H})}} q''(s', a') + \frac{1}{\alpha^{(\text{H})}} d_\infty(q', q''), \qquad s' \in S,$$

$$\alpha^{(\text{H})} \operatorname*{logsumexp}_{a'} \frac{1}{\alpha^{(\text{H})}} q'(s', a') - \alpha^{(\text{H})} \operatorname*{logsumexp}_{a'} \frac{1}{\alpha^{(\text{H})}} q''(s', a') \leq d_\infty(q', q''), s' \in S.$$

So for any $s \in S, a \in \mathcal{A}(s)$, we have

$$\mathfrak{b}_*^{(\text{soft})}(q')(s, a) - \mathfrak{b}_*^{(\text{soft})}(q')(s, a)$$

$$= \gamma \sum_{s'} p(s'|s, a)\left[\alpha^{(\text{H})} \operatorname*{logsumexp}_{a'} \frac{1}{\alpha^{(\text{H})}} q'(s', a') - \alpha^{(\text{H})} \operatorname*{logsumexp}_{a'} \frac{1}{\alpha^{(\text{H})}} q''(s', a')\right]$$

$$\leq \gamma d_\infty(q', q'').$$

Therefore,

$$d_\infty\left(\mathfrak{b}_*^{(\text{soft})}(q'), \mathfrak{b}_*^{(\text{soft})}(q'')\right) \leq \gamma d_\infty(q', q'').$$

The proof completes.)

Since the soft Bellman operator is a contraction mapping in a complete metric space, iterations using soft Bellman operator will converge to a unique optimal solution.

10.1.4 Optimal Values

The previous section introduced an iterative algorithm that can converge to a unique optimal solution. According to the soft optimal policy improvement theorem, at the optimal point, KL divergence should equal 0. Therefore, the optimal solution $\pi_*^{(H)}$ satisfies

$$\pi_*^{(H)}(a|s) \overset{\text{a.e.}}{=} \exp\left(\frac{1}{\alpha^{(H)}}\left(q_*^{(\text{soft})}(s,a) - v_*^{(\text{soft})}(s)\right)\right), \quad s \in \mathcal{S}, a \in \mathcal{A}(s),$$

where $q_*^{(\text{soft})}(s,a)$ and $v_*^{(\text{soft})}(s)$ are the soft action values and soft state values of the policy $\pi_*^{(H)}$ respectively. We call $q_*^{(\text{soft})}(s,a)$ **optimal soft action value**, and call $v_*^{(\text{soft})}(s)$ **optimal soft state value**.

The relationship between optimal soft values includes the following:

- Use optimal soft action values to back up optimal soft action values:

$$q_*^{(\text{soft})}(s,a)$$

$$= \sum_{s',r} p(s',r|s,a)\left[r + \gamma\left(\sum_{a'} \pi_*^{(H)}(a'|s')q_*^{(\text{soft})}(s',a') + \alpha^{(H)}\mathrm{H}\left[\pi_*^{(H)}(\cdot|s')\right]\right)\right]$$

$$= r(s,a) + \gamma\sum_{s'} p(s'|s,a)\left[\sum_{a'} \pi_*^{(H)}(a'|s')q_*^{(\text{soft})}(s',a') + \alpha^{(H)}\mathrm{H}\left[\pi_*^{(H)}(\cdot|s')\right]\right]$$

$$s \in \mathcal{S}, a \in \mathcal{A}(s).$$

This can also be written as

$$q_*^{(\text{soft})}(S_t, A_t) = \mathrm{E}\left[R_{t+1} + \gamma\left(\mathrm{E}_{\pi_*^{(H)}}\left[q_*^{(\text{soft})}(S_{t+1}, A_{t+1})\right]\right) + \alpha^{(H)}\mathrm{H}\left[\pi_*^{(H)}(\cdot|S_{t+1})\right]\right].$$

(Proof: The correctness is provided by the Bellman equation of using soft action values to back up soft action values for general policies.)

- Use the optimal action values to back up the optimal state values

$$\mathrm{E}_{\pi_*^{(H)}}\left[v_*^{(\text{soft})}(S_t)\right] = \mathrm{E}_{\pi_*^{(H)}}\left[q_*^{(\text{soft})}(S_t, A_t)\right] + \alpha^{(H)}\mathrm{H}\left[\pi_*^{(H)}(\cdot|S_t)\right].$$

(Proof: Since the optimal soft policy satisfies

$$\pi_*^{(H)}(a|s) = \exp\left(\frac{1}{\alpha^{(H)}}\left(q_*^{(\text{soft})}(s,a) - v_*^{(\text{soft})}(s)\right)\right), \quad s \in \mathcal{S}, a \in \mathcal{A}(s).$$

Therefore,

$$v_*^{(\text{soft})}(s) = q_*^{(\text{soft})}(s,a) - \alpha^{(H)}\ln\pi_*^{(H)}(a|s), \quad s \in \mathcal{S}, a \in \mathcal{A}(s).$$

Furthermore,

$$\sum_a \pi_*^{(\mathrm{H})}(a|s) v_*^{(\mathrm{soft})}(s)$$

$$= \sum_a \pi_*^{(\mathrm{H})}(a|s) q_*^{(\mathrm{soft})}(s,a) - \alpha^{(\mathrm{H})} \sum_a \pi_*^{(\mathrm{H})}(a|s) \ln \pi_*^{(\mathrm{H})}(a|s), \quad s \in \mathcal{S}$$

which is exactly what we want to prove. The proof completes.)

- Use the optimal state values to back up optimal soft action values:

$$q_*^{(\mathrm{soft})}(s,a) = \mathrm{E}\Big[R_{t+1} + \gamma v_*^{(\mathrm{soft})}(S_{t+1}) \Big| S_t = s, A_t = a \Big], \quad s \in \mathcal{S}, a \in \mathcal{A}(s).$$

It can also be written as

$$q_*^{(\mathrm{soft})}(S_t, A_t) = \mathrm{E}\Big[R_{t+1} + \gamma v_*^{(\mathrm{soft})}(S_{t+1}) \Big].$$

(Proof: Plugging

$$\mathrm{E}_{\pi_*^{(\mathrm{H})}}\Big[q_*^{(\mathrm{soft})}(S_t, A_t) \Big] + \alpha^{(\mathrm{H})} \mathrm{H}\Big[\pi_*^{(\mathrm{H})}(\cdot|S_t) \Big] = \mathrm{E}_{\pi_*^{(\mathrm{H})}}\Big[v_*^{(\mathrm{soft})}(S_t) \Big]$$

to

$$q_*^{(\mathrm{soft})}(S_t, A_t) = \mathrm{E}\Big[R_{t+1} + \gamma \Big(\mathrm{E}_{\pi_*^{(\mathrm{H})}} \Big[q_*^{(\mathrm{soft})}(S_t, A_t) \Big] + \alpha^{(\mathrm{H})} \mathrm{H}\Big[\pi_*^{(\mathrm{H})}(\cdot|S_t) \Big] \Big) \Big]$$

completes the proof.)

10.1.5 Soft Policy Gradient Theorem

The previous subsections analyzed from the aspect of values. This section considers the policy gradient.

Consider the policy $\pi(\mathbf{\theta})$, where $\mathbf{\theta}$ is the parameter. **Soft policy gradient theorem** gives the gradient of objective $\mathrm{E}_{\pi(\mathbf{\theta})}\Big[G_0^{(\mathrm{H})} \Big]$ with respect to the policy parameter $\mathbf{\theta}$:

$$\nabla \mathrm{E}_{\pi(\mathbf{\theta})}\Big[G_0^{(\mathrm{H})} \Big] = \mathrm{E}_{\pi(\mathbf{\theta})}\Bigg[\sum_{t=0}^{T-1} \gamma^t \Big(G_t^{(\mathrm{H})} - \alpha^{(\mathrm{H})} \ln \pi(A_t|S_t; \mathbf{\theta}) \Big) \nabla \ln \pi(A_t|S_t; \mathbf{\theta}) \Bigg].$$

Let us prove it here. Similar to the proof in Sect. 7.1.2, we will use two different methods to prove it. Both methods use the gradient of entropy:

$$\nabla \mathrm{H}\big[\pi(\cdot|s; \mathbf{\theta}) \big] = \nabla \sum_a -\pi(a|s; \mathbf{\theta}) \ln \pi(a|s; \mathbf{\theta})$$

$$= \sum_a -\big(\ln \pi(a|s; \boldsymbol{\theta}) + 1\big) \nabla \pi(a|s; \boldsymbol{\theta}).$$

Proof method 1: trajectory method. Calculating the gradient of the objective with respect to the policy parameter $\boldsymbol{\theta}$, we have

$$
\begin{aligned}
\nabla \mathrm{E}_{\pi(\boldsymbol{\theta})}\Big[G_0^{(\mathrm{H})}\Big] &= \nabla \sum_t \pi(t; \boldsymbol{\theta}) \nabla g_0^{(\mathrm{H})} \\
&= \sum_t g_0^{(\mathrm{H})} \nabla \pi(t; \boldsymbol{\theta}) + \sum_t \pi(t; \boldsymbol{\theta}) \nabla g_0^{(\mathrm{H})} \\
&= \sum_t g_0^{(\mathrm{H})} \pi(t; \boldsymbol{\theta}) \nabla \ln \pi(t; \boldsymbol{\theta}) + \sum_t \pi(t; \boldsymbol{\theta}) \nabla g_0^{(\mathrm{H})} \\
&= \mathrm{E}_{\pi(\boldsymbol{\theta})}\Big[G_0^{(\mathrm{H})} \nabla \ln \pi(T; \boldsymbol{\theta}) + \nabla G_0^{(\mathrm{H})}\Big],
\end{aligned}
$$

where $g_0^{(\mathrm{H})}$ is the sample value of $G_0^{(\mathrm{H})}$, it depends on both the trajectory t and the policy $\pi(\boldsymbol{\theta})$. Therefore, it has gradient with respect to the policy parameter $\boldsymbol{\theta}$. Since

$$\pi(T; \boldsymbol{\theta}) = p_{S_0}(S_0) \prod_{t=0}^{T-1} \pi(A_t|S_t; \boldsymbol{\theta}) p(S_{t+1}|S_t, A_t),$$

$$\ln \pi(T; \boldsymbol{\theta}) = \ln p_{S_0}(S_0) + \sum_{t=0}^{T-1} \big[\ln \pi(A_t|S_t; \boldsymbol{\theta}) + \ln p(S_{t+1}|S_t, A_t)\big],$$

$$\nabla \ln \pi(T; \boldsymbol{\theta}) = \sum_{t=0}^{T-1} \nabla \ln \pi(A_t|S_t; \boldsymbol{\theta}),$$

and

$$
\begin{aligned}
\nabla G_0^{(\mathrm{H})} &= \nabla \sum_{t=0}^{T-1} \gamma^t R_{t+1}^{(\mathrm{H})} \\
&= \nabla \sum_{t=0}^{T-1} \gamma^t \Big(R_{t+1} + \alpha^{(\mathrm{H})} \mathrm{H}\big[\pi(\cdot|S_t; \boldsymbol{\theta})\big]\Big) \\
&= \nabla \sum_{t=0}^{T-1} \gamma^t \alpha^{(\mathrm{H})} \mathrm{H}\big[\pi(\cdot|S_t; \boldsymbol{\theta})\big] \\
&= \sum_{t=0}^{T-1} \gamma^t \alpha^{(\mathrm{H})} \sum_{a \in \mathcal{A}(S_t)} -\big(\ln \pi(a|S_t; \boldsymbol{\theta}) + 1\big) \nabla \pi(a|S_t; \boldsymbol{\theta}) \\
&= \sum_{t=0}^{T-1} \gamma^t \alpha^{(\mathrm{H})} \mathrm{E}_{\pi(\boldsymbol{\theta})}\big[-\big(\ln \pi(A_t|S_t; \boldsymbol{\theta}) + 1\big) \nabla \ln \pi(A_t|S_t; \boldsymbol{\theta})\big],
\end{aligned}
$$

we have

$$\nabla E_{\pi(\theta)}\left[G_0^{(\mathrm{H})}\right]$$

$$= E_{\pi(\theta)}\left[G_0^{(\mathrm{H})}\nabla \ln\pi(T;\theta) + \nabla G_0^{(\mathrm{H})}\right]$$

$$= E_{\pi(\theta)}\left[G_0^{(\mathrm{H})}\sum_{t=0}^{T-1}\nabla \ln\pi(A_t|S_t;\theta)\right.$$

$$\left. + \sum_{t=0}^{T-1}\gamma^t\alpha^{(\mathrm{H})}E_{\pi(\theta)}\left[-\left(\ln\pi(A_t|S_t;\theta)+1\right)\nabla \ln\pi(A_t|S_t;\theta)\right]\right]$$

$$= E_{\pi(\theta)}\left[\sum_{t=0}^{T-1}\left(G_0^{(\mathrm{H})} - \gamma^t\alpha^{(\mathrm{H})}\left(\ln\pi(A_t|S_t;\theta)+1\right)\right)\nabla \ln\pi(A_t|S_t;\theta)\right].$$

Introducing the baseline, we get

$$\nabla E_{\pi(\theta)}\left[G_0^{(\mathrm{H})}\right] = E_{\pi(\theta)}\left[\sum_{t=0}^{T-1}\gamma^t\left(G_t^{(\mathrm{H})} - \alpha^{(\mathrm{H})}\left(\ln\pi(A_t|S_t;\theta)+1\right)\right)\nabla \ln\pi(A_t|S_t;\theta)\right].$$

Proof completes.

Proof method 2: recursive method. Considering the gradient of the following three formulas with respect to the policy parameter θ:

$$v_{\pi(\theta)}^{(\mathrm{H})}(s) = \sum_a \pi(a|s;\theta)q_{\pi(\theta)}^{(\mathrm{H})}(s,a), \qquad\qquad s \in \mathcal{S}$$

$$q_{\pi(\theta)}^{(\mathrm{H})}(s,a) = r^{(\mathrm{H})}(s,a;\theta) + \gamma\sum_{s'}p(s'|s,a)v_{\pi(\theta)}^{(\mathrm{H})}(s'), \quad s \in \mathcal{S}, a \in \mathcal{A}(s)$$

$$r^{(\mathrm{H})}(s,a;\theta) = r(s,a) + \alpha^{(\mathrm{H})}\mathrm{H}\left[\pi(\cdot|s;\theta)\right], \qquad\qquad s \in \mathcal{S}, a \in \mathcal{A}(s)$$

we have

$$\nabla v_{\pi(\theta)}^{(\mathrm{H})}(s)$$

$$= \nabla\sum_a \pi(a|s;\theta)q_{\pi(\theta)}^{(\mathrm{H})}(s,a)$$

$$= \sum_a q_{\pi(\theta)}^{(\mathrm{H})}(s,a)\nabla\pi(a|s;\theta) + \sum_a \pi(a|s;\theta)\nabla q_{\pi(\theta)}^{(\mathrm{H})}(s,a)$$

$$= \sum_a q_{\pi(\theta)}^{(\mathrm{H})}(s,a)\nabla\pi(a|s;\theta)$$

$$\quad + \sum_a \pi(a|s;\theta)\nabla\left(r^{(\mathrm{H})}(s,a;\theta) + \gamma\sum_{s'}p(s'|s,a)v_{\pi(\theta)}^{(\mathrm{H})}(s')\right)$$

$$= \sum_a q_{\pi(\theta)}^{(\mathrm{H})}(s,a)\nabla\pi(a|s;\theta)$$

$$+ \sum_a \pi(a|s; \boldsymbol{\theta}) \left(\nabla \left(\alpha^{(H)} H[\pi(\cdot|s; \boldsymbol{\theta})] \right) + \gamma \sum_{s'} p(s'|s, a) v_{\pi(\boldsymbol{\theta})}^{(H)}(s') \right)$$

$$= \sum_a q_{\pi(\boldsymbol{\theta})}^{(H)}(s, a) \nabla \pi(a|s; \boldsymbol{\theta}) + \nabla \left(\alpha^{(H)} H[\pi(\cdot|s; \boldsymbol{\theta})] \right)$$

$$+ \gamma \sum_a \pi(a|s; \boldsymbol{\theta}) \sum_{s'} p(s'|s, a) \nabla v_{\pi(\boldsymbol{\theta})}^{(H)}(s')$$

$$= \sum_a q_{\pi(\boldsymbol{\theta})}^{(H)}(s, a) \nabla \pi(a|s; \boldsymbol{\theta}) - \alpha^{(H)} \sum_a (\ln \pi(a|s; \boldsymbol{\theta}) + 1) \nabla \pi(a|s; \boldsymbol{\theta})$$

$$+ \gamma \sum_{s'} p_{\pi(\boldsymbol{\theta})}(s'|s) \nabla v_{\pi(\boldsymbol{\theta})}^{(H)}(s')$$

$$= \sum_a \left[q_{\pi(\boldsymbol{\theta})}^{(H)}(s, a) - \alpha^{(H)} (\ln \pi(a|s; \boldsymbol{\theta}) + 1) \right] \nabla \pi(a|s; \boldsymbol{\theta})$$

$$+ \gamma \sum_{s'} p_{\pi(\boldsymbol{\theta})}(s'|s) \nabla v_{\pi(\boldsymbol{\theta})}^{(H)}(s')$$

$$= \sum_a \pi(a|s; \boldsymbol{\theta}) \left[q_{\pi(\boldsymbol{\theta})}^{(H)}(s, a) - \alpha^{(H)} (\ln \pi(a|s; \boldsymbol{\theta}) + 1) \right] \nabla \ln \pi(a|s; \boldsymbol{\theta})$$

$$+ \gamma \sum_{s'} p_{\pi(\boldsymbol{\theta})}(s'|s) \nabla v_{\pi(\boldsymbol{\theta})}^{(H)}(s')$$

$$= E_{A \sim \pi(s; \boldsymbol{\theta})} \left[q_{\pi(\boldsymbol{\theta})}^{(H)}(s, a) - \alpha^{(H)} (\ln \pi(a|s; \boldsymbol{\theta}) + 1) \right]$$

$$+ \gamma \sum_{s'} p_{\pi(\boldsymbol{\theta})}(s'|s) \nabla v_{\pi(\boldsymbol{\theta})}^{(H)}(s')$$

Taking the expectation of the above equation over the policy $\pi(\boldsymbol{\theta})$ leads to the recursive form

$$E_{\pi(\boldsymbol{\theta})} \left[\nabla v_{\pi(\boldsymbol{\theta})}^{(H)}(S_t) \right]$$

$$= E_{\pi(\boldsymbol{\theta})} \left[\left[q_{\pi(\boldsymbol{\theta})}^{(H)}(S_t, A_t) - \alpha^{(H)} (\ln \pi(A_t|S_t; \boldsymbol{\theta}) + 1) \right] \nabla \ln \pi(A_t|S_t; \boldsymbol{\theta}) \right]$$

$$+ \gamma E_{\pi(\boldsymbol{\theta})} \left[\nabla v_{\pi(\boldsymbol{\theta})}^{(H)}(S_{t+1}) \right].$$

Therefore,

$$E_{\pi(\boldsymbol{\theta})} \left[\nabla v_{\pi(\boldsymbol{\theta})}^{(H)}(S_0) \right]$$

$$= \sum_{t=0}^{+\infty} \gamma^t E_{\pi(\boldsymbol{\theta})} \left[\left[q_{\pi(\boldsymbol{\theta})}^{(H)}(S_t, A_t) - \alpha^{(H)} (\ln \pi(A_t|S_t; \boldsymbol{\theta}) + 1) \right] \nabla \ln \pi(A_t|S_t; \boldsymbol{\theta}) \right].$$

So

$$\nabla E_{\pi(\boldsymbol{\theta})} \left[G_0^{(H)} \right]$$

$$= \mathrm{E}_{\pi(\boldsymbol{\theta})} \left[\nabla v_{\pi(\boldsymbol{\theta})}^{(\mathrm{H})} (S_0) \right]$$

$$= \sum_{t=0}^{+\infty} \gamma^t \mathrm{E}_{\pi(\boldsymbol{\theta})} \left[\left(q_{\pi(\boldsymbol{\theta})}^{(\mathrm{H})} (S_t, A_t) - \alpha^{(\mathrm{H})} \left(\ln \pi(A_t|S_t; \boldsymbol{\theta}) + 1 \right) \right) \nabla \ln \pi(A_t|S_t; \boldsymbol{\theta}) \right]$$

$$= \mathrm{E}_{\pi(\boldsymbol{\theta})} \left[\sum_{t=0}^{+\infty} P q_{\pi(\boldsymbol{\theta})}^{(\mathrm{H})} (S_t, A_t) - \alpha^{(\mathrm{H})} \left(\ln \pi(A_t|S_t; \boldsymbol{\theta}) + 1 \right) \nabla \ln \pi(A_t|S_t; \boldsymbol{\theta}) \right].$$

Use a baseline that is independent of the policy π, and then we have

$$\nabla \mathrm{E}_{\pi(\boldsymbol{\theta})} \left[G_0^{(\mathrm{H})} \right]$$

$$= \mathrm{E}_{\pi(\boldsymbol{\theta})} \left[\sum_{t=0}^{+\infty} \gamma^t \left(q_{\pi(\boldsymbol{\theta})}^{(\mathrm{soft})} (S_t, A_t) - \alpha^{(\mathrm{H})} \left(\ln \pi(A_t|S_t; \boldsymbol{\theta}) + 1 \right) \right) \nabla \ln \pi(A_t|S_t; \boldsymbol{\theta}) \right].$$

Using different baselines, the policy gradient can also be expressed as

$$\nabla \mathrm{E}_{\pi(\boldsymbol{\theta})} \left[G_0^{(\mathrm{H})} \right]$$

$$= \mathrm{E}_{\pi(\boldsymbol{\theta})} \left[\sum_{t=0}^{T-1} \left(G_t^{(\mathrm{H})} - \alpha^{(\mathrm{H})} \left(\ln \pi(A_t|S_t; \boldsymbol{\theta}) + b(S_t) \right) \right) \nabla \ln \pi(A_t|S_t; \boldsymbol{\theta}) \right]$$

$$= \mathrm{E}_{\pi(\boldsymbol{\theta})} \left[\sum_{t=0}^{T-1} \left(G_t^{(\mathrm{soft})} - \alpha^{(\mathrm{H})} \left(\ln \pi(A_t|S_t; \boldsymbol{\theta}) + b(S_t) \right) \right) \nabla \ln \pi(A_t|S_t; \boldsymbol{\theta}) \right].$$

Now we have learned the theoretical foundation of maximum entropy RL.

10.2 Soft RL Algorithms

This section will introduce maximum-entropy RL algorithms, including Soft Q Learning algorithm and Soft Actor–Critic algorithm.

10.2.1 SQL: Soft Q Learning

This section introduces the **Soft Q Learning (SQL)** algorithm.

Recap that Q learning algorithm (see Algo. 6.9 in Sect. 6.4) updates action value estimates using

$$\mathbf{w} \leftarrow \mathbf{w} + \alpha \left[U - q(S, A; \mathbf{w}) \right] \nabla q(S, A; \mathbf{w}),$$

where TD return is

$$U \leftarrow R + \gamma(1 - D') \max_a q\left(S', a; \mathbf{w}_{\text{target}}\right).$$

The TD return is due to the relationship between the optimal state values and the optimal action values $v(S'; \mathbf{w}) = \max_a q(S', a; \mathbf{w})$. For maximum-entropy RL, the relationship between soft action values and soft state values are

$$v\left(S'; \mathbf{w}\right) = \alpha^{(\text{H})} \operatorname*{logsumexp}_a \frac{1}{\alpha^{(\text{H})}} q(S', a; \mathbf{w}).$$

Therefore, we modify the TD return to

$$U \leftarrow R + \gamma\left(1 - D'\right) \alpha^{(\text{H})} \operatorname*{logsumexp}_a \frac{1}{\alpha^{(\text{H})}} q\left(S', a; \mathbf{w}_{\text{target}}\right)$$

and use this TD return to replace the update in Q learning. Note that, in the iteration, q stores the estimate of soft action values rather than the raw action values.

Replacing the TD return on Algo. 6.9 results in the SQL algorithm (Algo. 10.1).

Algorithm 10.1   SQL.

Input: environment (without mathematical model).
Output: optimal action value estimates $q(s, a; \mathbf{w})$ ($s \in \mathcal{S}, a \in \mathcal{A}$). We can use these optimal action value estimates to generate optimal policy estimates if needed.
Parameters: parameters of DQN, and weight of entropy $\alpha^{(\text{H})}$.

1. Initialize:

   1.1. (Initialize parameters) Initialize the parameter of the evaluation network $\mathbf{w}$. Initialize the parameters of the target network $\mathbf{w}_{\text{target}}$.
   1.2. (Initialize experience storage) $\mathcal{D} \leftarrow \varnothing$.

2. For each episode:

   2.1. (Initialize state) Choose the initial state S.
   2.2. Loop until the episode ends:
       2.2.1. (Collect experiences) Do the following once or multiple times:
           2.2.1.1. (Decide) Use the policy derived from $q(S, \cdot; \mathbf{w})$ (for example, ε-greedy policy) to determine the action A.
           2.2.1.2. (Sample) Execute the action A, and then observe the reward R, the next state S', and the indicator of episode end D'.
           2.2.1.3. Save the experience (S, A, R, S', D') in the storage $\mathcal{D}$.
           2.2.1.4. $S \leftarrow S'$.
       2.2.2. (Use experiences) Do the following once or multiple times:

2.2.2.1. (Replay) Sample a batch of experience $\mathcal{B}$ from the storage $\mathcal{D}$. Each entry is in the form of (S, A, R, S', D').

2.2.2.2. (Calculate TD returns) $U \leftarrow R + \gamma(1 - D')\alpha^{(\text{H})} \text{logsumexp}_a \frac{1}{\alpha^{(\text{H})}} q\left(S', a; \mathbf{w}_{\text{target}}\right)$ $((S, A, R, S', D') \in \mathcal{B})$.

2.2.2.3. (Update action value parameter) Update $\mathbf{w}$ to reduce $\frac{1}{|\mathcal{B}|} \sum_{(S,A,R,S',D') \in \mathcal{B}} \left[U - q(S, A; \mathbf{w})\right]^2$. (For example, $\mathbf{w} \leftarrow \mathbf{w} + \alpha \frac{1}{|\mathcal{B}|} \sum_{(S,A,R,S',D') \in \mathcal{B}} \left[U - q(S, A; \mathbf{w})\right] \nabla q(S, A; \mathbf{w})$.)

2.2.2.4. (Update target network) Under some condition (say, every several updates), update the parameters of target network $\mathbf{w}_{\text{target}} \leftarrow \left(1 - \alpha_{\text{target}}\right) \mathbf{w}_{\text{target}} + \alpha_{\text{target}} \mathbf{w}$.

10.2.2 SAC: Soft Actor–Critic

Soft Actor–Critic (SAC) algorithm is the actor–critic counterpart for soft learning, usually with experience replay. It was proposed by (Haarnoja, 2018: SAC1).

In order to make the training more stable, SAC algorithm uses different parametric functions to approximate optimal action values and optimal state values, while action values use the double learning trick and state values use the target network trick.

Let us look into the approximation of action values. SAC uses neural networks to approximate the soft action values $q_\pi^{(\text{soft})}$. Moreover, it uses double learning, who uses two parametric functions $q\left(\mathbf{w}^{(0)}\right)$ and $q\left(\mathbf{w}^{(1)}\right)$ of the same form. Previously we have known double Q learning can help reduce the maximum bias. Tabular double Q learning uses two suites of action value estimates $q\left(\mathbf{w}^{(0)}\right)$ and $q\left(\mathbf{w}^{(1)}\right)$, where one suite is used for estimating optimal actions (i.e. $A' = \arg\max_a q^{(0)}(S', a)$), and another one is used for calculating TD return (such as $q^{(1)}(S', A')$). In double DQN, since there are already two neural networks with parameters $\mathbf{w}$ and $\mathbf{w}_{\text{target}}$ respectively, we can use one parameter $\mathbf{w}$ to calculate optimal action (i.e. $A' = \arg\max_a q(S', a; \mathbf{w})$), and use the parameters of the target network $\mathbf{w}_{\text{target}}$ to estimate target (such as $q\left(S', A'; \mathbf{w}_{\text{target}}\right)$). However, these methods can not be directly applied to actor–critic algorithms, since actors have designated the actions. A usual way to reduce the maximum bias for actor–critic algorithms is as follows: We still maintain two parameters of action value networks $\mathbf{w}^{(i)}$ ($i = 0, 1$). When calculating return, we use the minimum values between the outputs of these two networks, i.e. $\min_{i=0,1} q\left(\cdot, \cdot; \mathbf{w}^{(i)}\right)$. General actor–critic algorithms can use this trick. Particularly, SAC algorithm uses this trick.

Next, we consider the approximation for state values. SAC algorithm uses the target network trick that uses two parameters $\mathbf{w}^{(v)}$ and $\mathbf{w}^{(v)}_{\text{target}}$ for evaluation net and target network respectively. The learning rate of the target network is denoted as α_{target}.

Therefore, approximation for the values uses four networks, whose parameters are $\mathbf{w}^{(0)}$, $\mathbf{w}^{(1)}$, $\mathbf{w}^{(v)}$, and $\mathbf{w}^{(v)}_{\text{target}}$. The soft action values $q_\pi^{(\text{soft})}(\cdot, \cdot)$ are approximated by $q\left(\cdot, \cdot; \mathbf{w}^{(i)}\right)$ ($i = 0, 1$), and the soft state values $v_\pi^{(\text{soft})}(\cdot)$ are approximated by $v\left(\cdot, \mathbf{w}^{(v)}\right)$. Their update targets are as follows:

- When updating $q\left(\cdot, \cdot; \mathbf{w}^{(i)}\right)$ ($i = 0, 1$), try to minimize

$$\mathrm{E}_\mathcal{D}\left[\left(q\left(S, A; \mathbf{w}^{(i)}\right) - U_t^{(q)}\right)^2\right],$$

where the TD return $U_t^{(q)} = R_{t+1} + \gamma v\left(S'; \mathbf{w}^{(v)}_{\text{target}}\right)$.

- When updating $v\left(\cdot; \mathbf{w}^{(v)}\right)$, try to minimize

$$\mathrm{E}_{S \sim \mathcal{D}}\left[\left(v\left(S; \mathbf{w}^{(v)}\right) - U_t^{(v)}\right)^2\right],$$

where the target

$$U_t^{(v)} = \mathrm{E}_{A' \sim \pi(\cdot|S;\boldsymbol{\theta})}\left[\min_{i=0,1} q\left(S, A'; \mathbf{w}^{(i)}\right)\right] + \alpha^{(\mathrm{H})}\mathrm{H}[\pi(\cdot|S; \boldsymbol{\theta})]$$

$$= \mathrm{E}_{A' \sim \pi(\cdot|S;\boldsymbol{\theta})}\left[\min_{i=0,1} q\left(S, A'; \mathbf{w}^{(i)}\right) - \alpha^{(\mathrm{H})}\ln\pi\left(A'|S; \boldsymbol{\theta}\right)\right].$$

SAC only uses one network $\pi(\boldsymbol{\theta})$ to approximate policy. It tries to maximize

$$\mathrm{E}_{A' \sim \pi(\cdot|S;\boldsymbol{\theta})}\left[q\left(S, A'; \mathbf{w}^{(0)}\right)\right] + \alpha^{(\mathrm{H})}\mathrm{H}[\pi(\cdot|S; \boldsymbol{\theta})]$$

$$= \mathrm{E}_{A' \sim \pi(\cdot|S;\boldsymbol{\theta})}\left[q\left(S, A'; \mathbf{w}^{(0)}\right) - \alpha^{(\mathrm{H})}\ln\pi\left(A'|S; \boldsymbol{\theta}\right)\right]$$

to update the policy parameter $\boldsymbol{\theta}$.

Using the above analysis, we can get the SAC algorithm in Algo. 10.2.

Algorithm 10.2   SAC.

Input: environment (without mathematical model).
Output: optimal policy estimates $\pi(\boldsymbol{\theta})$.

Parameters: optimizers, discount factor γ, parameters that control the episode and steps, learning rate of the target network α_{target}, and weight of entropy $\alpha^{(\mathrm{H})}$.

1. Initialize:

   1.1. (Initialize parameters) Initialize parameter $\boldsymbol{\theta}$, and initialize parameter $\mathbf{w}^{(0)}$ and $\mathbf{w}^{(v)}$. Set $\mathbf{w}^{(1)} \leftarrow \mathbf{w}^{(0)}$, and $\mathbf{w}^{(v)}_{\text{target}} \leftarrow \mathbf{w}^{(v)}$.

   1.2. (Initialize experience storage) $\mathcal{D} \leftarrow \varnothing$.

2. For each episode:

   2.1. (Initialize state) Choose the initial state S.

   2.2. Loop until the episode ends:

   2.2.1. (Collect experiences) Do the following once or multiple times:

   2.2.1.1. (Decide) Use the policy $\pi(\cdot|S; \boldsymbol{\theta})$ to determine the action A.

   2.2.1.2. (Sample) Execute the action A, and then observe the reward R, the next state S', and the indicator of episode end D'.

   2.2.1.3. Save the experience (S, A, R, S', D') in the storage $\mathcal{D}$.

   2.2.1.4. $S \leftarrow S'$.

   2.2.2. (Use experiences) Do the following once or multiple times:

   2.2.2.1. (Replay) Sample a batch of experience $\mathcal{B}$ from the storage $\mathcal{D}$. Each entry is in the form of (S, A, R, S', D').

   2.2.2.2. (Calculate TD returns) $U^{(q)} \leftarrow R + \gamma(1 - D')v\left(S'; \mathbf{w}^{(v)}_{\text{target}}\right)$ $((S, A, R, S', D') \in \mathcal{B})$,

   $$U^{(v)} \leftarrow \mathrm{E}_{A' \sim \pi(\cdot|S;\boldsymbol{\theta})} \left[\min_{i=0,1} q\left(S, A'; \mathbf{w}^{(i)}\right) - \alpha^{(\mathrm{H})} \ln \pi(A'|S; \boldsymbol{\theta}) \right] ((S, A, R, S', D') \in \mathcal{B}).$$

   2.2.2.3. (Update action value parameter) Update $\mathbf{w}^{(i)}$ ($i = 0, 1$) to reduce

   $$\frac{1}{|\mathcal{B}|} \sum_{(S,A,R,S',D') \in \mathcal{B}} \left[U^{(q)} - q\left(S, A; \mathbf{w}^{(i)}\right) \right]^2.$$

   and update $\mathbf{w}^{(v)}$ to reduce

   $$\frac{1}{|\mathcal{B}|} \sum_{(S,A,R,S',D') \in \mathcal{B}} \left[U^{(v)} - v\left(S; \mathbf{w}^{(v)}\right) \right]^2.$$

2.2.2.4. (Update policy parameter) Update $\boldsymbol{\theta}$ to reduce

$$-\tfrac{1}{|\mathcal{B}|} \Sigma_{(S,A,R,S',D')\in\mathcal{B}} \, \mathrm{E}_{A'\sim\pi(\cdot|S;\boldsymbol{\theta})} \left[q\Big(S, A'; \mathbf{w}^{(0)}\Big) - \right.$$

$$\left. \alpha^{(\mathrm{H})} \ln \pi(A'|S; \boldsymbol{\theta}) \right].$$

2.2.2.5. (Update target network) Under some condition (say, every several updates), update the parameters of target network $\mathbf{w}_{\text{target}} \leftarrow \Big(1 - \alpha_{\text{target}}\Big)\mathbf{w}_{\text{target}} + \alpha_{\text{target}}\mathbf{w}$.

Remarkably, both updating action value parameters and updating policy parameters use the expectation over the action $A' \sim \pi(\cdot|S; \boldsymbol{\theta})$. For discrete action space, this expectation can be calculated using

$$\mathrm{E}_{A'\sim\pi(\cdot|S;\boldsymbol{\theta})} \left[q(S, A'; \mathbf{w}) - \alpha^{(\mathrm{H})} \ln \pi(A'|S; \boldsymbol{\theta}) \right]$$

$$= \sum_{a\in\mathcal{A}(S)} \pi(a|S; \boldsymbol{\theta}) \left[q(S, a; \mathbf{w}) - \alpha^{(\mathrm{H})} \ln \pi(a|S; \boldsymbol{\theta}) \right]$$

Therefore, when the action space is discrete, action value network and policy network usually output vectors, whose length is the number of actions in the action space. For the tasks with continuous action space, we usually assume the policy obeys some parametric distributions, such as Gaussian distribution. We let the policy network outputs the distribution parameters (such as the mean and standard deviation of Gaussian distribution.) The action value network uses state–action pairs as inputs.

10.3 Automatic Entropy Adjustment

The definition of reward entropy uses the parameter $\alpha^{(\mathrm{H})}$ to tradeoff between exploitation and exploration. Large $\alpha^{(\mathrm{H})}$ means the entropy is important, and we care about exploration; while small $\alpha^{(\mathrm{H})}$ means the entropy is not that important, and we care about exploitation. We may care more about exploration at the beginning of learning, so we set a large $\alpha^{(\mathrm{H})}$. Then gradually, we care more about exploitation and reduce $\alpha^{(\mathrm{H})}$.

(Haarnoja, 2018: SAC2) proposed a method called automatically entropy adjustment to determine $\alpha^{(\mathrm{H})}$. It designates a reference value for the entropy component (denoted as $\bar{h}$). If the actual entropy is greater than this reference value, it thinks that $\alpha^{(\mathrm{H})}$ is too large, so it reduces $\alpha^{(\mathrm{H})}$. If the actual entropy is smaller than this reference value, it thinks that $\alpha^{(\mathrm{H})}$ is too small, so it increases $\alpha^{(\mathrm{H})}$. Therefore, we can minimize the loss function in the following form

$$f\Big(\alpha^{(\mathrm{H})}\Big)\Big(\mathrm{E}_S\Big[\mathrm{H}[\pi(\cdot|S)]\Big] - \bar{h}\Big),$$

where $f\left(\alpha^{(\mathrm{H})}\right)$ can be an arbitrary monotonically increasing function over $\alpha^{(\mathrm{H})}$, such as $f\left(\alpha^{(\mathrm{H})}\right) = \alpha^{(\mathrm{H})}$ or $f\left(\alpha^{(\mathrm{H})}\right) = \ln \alpha^{(\mathrm{H})}$.

If the adjustment is implemented using software libraries such as TensorFlow and PyTorch, we usually set the leaf variable in the form of $\ln \alpha^{(\mathrm{H})}$ to ensure $\alpha^{(\mathrm{H})} > 0$. In this case, we usually set $f\left(\alpha^{(\mathrm{H})}\right) = \ln \alpha^{(\mathrm{H})}$ for simplicity, and the update tries to minimize

$$\ln \alpha^{(\mathrm{H})}\left(\mathrm{E}_S\left[\mathrm{H}[\pi(\cdot|S)]\right] - \bar{h}\right).$$

How to select a suitable reference entropy value $\bar{h}$: If the action space is a finite set such as $\{0, 1, \ldots, n-1\}$, the range of entropy is $[0, \ln n]$. In this case, we can use a positive number for $\bar{h}$ such as $\frac{1}{4} \ln n$. If the action space is $\mathbb{R}^n$, the entropy can be either positive or negative, so we can use the reference value $\bar{h} = -n$.

Algorithm 10.3 shows SAC algorithm with automatic entropy adjustment.

Algorithm 10.3   SAC with automatic entropy adjustment.

Input: environment (without mathematical model).
Output: optimal policy estimates $\pi(\boldsymbol{\theta})$.
Parameters: optimizers (including the learning rate for automatic entropy adjustment), entropy reference value $\bar{h}$, et al.

1. Initialize:

   1.1. (Initialize parameters) Initialize parameter $\boldsymbol{\theta}$, and initialize parameter $\mathbf{w}^{(0)}$ and $\mathbf{w}^{(v)}$. Set $\mathbf{w}^{(1)} \leftarrow \mathbf{w}^{(0)}$, and $\mathbf{w}_{\mathrm{target}}^{(v)} \leftarrow \mathbf{w}^{(v)}$.
   1.2. (Initial weight for entropy) $\ln \alpha^{(\mathrm{H})} \leftarrow$ arbitrary value.
   1.3. (Initialize experience storage) $\mathcal{D} \leftarrow \varnothing$.

2. For each episode:

   2.1. (Initialize state) Choose the initial state S.
   2.2. Loop until the episode ends:
      2.2.1. (Collect experiences) Do the following once or multiple times:
         2.2.1.1. (Decide) Use the policy $\pi(\cdot|S; \boldsymbol{\theta})$ to determine the action A.
         2.2.1.2. (Sample) Execute the action A, and then observe the reward R, the next state S', and the indicator of episode end D'.
         2.2.1.3. Save the experience (S, A, R, S', D') in the storage $\mathcal{D}$.
         2.2.1.4. $S \leftarrow S'$.
      2.2.2. (Use experiences) Do the following once or multiple times:

2.2.2.1. (Replay) Sample a batch of experience $\mathcal{B}$ from the storage $\mathcal{D}$. Each entry is in the form of (S, A, R, S', D').

2.2.2.2. (Update entropy parameter) Calculate the average entropy $\bar{H} \leftarrow \frac{1}{|\mathcal{B}|} \sum_{(S,A,R,S',D') \in \mathcal{B}} H[\pi(\cdot|S)]$. Update $\alpha^{(\mathrm{H})}$ to reduce $\ln \alpha^{(\mathrm{H})} (\bar{H} - \bar{h})$.

2.2.2.3. (Calculate TD returns) $U^{(q)} \leftarrow R + \gamma(1 - D')v\left(S'; \mathbf{w}_{\mathrm{target}}^{(v)}\right)$ $((S, A, R, S', D') \in \mathcal{B})$,

$$U^{(v)} \leftarrow \mathrm{E}_{A' \sim \pi(\cdot|S;\boldsymbol{\theta})} \left[\min_{i=0,1} q\left(S, A'; \mathbf{w}^{(i)}\right) - \alpha^{(\mathrm{H})} \ln \pi(A'|S; \boldsymbol{\theta}) \right] \, ((S, A, R, S', D') \in \mathcal{B}).$$

2.2.2.4. (Update action value parameter) Update $\mathbf{w}^{(i)}$ ($i = 0, 1$) to reduce

$$\frac{1}{|\mathcal{B}|} \sum_{(S,A,R,S',D') \in \mathcal{B}} \left[U^{(q)} - q\left(S, A; \mathbf{w}^{(i)}\right) \right]^2.$$

and update $\mathbf{w}^{(q)}$ to reduce

$$\frac{1}{|\mathcal{B}|} \sum_{(S,A,R,S',D') \in \mathcal{B}} \left[U^{(v)} - v\left(S; \mathbf{w}^{(v)}\right) \right]^2.$$

2.2.2.5. (Update policy parameter) Update $\boldsymbol{\theta}$ to reduce

$$-\frac{1}{|\mathcal{B}|} \sum_{(S,A,R,S',D') \in \mathcal{B}} \mathrm{E}_{A' \sim \pi(\cdot|S;\boldsymbol{\theta})} \left[q\left(S, A'; \mathbf{w}^{(0)}\right) - \alpha^{(\mathrm{H})} \ln \pi(A'|S; \boldsymbol{\theta}) \right].$$

2.2.2.6. (Update target network) Under some condition (say, every several updates), update the parameters of target network $\mathbf{w}_{\mathrm{target}} \leftarrow \left(1 - \alpha_{\mathrm{target}}\right) \mathbf{w}_{\mathrm{target}} + \alpha_{\mathrm{target}} \mathbf{w}$.

10.4 Case Study: Lunar Lander

This section considers the problem Lunar Lander, where a spacecraft wants to land a platform on the moon. The Box2D subpackage of Gym implements two environments for this problem: the environment LunarLander-v2 has a finite action space, while the environment LunarLanderContinuous-v2 has a continuous action space. This section will first introduce how to install the subpackage Box2D of Gym, and then introduce how to use these environments. Finally, we use soft RL agents to solve these two tasks.

10.4.1 Install Environment

This subsection considers the installation of subpackage Box2D of Gym. The installation consists of two steps: install SWIG, and then install Python package gym[box2d].

Install SWIG: You can visit the following website to get the installer of SWIG:

```
http://www.swig.org/download.html
```

The URL of installer may be

```
http://prdownloads.sourceforge.net/swig/swigwin-4.2.1.zip
```

The size of installer is about 11MB. Please unzip it into a permanent location, such as %PROGRAMFILE%/swig (this location requires administrator permission), and then add the unzipped directory containing the file swig.exe, such as %PROGRAMFILE%/swig/swigwin-4.2.1, to the path of system variable.

The way to set environment variables in Windows is as follows: Press "Windows+R" to open the "Run" window, and then type sysdm.cpl and press Enter to open "System Properties". Go to the "Advanced" tab and click "Environment Variables", and select "PATH" in it and add the location to it. After setting, re-login Windows to make sure the change is in effect.

After the installation of SWIG, we can execute the following command in Anaconda Prompt as an administrator to install gym[box2d]:

```
pip install gym[box2d]
```

❗ Note

The installation of Box2D may fail if SWIG has not been correctly installed. In this case, install SWIG correctly, and relogin Windows, and retry the pip command will work.

10.4.2 Use Environment

In the Lunar Lander problem, a spacecraft wants to land a platform on the moon. The spacecraft has a left orientation engine, a main engine, and a right orientation engine. For the environment LunarLander-v2, the action space is Discrete(4) and the meaning of actions is shown in Table 10.1. For the environment LunarLanderContinous-v2, the action space is Box(2,), and all components in the action vector are in range $[-1, 1]$. The meaning of actions is shown in Table 10.2.

If the spacecraft crashes, the episode ends with reward -100. If the spacecraft lands successfully and shuts down all engines, the episode ends with reward 200.

Table 10.1   Actions in the task `LunarLander-v2`.

action	description
0	do nothing
1	fire the left orientation engine
2	fire the main engine
3	fire the right orientation engine

Table 10.2   Actions in the task `LunarLanderContinuous-v2`.

action		description
element 0	$[-1, 0]$	shut down the main engine
	$(0, +1]$	fire the main engine
element 1	$[-1, -0.5)$	fire the left orientation engine
	$[-0.5, +0.5]$	shut down the orientation engines
	$(+0.5, +1]$	fire the right orientation engine

• When the engine is used, the absolute value of the component decides the throttle.

Firing the main engine costs −0.3 per step. The spacecraft has two legs. Each leg touching gets the reward +10. If the spacecraft leaves the platform after it landed, it will lose the reward that it has obtained. Each episode has 1000 steps at most.

The task is solved if the average episode reward over 100 successive episodes exceeds 200. We need to land successfully for most cases in order to solve the task.

Codes 10.1 and 10.2 give the closed-form solutions of these two environments.

Code 10.1   Closed-form solution of `LunarLander-v2`.
LunarLander-v2_ClosedForm.ipynb

```
class ClosedFormAgent:
    def __init__(self, _):
        pass

    def reset(self, mode=None):
        pass

    def step(self, observation, reward, terminated):
        x, y, v_x, v_y, angle, v_angle, contact_left, contact_right = \
                observation

        if contact_left or contact_right:  # legs have contact
            f_y = -10. * v_y - 1.
            f_angle = 0.
        else:
            f_y = 5.5 * np.abs(x) - 10. * y - 10. * v_y - 1.
            f_angle = -np.clip(5. * x + 10. * v_x, -4, 4) + 10. * angle \
                    + 20. * v_angle

        if np.abs(f_angle) <= 1 and f_y <= 0:
            action = 0 # do nothing
        elif np.abs(f_angle) < f_y:
            action = 2 # main engine
        elif f_angle < 0.:
            action = 1 # left engine
        else:
```

```
27          action = 3 # right engine
28        return action
29
30    def close(self):
31        pass
32
33
34  agent = ClosedFormAgent(env)
```

Code 10.2   Closed-form solution of `LunarLanderContinuous-v2`
`LunarLanderContinuous-v2_ClosedForm.ipynb`

```
1  class ClosedFormAgent:
2      def __init__(self, _):
3          pass
4
5      def reset(self, mode=None):
6          pass
7
8      def step(self, observation, reward, terminated):
9          x, y, v_x, v_y, angle, v_angle, contact_left, contact_right = \
10                 observation
11
12          if contact_left or contact_right:  # legs have contact
13              f_y = -10. * v_y - 1.
14              f_angle = 0.
15          else:
16              f_y = 5.5 * np.abs(x) - 10. * y - 10. * v_y - 1.
17              f_angle = -np.clip(5. * x + 10. * v_x, -4, 4) + 10. * angle \
18                      + 20. * v_angle
19
20          action = np.array([f_y, f_angle])
21          return action
22
23      def close(self):
24          pass
25
26
27  agent = ClosedFormAgent(env)
```

10.4.3 Use SQL to Solve LunarLander

This section uses SQL to solve `LunarLander-v2`. Codes 10.3 and 10.4 implement this algorithm. Code 10.3 shows the TensorFlow version of the agent class. When its member function `step()` decides the action, it first calculates the soft values divided by `self.alpha`, i.e. `q_div_alpha` and `v_div_alpha`, and then calculates the policy `prob = np.exp(q_div_alpha - v_div_alpha)`. In theory, the sum of all elements of prob should be 1. However, due to the calculation precision, the sum of all elements in prob may slightly differ from 1, which may make the function `np.random.choice()` raise error. Therefore, is used to make sure the sum of all elements of prob equals 1.

Code 10.3   SQL agent (TensorFlow version).

`LunarLander-v2_SQL_tf.ipynb`

```python
class SQLAgent:
    def __init__(self, env):
        self.action_n = env.action_space.n
        self.gamma = 0.99

        self.replayer = DQNReplayer(10000)

        self.alpha = 0.02

        self.evaluate_net = self.build_net(
                input_size=env.observation_space.shape[0],
                hidden_sizes=[64, 64], output_size=self.action_n)
        self.target_net = models.clone_model(self.evaluate_net)

    def build_net(self, input_size, hidden_sizes, output_size):
        model = keras.Sequential()
        for layer, hidden_size in enumerate(hidden_sizes):
            kwargs = dict(input_shape=(input_size,)) if not layer else {}
            model.add(layers.Dense(units=hidden_size,
                    activation=nn.relu, **kwargs))
        model.add(layers.Dense(units=output_size))
        optimizer = optimizers.Adam(0.001)
        model.compile(loss=losses.mse, optimizer=optimizer)
        return model

    def reset(self, mode=None):
        self.mode = mode
        if self.mode == 'train':
            self.trajectory = []
            self.target_net.set_weights(self.evaluate_net.get_weights())

    def step(self, observation, reward, terminated):
        qs = self.evaluate_net.predict(
                observation[np.newaxis], verbose=0)
        q_div_alpha = qs[0] / self.alpha
        v_div_alpha = scipy.special.logsumexp(q_div_alpha)
        prob = np.exp(q_div_alpha - v_div_alpha)
        prob /= prob.sum()  # work around for np.random.choice
        action = np.random.choice(self.action_n, p=prob)
        if self.mode == 'train':
            self.trajectory += [observation, reward, terminated, action]
            if len(self.trajectory) >= 8:
                state, _, _, act, next_state, reward, terminated, _ = \
                        self.trajectory[-8:]
                self.replayer.store(state, act, reward, next_state, terminated)
            if self.replayer.count >= 500:
                self.learn()
        return action

    def close(self):
        pass

    def learn(self):
        # replay
        states, actions, rewards, next_states, terminateds = \
                self.replayer.sample(128)

        # update value net
        next_qs = self.target_net.predict(next_states, verbose=0)
        next_vs = self.alpha * scipy.special.logsumexp(next_qs / self.alpha,
                axis=-1)
        us = rewards + self.gamma * (1. - terminateds) * next_vs
```

```
63            targets = self.evaluate_net.predict(states, verbose=0)
64            targets[np.arange(us.shape[0]), actions] = us
65            self.evaluate_net.fit(states, targets, verbose=0)
66
67
68    agent = SQLAgent(env)
```

Code 10.4   SQL agent (PyTorch version).
LunarLander-v2_SQL_torch.ipynb

```
1    class SQLAgent:
2        def __init__(self, env):
3            self.action_n = env.action_space.n
4            self.gamma = 0.99
5
6            self.replayer = DQNReplayer(10000)
7
8            self.alpha = 0.02
9
10           self.evaluate_net = self.build_net(
11                   input_size=env.observation_space.shape[0],
12                   hidden_sizes=[256, 256], output_size=self.action_n)
13           self.optimizer = optim.Adam(self.evaluate_net.parameters(), lr=3e-4)
14           self.loss = nn.MSELoss()
15
16       def build_net(self, input_size, hidden_sizes, output_size):
17           layers = []
18           for input_size, output_size in zip([input_size,] + hidden_sizes,
19                   hidden_sizes + [output_size,]):
20               layers.append(nn.Linear(input_size, output_size))
21               layers.append(nn.ReLU())
22           layers = layers[:-1]
23           model = nn.Sequential(*layers)
24           return model
25
26       def reset(self, mode=None):
27           self.mode = mode
28           if self.mode == 'train':
29               self.trajectory = []
30               self.target_net = copy.deepcopy(self.evaluate_net)
31
32       def step(self, observation, reward, terminated):
33           state_tensor = torch.as_tensor(observation,
34                   dtype=torch.float).squeeze(0)
35           q_div_alpha_tensor = self.evaluate_net(state_tensor) / self.alpha
36           v_div_alpha_tensor = torch.logsumexp(
37                   q_div_alpha_tensor, dim=-1, keepdim=True)
38           prob_tensor = (q_div_alpha_tensor - v_div_alpha_tensor).exp()
39           action_tensor = distributions.Categorical(prob_tensor).sample()
40           action = action_tensor.item()
41           if self.mode == 'train':
42               self.trajectory += [observation, reward, terminated, action]
43               if len(self.trajectory) >= 8:
44                   state, _, _, act, next_state, reward, terminated, _ = \
45                           self.trajectory[-8:]
46                   self.replayer.store(state, act, reward, next_state, terminated)
47               if self.replayer.count >= 500:
48                   self.learn()
49           return action
50
51       def close(self):
52           pass
53
54       def learn(self):
55           # replay
```

```
56    states, actions, rewards, next_states, terminateds \
57            = self.replayer.sample(128)
58    state_tensor = torch.as_tensor(states, dtype=torch.float)
59    action_tensor = torch.as_tensor(actions, dtype=torch.long)
60    reward_tensor = torch.as_tensor(rewards, dtype=torch.float)
61    next_state_tensor = torch.as_tensor(next_states, dtype=torch.float)
62    terminated_tensor = torch.as_tensor(terminateds, dtype=torch.float)
63
64    # update value net
65    next_q_tensor = self.target_net(next_state_tensor)
66    next_v_tensor = self.alpha * torch.logsumexp(
67            next_q_tensor / self.alpha, dim=-1)
68    target_tensor = reward_tensor + self.gamma * (1. -
69            terminated_tensor) * next_v_tensor
70    pred_tensor = self.evaluate_net(state_tensor)
71    q_tensor = pred_tensor.gather(1, action_tensor.unsqueeze(1)).squeeze(1)
72    loss_tensor = self.loss(q_tensor, target_tensor.detach())
73    self.optimizer.zero_grad()
74    loss_tensor.backward()
75    self.optimizer.step()
76
77
78  agent = SQLAgent(env)
```

Interaction between this agent and the environment uses Code 1.3.

We can observe the change in episode rewards and episode steps during the learning process. In the beginning, the spacecraft can neither fly nor land. In this case, the episode reward is usually within the range of −300 to 100, and each episode usually has hundreds of steps. Then the spacecraft knows how to fly, but it still can not land. In this case, the episode step usually reaches the upper bound 1000, but the episode reward is still negative. Finally, the spacecraft can both fly and land. The number of steps of each episode becomes hundreds of steps again, and the episode reward exceeds 200.

10.4.4 Use SAC to Solve LunarLander

Codes 10.5 and 10.6 implement SAC algorithm. These implementations fix the weight of entropy $\alpha^{(H)}$ as 0.02. Code 10.5 is the TensorFlow version, and it uses Keras APIs to construct neural networks. In order to train the actor network using Keras API, the function sac_loss() is defined and passed as loss. The function sac_loss() has two parameters. The first parameter provides action value estimates $q\left(S, a; \mathbf{w}^{(0)}\right)$ ($a \in \mathcal{A}$), and the second parameter provides optimal policy estimate $\pi(a|S; \boldsymbol{\theta})$ ($a \in \mathcal{A}$). The function sac_loss() then calculates $\sum_a\left[\alpha^{(H)}\pi(a|S; \boldsymbol{\theta}) \ln \pi(a|S; \boldsymbol{\theta}) - q\left(S, a; \mathbf{w}^{(0)}\right)\pi(a|S; \boldsymbol{\theta})\right]$, which is exactly the loss for this sample entry $-E_{A' \sim \pi(\cdot|S;\boldsymbol{\theta})}\left[q\left(S, A'; \mathbf{w}^{(0)}\right) - \alpha^{(H)} \ln \pi(A'|S; \boldsymbol{\theta})\right]$.

Code 10.5 SAC agent (TensorFlow version).
LunarLander-v2_SACwoA_tf.ipynb

```
class SACAgent:
    def __init__(self, env):
        self.action_n = env.action_space.n
        self.gamma = 0.99

        self.replayer = DQNReplayer(100000)

        self.alpha = 0.02

        # create actor
        def sac_loss(y_true, y_pred):
            """
            y_true is Q(*, action_n),
            y_pred is pi(*, action_n)
            """
            qs = self.alpha * tf.math.xlogy(y_pred, y_pred) - y_pred * y_true
            return tf.reduce_sum(qs, axis=-1)
        self.actor_net = self.build_net(hidden_sizes=[256, 256],
                output_size=self.action_n, output_activation=nn.softmax,
                loss=sac_loss)

        # create Q critic
        self.q0_net = self.build_net(hidden_sizes=[256, 256],
                output_size=self.action_n)
        self.q1_net = self.build_net(hidden_sizes=[256, 256],
                output_size=self.action_n)

        # create V critic
        self.v_evaluate_net = self.build_net(hidden_sizes=[256, 256])
        self.v_target_net = models.clone_model(self.v_evaluate_net)

    def build_net(self, hidden_sizes, output_size=1, activation=nn.relu,
            output_activation=None, loss=losses.mse, learning_rate=0.0003):
        model = keras.Sequential()
        for hidden_size in hidden_sizes:
            model.add(layers.Dense(units=hidden_size, activation=activation))
        model.add(layers.Dense(units=output_size,
                activation=output_activation))
        optimizer = optimizers.Adam(learning_rate)
        model.compile(optimizer=optimizer, loss=loss)
        return model

    def reset(self, mode=None):
        self.mode = mode
        if self.mode == 'train':
            self.trajectory = []

    def step(self, observation, reward, terminated):
        probs = self.actor_net.predict(observation[np.newaxis], verbose=0)[0]
        action = np.random.choice(self.action_n, p=probs)
        if self.mode == 'train':
            self.trajectory += [observation, reward, terminated, action]
            if len(self.trajectory) >= 8:
                state, _, _, action, next_state, reward, terminated, _ = \
                        self.trajectory[-8:]
                self.replayer.store(state, action, reward, next_state,
                        terminated)
                if self.replayer.count >= 500:
                    self.learn()
        return action

    def close(self):
```

```
63              pass
64
65          def update_net(self, target_net, evaluate_net, learning_rate=0.005):
66              average_weights = [(1. - learning_rate) * t + learning_rate * e for
67                      t, e in zip(target_net.get_weights(),
68                      evaluate_net.get_weights())]
69              target_net.set_weights(average_weights)
70
71          def learn(self):
72              states, actions, rewards, next_states, terminateds = \
73                      self.replayer.sample(128)
74
75              # update actor
76              q0s = self.q0_net.predict(states, verbose=0)
77              q1s = self.q1_net.predict(states, verbose=0)
78              self.actor_net.fit(states, q0s, verbose=0)
79
80              # update V critic
81              q01s = np.minimum(q0s, q1s)
82              pis = self.actor_net.predict(states, verbose=0)
83              entropic_q01s = pis * q01s - self.alpha * scipy.special.xlogy(pis, pis)
84              v_targets = entropic_q01s.sum(axis=-1)
85              self.v_evaluate_net.fit(states, v_targets, verbose=0)
86
87              # update Q critic
88              next_vs = self.v_target_net.predict(next_states, verbose=0)
89              q_targets = rewards[:, np.newaxis] + self.gamma * (1. -
90                      terminateds[:, np.newaxis]) * next_vs
91              np.put_along_axis(q0s, actions.reshape(-1, 1), q_targets, -1)
92              np.put_along_axis(q1s, actions.reshape(-1, 1), q_targets, -1)
93              self.q0_net.fit(states, q0s, verbose=0)
94              self.q1_net.fit(states, q1s, verbose=0)
95
96              # update v network
97              self.update_net(self.v_target_net, self.v_evaluate_net)
98
99
100  agent = SACAgent(env)
```

Code 10.6   SAC agent (PyTorch version).
LunarLander-v2_SACwoA_torch.ipynb

```
1   class SACAgent:
2       def __init__(self, env):
3           state_dim = env.observation_space.shape[0]
4           self.action_n = env.action_space.n
5           self.gamma = 0.99
6           self.replayer = DQNReplayer(10000)
7
8           self.alpha = 0.02
9
10          # create actor
11          self.actor_net = self.build_net(input_size=state_dim,
12                  hidden_sizes=[256, 256], output_size=self.action_n,
13                  output_activator=nn.Softmax(-1))
14          self.actor_optimizer = optim.Adam(self.actor_net.parameters(),
15                  lr=3e-4)
16
17          # create V critic
18          self.v_evaluate_net = self.build_net(input_size=state_dim,
19                  hidden_sizes=[256, 256])
20          self.v_target_net = copy.deepcopy(self.v_evaluate_net)
21          self.v_optimizer = optim.Adam(self.v_evaluate_net.parameters(),
22                  lr=3e-4)
23          self.v_loss = nn.MSELoss()
```

```
24
25        # create Q critic
26        self.q0_net = self.build_net(input_size=state_dim,
27                hidden_sizes=[256, 256], output_size=self.action_n)
28        self.q1_net = self.build_net(input_size=state_dim,
29                hidden_sizes=[256, 256], output_size=self.action_n)
30        self.q0_loss = nn.MSELoss()
31        self.q1_loss = nn.MSELoss()
32        self.q0_optimizer = optim.Adam(self.q0_net.parameters(), lr=3e-4)
33        self.q1_optimizer = optim.Adam(self.q1_net.parameters(), lr=3e-4)
34
35    def build_net(self, input_size, hidden_sizes, output_size=1,
36            output_activator=None):
37        layers = []
38        for input_size, output_size in zip([input_size,] + hidden_sizes,
39                hidden_sizes + [output_size,]):
40            layers.append(nn.Linear(input_size, output_size))
41            layers.append(nn.ReLU())
42        layers = layers[:-1]
43        if output_activator:
44            layers.append(output_activator)
45        net = nn.Sequential(*layers)
46        return net
47
48    def reset(self, mode=None):
49        self.mode = mode
50        if self.mode == 'train':
51            self.trajectory = []
52
53    def step(self, observation, reward, terminated):
54        state_tensor = torch.as_tensor(observation,
55                dtype=torch.float).unsqueeze(0)
56        prob_tensor = self.actor_net(state_tensor)
57        action_tensor = distributions.Categorical(prob_tensor).sample()
58        action = action_tensor.numpy()[0]
59        if self.mode == 'train':
60            self.trajectory += [observation, reward, terminated, action]
61            if len(self.trajectory) >= 8:
62                state, _, _, action, next_state, reward, terminated, _ = \
63                        self.trajectory[-8:]
64                self.replayer.store(state, action, reward, next_state,
65                        terminated)
66            if self.replayer.count >= 500:
67                self.learn()
68        return action
69
70    def close(self):
71        pass
72
73    def update_net(self, target_net, evaluate_net, learning_rate=0.0025):
74        for target_param, evaluate_param in zip(
75                target_net.parameters(),
76                evaluate_net.parameters()):
77            target_param.data.copy_(learning_rate * evaluate_param.data +
78                    (1 - learning_rate) * target_param.data)
79
80    def learn(self):
81        states, actions, rewards, next_states, terminateds = \
82                self.replayer.sample(128)
83        state_tensor = torch.as_tensor(states, dtype=torch.float)
84        action_tensor = torch.as_tensor(actions, dtype=torch.long)
85        reward_tensor = torch.as_tensor(rewards, dtype=torch.float)
86        next_state_tensor = torch.as_tensor(next_states, dtype=torch.float)
87        terminated_tensor = torch.as_tensor(terminateds, dtype=torch.float)
88
89        # update Q critic
90        next_v_tensor = self.v_target_net(next_state_tensor)
```

```
 91      q_target_tensor = reward_tensor.unsqueeze(1) + self.gamma * \
 92              (1. - terminated_tensor.unsqueeze(1)) * next_v_tensor
 93
 94      all_q0_pred_tensor = self.q0_net(state_tensor)
 95      q0_pred_tensor = torch.gather(all_q0_pred_tensor, 1,
 96              action_tensor.unsqueeze(1))
 97      q0_loss_tensor = self.q0_loss(q0_pred_tensor, q_target_tensor.detach())
 98      self.q0_optimizer.zero_grad()
 99      q0_loss_tensor.backward()
100      self.q0_optimizer.step()
101
102      all_q1_pred_tensor = self.q1_net(state_tensor)
103      q1_pred_tensor = torch.gather(all_q1_pred_tensor, 1,
104              action_tensor.unsqueeze(1))
105      q1_loss_tensor = self.q1_loss(q1_pred_tensor, q_target_tensor.detach())
106      self.q1_optimizer.zero_grad()
107      q1_loss_tensor.backward()
108      self.q1_optimizer.step()
109
110      # update V critic
111      q0_tensor = self.q0_net(state_tensor)
112      q1_tensor = self.q1_net(state_tensor)
113      q01_tensor = torch.min(q0_tensor, q1_tensor)
114      prob_tensor = self.actor_net(state_tensor)
115      ln_prob_tensor = torch.log(prob_tensor.clamp(1e-6, 1.))
116      entropic_q01_tensor = prob_tensor * (q01_tensor - self.alpha *
117              ln_prob_tensor)
118      # OR entropic_q01_tensor = prob_tensor * (q01_tensor - self.alpha * \
119      #         torch.xlogy(prob_tensor, prob_tensor)
120      v_target_tensor = torch.sum(entropic_q01_tensor, dim=-1, keepdim=True)
121      v_pred_tensor = self.v_evaluate_net(state_tensor)
122      v_loss_tensor = self.v_loss(v_pred_tensor, v_target_tensor.detach())
123      self.v_optimizer.zero_grad()
124      v_loss_tensor.backward()
125      self.v_optimizer.step()
126
127      self.update_net(self.v_target_net, self.v_evaluate_net)
128
129      # update actor
130      prob_q_tensor = prob_tensor * (self.alpha * ln_prob_tensor - q0_tensor)
131      actor_loss_tensor = prob_q_tensor.sum(axis=-1).mean()
132      self.actor_optimizer.zero_grad()
133      actor_loss_tensor.backward()
134      self.actor_optimizer.step()
135
136
137  agent = SACAgent(env)
```

Interaction between this agent and the environment once again uses Code 1.3.

10.4.5 Use Automatic Entropy Adjustment to Solve LunarLander

This section considers automatic entropy adjustment.

Codes 10.7 and 10.8 implement the SAC with automatic entropy adjustment with TensorFlow and PyTorch respectively. The constructor of the class SACAgent defines the training variable self.ln_alpha_tensor for $\ln \alpha^{(H)}$, an Adam optimizer to train this variable, and the reference entropy value $\bar{h} = \frac{1}{4} \ln|\mathcal{A}|$.

Code 10.7   SAC with automatic entropy adjustment (TensorFlow version).
LunarLander-v2_SACwA_tf.ipynb

```python
class SACAgent:
    def __init__(self, env):
        state_dim = env.observation_space.shape[0]
        self.action_n = env.action_space.n
        self.gamma = 0.99

        self.replayer = DQNReplayer(100000)

        # create alpha
        self.target_entropy = np.log(self.action_n) / 4.
        self.ln_alpha_tensor = tf.Variable(0., dtype=tf.float32)
        self.alpha_optimizer = optimizers.Adam(0.0003)

        # create actor
        self.actor_net = self.build_net(hidden_sizes=[256, 256],
                output_size=self.action_n, output_activation=nn.softmax)

        # create Q critic
        self.q0_net = self.build_net(hidden_sizes=[256, 256],
                output_size=self.action_n)
        self.q1_net = self.build_net(hidden_sizes=[256, 256],
                output_size=self.action_n)

        # create V critic
        self.v_evaluate_net = self.build_net(input_size=state_dim,
                hidden_sizes=[256, 256])
        self.v_target_net = models.clone_model(self.v_evaluate_net)

    def build_net(self, hidden_sizes, output_size=1, activation=nn.relu,
            output_activation=None, input_size=None, loss=losses.mse,
            learning_rate=0.0003):
        model = keras.Sequential()
        for layer_idx, hidden_size in enumerate(hidden_sizes):
            kwargs = {'input_shape': (input_size,)} if layer_idx == 0 \
                    and input_size is not None else {}
            model.add(layers.Dense(units=hidden_size,
                    activation=activation, **kwargs))
        model.add(layers.Dense(units=output_size,
                activation=output_activation))
        optimizer = optimizers.Adam(learning_rate)
        model.compile(optimizer=optimizer, loss=loss)
        return model

    def reset(self, mode=None):
        self.mode = mode
        if self.mode == 'train':
            self.trajectory = []

    def step(self, observation, reward, terminated):
        probs = self.actor_net.predict(
                observation[np.newaxis], verbose=0)[0]
        action = np.random.choice(self.action_n, p=probs)
        if self.mode == 'train':
            self.trajectory += [observation, reward,
                    terminated, action]
            if len(self.trajectory) >= 8:
                state, _, _, action, next_state, reward, terminated, _ = \
                        self.trajectory[-8:]
                self.replayer.store(state, action, reward, next_state,
                        terminated)
            if self.replayer.count >= 500:
                self.learn()
```

```
63          return action
64
65      def close(self):
66          pass
67
68      def update_net(self, target_net, evaluate_net,
69              learning_rate=0.005):
70          average_weights = [(1. - learning_rate) * t + learning_rate * e for
71                  t, e in zip(target_net.get_weights(),
72                  evaluate_net.get_weights())]
73          target_net.set_weights(average_weights)
74
75      def learn(self):
76          states, actions, rewards, next_states, terminateds = \
77                  self.replayer.sample(128)
78
79          # update alpha
80          all_probs = self.actor_net.predict(states, verbose=0)
81          probs = np.take_along_axis(all_probs, actions[np.newaxis, :], axis=-1)
82          ln_probs = np.log(probs.clip(1e-6, 1.))
83          mean_ln_prob = ln_probs.mean()
84          with tf.GradientTape() as tape:
85              alpha_loss_tensor = -self.ln_alpha_tensor \
86                      * (mean_ln_prob + self.target_entropy)
87          grads = tape.gradient(alpha_loss_tensor, [self.ln_alpha_tensor,])
88          self.alpha_optimizer.apply_gradients(zip(grads,
89                  [self.ln_alpha_tensor,]))
90
91          # update V critic
92          q0s = self.q0_net.predict(states, verbose=0)
93          q1s = self.q1_net.predict(states, verbose=0)
94          q01s = np.minimum(q0s, q1s)
95          pis = self.actor_net.predict(states, verbose=0)
96          alpha = tf.exp(self.ln_alpha_tensor).numpy()
97          entropic_q01s = pis * q01s - alpha * scipy.special.xlogy(pis, pis)
98          v_targets = entropic_q01s.sum(axis=-1)
99          self.v_evaluate_net.fit(states, v_targets, verbose=0)
100         self.update_net(self.v_target_net, self.v_evaluate_net)
101
102         # update Q critic
103         next_vs = self.v_target_net.predict(next_states, verbose=0)
104         q_targets = rewards[:, np.newaxis] + self.gamma * (1. -
105                 terminateds[:, np.newaxis]) * next_vs
106         np.put_along_axis(q0s, actions.reshape(-1, 1), q_targets, -1)
107         np.put_along_axis(q1s, actions.reshape(-1, 1), q_targets, -1)
108         self.q0_net.fit(states, q0s, verbose=0)
109         self.q1_net.fit(states, q1s, verbose=0)
110
111         # update actor
112         state_tensor = tf.convert_to_tensor(states, dtype=tf.float32)
113         q0s_tensor = self.q0_net(state_tensor)
114         with tf.GradientTape() as tape:
115             probs_tensor = self.actor_net(state_tensor)
116             alpha_tensor = tf.exp(self.ln_alpha_tensor)
117             losses_tensor = alpha_tensor * tf.math.xlogy(
118                     probs_tensor, probs_tensor) - probs_tensor * q0s_tensor
119             actor_loss_tensor = tf.reduce_sum(losses_tensor, axis=-1)
120         grads = tape.gradient(actor_loss_tensor,
121                 self.actor_net.trainable_variables)
122         self.actor_net.optimizer.apply_gradients(zip(grads,
123                 self.actor_net.trainable_variables))
124
125
126 agent = SACAgent(env)
```

Code 10.8   SAC with automatic entropy adjustment (PyTorch version).

LunarLander-v2_SACwA_torch.ipynb

```python
class SACAgent:
    def __init__(self, env):
        state_dim = env.observation_space.shape[0]
        self.action_n = env.action_space.n
        self.gamma = 0.99

        self.replayer = DQNReplayer(10000)

        # create alpha
        self.target_entropy = np.log(self.action_n) / 4.
        self.ln_alpha_tensor = torch.zeros(1, requires_grad=True)
        self.alpha_optimizer = optim.Adam([self.ln_alpha_tensor,], lr=3e-4)

        # create actor
        self.actor_net = self.build_net(input_size=state_dim,
                hidden_sizes=[256, 256], output_size=self.action_n,
                output_activator=nn.Softmax(-1))
        self.actor_optimizer = optim.Adam(self.actor_net.parameters(), lr=3e-4)

        # create V critic
        self.v_evaluate_net = self.build_net(input_size=state_dim,
                hidden_sizes=[256, 256])
        self.v_target_net = copy.deepcopy(self.v_evaluate_net)
        self.v_optimizer = optim.Adam(self.v_evaluate_net.parameters(),
                lr=3e-4)
        self.v_loss = nn.MSELoss()

        # create Q critic
        self.q0_net = self.build_net(input_size=state_dim,
                hidden_sizes=[256, 256], output_size=self.action_n)
        self.q1_net = self.build_net(input_size=state_dim,
                hidden_sizes=[256, 256], output_size=self.action_n)
        self.q0_loss = nn.MSELoss()
        self.q1_loss = nn.MSELoss()
        self.q0_optimizer = optim.Adam(self.q0_net.parameters(), lr=3e-4)
        self.q1_optimizer = optim.Adam(self.q1_net.parameters(), lr=3e-4)

    def build_net(self, input_size, hidden_sizes, output_size=1,
            output_activator=None):
        layers = []
        for input_size, output_size in zip([input_size,] + hidden_sizes,
                hidden_sizes + [output_size,]):
            layers.append(nn.Linear(input_size, output_size))
            layers.append(nn.ReLU())
        layers = layers[:-1]
        if output_activator:
            layers.append(output_activator)
        net = nn.Sequential(*layers)
        return net

    def reset(self, mode=None):
        self.mode = mode
        if self.mode == 'train':
            self.trajectory = []

    def step(self, observation, reward, terminated):
        state_tensor = torch.as_tensor(observation,
                dtype=torch.float).unsqueeze(0)
        prob_tensor = self.actor_net(state_tensor)
        action_tensor = distributions.Categorical(prob_tensor).sample()
        action = action_tensor.numpy()[0]
        if self.mode == 'train':
```

```
63        self.trajectory += [observation, reward, terminated, action]
64        if len(self.trajectory) >= 8:
65            state, _, _, action, next_state, reward, terminated, _ = \
66                    self.trajectory[-8:]
67            self.replayer.store(state, action, reward, next_state,
68                    terminated)
69            if self.replayer.count >= 500:
70                self.learn()
71    return action
72
73    def close(self):
74        pass
75
76    def update_net(self, target_net, evaluate_net,
77            learning_rate=0.0025):
78        for target_param, evaluate_param in zip(
79                target_net.parameters(), evaluate_net.parameters()):
80            target_param.data.copy_(learning_rate * evaluate_param.data
81                    + (1 - learning_rate) * target_param.data)
82
83    def learn(self):
84        states, actions, rewards, next_states, terminateds = \
85                self.replayer.sample(128)
86        state_tensor = torch.as_tensor(states, dtype=torch.float)
87        action_tensor = torch.as_tensor(actions, dtype=torch.long)
88        reward_tensor = torch.as_tensor(rewards, dtype=torch.float)
89        next_state_tensor = torch.as_tensor(next_states, dtype=torch.float)
90        terminated_tensor = torch.as_tensor(terminateds, dtype=torch.float)
91
92        # update alpha
93        prob_tensor = self.actor_net(state_tensor)
94        ln_prob_tensor = torch.log(prob_tensor.clamp(1e-6, 1))
95        neg_entropy_tensor = (prob_tensor * ln_prob_tensor).sum()
96        # OR neg_entropy_tensor = torch.xlogy(prob_tensor, prob_tensor).sum()
97        grad_tensor = neg_entropy_tensor + self.target_entropy
98        alpha_loss_tensor = -self.ln_alpha_tensor * grad_tensor.detach()
99        self.alpha_optimizer.zero_grad()
100       alpha_loss_tensor.backward()
101       self.alpha_optimizer.step()
102
103       # update Q critic
104       next_v_tensor = self.v_target_net(next_state_tensor)
105       q_target_tensor = reward_tensor.unsqueeze(1) + self.gamma * (1. -
106               terminated_tensor.unsqueeze(1)) * next_v_tensor
107
108       all_q0_pred_tensor = self.q0_net(state_tensor)
109       q0_pred_tensor = torch.gather(all_q0_pred_tensor, 1,
110               action_tensor.unsqueeze(1))
111       q0_loss_tensor = self.q0_loss(q0_pred_tensor,
112               q_target_tensor.detach())
113       self.q0_optimizer.zero_grad()
114       q0_loss_tensor.backward()
115       self.q0_optimizer.step()
116
117       all_q1_pred_tensor = self.q1_net(state_tensor)
118       q1_pred_tensor = torch.gather(all_q1_pred_tensor, 1,
119               action_tensor.unsqueeze(1))
120       q1_loss_tensor = self.q1_loss(q1_pred_tensor,
121               q_target_tensor.detach())
122       self.q1_optimizer.zero_grad()
123       q1_loss_tensor.backward()
124       self.q1_optimizer.step()
125
126       # update V critic
127       q0_tensor = self.q0_net(state_tensor)
128       q1_tensor = self.q1_net(state_tensor)
129       q01_tensor = torch.min(q0_tensor, q1_tensor)
```

```
130    prob_tensor = self.actor_net(state_tensor)
131    ln_prob_tensor = torch.log(prob_tensor.clamp(1e-6, 1.))
132    alpha = self.ln_alpha_tensor.exp().detach().item()
133    entropic_q01_tensor = prob_tensor * (q01_tensor -
134            alpha * ln_prob_tensor)
135    # OR entropic_q01_tensor = prob_tensor * (q01_tensor −
136    #          alpha * torch.xlogy(prob_tensor, prob_tensor)
137    v_target_tensor = torch.sum(entropic_q01_tensor, dim=-1, keepdim=True)
138    v_pred_tensor = self.v_evaluate_net(state_tensor)
139    v_loss_tensor = self.v_loss(v_pred_tensor, v_target_tensor.detach())
140    self.v_optimizer.zero_grad()
141    v_loss_tensor.backward()
142    self.v_optimizer.step()
143
144    self.update_net(self.v_target_net, self.v_evaluate_net)
145
146    # update actor
147    prob_q_tensor = prob_tensor * (alpha * ln_prob_tensor - q0_tensor)
148    actor_loss_tensor = prob_q_tensor.sum(axis=-1).mean()
149    self.actor_optimizer.zero_grad()
150    actor_loss_tensor.backward()
151    self.actor_optimizer.step()
152
153
154 agent = SACAgent(env)
```

Interaction between this agent and the environment once again uses Code 1.3.

10.4.6 Solve LunarLanderContinuous

This section considers the environment LunarLanderContinuous-v2, the version with continuous action space.

Codes 10.9 and 10.10 use the SAC with automatic entropy adjustment to solve LunarLanderContinuous-v2. On policy approximation, it uses the Gaussian distribution with independent variables to approximate optimal policy, and the actor network outputs the mean vector and logarithm standard deviation vector. Then we use the Gaussian distribution with the output mean and standard deviation to generate actions. On the automatic entropy adjustment, the reference entropy value is $\bar{h} = -\ln \dim \mathcal{A}$, where $\dim \mathcal{A}$ is the dimension of the action space $\mathcal{A}$.

Code 10.9 SAC with automatic entropy adjustment for continuous action space (TensorFlow version).
LunarLanderContinuous-v2_SACwA_tf.ipynb

```
1  class SACAgent:
2      def __init__(self, env):
3          state_dim = env.observation_space.shape[0]
4          action_dim = env.action_space.shape[0]
5          self.action_low = env.action_space.low
6          self.action_high = env.action_space.high
7          self.gamma = 0.99
8
9          self.replayer = DQNReplayer(100000)
10
11         # create alpha
12         self.target_entropy = -action_dim
13         self.ln_alpha_tensor = tf.Variable(0., dtype=tf.float32)
```

```
14          self.alpha_optimizer = optimizers.Adam(3e-4)
15
16          # create actor
17          self.actor_net = self.build_net(input_size=state_dim,
18                  hidden_sizes=[256, 256], output_size=action_dim*2,
19                  output_activation=tf.tanh)
20
21          # create V critic
22          self.v_evaluate_net = self.build_net(input_size=state_dim,
23                  hidden_sizes=[256, 256])
24          self.v_target_net = models.clone_model(self.v_evaluate_net)
25
26          # create Q critic
27          self.q0_net = self.build_net(input_size=state_dim+action_dim,
28                  hidden_sizes=[256, 256])
29          self.q1_net = self.build_net(input_size=state_dim+action_dim,
30                  hidden_sizes=[256, 256])
31
32      def build_net(self, input_size, hidden_sizes, output_size=1,
33              activation=nn.relu, output_activation=None, loss=losses.mse,
34              learning_rate=3e-4):
35          model = keras.Sequential()
36          for layer, hidden_size in enumerate(hidden_sizes):
37              kwargs = {'input_shape' : (input_size,)} \
38                      if layer == 0 else {}
39              model.add(layers.Dense(units=hidden_size,
40                      activation=activation, **kwargs))
41          model.add(layers.Dense(units=output_size,
42                  activation=output_activation))
43          optimizer = optimizers.Adam(learning_rate)
44          model.compile(optimizer=optimizer, loss=loss)
45          return model
46
47      def get_action_ln_prob_tensors(self, state_tensor):
48          mean_ln_std_tensor = self.actor_net(state_tensor)
49          mean_tensor, ln_std_tensor = tf.split(mean_ln_std_tensor, 2, axis=-1)
50          if self.mode == 'train':
51              std_tensor = tf.math.exp(ln_std_tensor)
52              normal_dist = distributions.Normal(mean_tensor, std_tensor)
53              sample_tensor = normal_dist.sample()
54              action_tensor = tf.tanh(sample_tensor)
55              ln_prob_tensor = normal_dist.log_prob(sample_tensor) - \
56                      tf.math.log1p(1e-6 - tf.pow(action_tensor, 2))
57              ln_prob_tensor = tf.reduce_sum(ln_prob_tensor, axis=-1,
58                      keepdims=True)
59          else:
60              action_tensor = tf.tanh(mean_tensor)
61              ln_prob_tensor = tf.ones_like(action_tensor)
62          return action_tensor, ln_prob_tensor
63
64      def reset(self, mode):
65          self.mode = mode
66          if self.mode == 'train':
67              self.trajectory = []
68
69      def step(self, observation, reward, terminated):
70          if self.mode == 'train' and self.replayer.count < 5000:
71              action = np.random.uniform(self.action_low, self.action_high)
72          else:
73              state_tensor = tf.convert_to_tensor(observation[np.newaxis, :],
74                      dtype=tf.float32)
75              action_tensor, _ = self.get_action_ln_prob_tensors(state_tensor)
76              action = action_tensor[0].numpy()
77          if self.mode == 'train':
78              self.trajectory += [observation, reward, terminated, action]
79              if len(self.trajectory) >= 8:
80                  state, _, _, act, next_state, reward, terminated, _ = \
```

```
81                              self.trajectory[-8:]
82                  self.replayer.store(state, act, reward, next_state, terminated)
83              if self.replayer.count >= 120:
84                  self.learn()
85          return action
86
87      def close(self):
88          pass
89
90      def update_net(self, target_net, evaluate_net, learning_rate=0.005):
91          average_weights = [(1. - learning_rate) * t + learning_rate * e for
92                  t, e in zip(target_net.get_weights(),
93                  evaluate_net.get_weights())]
94          target_net.set_weights(average_weights)
95
96      def learn(self):
97          states, actions, rewards, next_states, terminateds = \
98                  self.replayer.sample(128)
99          state_tensor = tf.convert_to_tensor(states, dtype=tf.float32)
100
101         # update alpha
102         act_tensor, ln_prob_tensor = \
103                 self.get_action_ln_prob_tensors(state_tensor)
104         with tf.GradientTape() as tape:
105             alpha_loss_tensor = -self.ln_alpha_tensor * (tf.reduce_mean(
106                 ln_prob_tensor, axis=-1) + self.target_entropy)
107         grads = tape.gradient(alpha_loss_tensor, [self.ln_alpha_tensor,])
108         self.alpha_optimizer.apply_gradients(zip(grads,
109                 [self.ln_alpha_tensor,]))
110
111         # update Q critic
112         state_actions = np.concatenate((states, actions), axis=-1)
113         next_vs = self.v_target_net.predict(next_states, verbose=0)
114         q_targets = rewards[:, np.newaxis] + self.gamma * (1. -
115                 terminateds[:, np.newaxis]) * next_vs
116         self.q0_net.fit(state_actions, q_targets, verbose=False)
117         self.q1_net.fit(state_actions, q_targets, verbose=False)
118
119         # update V critic
120         state_act_tensor = tf.concat((state_tensor, act_tensor), axis=-1)
121         q0_pred_tensor = self.q0_net(state_act_tensor)
122         q1_pred_tensor = self.q1_net(state_act_tensor)
123         q_pred_tensor = tf.minimum(q0_pred_tensor, q1_pred_tensor)
124         alpha_tensor = tf.exp(self.ln_alpha_tensor)
125         v_target_tensor = q_pred_tensor - alpha_tensor * ln_prob_tensor
126         v_targets = v_target_tensor.numpy()
127         self.v_evaluate_net.fit(states, v_targets, verbose=False)
128         self.update_net(self.v_target_net, self.v_evaluate_net)
129
130         # update actor
131         with tf.GradientTape() as tape:
132             act_tensor, ln_prob_tensor = \
133                     self.get_action_ln_prob_tensors(state_tensor)
134             state_act_tensor = tf.concat((state_tensor, act_tensor), axis=-1)
135             q0_pred_tensor = self.q0_net(state_act_tensor)
136             alpha_tensor = tf.exp(self.ln_alpha_tensor)
137             actor_loss_tensor = tf.reduce_mean(
138                     alpha_tensor * ln_prob_tensor - q0_pred_tensor)
139         grads = tape.gradient(actor_loss_tensor,
140                 self.actor_net.trainable_variables)
141         self.actor_net.optimizer.apply_gradients(zip(grads,
142                 self.actor_net.trainable_variables))
143
144
145 agent = SACAgent(env)
```

Code 10.10   SAC with automatic entropy adjustment for continuous action space (PyTorch version).
LunarLanderContinuous-v2_SACwA_torch.ipynb

```
1   class SACAgent:
2       def __init__(self, env):
3           state_dim = env.observation_space.shape[0]
4           self.action_dim = env.action_space.shape[0]
5           self.action_low = env.action_space.low
6           self.action_high = env.action_space.high
7           self.gamma = 0.99
8
9           self.replayer = DQNReplayer(100000)
10
11          # create alpha
12          self.target_entropy = -self.action_dim
13          self.ln_alpha_tensor = torch.zeros(1, requires_grad=True)
14          self.alpha_optimizer = optim.Adam([self.ln_alpha_tensor,], lr=0.0003)
15
16          # create actor
17          self.actor_net = self.build_net(input_size=state_dim,
18                  hidden_sizes=[256, 256], output_size=self.action_dim*2,
19                  output_activator=nn.Tanh())
20          self.actor_optimizier = optim.Adam(self.actor_net.parameters(),
21                  lr=0.0003)
22
23          # create V critic
24          self.v_evaluate_net = self.build_net(input_size=state_dim,
25                  hidden_sizes=[256, 256])
26          self.v_target_net = copy.deepcopy(self.v_evaluate_net)
27          self.v_loss = nn.MSELoss()
28          self.v_optimizer = optim.Adam(self.v_evaluate_net.parameters(),
29                  lr=0.0003)
30
31          # create Q critic
32          self.q0_net = self.build_net(input_size=state_dim+self.action_dim,
33                  hidden_sizes=[256, 256])
34          self.q1_net = self.build_net(input_size=state_dim+self.action_dim,
35                  hidden_sizes=[256, 256])
36          self.q0_loss = nn.MSELoss()
37          self.q1_loss = nn.MSELoss()
38          self.q0_optimizer = optim.Adam(self.q0_net.parameters(), lr=0.0003)
39          self.q1_optimizer = optim.Adam(self.q1_net.parameters(), lr=0.0003)
40
41      def build_net(self, input_size, hidden_sizes, output_size=1,
42              output_activator=None):
43          layers = []
44          for input_size, output_size in zip([input_size,] + hidden_sizes,
45                  hidden_sizes + [output_size,]):
46              layers.append(nn.Linear(input_size, output_size))
47              layers.append(nn.ReLU())
48          layers = layers[:-1]
49          if output_activator:
50              layers.append(output_activator)
51          net = nn.Sequential(*layers)
52          return net
53
54      def get_action_ln_prob_tensors(self, state_tensor):
55          mean_ln_std_tensor = self.actor_net(state_tensor)
56          mean_tensor, ln_std_tensor = torch.split(mean_ln_std_tensor,
57                  self.action_dim, dim=-1)
58          if self.mode == 'train':
59              std_tensor = torch.exp(ln_std_tensor)
60              normal_dist = distributions.Normal(mean_tensor, std_tensor)
```

```
61          rsample_tensor = normal_dist.rsample()
62          action_tensor = torch.tanh(rsample_tensor)
63          ln_prob_tensor = normal_dist.log_prob(rsample_tensor) - \
64                  torch.log1p(1e-6 - action_tensor.pow(2))
65          ln_prob_tensor = ln_prob_tensor.sum(-1, keepdim=True)
66        else:
67          action_tensor = torch.tanh(mean_tensor)
68          ln_prob_tensor = torch.ones_like(action_tensor)
69        return action_tensor, ln_prob_tensor
70
71    def reset(self, mode):
72        self.mode = mode
73        if self.mode == 'train':
74            self.trajectory = []
75
76    def step(self, observation, reward, terminated):
77        if self.mode == 'train' and self.replayer.count < 5000:
78            action = np.random.uniform(self.action_low, self.action_high)
79        else:
80            state_tensor = torch.as_tensor(observation,
81                    dtype=torch.float).unsqueeze(0)
82            action_tensor, _ = self.get_action_ln_prob_tensors(state_tensor)
83            action = action_tensor[0].detach().numpy()
84        if self.mode == 'train':
85            self.trajectory += [observation, reward, terminated, action]
86            if len(self.trajectory) >= 8:
87                state, _, _, act, next_state, reward, terminated, _ = \
88                        self.trajectory[-8:]
89                self.replayer.store(state, act, reward, next_state, terminated)
90            if self.replayer.count >= 128:
91                self.learn()
92        return action
93
94    def close(self):
95        pass
96
97    def update_net(self, target_net, evaluate_net, learning_rate=0.005):
98        for target_param, evaluate_param in zip(target_net.parameters(),
99                evaluate_net.parameters()):
100            target_param.data.copy_(learning_rate * evaluate_param.data +
101                    (1 - learning_rate) * target_param.data)
102
103    def learn(self):
104        states, actions, rewards, next_states, terminateds = \
105                self.replayer.sample(128)
106        state_tensor = torch.as_tensor(states, dtype=torch.float)
107        action_tensor = torch.as_tensor(actions, dtype=torch.float)
108        reward_tensor = torch.as_tensor(rewards, dtype=torch.float)
109        next_state_tensor = torch.as_tensor(next_states, dtype=torch.float)
110        terminated_tensor = torch.as_tensor(terminateds, dtype=torch.float)
111
112        # update alpha
113        act_tensor, ln_prob_tensor = \
114                self.get_action_ln_prob_tensors(state_tensor)
115        alpha_loss_tensor = (-self.ln_alpha_tensor * (ln_prob_tensor +
116                self.target_entropy).detach()).mean()
117
118        self.alpha_optimizer.zero_grad()
119        alpha_loss_tensor.backward()
120        self.alpha_optimizer.step()
121
122        # update Q critic
123        states_action_tensor = torch.cat((state_tensor, action_tensor), dim=-1)
124        q0_tensor = self.q0_net(states_action_tensor)
125        q1_tensor = self.q1_net(states_action_tensor)
126        next_v_tensor = self.v_target_net(next_state_tensor)
127        q_target = reward_tensor.unsqueeze(1) + self.gamma * \
```

```
128                next_v_tensor * (1. - terminated_tensor.unsqueeze(1))
129            q0_loss_tensor = self.q0_loss(q0_tensor, q_target.detach())
130            q1_loss_tensor = self.q1_loss(q1_tensor, q_target.detach())
131
132            self.q0_optimizer.zero_grad()
133            q0_loss_tensor.backward()
134            self.q0_optimizer.step()
135
136            self.q1_optimizer.zero_grad()
137            q1_loss_tensor.backward()
138            self.q1_optimizer.step()
139
140            # update V critic
141            state_act_tensor = torch.cat((state_tensor, act_tensor), dim=-1)
142            v_pred_tensor = self.v_evaluate_net(state_tensor)
143            q0_pred_tensor = self.q0_net(state_act_tensor)
144            q1_pred_tensor = self.q1_net(state_act_tensor)
145            q_pred_tensor = torch.min(q0_pred_tensor, q1_pred_tensor)
146            alpha_tensor = self.ln_alpha_tensor.exp()
147            v_target_tensor = q_pred_tensor - alpha_tensor * ln_prob_tensor
148            v_loss_tensor = self.v_loss(v_pred_tensor, v_target_tensor.detach())
149
150            self.v_optimizer.zero_grad()
151            v_loss_tensor.backward()
152            self.v_optimizer.step()
153
154            self.update_net(self.v_target_net, self.v_evaluate_net)
155
156            # update actor
157            actor_loss_tensor = (alpha_tensor * ln_prob_tensor
158                    - q0_pred_tensor).mean()
159
160            self.actor_optimizier.zero_grad()
161            actor_loss_tensor.backward()
162            self.actor_optimizier.step()
163
164
165  agent = SACAgent(env)
```

Interaction between this agent and the environment once again uses Code 1.3.

10.5 Summary

- Reward engineering modifies the definition of reward in the original task and trains on the modified task in order to solve the original problem.
- Maximum-entropy RL tries to maximize the linear combination of return and entropy.
- The relationship between soft action values and soft state values is

$$v_\pi^{(\text{soft})}(s) = \alpha^{(\text{H})} \operatorname*{logsumexp}_{a} \left(\frac{1}{\alpha^{(\text{H})}} q_\pi^{(\text{soft})}(s, a) \right), \quad s \in \mathcal{S}.$$

- Soft policy gradient theorem uses the following way to construct the new policy

$$\tilde{\pi}(a|s) = \exp\left(\frac{1}{\alpha^{(\text{H})}} a_\pi^{(\text{soft})}(s, a) \right), \quad s \in \mathcal{S}, a \in \mathcal{A}.$$

- Soft policy gradient theorem

$$\nabla \mathrm{E}_{\pi(\boldsymbol{\theta})}\left[G_0^{(\mathrm{H})}\right] = \mathrm{E}_{\pi(\boldsymbol{\theta})}\left[\sum_{t=0}^{T-1} \gamma^t \left(G_t^{(\mathrm{H})} - \alpha^{(\mathrm{H})} \ln \pi(A_t|S_t;\boldsymbol{\theta})\right) \nabla \ln\pi(A_t|S_t;\boldsymbol{\theta})\right]$$

- Automatic entropy adjustment tries to reduce

$$\ln \alpha^{(\mathrm{H})}\left(\mathrm{E}_S\left[\mathrm{H}[\pi(\cdot|S)]\right] - \bar{h}\right).$$

- The Gym library has sub-packages such as Box2d and Atari. We can install a complete version of Gym.

10.6 Exercises

10.6.1 Multiple Choices

10.1 On the reward with entropy $R_{t+1}^{(\mathrm{H})} = R_{t+1} + \alpha^{(\mathrm{H})}\mathrm{H}[\pi(\cdot|S_t)]$, choose the correct one: (   )

A. Reducing $\alpha^{(\mathrm{H})}$ can increase exploration.
B. Increasing $\alpha^{(\mathrm{H})}$ can increase exploration.
C. Neither reducing nor increasing $\alpha^{(\mathrm{H})}$ can increase exploration.

10.2 On soft values, choose the correct one: (   )

A. $v_\pi^{(\mathrm{soft})}(s) = \mathrm{E}_\pi\left[q_\pi^{(\mathrm{soft})}(s,a)\right]$ $(s \in \mathcal{S})$.
B. $v_\pi^{(\mathrm{soft})}(s) = \max_{a \in \mathcal{A}(s)} q_\pi^{(\mathrm{soft})}(s,a)$ $(s \in \mathcal{S})$.
C. $\frac{v_\pi^{(\mathrm{soft})}(s)}{\alpha^{(\mathrm{H})}} = \mathrm{logsumexp}_{a \in \mathcal{A}(s)} \frac{q_\pi^{(\mathrm{soft})}(s,a)}{\alpha^{(\mathrm{H})}}$ $(s \in \mathcal{S})$.

10.6.2 Programming

10.3 Use SAC algorithm to solve the task `Acrobat-v1`.

10.6.3 Mock Interview

10.4 What is reward engineering?

10.5 Why does maximum-entropy RL consider the reward with entropy?

10.6 Can maximum-entropy RL find the optimal policy of the original problem? Why?

Chapter 11
Policy-Based Gradient-Free Algorithms

This chapter covers

- Evolution Strategy (ES) algorithm
- Augmented Random Search (ARS) algorithm
- comparison between gradient-free algorithms and policy-gradient algorithms

So far, we have used the parametric function $\pi(\boldsymbol{\theta})$ to approximate optimal policy, and try to find suitable policy parameter $\boldsymbol{\theta}$ to maximize the return expectation. The parameter finding process usually changes the policy parameter according to the guidance of gradients, which can be determined by policy gradient theorem or performance difference lemma. In contrast, this chapter introduces a method that finds suitable policy parameters without using gradients. This type of algorithms is called **gradient-free algorithm**.

11.1 Gradient-Free Algorithms

This section will introduce two kinds of gradient-free algorithms.

11.1.1 ES: Evolution Strategy

This subsection introduces **Evolution Strategy (ES)** algorithm (Salimans, 2017).

© The Author(s), under exclusive license to Springer Nature Singapore Pte Ltd. 2024
Z. Xiao, *Reinforcement Learning*, https://doi.org/10.1007/978-981-19-4933-3_11

The idea of ES is as follows: Consider a deterministic policy $\pi(\boldsymbol{\theta})$. We try to update its policy parameter $\boldsymbol{\theta}$ iteratively. Let $\boldsymbol{\theta}_k$ denote the policy parameter before the k-th iteration. In order to improve policy, we choose n random directional parameters $\boldsymbol{\delta}_k^{(0)}, \boldsymbol{\delta}_k^{(1)}, \ldots, \boldsymbol{\delta}_k^{(n-1)}$, whose shapes are all the same as that of the policy parameter $\boldsymbol{\theta}_k$. Then change the policy parameter $\boldsymbol{\theta}_k$ in these n directions, and we can get n new policies $\pi\left(\boldsymbol{\theta}_k + \sigma\boldsymbol{\delta}_k^{(0)}\right), \pi\left(\boldsymbol{\theta}_k + \sigma\boldsymbol{\delta}_k^{(1)}\right), \ldots, \pi\left(\boldsymbol{\theta}_k + \sigma\boldsymbol{\delta}_k^{(n-1)}\right)$, where $\sigma > 0$ is the amplitude of changes. Then new policies interact with the environments, and we get the return estimates for these policies, denoted as $G_k^{(0)}, G_k^{(1)}, \ldots, G_k^{(n-1)}$. If one return estimate, say $G_k^{(i)}$ $(0 \leq i < n)$, is greater than all other return estimates, it indicates that the policy $\pi\left(\boldsymbol{\theta}_k + \sigma\boldsymbol{\delta}_k^{(i)}\right)$ may be smarter than all other policies, and the direction $\boldsymbol{\delta}_k^{(i)}$ is sensible. Therefore, we may consider changing the policy parameter $\boldsymbol{\theta}_k$ in the direction of $\boldsymbol{\delta}_k^{(i)}$. If one return sample, say $G_k^{(i)}$ $(0 \leq i < n)$ is smaller than other return samples, it indicates that the policy $\pi\left(\boldsymbol{\theta}_k + \sigma\boldsymbol{\delta}_k^{(i)}\right)$ may be more stupid than all other policies, and the direction $\boldsymbol{\delta}_k^{(i)}$ is stupid. Therefore, we may consider changing the policy parameter $\boldsymbol{\theta}_k$ in the opposite direction of $\boldsymbol{\delta}_k^{(i)}$. Based on this idea, we can standardize all return estimates $G_k^{(0)}, G_k^{(1)}, \ldots, G_k^{(n-1)}$ to a set of fitness score

$$F_k^{(i)} = \frac{G_k^{(i)} - \text{mean}_{0 \leq j < n} \, G_k^{(j)}}{\text{std}_{0 \leq j < n} \, G_k^{(j)}}, \quad 0 \leq i < n,$$

whose mean is 0, and standard deviation is 1. In this way, some $F_k^{(i)}$ are positive, while some $F_k^{(i)}$ are negative. Then we average $\boldsymbol{\delta}_k^{(i)}$ $(0 \leq i < n)$ with the weight $F_k^{(i)}$, resulting in the weighted average direction $\sum_{i=0}^{n-1} F_k^{(i)} \boldsymbol{\delta}_k^{(i)}$. This direction is likely to improve the policy, so we use this direction to update the policy parameter:

$$\boldsymbol{\theta}_{k+1} = \boldsymbol{\theta}_k + \alpha \sum_{i=0}^{n-1} F_k^{(i)} \boldsymbol{\delta}_k^{(i)}.$$

Algorithm 11.1 shows the ES algorithm.

Algorithm 11.1   ES.

Input: environment (without mathematical model).
Output: optimal policy estimate $\pi(\boldsymbol{\theta})$.
Parameters: number of policies in each iteration n, amplitude of change σ, learning rate α, and parameters that control the end of iterations.

1. (Initialize) Initialize the policy parameter $\boldsymbol{\theta}$.

2. Loop over generations, until the breaking conditions are met (for example, the number of generations reaches the threshold):

    2.1. (Generate policies) For $i = 0, 1, \ldots, n - 1$, randomly select direction $\boldsymbol{\delta}^{(i)} \sim \text{normal}(\mathbf{0}, \mathbf{I})$, and get the policy $\pi\left(\boldsymbol{\theta} + \sigma \boldsymbol{\delta}^{(i)}\right)$.

    2.2. (Evaluate performance of policies) For $i = 0, 1, \ldots, n - 1$, policy $\pi\left(\boldsymbol{\theta} + \sigma \boldsymbol{\delta}^{(i)}\right)$ interacts with the environments, and gets return estimate $G^{(i)}$.

    2.3. (Calculate fitness score) Determine the fitness score $\left\{ F^{(i)} : i = 0, 1, \ldots, n - 1 \right\}$ using $\left\{ G^{(i)} : i = 0, 1, \ldots, n - 1 \right\}$. (For example, $F^{(i)} \leftarrow \frac{G^{(i)} - \text{mean}_j\, G^{(j)}}{\text{std}_j\, G^{(j)}}$.)

    2.4. (Update policy parameter) $\boldsymbol{\theta} \leftarrow \boldsymbol{\theta} + \alpha \sum_i F^{(i)} \boldsymbol{\delta}^{(i)}$.

11.1.2 ARS: Augmented Random Search

ES algorithm has many variants, and **Augmented Random Search** (**ARS**) is a type of algorithms. This section introduces one of ARS algorithms.

The idea of the ARS algorithm (Mania, 2018) is as follows: We once again assume that the policy parameter before the k-th iteration is $\boldsymbol{\theta}_k$, and randomly select n direction parameters $\boldsymbol{\delta}_k^{(0)}, \boldsymbol{\delta}_k^{(1)}, \ldots, \boldsymbol{\delta}_k^{(n-1)}$ that share the same shape with the parameter $\boldsymbol{\theta}_k$. However, we use each direction $\boldsymbol{\delta}_k^{(i)}$ to determine two different policies: $\pi\left(\boldsymbol{\theta}_k + \sigma \boldsymbol{\delta}_k^{(i)}\right)$ and $\pi\left(\boldsymbol{\theta}_k - \sigma \boldsymbol{\delta}_k^{(i)}\right)$. Therefore, there are totally $2n$ policies. Then these $2n$ policies interact with the environment, and get a return estimate for each policy. Let $G_{+,k}^{(i)}$ and $G_{-,k}^{(i)}$ denote the return estimates for the policy $\pi\left(\boldsymbol{\theta}_k + \sigma \boldsymbol{\delta}_k^{(i)}\right)$ and policy $\pi\left(\boldsymbol{\theta}_k - \sigma \boldsymbol{\delta}_k^{(i)}\right)$ respectively. If $G_{+,k}^{(i)}$ is much larger than $G_{-,k}^{(i)}$, the direction $\boldsymbol{\delta}_k^{(i)}$ is probably smart, and we may change the parameter in the direction of $\boldsymbol{\delta}_k^{(i)}$; if $G_{+,k}^{(i)}$ is much smaller than $G_{-,k}^{(i)}$, the direction $\boldsymbol{\delta}_k^{(i)}$ is probably stupid, and we may change the parameter in the opposite direction of $\boldsymbol{\delta}_k^{(i)}$. Therefore, the fitness of the direction $\boldsymbol{\delta}_k^{(i)}$ can be defined as the difference between two return estimates, i.e.

$$F_k^{(i)} = G_{+,k}^{(i)} - G_{-,k}^{(i)}.$$

Algorithm 11.2 shows this version of ARS algorithm.

Algorithm 11.2   ARS.

Input: environment (without mathematical model).
Output: optimal policy estimate $\pi(\boldsymbol{\theta})$.
Parameters: number of policies in each direction n, amplitude of change σ, learning rate α, and parameters that control the end of iterations.

1. (Initialize) Initialize the policy parameter $\boldsymbol{\theta}$.
2. Loop over generations, until the breaking conditions are met (for example, the number of generations reaches the threshold):

    2.1. (Generate policies) For $i = 0, 1, \ldots, n-1$, randomly select direction $\boldsymbol{\delta}^{(i)} \sim \text{normal}(\mathbf{0}, \mathbf{I})$, and get the policy $\pi\left(\boldsymbol{\theta} + \sigma\boldsymbol{\delta}^{(i)}\right)$ and $\pi\left(\boldsymbol{\theta} - \sigma\boldsymbol{\delta}^{(i)}\right)$.

    2.2. (Evaluate performance of policies) For $i = 0, 1, \ldots, n-1$, policy $\pi\left(\boldsymbol{\theta} + \sigma\boldsymbol{\delta}^{(i)}\right)$ interacts with the environments, and get return estimate $G_+^{(i)}$; policy $\pi\left(\boldsymbol{\theta} - \sigma\boldsymbol{\delta}^{(i)}\right)$ interacts with the environments, and get return estimate $G_-^{(i)}$.

    2.3. (Calculate fitness score) Determine the fitness score $\left\{F^{(i)} : i = 0, 1, \ldots, n-1\right\}$ using $\left\{G_\pm^{(i)} : i = 0, 1, \ldots, n-1\right\}$. (For example, $F^{(i)} \leftarrow G_+^{(i)} - G_-^{(i)}$.)

    2.4. (Update policy parameter) $\boldsymbol{\theta} \leftarrow \boldsymbol{\theta} + \alpha \sum_i F^{(i)} \boldsymbol{\delta}^{(i)}$.

11.2 Compare Gradient-Free Algorithms and Policy Gradient Algorithms

Both policy-based gradient-free algorithms and policy-gradient algorithms try to adjust policy parameters to find an optimal policy. This section compares these two categories and gives their advantages and disadvantages.

Since RL can be viewed as an optimization problem, the comparison between the gradient-free algorithms and policy-gradient algorithms is alike the comparison between gradient-free optimization and gradient-based optimization. Nevertheless, we have additional insights for RL tasks.

Gradient-free algorithms have the following advantages:

- Gradient-free algorithms explore better. When gradient-free algorithms generate new policies, the directions of parameter changes are arbitrary, without preference for some particular direction. Therefore, the exploration is more impartial.

- Gradient-free algorithms can cope with parallel computing better. The main resource consumption of gradient-free algorithms is to estimate the return of each policy, which can be computed parallelly by assigning each policy to a different machine. Therefore, gradient-free algorithms are especially suitable for parallel computation.
- Gradient-free algorithms are more robust and less impacted by random seeds.
- Gradient-free algorithms do not involve gradient, and can be used for the scenario that gradients are not available.

Gradient-free algorithms have the following disadvantage:

- Gradient-free algorithms usually have low sample efficiency, so they are not suitable for the environments that are expensive to interact with. The low sample efficiency is because the change direction of policy parameters is quite random and lack of guidance, so many unsuitable policies are generated and evaluated.

11.3 Case Study: BipedalWalker

This section considers the task `BipedalWalker-v3` in the subpackage Box2D, where we try to teach a robot to walk. As in Fig. 1.4, The robot has a hull and two legs. Each leg has two segments. Two hips connect the hull and two legs, while two knees connect the two segments of the legs. We will control the hips and knees to make it walk.

The observation space of this environment is `Box(24,)`, and the meaning of observation is shown in Table 11.1. The action space of this environment is `Box(4,)`, and the meaning of action is shown in Table 11.2.

Table 11.1   Observations in the task `BipedalWalker-v3`.

index	name
0	hull_angle
1	hull_angle_velocity
2	hull_x_velocity
3	hull_y_velocity
4	hip_angle0
5	hip_speed0
6	knee_angle0
7	knee_speed0
8	contact0
9	hip_angle1
10	hip_speed1
11	knee_angle1
12	knee_speed1
13	contact1
14–23	lidar0–lidar9

Table 11.2   Actions in the task `BipedalWalker-v3`.

index	name
0	hip_speed0
1	knee_torque0
2	hip_speed1
3	knee_torque1

When the robot moves forwards, we get positive rewards. We get some little negative rewards when the robot is in a bad shape. If the robot falls to the ground, we get the reward −100, and the episode ends. The threshold of episode reward is 300, and the maximum number of steps is 1600.

Code 11.1 shows a linear-approximated deterministic policy that can solve this task.

Code 11.1   Closed-form solution of `BipedalWalker-v3`.

`BipedalWalker-v3_ClosedForm.ipynb`

```
class ClosedFormAgent:
    def __init__(self, env):
        self.weights = np.array([
            [ 0.9, -0.7,  0.0, -1.4],
            [ 4.3, -1.6, -4.4, -2.0],
            [ 2.4, -4.2, -1.3, -0.1],
            [-3.1, -5.0, -2.0, -3.3],
            [-0.8,  1.4,  1.7,  0.2],
            [-0.7,  0.2, -0.2,  0.1],
            [-0.6, -1.5, -0.6,  0.3],
            [-0.5, -0.3,  0.2,  0.1],
            [ 0.0, -0.1, -0.1,  0.1],
            [ 0.4,  0.8, -1.6, -0.5],
            [-0.4,  0.5, -0.3, -0.4],
            [ 0.3,  2.0,  0.9, -1.6],
            [ 0.0, -0.2,  0.1, -0.3],
            [ 0.1,  0.2, -0.5, -0.3],
            [ 0.7,  0.3,  5.1, -2.4],
            [-0.4, -2.3,  0.3, -4.0],
            [ 0.1, -0.8,  0.3,  2.5],
            [ 0.4, -0.9, -1.8,  0.3],
            [-3.9, -3.5,  2.8,  0.8],
            [ 0.4, -2.8,  0.4,  1.4],
            [-2.2, -2.1, -2.2, -3.2],
            [-2.7, -2.6,  0.3,  0.6],
            [ 2.0,  2.8,  0.0, -0.9],
            [-2.2,  0.6,  4.7, -4.6],
            ])
        self.bias = np.array([3.2, 6.1, -4.0, 7.6])

    def reset(self, mode=None):
        pass

    def step(self, observation, reward, terminated):
        action = np.matmul(observation, self.weights) + self.bias
        return action

    def close(self):
        pass

```

```
42  agent = ClosedFormAgent(env)
```

This remaining of this section will use gradient-free algorithms to solve this task.

11.3.1 Reward Shaping and Reward Clipping

A reward shaping tricked call reward clipped can be used for solving the task BipedalWalker-v3.

Reward shaping is a trick that changes the reward to make the training easier. It modifies the reward to make the learning easier and faster, but it may impact the performance of learned agent since the shaped reward is no longer the ultimate goal. A good reward shaping application should be able to both benefit the learning process and avoid unacceptable performance degradation, and design a good reward shaping trick usually requires the domain knowledge of the task.

For the task BipedalWalker-v3, we will observe the change of episode rewards and episode steps during the training process. If the robot dies soon after the episode starts, the episode reward is about -90. If the episode reward obviously exceeds this value, it means that the robot has learned something. If the robot just learns to walk, the episode reward is about 200. But the robot is not fast enough so that it can not finish before the maximum number of steps. If the robot moves faster, the episode reward can be larger, say 250. The robot needs to be quite fast to finish in order to exceed the episode reward threshold 300.

Accordingly, the training of task BipedalWalker-v3 can use a reward shaping trick called **reward clipping**. This trick bounds the reward in the range of $[-1, +1]$ during training, while testing is not impacted. This trick is not suitable for all tasks, but it is beneficial to the task BipedalWalker-v3. The reasons why the trick is suitable for this task are:

- In order to make the episode reward exceed 300 and solve this task successfully, the robot needs to move forward at a certain speed, and the robot must not fall to the ground. In the original task, the reward exceeds the range $[-1, +1]$ only when the robot falls. Therefore, for the agent that never makes the robot fall to the ground, the reward clipping is invalid. Therefore, the reward clipping does not degrade the final performance of the agent.
- In the original task, we get the reward -100 when the robot falls to the ground. This is a very large penalty, and the agent will always try to avoid it. A possible way to avoid that is to adjust the posture of the robot so that the robot can neither move nor fall. This way can avoid the large penalty, but the stuck posture is not what we want. Therefore, we bound the reward in the range of $[-1, +1]$ to avoid such large penalty in the training, and encourage the robot to move forward.

Reward clipping can be implemented with the help of the class gym.wrappers. TransformReward. The codes to use the class gym.wrappers.TransformReward to wrap are as follows:

```
1  def clip_reward(reward):
2      return np.clip(reward, -1., 1.)
3  reward_clipped_env = gym.wrappers.TransformReward(env, clip_reward)
```

Online Contents

Advanced readers can check the explanation of the class gym.wrappers.
TransformReward in GitHub repo of this book.

11.3.2 ES

Code 11.2 implements the ES algorithm. The policy is linear approximated, and the
policy parameters are weights and bias. The agent class ESAgent has a member
function train(), which accepts the parameter env, an environment with reward
clipping. The member function train() will create lots of new ESAgent objects.

Code 11.2   ES agent.
BipedalWalker-v3_ES.ipynb

```
1   class ESAgent:
2       def __init__(self, env=None, weights=None, bias=None):
3           if weights is not None:
4               self.weights = weights
5           else:
6               self.weights = np.zeros((env.observation_space.shape[0],
7                       env.action_space.shape[0]))
8           if bias is not None:
9               self.bias = bias
10          else:
11              self.bias = np.zeros(env.action_space.shape[0])
12
13      def reset(self, mode=None):
14          pass
15
16      def close(self):
17          pass
18
19      def step(self, observation, reward, terminated):
20          action = np.matmul(observation, self.weights)
21          return action
22
23      def train(self, env, scale=0.05, learning_rate=0.2, population=16):
24          # permulate weights
25          weight_deltas = [scale * np.random.randn(*agent.weights.shape) for _ in
26                  range(population)]
27          bias_deltas = [scale * np.random.randn(*agent.bias.shape) for _ in
28                  range(population)]
29
30          # calculate rewards
31          agents = [ESAgent(weights=self.weights + weight_delta,
32                  bias=self.bias + bias_delta) for weight_delta, bias_delta in
33                  zip(weight_deltas, bias_deltas)]
34          rewards = np.array([play_episode(env, agent)[0] for agent in agents])
35
36          # standardize the rewards
```

```
37      std = rewards.std()
38      if np.isclose(std, 0):
39          coeffs = np.zeros(population)
40      else:
41          coeffs = (rewards - rewards.mean()) / std
42
43      # update weights
44      weight_updates = sum([coeff * weight_delta for coeff, weight_delta in
45              zip(coeffs, weight_deltas)])
46      bias_updates = sum([coeff * bias_delta for coeff, bias_delta in
47              zip(coeffs, bias_deltas)])
48      self.weights += learning_rate * weight_updates / population
49      self.bias += learning_rate * bias_updates / population
50
51
52  agent = ESAgent(env=env)
```

Code 11.3 shows the codes to train and test the agent. It still uses the function play_episode() in Code 1.3, but its training uses reward_clipped_env, an environment object wrapped by the class TransformReward.

Code 11.3 Train and test ES agent.

BipedalWalker-v3_ES.ipynb

```
1   logging.info('==== train & evaluate ====')
2   episode_rewards = []
3   for generation in itertools.count():
4       agent.train(reward_clipped_env)
5       episode_reward, elapsed_steps = play_episode(env, agent)
6       episode_rewards.append(episode_reward)
7       logging.info('evaluate generation %d: reward = %.2f, steps = %d',
8               generation, episode_reward, elapsed_steps)
9       if np.mean(episode_rewards[-10:]) > env.spec.reward_threshold:
10          break
11  plt.plot(episode_rewards)
12
13
14  logging.info('==== test ====')
15  episode_rewards = []
16  for episode in range(100):
17      episode_reward, elapsed_steps = play_episode(env, agent)
18      episode_rewards.append(episode_reward)
19      logging.info('test episode %d: reward = %.2f, steps = %d',
20              episode, episode_reward, elapsed_steps)
21  logging.info('average episode reward = %.2f ± %.2f',
22          np.mean(episode_rewards), np.std(episode_rewards))
```

11.3.3 ARS

Code 11.4 implements one of ARS algorithms. The agent class ARSAgent differs from the class ESAgent only in the member function train(), which still accepts the parameter env, an environment with reward clipping.

Code 11.4   ARS agent.

BipedalWalker-v3_ARS.ipynb

```
 1  class ARSAgent:
 2      def __init__(self, env=None, weights=None, bias=None):
 3          if weights is not None:
 4              self.weights = weights
 5          else:
 6              self.weights = np.zeros((env.observation_space.shape[0],
 7                      env.action_space.shape[0]))
 8          if bias is not None:
 9              self.bias = bias
10          else:
11              self.bias = np.zeros(env.action_space.shape[0])
12
13      def reset(self, mode=None):
14          pass
15
16      def close(self):
17          pass
18
19      def step(self, observation, reward, terminated):
20          action = np.matmul(observation, self.weights)
21          return action
22
23      def train(self, env, scale=0.06, learning_rate=0.09, population=16):
24          weight_updates = np.zeros_like(self.weights)
25          bias_updates = np.zeros_like(self.bias)
26          for _ in range(population):
27              weight_delta = scale * np.random.randn(*agent.weights.shape)
28              bias_delta = scale * np.random.randn(*agent.bias.shape)
29              pos_agent = ARSAgent(weights=self.weights + weight_delta,
30                      bias=self.bias + bias_delta)
31              pos_reward, _ = play_episode(env, pos_agent)
32              neg_agent = ARSAgent(weights=self.weights - weight_delta,
33                      bias=self.bias - bias_delta)
34              neg_reward, _ = play_episode(env, neg_agent)
35              weight_updates += (pos_reward - neg_reward) * weight_delta
36              bias_updates += (pos_reward - neg_reward) * bias_delta
37          self.weights += learning_rate * weight_updates / population
38          self.bias += learning_rate * bias_updates / population
39
40
41  agent = ARSAgent(env=env)
```

11.4 Summary

- In each generation, gradient-free algorithms generate multiple direction
  parameters $\delta^{(i)}$ $(0 \leq i < n)$ and calculate their fitness $F^{(i)}$. Then update the
  policy parameters using

$$\theta \leftarrow \theta + \alpha \sum_i F^{(i)} \delta^{(i)}.$$

- The fitness of Evolution Strategy (ES) algorithm is

$$F^{(i)} = \frac{G^{(i)} - \text{mean}_{0 \le j < n}\, G^{(i)}}{\text{std}_{0 \le j < n}\, G^{(i)}}, \quad 0 \le i < n,$$

where $G^{(i)}$ is the return estimate for the policy $\pi\left(\boldsymbol{\theta} + \sigma\boldsymbol{\theta}^{(i)}\right)$.

- The fitness of Augmented Random Search (ARS) algorithm is

$$F^{(i)} = G_+^{(i)} - G_-^{(i)}, \quad 0 \le i < n,$$

where $G_+^{(i)}$ and $G_-^{(i)}$ are the return estimates for the policy $\pi\left(\boldsymbol{\theta} + \sigma\boldsymbol{\theta}^{(i)}\right)$ and $\pi\left(\boldsymbol{\theta} - \sigma\boldsymbol{\theta}^{(i)}\right)$ respectively.

- Gradient-free algorithms tend to explore better, and are more suitable for parallel computing. They can be used when the gradients are not available.
- Gradient-free algorithms usually have low sample efficiency.

11.5 Exercises

11.5.1 Multiple Choices

11.1 On ES algorithm, choose the correct one: ( )

A. ES algorithm is a value-based algorithm.
B. ES algorithm is a policy-based algorithm.
C. ES algorithm is an actor–critic algorithm.

11.2 On ES algorithm, choose the correct one: ( )

A. ES algorithm generates multiple $\boldsymbol{\delta}^{(i)}$ $(0 \le i < n)$ as the change of policy parameter $\boldsymbol{\theta}$ in each generation.
B. ES algorithm generates multiple $\boldsymbol{\delta}^{(i)}$ $(0 \le i < n)$ as the change of value parameter $\mathbf{w}$ in each generation.
C. ES algorithm generates multiple $\boldsymbol{\delta}_{\boldsymbol{\theta}}^{(i)}$ and $\boldsymbol{\delta}_{\mathbf{w}}^{(i)}$ $(0 \le i < n)$ as the change of policy parameter $\boldsymbol{\theta}$ and value parameter $\mathbf{w}$ in each generation.

11.3 Compared to the policy-based gradient-free algorithm, choose the correct one: ( )

A. Policy-gradient algorithms tend to explore more thoroughly.
B. Policy-gradient algorithms tend to have better sample efficiency.
C. Policy-gradient algorithms are more suitable for parallel computing.

11.5.2 Programming

11.4 Use ES algorithm or ARS algorithm to solve the task `CartPole-v0`. The form of policy is linear policy.

11.5.3 Mock Interview

11.5 What are the advantages and disadvantages of gradient-free algorithms, compared to policy gradient algorithms?

Chapter 12
Distributional RL

This chapter covers

- distributional RL
- maximum utility RL
- distortion function and distorted expectation
- Categorical Deep Q Network (C51) algorithm
- Quantile Regression Deep Q Network (QR-DQN) algorithm
- Implicit Quantile Network (IQN) algorithm

This chapter considers a variant of optimal-value RL: distributional RL.

Chapter 2 told us that the return on the condition of state or state–action pair is a random variable, and value is the expectation of the random variable. Optimal-value algorithms try to maximize the values. However, in some tasks, only expectations do not suffice for considering the task in its entirety, and the entire distribution can help make the decision smarter. Specifically, some tasks not only want to maximize the expectation of the episode rewards, but also want to optimize the utility or statistics risk that is determined by the entire distribution (for example, try to minimize the standard deviation). In those cases, considering the entire distribution has advantages.

12.1 Value Distribution and its Properties

Given the policy π, define the state value random variable and action value random variable as conditional return:

© The Author(s), under exclusive license to Springer Nature Singapore Pte Ltd. 2024
Z. Xiao, *Reinforcement Learning*, https://doi.org/10.1007/978-981-19-4933-3_12

$$V_\pi(s) \overset{d}{=} [G_t | S_t = s; \pi], \qquad\qquad s \in \mathcal{S}$$

$$Q_\pi(s, a) \overset{d}{=} [G_t | S_t = s, A_t = a; \pi], \quad s \in \mathcal{S}, a \in \mathcal{A}(s),$$

where $X \overset{d}{=} Y$ means two random variables X and Y share the same distribution. The relationship between value random variables and the values is as follows.

- The relationship between state values and state value random variables:

$$v_\pi(s) = E_\pi[V_\pi(s)], \quad s \in \mathcal{S}.$$

- The relationship between action values and action value random variables:

$$q_\pi(s, a) = E_\pi[Q_\pi(s, a)], \quad s \in \mathcal{S}, a \in \mathcal{A}(s).$$

Since optimal-value algorithms mainly consider action value, the distributional RL algorithms in this chapter mainly consider action value random variables.

The Bellman equation to use action value random variables to back up action value random variables is:

$$Q_\pi(S_t, A_t) \overset{d}{=} R_{t+1} + \gamma Q_\pi(S_{t+1}, A_{t+1}).$$

Accordingly, we can define the Bellman operator of the policy π as $\mathfrak{B}_\pi : Q \mapsto \mathfrak{B}_\pi(Q)$:

$$\mathfrak{B}_\pi(Q)(s, a) \overset{def}{=} [R_{t+1} + \gamma Q(S_{t+1}, A_{t+1}) | S_t = s, A_t = a], \quad s \in \mathcal{S}, a \in \mathcal{A}(s).$$

This operator is a contraction mapping in the max-form of Wasserstein metric. In the sequel, we will introduce the Wasserstein metric, and prove this operator is a contraction mapping.

Interdisciplinary Reference 12.1
Probability Theory: Quantile Function

The quantile function of a random variable X is defined as

$$\phi_X(\omega) \overset{def}{=} \inf \{x \in \mathcal{X} : \omega \le \Pr[X \le x]\}, \quad \omega \in [0, 1].$$

Let $\Omega \sim$ uniform $[0, 1]$ be a uniformly distributed random variable, and then $\phi_X(\Omega) \overset{d}{=} X$. Therefore,

$$E[X] = E_{\Omega \sim \text{uniform}[0,1]}[\phi_X(\Omega)] = \int_0^1 \phi_X(\omega) d\omega.$$

Interdisciplinary Reference 12.2
Metric Geometry: Wasserstein Metric

Consider two random variables X and Y, whose quantile functions are ϕ_X and ϕ_Y respectively. The p-Wasserstein metric between these two variables is defined as

$$d_{\mathrm{W},p}(X,Y) \stackrel{\text{def}}{=} \sqrt[p]{\int_0^1 |\phi_X(\omega) - \phi_Y(\omega)|^p \, d\omega}, \quad \omega \in [0,1],$$

and the ∞-Wasserstein metric between these two variables is defined as

$$d_{\mathrm{W},\infty}(X,Y) \stackrel{\text{def}}{=} \sup_{\omega \in [0,1]} |\phi_X(\omega) - \phi_Y(\omega)|.$$

Consider the set of all possible action value random variables, or their distributions equivalently, as

$$\mathcal{Q}_p \stackrel{\text{def}}{=} \left\{ Q : \forall s \in \mathcal{S}, a \in \mathcal{A}(s), \mathrm{E}\left[|Q(s,a)|^p\right] < +\infty \right\}.$$

We can define the max-form of Wasserstein metric as

$$d_{\mathrm{supW},p}(Q',Q'') \stackrel{\text{def}}{=} \sup_{s,a} d_{\mathrm{W},p}(Q'(s,a), Q''(s,a)), \quad Q', Q'' \in \mathcal{Q}_p.$$

We can verify that $d_{\mathrm{supW},p}$ is a metric on $\mathcal{Q}_p$. (There we verify the triangle inequality: For any $Q', Q'', Q''' \in \mathcal{Q}_p$, since $d_{\mathrm{W},p}$ is a metric that satisfies the triangle inequality, we have

$$
\begin{aligned}
&d_{\mathrm{supW},p}(Q',Q'') \\
&= \sup_{s,a} d_{\mathrm{W},p}(Q'(s,a), Q''(s,a)) \\
&\leq \sup_{s,a}\left(d_{\mathrm{W},p}(Q'(s,a), Q'''(s,a)) + d_{\mathrm{W},p}(Q'''(s,a), Q''(s,a)) \right) \\
&\leq \sup_{s,a} d_{\mathrm{W},p}(Q'(s,a), Q'''(s,a)) + \sup_{s,a} d_{\mathrm{W},p}(Q'''(s,a), Q''(s,a)) \\
&= d_{\mathrm{supW},p}(Q',Q''') + d_{\mathrm{supW},p}(Q''',Q'').
\end{aligned}
$$

Proof completes.)

Now we prove that the Bellman operator $\mathfrak{B}_\pi$ is a contraction mapping on $\left(\mathcal{Q}_p, d_{\mathrm{supW},p}\right)$, i.e.

$$d_{\mathrm{supW},p}\left(\mathfrak{B}_\pi(Q'), \mathfrak{B}_\pi(Q'')\right) \leq \gamma d_{\mathrm{supW},p}(Q',Q''), \quad Q', Q'' \in \mathcal{Q}_p.$$

(Proof: Since

$$
d_{W,p}\Big(\mathcal{B}_\pi(Q')(s,a), \mathcal{B}_\pi(Q'')(s,a)\Big)
$$

$$
= d_{W,p}\Big(\big[R_{t+1} + \gamma Q'(S_{t+1}, A_{t+1})\big|S_t = s, A_t = a; \pi\big],
$$

$$
\big[R_{t+1} + \gamma Q''(S_{t+1}, A_{t+1})\big|S_t = s, A_t = a; \pi\big]\Big)
$$

$$
\leq \gamma\big[d_{W,p}(Q'(S_{t+1}, A_{t+1}), Q''(S_{t+1}, A_{t+1}))\big|S_t = s, A_t = a\big]
$$

$$
\leq \gamma \sup_{s',a'} d_{W,p}\Big(Q'(s',a'), Q''(s',a')\Big),
$$

$$
\leq d_{\sup W,p}(Q', Q''), \qquad\qquad s \in \mathcal{S}, a \in \mathcal{A}(s),
$$

we have

$$
d_{\sup W,p}\Big(\mathcal{B}_\pi(Q'), \mathcal{B}_\pi(Q'')\Big)
$$

$$
= \sup_{s,a} d_{W,p}\Big(\mathcal{B}_\pi(Q')(s,a), \mathcal{B}_\pi(Q'')(s,a)\Big)
$$

$$
\leq \sup_{s,a} \gamma d_{\sup W,p}(Q', Q'')
$$

$$
\leq \gamma d_{\sup W,p}(Q', Q'').
$$

The proof completes.)

For the task that tries to maximize the expectation, the greedy policy π_* derived from the action value distribution chooses the action that maximizes the action value expectation, i.e.

$$
\sum_{a \in \mathcal{A}(s)} \pi(a|s)\mathrm{E}\big[Q(s,a)\big] = \max_{a \in \mathcal{A}(s)} \mathrm{E}\big[Q(s,a)\big], \quad s \in \mathcal{S},
$$

or we can write

$$
\pi_*(s) = \arg\max_{a \in \mathcal{A}(s)} \mathrm{E}\big[Q(s,a)\big], \quad s \in \mathcal{S}.
$$

Here we assume that the maximum can be obtained. If multiple actions can lead to the same maximum value, we can pick an arbitrary action.

In this case, the Bellman optimal operator that uses action value random variable to back up action value random variable is

$$
Q_*(S_t, A_t) \overset{\mathrm{d}}{=} R_{t+1} + \gamma Q_*\Big(S_{t+1}, \arg\max_{a' \in \mathcal{A}(S_{t+1})} \mathrm{E}\big[Q_*(S_{t+1}, a')\big]\Big).
$$

Furthermore, we can define Bellman optimal operator $\mathcal{B}_*$ as

$$\mathfrak{B}_*(Q)(s, a) \stackrel{\text{def}}{=} \left[R_{t+1} + \gamma Q\left(S_{t+1}, \underset{a' \in \mathcal{A}(S_{t+1})}{\arg\max} \mathrm{E}\left[Q(S_{t+1}, a') \right] \right) \middle| S_t = s, A_t = a \right],$$

$$s \in \mathcal{S}, a \in \mathcal{A}(s).$$

Distributed VI algorithm: Starting from an action value distribution Q_0, apply the Bellman optimal operator $\mathfrak{B}_*$ repeatedly. Then we may get the optimal value random variable, and then get the optimal policy.

If the action space is a finite set, and we use a deterministic order to select an action when multiple actions are encountered in the argmax operator of the greedy option, (for example, the action space is an order set, and we always choose the smallest action among actions that can obtain the maximum), there exists a unique fixed point. The idea of the proof is similar to the proof in the expectation version in Sect. 3.1, but the details are much more complex. Here I only prove that $\mathfrak{B}_*$ is a contraction mapping in the expectation aspect, i.e.

$$d_\infty\left(\mathrm{E}\left[\mathfrak{B}_*(Q') \right], \mathrm{E}\left[\mathfrak{B}_*(Q'') \right] \right) \le \gamma d_\infty\left(\mathrm{E}[Q'], \mathrm{E}[Q''] \right), \quad Q', Q'' \in Q_\infty,$$

where d_∞ is defined in Sect. 3.1, i.e.

$$d_\infty(q', q'') = \max_{s, a} \left| q'(s, a) - q''(s, a) \right|, \quad q', q'' \in Q.$$

(Proof: Since the Bellman optimal operator $\mathfrak{b}_*$ in the expectation form is a contraction mapping, we have

$$d_\infty\left(\mathfrak{b}_*(q'), \mathfrak{b}_*(q'') \right) \le \gamma d_\infty(q', q''), \quad q', q'' \in Q.$$

Since $\mathrm{E}\left[\mathfrak{B}_*(Q) \right] = \mathfrak{b}_*(q)$ holds for $Q \in Q$, and $q = \mathrm{E}[Q]$, plugging it into the above inequality finishes the proof.)

12.2 Maximum Utility RL

In the previous section, the results related to the optimal policy are aimed to maximize the expectation of rewards. However, some tasks not only want to maximize rewards or minimize costs, but also consider other utility and risk measures. For example, a task may want to both maximize the expectation and minimize the standard deviation. It is difficult for the optimal value algorithms that maintain expectation only. However, the distributional RL algorithms in this chapter can do that since they maintain the whole distribution of action value random variables.

Maximum utility RL tries to maximize the utility.

Interdisciplinary Reference 12.3
Utility Theory: von Neumann Morgenstern Utility

The utility is a mapping from a random value to a random variable, so that the statistic of the resulting random variable can have some meaning. For example, a utility u can make the utility expectation $\mathrm{E}[u(X)]$ meaningful, where X is a random event. This description is very abstract. In fact, utility has many different mathematical definitions.

Among various definitions of utility, von Neumann Morgenstern (VNM) utility is the most common one (Neumann, 1947). The definition of VNM utility is based on the VNM utility theorem. VNM utility theorem says if the decision-maker follows the set of VNM axioms (will be introduced in the sequel), the decision-maker de facto tries to maximize the expectation of a function. This theorem should be understood as follows: When the decision-maker picks an event from two different random events X' and X'', (1) $X' < X''$ means that the decision-maker prefers X'' than X'; (2) $X' \sim X''$ means that the decision-maker is indifferent with these two random events; (3) $X' > X''$ means that the decision-maker prefers X' than X''. If the decision-maker follows the set of VNM axioms, there exists a functional $u : X \to \mathbb{R}$ such that

$$X' < X'' \iff \mathrm{E}[u(X')] < \mathrm{E}[u(X'')]$$
$$X' \sim X'' \iff \mathrm{E}[u(X')] = \mathrm{E}[u(X'')]$$
$$X' > X'' \iff \mathrm{E}[u(X')] > \mathrm{E}[u(X'')].$$

The expectation utility $\mathrm{E}[u(\ )]$ is called VNM utility.

Then what are VNM axioms? Let us first learn a core operation of VNM axioms: the mixture of events. Consider two random events X' and X''. I can create a new random event, so that I pick event X' with probability p $(0 \le p \le 1)$, and pick event X'' with probability $1 - p$. Such mixture is denoted as $pX' + (1 - p)X''$. Note that $pX' + (1 - p)X''$ does not mean multiplying the value of X' with p and multiplying the value of X'' with $(1 - p)$. The random events X' and X'' are not necessarily numbers, so the multiplication and addition are not well defined. Specifically, when both X' and X'' and random variables, the Cumulated Distribution Function (CDF) of $pX' + (1 - p)X''$ is the weighted average of the CDF of X' and CDF of X''. Now we know the mixture operation. The VNM axioms are:

- Completeness: For two random events X' and X'', one and only one of the following holds: (1) $X' < X''$; (2) $X' \sim X''$; (3) $X' > X''$.
- Transitivity: For three random events X', X'', X''', (1) $X' < X'''$ and $X''' < X''$ lead to $X' < X''$; (2) $X' \sim X'''$ and $X''' \sim X''$ lead to $X' \sim X''$; (3) $X' > X'''$ and $X''' > X''$ lead to $X' > X''$.
- Continuity: For three random events X', X'', X''' such that $X' < X'' < X'''$, there exists a probability $p \in [0, 1]$ such that $pX' + (1 - p)X''' \sim X''$.

- Independence: For three random events X', X'', X''' and the probability value $p \in [0, 1]$, $X' \prec X''$ holds if and only if $pX + (1-p)X' \prec X + (1-p)X''$.

Example 12.1 (Exponential Utility) The utility function of exponential utility is

$$u(x) = \begin{cases} -\mathrm{e}^{-ax}, & a \neq 0 \\ x, & a = 0 \end{cases}$$

where a is the parameter that indicates risk preference. $a > 0$ means risk-aversion; $a = 0$ means risk-neutral; and $a < 0$ means risk-seeking. $a > 0$ somehow tries to maximize expectation and minimize standard deviation at the same time. Especially, the exponential utility of a normal distributed random variable $X \sim \mathrm{normal}(\mu, \sigma^2)$ is $\mathrm{E}[u(X)] = -\exp\left(-a\left(\mu - \frac{1}{2}a\sigma^2\right)\right)$, so maximizing its exponential utility in fact maximizes $\mu - \frac{1}{2}a\sigma^2$, or equivalently maximizes the linear combination of the mean μ and variance σ^2. Therefore, adjusting the parameter a can trade off between mean and variance.

Maximum utility RL can try to maximize the VNM utility. The idea is as follows: Apply the utility function u on the action value random variable or its samples, and get $u(Q(\cdot, \cdot))$. Then we try to maximize the expectation or sample mean of $u(Q(\cdot, \cdot))$. That is, the optimal strategy maximizes the utility, i.e.

$$\pi(s) = \arg\max_{a} \mathrm{E}\big[u(Q(s, a))\big], \quad s \in \mathcal{S}.$$

Although VNM utility is the most common utility, it also has defects. For example, Allais paradox is a usual example to show the defect of VNM utility. There are other definitions of utility. Some maximum utility RL researches also consider Yarri utility.

Interdisciplinary Reference 12.4
Utility Theory: Yarri Utility

Recall that the axiom of independent in VNM utility considers the CDF of event mixture. Due to the defect of VNM utility, (Yarri, 1987) considers changing the axiom of independence from the mixture of CDF function to the mixture of quantile functions, i.e.

$$p\phi_X + (1-p)\phi_{X'} \leq p\phi_X + (1-p)\phi_{X''}.$$

It equivalently defines a distortion function $\beta : [0, 1] \to [0, 1]$ such that β is a strictly monotonously increasing function with $\beta(0) = 0$ and $\beta(1) = 1$. Let β^{-1} denote the inverse function of the distortion function. If the distortion function β is an identity function (i.e. $\beta(\omega) = \omega$ for all $\omega \in [0, 1]$), the utility is risk-neutral. If the distortion function is convex, i.e. the graph is beneath the identity function, we emphasize more on the worse case, so it is risk-averse. If the distortion function is concave, i.e. the graph is above the identity function, we emphasize on better case, so it is risk-seeking.

Given the distortion function β, the distorted expectation is defined as

$$E^{\langle\beta\rangle}[X] \overset{\text{def}}{=} E_{\Omega\sim\text{uniform}[0,1]}\left[\phi_X\big(\beta(\Omega)\big)\right].$$

We can easily get

$$E^{\langle\beta\rangle}[X] = \int_0^1 \phi_X\big(\beta(\omega)\big)d\omega = \int_0^1 \phi_X(\omega)d\beta^{-1}(\omega).$$

We can also understand the distortion function as reparameterizing the uniform distributed random variable over uniform $[0, 1]$ to another random variable with a new distribution. We can use uniform$^{\langle\beta\rangle}$ $[0, 1]$ to denote the new distribution. Then the distorted expectation can be represented as

$$E^{\langle\beta\rangle}[X] = E_{\Omega\sim\text{uniform}^{\langle\beta\rangle}[0,1]}\left[\phi_X(\Omega)\right].$$

If the distortion function β is the identity function, the distorted expectation is the ordinary expectation.

Some maximum utility RL algorithms can try to maximize the distorted expectation. The idea is, after getting the quantile function, calculate the distorted action value

$$q^{\langle\beta\rangle}(s, a) = E_{\Omega\sim\text{uniform}[0,1]}\left[\phi_{Q(s,a)}\big(\beta(\Omega)\big)\right].$$

Then choose the action that maximizes the distorted action value, i.e.

$$\pi^{\langle\beta\rangle}(s) = \arg\max_a q^{\langle\beta\rangle}(s, a).$$

12.3 Probability-Based Algorithm

This section introduces the first type of distributed RL algorithm: probability-based algorithm. Based on the optimal-value algorithm such as Q learning, this type of algorithm maintains the probability distribution of action value, so the algorithm becomes a distributed RL algorithm. Remarkably, such algorithms do not model the action value distribution as a parametric distribution such as Gaussian distribution. This is because, if the action value distribution is modeled as a parametric distribution, we can directly learn the distribution parameters, and then use the parameters to get statistics. In such cases, we do not need to learn the entire distribution.

12.3.1 C51: Categorical DQN

This section introduces Categorical Deep Q Network algorithm. This algorithm is called as such because it uses categorical distribution to approximate the distribution of action value random variables.

Interdisciplinary Reference 12.5
Probably Theory: Categorical Distribution

Categorical distribution, a.k.a. generalized Bernoulli distribution, is a discrete probability distribution defined by its support list $(x : x \in X)$ and probability in each support value $(p(x) : x \in X)$, i.e.

$$\Pr[X = x] = p(x), \quad x \in X,$$

where $(p(x) : x \in X)$ should satisfy $p(x) \geq 0$ $(x \in X)$ and $\sum_x p(x) = 1$.

Categorical Deep Q Network algorithm is sometimes called C51 algorithm in the original paper when the number of elements in the support set is 51. The original paper also considers other numbers of supporting elements.

The idea of **Categorical Deep Q Network (Categorical DQN)** is as follows (Bellemare, 2017): Approximate the distribution of $Q(s, a)$ as the categorical distribution with the support set $\left\{ q^{(i)} : i \in \mathcal{I} \right\}$, where $q^{(0)} < q^{(1)} < \cdots < q^{(|\mathcal{I}|-1)}$ are pre-defined. We use a neural network $p^{(i)}(s, a; \mathbf{w})$ $(i \in \mathcal{I})$ to output $\Pr\left[Q(s, a) = q^{(i)} \right]$. Note that the output of the neural network $p^{(i)}(s, a; \mathbf{w})$ should be in the form of Probability Mass Function (PMF) of the categorical distribution, so it needs to satisfy $p^{(i)}(s, a; \mathbf{w}) \geq 0$ $(i \in \mathcal{I})$ and $\sum_{i \in \mathcal{I}} p^{(i)}(s, a; \mathbf{w}) = 1$. This can be achieved by adding a softmax layer in the neural network. For example, applying the softmax operation on the output of a layer of the neural network $\xi^{(i)}(s, a; \mathbf{w})$ becomes $p^{(i)}(s, a; \mathbf{w}) = \frac{\exp \xi^{(i)}(s, a; \mathbf{w})}{\sum_{i \in \mathcal{I}} \xi^{(i)}(s, a; \mathbf{w})}$ $(i \in \mathcal{I})$, which can satisfy the constraint.

The training objective of Categorical DQN algorithm is to minimize the cross entropy loss between the output distribution of neural network $p^{(\cdot)}(\cdot, \cdot; \mathbf{w})$ and the target distribution $p^{(\cdot)}_{\text{target}}(\cdot, \cdot)$ obtained by bootstrapping. That is, we want to minimize

$$-\sum_{i \in \mathcal{I}} p^{(i)}(s, a; \mathbf{w}) \ln p^{(i)}_{\text{target}}(s, a), \quad s \in \mathcal{S}, a \in \mathcal{A}.$$

The way to calculate the target probability $p^{(i)}_{\text{target}}(s, a)$ for a state–action pair (s, a) is as follows: Upon we get a transition (s, a, r, s', d'), we can use a greedy policy to choose the action after the state s' (denoted the action as a'). For the tasks that try to maximize the expectation of rewards, the greedy policy tries to maximize the action value expectation. In order to achieve this, we need to calculate the action value expectation estimate $q(s', a')$ for each action candidate $a' \in \mathcal{A}(s')$. This can be obtained by first obtaining the categorical distribution probability $p^{(j)}(s', a'; \mathbf{w})$ ($j \in I$) using the neural network and then calculating the expectation using $q(s', a'; \mathbf{w}) = \sum_{i \in I} q^{(i)} p^{(i)}(s', a'; \mathbf{w})$. After choosing the next action a', we bootstrap using $p^{(j)}\left(s', a', \mathbf{w}_{\text{target}}\right)$ ($j \in I$). Each $q^{(j)}$ ($j \in I$) will be bootstrapped to $u^{(j)} = r + \gamma(1 - d')q^{(j)}$, with probability $p^{(j)}\left(s', a'; \mathbf{w}_{\text{target}}\right)$. Since $u^{(j)}$ is probably not within $\left\{q^{(i)} : i \in I\right\}$, we need to project the probability from $u^{(j)}$ ($j \in I$) to $q^{(i)}$ ($i \in I$) in some way. Let $\varsigma^{(i)}\left(u^{(j)}\right)$ ($i, j \in I$) denote the project probability ratio from $u^{(j)}$ to $q^{(i)}$. Then the total projected probability of $q^{(i)}$ ($i \in I$) from all $u^{(j)}$ ($j \in I$) is

$$p_{\text{target}}^{(i)}(s, a) = \sum_{j \in I} \varsigma^{(i)}\left(u^{(j)}\right) p^{(j)}\left(s', a'; \mathbf{w}_{\text{target}}\right), \quad i \in I.$$

The support set of categorical distribution is usually set to the form $q^{(i)} = q^{(0)} + i\Delta q$ ($i = 0, 1, \ldots, |I| - 1$), where $\Delta q > 0$ is a positive real number. In such setting, $q^{(i)}$ ($i \in I$) are $|I|$ values with equal intervals and $q^{(i)}$. This setting can be combined with the following rule to determine the ratio $\varsigma^{(i)}\left(u^{(j)}\right)$ ($i, j \in I$):

- If $u^{(j)} < q^{(0)}$, the probability is all projected to $q^{(0)}$. while ratios to other $q^{(i)}$ ($i > 0$) are 0.
- If $u^{(j)} > q^{(|I|-1)}$, the probability is all projected to $q^{(|I|-1)}$, while ratios to other $q^{(i)}$ ($i < |I| - 1$) are 0.
- If $u^{(j)}$ is identical to some $q^{(i)}$ ($i \in I$), the probability is all projected to $q^{(i)}$.
- Otherwise, $u^{(j)}$ must be between some $q^{(i)}$ and $q^{(i+1)}$ ($i < |I| - 1$) and the ratio to $q^{(i)}$ is $\varsigma^{(i)}\left(u^{(j)}\right) = \frac{q^{(i+1)} - u^{(j)}}{\Delta q}$ and the ratio to $q^{(i+1)}$ is $\varsigma^{(i+1)}\left(u^{(j)}\right) = \frac{u^{(j)} - q^{(i)}}{\Delta q}$.

The above support set and projection rule can be implemented in the following method: For each $u^{(j)}$ ($j \in I$), first calculate $u_{\text{clip}}^{(j)} \leftarrow \text{clip}\left(u^{(j)}, q^{(0)}, q^{(|I|-1)}\right)$ to convert the first two cases in the rule to the third case. Then we calculate $\left|u_{\text{clip}}^{(j)} - q^{(i)}\right|/\Delta q$ for each $i \in I$ to know how many Δq do current $u_{\text{clip}}^{(j)}$ and $q^{(i)}$ differ. If the difference is zero times Δq, the projection ratio will be 1; if the difference is ≥ 1 times Δq, the projection ratio will be 0. Therefore, the projection ratio to $q^{(i)}$ is

$$\varsigma^{(i)}\left(u^{(j)}\right) = 1 - \text{clip}\left(\frac{\left|u_{\text{clip}}^{(j)} - q^{(i)}\right|}{\Delta q}, 0, 1\right).$$

Algorithm 12.1 shows Categorical DQN algorithm that is used to maximize return expectation.

Algorithm 12.1 **Categorical DQN to find the optimal policy (to maximize expectation).**

Parameters: support values of action value $q^{(i)}$ ($i \in I$) (usually set as $q^{(i)} = q^{(0)} + i\triangle q$, $i = 0, 1, \ldots, |I| - 1$), et al.

1. Initialize:

    1.1. (Initialize parameters) Initialize the parameters of action value evaluation network $\mathbf{w}$. Initialize the action value target network $\mathbf{w}_{\text{target}} \leftarrow \mathbf{w}$.

    1.2. (Initialize experience storage) $\mathcal{D} \leftarrow \varnothing$.

2. For each episode:

    2.1. (Initialize state) Choose the initial state S.

    2.2. Loop until the episode ends:

        2.2.1. (Collect experiences) Do the following once or multiple times:

            2.2.1.1. (Decide) For each action $a \in \mathcal{A}(S)$, compute $p^{(i)}(S, a; \mathbf{w})$ ($i \in I$), and then compute $q(S, a) \leftarrow \sum_{i \in I} q^{(i)} p^{(i)}(S, a; \mathbf{w})$. Use the policy derived from $q(S, \cdot)$ (say ε-greedy policy) to determine the action A.

            2.2.1.2. (Sample) Execute the action A, and then observe the reward R, the next state S', and the indicator of episode end D'.

            2.2.1.3. Save the experience (S, A, R, S', D') in the storage $\mathcal{D}$.

            2.2.1.4. $S \leftarrow S'$.

        2.2.2. (Use experiences) Do the following once or multiple times:

            2.2.2.1. (Replay) Sample a batch of experience $\mathcal{B}$ from the storage $\mathcal{D}$. Each entry is in the form of (S, A, R, S', D').

            2.2.2.2. (Choose the next action for the next state) For each transition, when $D' = 0$, calculate the probability for each possible next action $a' \in \mathcal{A}(S')$ and each support value as $p^{(i)}(S', a'; \mathbf{w})$ ($i \in I$), and then calculate the action value $q(S', a') \leftarrow \sum_{i \in I} q^{(i)} p^{(i)}(S', a'; \mathbf{w})$. Choose the next action using $A' \leftarrow \arg\max_{a'} q(S', a')$. The values can be arbitrarily assigned when $D' = 1$.

            2.2.2.3. (Calculate TD returns) For each transition, calculate $U^{(j)} \leftarrow R + \gamma(1 - D')q^{(j)}$ ($j \in I$).

2.2.2.4. (Calculate probability of TD returns) For each transition, calculate the project ratio $\varsigma^{(j)}\!\left(U^{(i)}\right)$ from $U^{(j)}$ ($j \in \mathcal{I}$) to $q^{(i)}$ ($i \in \mathcal{I}$). (For example,

$$U^{(j)}_{\text{clip}} \leftarrow \text{clip}\!\left(U^{(j)}, q^{(0)}, q^{(|\mathcal{I}|-1)}\right),\ \varsigma^{(i)}\!\left(U^{(j)}\right)$$

$$\leftarrow 1 - \text{clip}\!\left(\frac{\left|U^{(j)}_{\text{clip}} - q^{(i)}\right|}{\triangle q}, 0, 1\right).)\ \text{Then}\ p^{(i)}_{\text{target}} \leftarrow$$

$\sum_{j \in \mathcal{I}} \varsigma^{(i)}\!\left(U^{(j)}\right) p^{(j)}\!\left(S', A'; \mathbf{w}_{\text{target}}\right)$ ($i \in \mathcal{I}$).

2.2.2.5. (Update action value parameter) Update $\mathbf{w}$ to reduce $\frac{1}{|\mathcal{B}|} \sum_{(S,A,R,S',D') \in \mathcal{B}} \left(-\sum_{i \in \mathcal{I}} p^{(i)}_{\text{target}} \ln p^{(i)}(s, a; \mathbf{w})\right)$.

2.2.2.6. (Update target network) Under some condition (say, every several updates), update the parameters of target network $\mathbf{w}_{\text{target}} \leftarrow \left(1 - \alpha_{\text{target}}\right) \mathbf{w}_{\text{target}} + \alpha_{\text{target}} \mathbf{w}$.

12.3.2 Categorical DQN with Utility

This section introduces how to use Categorical DQN to maximize VNM utility.

The Categorical DQN in the previous section tries to maximize the return expectation, and it chooses the action that maximizes the action value expectation, i.e. tries to choose an action $a \in \mathcal{A}(s)$ to maximize the sample $\sum_{i \in \mathcal{I}} q^{(i)} p^{(i)}(s, a; \mathbf{w})$. After introducing the VNM utility, we need to choose an action $a \in \mathcal{A}(s)$ to maximize the sample $\sum_{i \in \mathcal{I}} u\!\left(q^{(i)}\right) p^{(i)}(s, a; \mathbf{w})$. Accordingly, the Categorical DQN algorithm to maximize the utility is shown in Algo. 12.2.

Algorithm 12.2   Categorical DQN to find the optimal policy (to maximize VNM utility).

Parameters: utility function u, the support values of action value $q^{(i)}$ ($i \in \mathcal{I}$) (usually set as $q^{(i)} = q^{(0)} + i\triangle q$, $i = 0, 1, \ldots, |\mathcal{I}| - 1$), et al.

1. Initialize:

    1.1. (Initialize parameters) Initialize the parameters of action value evaluation network $\mathbf{w}$. Initialize the action value target network $\mathbf{w}_{\text{target}} \leftarrow \mathbf{w}$.

    1.2. (Initialize experience storage) $\mathcal{D} \leftarrow \varnothing$.

2. For each episode:

2.1. (Initialize state) Choose the initial state S.

2.2. Loop until the episode ends:

    2.2.1. (Collect experiences) Do the following once or multiple times:

        2.2.1.1. (Decide) For each action $a \in \mathcal{A}(S)$, compute $p^{(i)}(S, a; \mathbf{w})$ $(i \in \mathcal{I})$, and then compute $q(S, a) \leftarrow \sum_{i \in \mathcal{I}} u\big(q^{(i)}\big) p^{(i)}(S, a; \mathbf{w})$. Use the policy derived from $q(S, \cdot)$ (say ε-greedy policy) to determine the action A.

        2.2.1.2. (Sample) Execute the action A, and then observe the reward R, the next state S', and the indicator of episode end D'.

        2.2.1.3. Save the experience (S, A, R, S', D') in the storage $\mathcal{D}$.

        2.2.1.4. $S \leftarrow S'$.

    2.2.2. (Use experiences) Do the following once or multiple times:

        2.2.2.1. (Replay) Sample a batch of experience $\mathcal{B}$ from the storage $\mathcal{D}$. Each entry is in the form of (S, A, R, S', D').

        2.2.2.2. (Choose the next action for the next state) For each transition, when $D' = 0$, calculate the probability for each possible next action $a' \in \mathcal{A}(S')$ and each support value as $p^{(i)}(S', a'; \mathbf{w})$ $(i \in \mathcal{I})$, and then calculate the action value $q(S', a') \leftarrow \sum_{i \in \mathcal{I}} u\big(q^{(i)}\big) p^{(i)}(S', a'; \mathbf{w})$. Choose the next action using $A' \leftarrow \arg\max_{a'} q(S', a')$. The values can be arbitrarily assigned when $D' = 1$.

        2.2.2.3. (Calculate TD returns) For each transition, calculate $U^{(j)} \leftarrow R + \gamma(1 - D')q^{(j)}$ $(j \in \mathcal{I})$.

        2.2.2.4. (Calculate probability of TD returns) For each transition, calculate the project ratio $\varsigma^{(j)}\big(U^{(i)}\big)$ from $U^{(j)}$ $(j \in \mathcal{I})$ to $q^{(i)}$ $(i \in \mathcal{I})$. (For example, $U^{(j)}_{\text{clip}} \leftarrow \text{clip}\big(U^{(j)}, q^{(0)}, q^{(|\mathcal{I}|-1)}\big)$, $\varsigma^{(i)}\big(U^{(j)}\big) \leftarrow 1 - \text{clip}\left(\dfrac{\big|U^{(j)}_{\text{clip}} - q^{(i)}\big|}{\Delta q}, 0, 1\right)$.) Then $p^{(i)}_{\text{target}} \leftarrow \sum_{j \in \mathcal{I}} \varsigma^{(i)}\big(U^{(j)}\big) p^{(j)}\big(S', A'; \mathbf{w}_{\text{target}}\big)$ $(i \in \mathcal{I})$.

        2.2.2.5. (Update action value parameter) Update $\mathbf{w}$ to reduce $\frac{1}{|\mathcal{B}|} \sum_{(S, A, R, S', D') \in \mathcal{B}} \left(-\sum_{i \in \mathcal{I}} p^{(i)}_{\text{target}} \ln p^{(i)}(s, a; \mathbf{w})\right)$.

> **2.2.2.6.** (Update target network) Under some condition (say,
> every several updates), update the parameters of
> target network $\mathbf{w}_{\text{target}} \leftarrow \left(1 - \alpha_{\text{target}}\right)\mathbf{w}_{\text{target}} + \alpha_{\text{target}}\mathbf{w}$.

12.4 Quantile Based RL

This section considers quantile-based RL, another type of RL algorithms based
on value distribution. This type of algorithms maintains the estimate of quantile
functions, rather than estimates of PMF or PDF. They conduct quantile regression
during training, rather than minimizing cross-entropy loss. Let us review the quantile
regression below.

Interdisciplinary Reference 12.6
Machine Learning: Quantile Regression

Quantile Regression (QR) is an algorithm to regress quantile values.
Consider a random variable X and its quantile value $\phi_X(\omega)$ at a given cumulated
probability $\omega \in [0, 1]$. The probability of $X < \phi_X(\omega)$ is ω, or equivalently the
probability of $X - \phi_X(\omega) < 0$ is ω. Let $\hat{\phi}$ denote an estimate of $\phi_X(\omega)$. We can
judge the estimation accuracy by observing the empirical probability of $X - \hat{\phi} < 0$:
If $\Pr\left[X - \hat{\phi} < 0\right] < \omega$, or equivalently $\Pr\left[X < \hat{\phi}\right] < \omega$, it means that the estimate
$\hat{\phi}$ is too small, and we need to increase the estimate $\hat{\phi}$; if $\Pr\left[X - \hat{\phi} < 0\right] > \omega$,
or $\Pr\left[X < \hat{\phi}\right] > \omega$, it means that the estimate $\hat{\phi}$ is too large, and we need to
decrease the estimate $\hat{\phi}$. During training, we can get the quantile value by minimizing
$\mathrm{E}\left[\left(\omega - 1_{\left[X - \phi < 0\right]}\right)\left(X - \phi\right)\right]$. (Proof: For simplicity, here I only consider the case
that X is a continuous random variable. Note that

$$
\mathrm{E}\left[\left(\omega - 1_{\left[X - \phi < 0\right]}\right)\left(X - \phi\right)\right]
$$

$$
= \omega\mathrm{E}[X] - \omega\phi - \mathrm{E}\left[1_{\left[X - \phi < 0\right]}X\right] + \phi\mathrm{E}\left[1_{X - \phi < 0}\right]
$$

$$
= \omega\mathrm{E}[X] - \omega\phi - \int_{x \in \mathcal{X}: x < \phi} x p(x)\mathrm{d}x + \phi \int_{x \in \mathcal{X}: x < \phi} p(x)\mathrm{d}x,
$$

and we have

$$
\frac{\mathrm{d}}{\mathrm{d}\phi}\mathrm{E}\left[\left(\omega - 1_{\left[X - \phi < 0\right]}\right)\left(X - \phi\right)\right]
$$

$$
= \frac{\mathrm{d}}{\mathrm{d}\phi}\left[\omega\mathrm{E}[X] - \omega\phi - \int_{x \in \mathcal{X}: x < \phi} x p(x)\mathrm{d}x + \phi \int_{x \in \mathcal{X}: x < \phi} p(x)\mathrm{d}x\right]
$$

$$= 0 - \omega - \phi p(\phi) + \left[\int_{x \in \mathcal{X}: x < \phi} p(x) \mathrm{d}x + \phi p(\phi) \right]$$

$$= -\omega + \int_{x \in \mathcal{X}: x < \phi} p(x) \mathrm{d}x.$$

Setting $\frac{\mathrm{d}}{\mathrm{d}\phi} \mathrm{E} \left[\left(\omega - 1_{[X-\phi<0]} \right) (X - \phi) \right] = 0$ leads to $\int_{x \in \mathcal{X}: x < \phi} p(x) \mathrm{d}x = \omega$. Therefore, at the extreme point of the optimization objective, ϕ is the quantile value at the cumulated probability ω.) Therefore, during training, after obtaining the samples $x_0, x_1, x_2, \ldots, x_{c-1}$, we can try to minimize the **Quantile Regression loss (QR loss)**:

$$\frac{1}{c} \sum_{i=0}^{c-1} \ell_{\mathrm{QR}} (x_i - \phi),$$

where $\ell_{\mathrm{QR}}(\delta; \omega) \overset{\mathrm{def}}{=} \left(\omega - 1_{[\delta<0]} \right) \delta$ is the sample loss of each sample. Here $\ell_{\mathrm{QR}}(\delta; \omega) = \left(\omega - 1_{[\delta<0]} \right) \delta$ can also be written as $\ell_{\mathrm{QR}}(\delta; \omega) = \left| \omega - 1_{[\delta<0]} \right| |\delta|$. If the estimate $\hat{\phi}$ is too small, $\delta > 0$, and we increase ϕ with weight ω; if the estimate $\hat{\phi}$ is too large, $\delta < 0$, and we decrease δ with weight $(1 - \omega)$.

The QR loss is not smooth at $\delta = 0$, which may degrade performance sometimes. Therefore, some algorithms consider combining the QR loss with Huber loss, and then obtain the QR Huber loss. Huber loss is defined as

$$\ell_{\mathrm{Hubor}}(\delta; \kappa) \overset{\mathrm{def}}{=} \begin{cases} \frac{\delta^2}{2\kappa}, & |\delta| < \kappa \\ |\delta| - \frac{1}{2}\kappa, & |\delta| \geq \kappa. \end{cases}$$

Quantile Regression Huber loss (QR Huber loss) is defined as

$$\ell_{\mathrm{QRHubor}}(\delta; \omega, \kappa) \overset{\mathrm{def}}{=} \left| \omega - 1_{[\delta<0]} \right| \ell_{\mathrm{Hubor}}(\delta; \kappa)$$

$$= \begin{cases} \left| \omega - 1_{[\delta<0]} \right| \frac{\delta^2}{2\kappa}, & |\delta| < \kappa \\ \left| \omega - 1_{[\delta<0]} \right| \left(|\delta| - \frac{1}{2}\kappa \right), & |\delta| \geq \kappa. \end{cases}$$

When $\kappa = 0$, QR Huber loss falls back to Huber loss.

The algorithms in this section will try to minimize QR Huber loss.

12.4.1 QR-DQN: Quantile Regression Deep Q Network

For a general random variable, we can calculate its expectation in $|\mathcal{I}|$ segments:

$$E[X] = \sum_{i=0}^{|\mathcal{I}|-1} \int_{i/|\mathcal{I}|}^{(i+1)/|\mathcal{I}|} \phi_X(\omega)d\omega.$$

If we consider the quantile values on the following cumulated probability

$$\omega^{(i)} = \frac{i+0.5}{|\mathcal{I}|}, \quad i = 0, 1, \ldots, |\mathcal{I}| - 1,$$

we have

$$E[X] \approx \sum_{i=0}^{|\mathcal{I}|-1} \phi_X\left(\frac{i+0.5}{|\mathcal{I}|}\right)\frac{1}{|\mathcal{I}|}.$$

Therefore, the expectation can be approximated represented as

$$E[X] \approx \phi_X^\top \Delta\omega,$$

where

$$\phi \stackrel{\text{def}}{=} \left(\phi_X\left(\frac{0.5}{|\mathcal{I}|}\right), \phi_X\left(\frac{1.5}{|\mathcal{I}|}\right), \ldots, \phi_X\left(\frac{|\mathcal{I}| - 0.5}{|\mathcal{I}|}\right)\right)^\top$$

$$\Delta\omega \stackrel{\text{def}}{=} \frac{1}{|\mathcal{I}|}\mathbf{1}_{|\mathcal{I}|}$$

and $\mathbf{1}_{|\mathcal{I}|}$ is the all-one vector consisting of $|\mathcal{I}|$ elements.

Quantile Regression Deep Q Network (QR-DQN) algorithm (Dabney, 2018: QR-DQN) uses a neural network to approximate quantile functions, and makes decisions accordingly. Specifically, it uses a neural network $\phi(s, a; \mathbf{w})$ to approximate the quantile function of the action value random variable. The parameter of the network is $\mathbf{w}$. The inputs of the network are the state–action pair (s, a). The outputs of the network are $|\mathcal{I}|$ quantile value estimates that correspond to $|\mathcal{I}|$ cumulated probability values respectively. The action value expectation can be calculated using

$$q(s, a) \leftarrow \left[\phi(s, a; \mathbf{w})\right]^\top \Delta\omega,$$

where $\Delta\omega \stackrel{\text{def}}{=} \frac{1}{|\mathcal{I}|}\mathbf{1}_{|\mathcal{I}|}$.

Algorithm 12.3 shows the QR-DQN algorithm that maximizes the return expectation.

Algorithm 12.3   QR-DQN to Find the Optimal Policy (To Maximize Expectation).

Parameters: number of cumulated values for quantile function (furthermore we can get $\omega^{(i)} = (i+0.5)/|\mathcal{I}|$ ($i = 0, 1, \ldots, |\mathcal{I}| - 1$) and $\Delta\omega = \frac{1}{|\mathcal{I}|}\mathbf{1}_{|\mathcal{I}|}$), and parameter for QR Huber loss κ, et al.

1. Initialize:

1.1. (Initialize parameters) Initialize the parameters of action value evaluation network $\mathbf{w}$. Initialize the action value target network $\mathbf{w}_{\text{target}} \leftarrow \mathbf{w}$.

1.2. (Initialize experience storage) $\mathcal{D} \leftarrow \varnothing$.

2. For each episode:

2.1. (Initialize state) Choose the initial state S.

2.2. Loop until the episode ends:

2.2.1. (Collect experiences) Do the following once or multiple times:

2.2.1.1. (Decide) For each action $a \in \mathcal{A}(S)$, compute $\phi(S, a; \mathbf{w})$, and then compute $q(S, a) \leftarrow \left[\phi(S, a; \mathbf{w})\right]^{\top} \Delta \omega$. Use the policy derived from $q(S, \cdot)$ (say ε-greedy policy) to determine the action A.

2.2.1.2. (Sample) Execute the action A, and then observe the reward R, the next state S', and the indicator of episode end D'.

2.2.1.3. Save the experience (S, A, R, S', D') in the storage $\mathcal{D}$.

2.2.1.4. $S \leftarrow S'$.

2.2.2. (Use experiences) Do the following once or multiple times:

2.2.2.1. (Replay) Sample a batch of experience $\mathcal{B}$ from the storage $\mathcal{D}$. Each entry is in the form of (S, A, R, S', D').

2.2.2.2. (Choose the next action for the next state) For each transition, when $D' = 0$, calculate the quantile values for each next action $a' \in \mathcal{A}(S')$ as $\phi(S', a'; \mathbf{w})$, and then calculate the action value $q(S', a') \leftarrow \left[\phi(S', a'; \mathbf{w})\right]^{\top} \Delta \omega$. Choose the next action using $A' \leftarrow \arg\max_{a'} q(S', a')$. The values can be arbitrarily assigned when $D' = 1$.

2.2.2.3. (Calculate TD returns) For each transition and each $j \in \mathcal{I}$, when $D' = 0$, calculate the quantile value $\phi^{(j)}\left(S', A'; \mathbf{w}_{\text{target}}\right)$ for each cumulated probability $\omega^{(j)}$. The values can be arbitrarily set when $D' = 1$. Then calculate $U^{(j)} \leftarrow R + \gamma(1 - D')\phi^{(j)}\left(S', A'; \mathbf{w}_{\text{target}}\right)$.

2.2.2.4. (Update action value parameters) Update $\mathbf{w}$ to reduce $\frac{1}{|\mathcal{B}|} \sum_{\substack{(S, A, R, S', D') \in \mathcal{B} \\ i, j \in \mathcal{I}}} \ell_{\text{QRHubor}}\left(U^{(j)} - \phi^{(i)}(S_t, A_t; \mathbf{w}); \omega^{(i)}, \kappa\right)$.

2.2.2.5. (Update target network) Under some condition (say, every several updates), update the parameters of target network $\mathbf{w}_{\text{target}} \leftarrow \left(1 - \alpha_{\text{target}}\right)\mathbf{w}_{\text{target}} + \alpha_{\text{target}}\mathbf{w}$.

QR-DQN algorithm pre-defines the cumulated probability values for the quantile function, and only fits the quantile values on those cumulated probability values. This may lead to some unwanted behaviors. The next section will introduce an algorithm to resolve the problem.

12.4.2 IQN: Implicit Quantile Networks

This section introduces **Implicit Quantile Networks** (**IQN**) algorithm. This algorithm can be obtained from a slight modification of QR-DQN algorithm (Dabney 2018: IQN). The idea of IQN algorithm is as follows: Consider the tasks that try to maximize the return expectation. In order to estimate the action value expectation $Q(s, a)$ for a given state–action pair $Q(s, a)$, since $\mathrm{E}\big[Q(s, a)\big] = \mathrm{E}_{\Omega \sim \text{uniform}[0,1]}\big[\phi_{Q(s,a)}(\Omega)\big]$, we draw c samples of cumulated probability values $\omega^{(i)}$ ($i = 0, 1, \ldots, c - 1$) from the uniform distribution uniform $[0, 1]$, and then use a quantile network ϕ to get the quantiles values $\phi\big(s, a, \omega^{(i)}; \mathbf{w}\big)$, and use the average $\frac{1}{c}\sum_{i=0}^{c-1}\phi\big(s, a, \omega^{(i)}; \mathbf{w}\big)$ as the estimate of action value expectation. Algorithm 12.4 shows the algorithm.

Algorithm 12.4 IQN to find the optimal policy (to maximize expectation).

Parameters: number of cumulated value samples for each expectation evaluation c, parameter for QR Huber loss κ, et al.

1. Initialize:

   1.1. (Initialize parameters) Initialize the parameters of action value evaluation network $\mathbf{w}$. Initialize the action value target network $\mathbf{w}_{\text{target}} \leftarrow \mathbf{w}$.

   1.2. (Initialize experience storage) $\mathcal{D} \leftarrow \varnothing$.

2. For each episode:

   2.1. (Initialize state) Choose the initial state S.

   2.2. Loop until the episode ends:

   2.2.1. (Collect experiences) Do the following once or multiple times:

   2.2.1.1. (Decide) For each action $a \in \mathcal{A}(S)$, draw c random samples $\Omega^{(i)}$ ($i = 0, 1, \ldots, c - 1$), and compute

$\phi\left(S, a, \Omega^{(i)}; \mathbf{w}\right)$, and then compute $q(S, a)$ $\leftarrow$ $\frac{1}{c} \sum_{i=0}^{c-1} \phi\left(S, a, \Omega^{(i)}; \mathbf{w}\right)$. Use the policy derived from $q(S, \cdot)$ (say ε-greedy policy) to determine the action A.

2.2.1.2. (Sample) Execute the action A, and then observe the reward R, the next state S', and the indicator of episode end D'.

2.2.1.3. Save the experience (S, A, R, S', D') in the storage $\mathcal{D}$.

2.2.1.4. $S \leftarrow S'$.

2.2.2. (Use experiences) Do the following once or multiple times:

2.2.2.1. (Replay) Sample a batch of experience $\mathcal{B}$ from the storage $\mathcal{D}$. Each entry is in the form of (S, A, R, S', D').

2.2.2.2. (Choose the next action for the next state) For each transition, when $D' = 0$, draw c random samples $\Omega'^{(i)}$ $(i = 0, 1, \ldots, c - 1)$ from the uniform distribution uniform$[0, 1]$, and then compute $\phi\left(S', a', \Omega'^{(i)}; \mathbf{w}\right)$ and $q(S', a') \leftarrow \frac{1}{c} \sum_{i=0}^{c-1} \phi\left(S', a', \Omega'^{(i)}; \mathbf{w}\right)$. Choose the next action using $A' \leftarrow \arg\max_{a'} q(S', a')$. The values can be arbitrarily assigned when $D' = 1$.

2.2.2.3. (Calculate TD returns) For each transition and each $j \in I$, when $D' = 0$, draw $\Omega'^{(j)}_{\text{target}} \sim$ uniform$[0, 1]$, and calculate TD return $U^{(j)} \leftarrow R + \gamma\phi\left(S', A', \Omega'^{(j)}_{\text{target}}; \mathbf{w}_{\text{target}}\right)$. The values can be arbitrarily set when $D' = 1$.

2.2.2.4. (Update action value parameters) Update $\mathbf{w}$ to minimize cross-entropy loss $\frac{1}{|\mathcal{B}|} \sum_{\substack{(S, A, R, S', D') \in \mathcal{B} \\ i, j \in I}} \ell_{\text{QRHubor}}\left(U^{(j)} - \phi\left(S, A, \Omega^{(i)}; \mathbf{w}\right); \Omega^{(i)}, \kappa\right)$.

2.2.2.5. (Update target network) Under some condition (say, every several updates), update the parameters of target network $\mathbf{w}_{\text{target}} \leftarrow \left(1 - \alpha_{\text{target}}\right)\mathbf{w}_{\text{target}} + \alpha_{\text{target}}\mathbf{w}$.

12.4.3 QR Algorithms with Utility

We can apply both VNM utility and Yarri utility in the quantile regression based algorithms, including QR-DQN and IQN.

The way to apply VNM utility is similar to that in Sect. 12.3.2. That is, when we make the decision, we need to consider the utility. Specifically speaking, we apply the utility function u on the output of quantile function $\phi(\cdot)$, leading to $u(\phi(\cdot))$. Then we estimate expectation and decide the action accordingly. In Algos. 12.3 and 12.4, we need to modify Steps 2.2.1.1 and 2.2.2.2.

Now we consider applying Yarri utility. When we use Yarri utility, we decide according to distorted expectation

$$\pi^{\langle\beta\rangle}(s) = \arg\max_a q^{\langle\beta\rangle}(s, a), \quad s \in \mathcal{S}.$$

Since it only modifies the expectation, we only need to modify Steps 2.2.1.1 and 2.2.2.2 in Algos. 12.3 and 12.4.

QR-DQN algorithm can use segmental approximation to calculate distorted expectation:

$$\begin{aligned}
E^{\langle\beta\rangle}[X] &= \int_0^1 \phi_X(\omega)d\beta^{-1}(\omega) \\
&= \sum_{i=0}^{|\mathcal{I}|-1} \int_{i/|\mathcal{I}|}^{(i+1)/|\mathcal{I}|} \phi_X(\omega)d\beta^{-1}(\omega) \\
&\approx \sum_{i=0}^{|\mathcal{I}|-1} \phi_X\left(\frac{i+0.5}{|\mathcal{I}|}\right)\left[\beta^{-1}\left(\frac{i+1}{|\mathcal{I}|}\right) - \beta^{-1}\left(\frac{i}{|\mathcal{I}|}\right)\right].
\end{aligned}$$

Consider the cumulated probability values

$$\omega^{(i)} = \frac{i+0.5}{|\mathcal{I}|}, \quad i = 0, 1, \ldots, |\mathcal{I}| - 1.$$

Let

$$\phi_X \overset{\text{def}}{=} \left(\phi_X\left(\frac{0.5}{|\mathcal{I}|}\right), \phi_X\left(\frac{1.5}{|\mathcal{I}|}\right), \ldots, \phi_X\left(\frac{|\mathcal{I}| - 0.5}{|\mathcal{I}|}\right)\right)^{\mathsf{T}}$$

$$\Delta\omega^{\langle\beta\rangle} \overset{\text{def}}{=} \left(\beta^{-1}\left(\frac{1}{|\mathcal{I}|}\right), \beta^{-1}\left(\frac{2}{|\mathcal{I}|}\right) - \beta^{-1}\left(\frac{1}{|\mathcal{I}|}\right), \ldots, 1 - \beta^{-1}\left(1 - \frac{1}{|\mathcal{I}|}\right)\right)^{\mathsf{T}}.$$

So we can get the pdistorted expectation as the inner product of the above two vectors, i.e.

$$E^{\langle\beta\rangle}[X] \approx \phi_X^{\mathsf{T}}\Delta\omega^{\langle\beta\rangle}.$$

The complete algorithm of QR-DQN with Yarri utility can be found in Algo. 12.5.

Algorithm 12.5 Categorical DQN to find the optimal policy (use Yarri distortion function).

Parameters: distortion function β, number of cumulated values for quantile function (furthermore we can get $\omega^{(i)} = (i + 0.5)/|\mathcal{I}|$ $(i = 0, 1, \ldots, |\mathcal{I}| - 1)$ and $\Delta\omega^{\langle\beta\rangle}$), and parameter for QR Huber loss κ, et al.

1. Initialize:

   1.1. (Initialize parameters) Initialize the parameters of action value evaluation network $\mathbf{w}$. Initialize the action value target network $\mathbf{w}_{\text{target}} \leftarrow \mathbf{w}$.

   1.2. (Initialize experience storage) $\mathcal{D} \leftarrow \varnothing$.

2. For each episode:

   2.1. (Initialize state) Choose the initial state S.

   2.2. Loop until the episode ends:

   2.2.1. (Collect experiences) Do the following once or multiple times:

   2.2.1.1. (Decide) For each action $a \in \mathcal{A}(S)$, compute $\phi(S, a; \mathbf{w})$, and then compute $q^{\langle\beta\rangle}(S, a) \leftarrow \left[\phi(S, a; \mathbf{w})\right]^{\top} \Delta\omega^{\langle\beta\rangle}$. Use the policy derived from $q^{\langle\beta\rangle}(S, \cdot)$ (say ε-greedy policy) to determine the action A.

   2.2.1.2. (Sample) Execute the action A, and then observe the reward R, the next state S', and the indicator of episode end D'.

   2.2.1.3. Save the experience (S, A, R, S', D') in the storage $\mathcal{D}$.

   2.2.1.4. $S \leftarrow S'$.

   2.2.2. (Use experiences) Do the following once or multiple times:

   2.2.2.1. (Replay) Sample a batch of experience $\mathcal{B}$ from the storage $\mathcal{D}$. Each entry is in the form of (S, A, R, S', D').

   2.2.2.2. (Choose the next action for the next state) For each transition, when $D' = 0$, calculate the quantile values for each next action $a' \in \mathcal{A}(S')$ as $\phi(S', a'; \mathbf{w})$, and then calculate the action value $q^{\langle\beta\rangle}(S', a') \leftarrow \left[\phi(S', a'; \mathbf{w})\right]^{\top} \Delta\omega^{\langle\beta\rangle}$. Choose the next action using $A' \leftarrow \arg\max_{a'} q^{\langle\beta\rangle}(S', a')$. The values can be arbitrarily assigned when $D' = 1$.

   2.2.2.3. (Calculate TD returns) For each transition and each $j \in \mathcal{I}$, when $D' = 0$, calculate the quantile value

$\phi^{(j)}\left(S', A'; \mathbf{w}_{\text{target}}\right)$ for each cumulated probability
$\omega^{(j)}$. The values can be arbitrarily set when
$D' = 1$. Then calculate $U^{(j)} \leftarrow R + \gamma(1 - D')\phi^{(j)}\left(S', A'; \mathbf{w}_{\text{target}}\right)$.

2.2.2.4. (Update action value parameters) Update $\mathbf{w}$ to reduce
$$\frac{1}{|\mathcal{B}|} \sum_{\substack{(S,A,R,S',D')\in\mathcal{B} \\ i,j\in I}} \ell_{\text{QRHubor}}\left(U^{(j)} - \phi^{(i)}(S_t, A_t; \mathbf{w}); \omega^{(i)}, \kappa\right).$$

2.2.2.5. (Update target network) Under some condition (say,
every several updates), update the parameters of
target network $\mathbf{w}_{\text{target}} \leftarrow \left(1 - \alpha_{\text{target}}\right)\mathbf{w}_{\text{target}} + \alpha_{\text{target}}\mathbf{w}$.

For IQN algorithm, when it calculates the distorted expectation, it still draws cumulated probability samples from the uniform distribution, and applies the distortion function β to these samples. Then the distorted cumulated probability values are sent to neural networks to get quantile values, and then average those quantile values to get the distorted expectation and decide accordingly. Therefore, we only need to change the form of calculating distorted expectation in Algo. 12.4 to

$$q(S, a) \leftarrow \frac{1}{c} \sum_{i=0}^{c-1} \phi\left(S, a, \beta\left(\Omega^{(i)}\right); \mathbf{w}\right).$$

12.5 Compare Categorical DQN and QR Algorithms

This chapter introduced Categorical DQN algorithm, QR-DQN algorithm, and IQN algorithm. This section compares the three algorithms (Table 12.1).

Table 12.1   Compare Categorical DQN, QR-DQN, and IQN.

algorithm	meaning of neural network	training objective	compatible utility
Categorical DQN	PMF (vector output)	minimize cross-entropy loss	VNM utility
QR-DQN	quantile function (vector output)	minimize QR Huber loss	VNM utility and Yarri utility
IQN	quantile function (cumulated probability as input, scaler output)	minimize QR Huber loss	VNM utility and Yarri utility

Categorical DQN algorithm uses neural networks to approximate PMF or PDF, while QR-DQN algorithm and IQN algorithm use neural networks to approximate quantile function.

Since Categorical DQN maintains PMF, it trains using cross-entropy loss. Since QR-DQN and IQN maintain quantile functions, they train using QR Huber loss.

All three algorithms can be used to maximize VNM utility, but only the algorithms that maintain the quantile function (i.e. QR-DQN and IQN) are suitable for maximizing Yarri utility.

Categorical DQN predefines the support set of the categorical distribution, and QR-DQN predefines the cumulated probabilities. Therefore, the neural networks in these two algorithms output vectors. IQN generates cumulated probability values randomly, so the inputs of the quantile networks include the cumulated probabilities, and calculation is usually conducted in batch.

QR-DQN needs to predefine the cumulated probability values, so the learning processing only focuses on the quantile values at given cumulated probabilities, and completely ignores the quantile values at other cumulated probabilities. This may introduce some issues. Contrarily, IQN randomly samples cumulated probabilities among $[0, 1]$ so that all possible cumulated probability values can be considered. This can avoid the overfitting to some specific cumulated probability values, and have better generalization performance.

12.6 Case Study: Atari Game Pong

This section uses distributional RL algorithms to play Pong, which is an Atari video games.

12.6.1 Atari Game Environment

Atari games are a collection of games that ran in the game console Atari 2600, which was put on sale in 1977. Players can insert different game cards into the console to load different games, and connect the console with TV to play video, and use handles to control. Later B. Matt developed the simulator Stella, so that Atari games can be played in Windows, macOS, and Linux. Stella was further wrapped and packaged. Especially, OpenAI integrated these Atari games into Gym, so that we can use Gym API to access these games.

The subpackage gym[atari] provides the environments for Atari games, including about 60 Atari games such as Breakout and Pong (Fig. 12.1). Each game has its own screen size. The screen size of most of games is 210×160, while some other games use the screen size 230×160 or 250×160. Different games have different action candidates. Different games have different ranges of episode rewards too. For example, the episode rewards of the game Pong ranges from -21 to 21, while the episode rewards of MontezumaRevenge is non-negative and can be thousands.

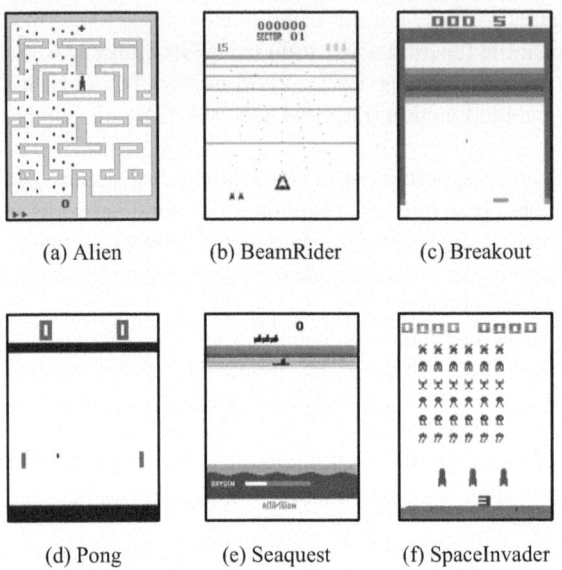

(a) Alien   (b) BeamRider   (c) Breakout

(d) Pong   (e) Seaquest   (f) SpaceInvader

Fig. 12.1   Some Atari games.
Animations can be found in https://www.gymlibrary.dev/environments/atari/complete_
list/.

Some Atari games are more friendly to humans, while others are more friendly to AI. For example, the game Pong is more friendly to AI. For the game Pong, a good AI player can attain the average episode reward 21, while a human player can only attain the average episode 14.6. MontezumaRevenge is more friendly to humans. Human players can attain 4753.3 on average, while most AI players can not gain any rewards, so the average episode rewards of an AI player is usually 0.

We can use the following command to install gym[atari]:

```
pip install --upgrade gym[atari,accept-rom-license,other]
```

Gym provides different versions for the same game. For example, there are 14 different versions for the same Pong (some of them are deprecated), shown in Table 12.2.

- Difference between the versions without ram and the versions with ram: Observations of the versions without ram is the RGB images. The observations usually can not fully determine the states. Observations of the versions with ram are the states in the memory. The observation spaces of these two categories are:

```
Box(low=0, high=255, shape=(length,width,3), dtype=np.uint8)  # without ram
Box(low=0, high=255, shape=(128,), dtype=np.uint8)  # with ram
```

- Difference among v0, v4, and v5: In the versions with v0, the agent has an additional constraint such that each time the agent picks an action, it is forced to use the same action with 25% probability in the next step. v5 is the latest version.
- Difference among the versions with Deterministic, the versions with NoFrameskip, and the versions without Deterministic or NoFrameskip: They differ on how many frames will proceed each time env.step() is called. For the versions with Deterministic, each call of env.step() proceeds 4 frames, gets the observation after 4 frames, and returns the reward summation of the 4 frames. For the versions with NoFrameskip, each call of env.step() proceeds only 1 frame, and gets the observation of the next frame. For the versions without either Deterministic and NoFrameskip, each call of env.step() proceeds τ frames, where τ is a random number among $\{2, 3, 4\}$. This can be understood as Semi-Markov Decision Process, which will be introduced in Sect. 15.4. If it is a v5 environment, $\tau = 4$.

Table 12.2 Different versions of the Pong game.

images as observations	memory as observations
Pong-v5	Pong-ram-v5
Pong-v0	Pong-ram-v0
Pong-v4	Pong-ram-v4
PongDeterministic-v0	Pong-ramDeterministic-v0
PongDeterministic-v4	Pong-ramDeterministic-v4
PongNoFrameskip-v0	Pong-ramNoFrameskip-v0
PongNoFrameskip-v4	Pong-ramNoFrameskip-v4

The usage of these environments is almost the same as other environments in Gym. We can use reset() to start a new episode, use step() for the next step. We can also check the maximum number of steps using env.spec.max_episode_steps. Additionally, there are some extra conventional when using the Atari environments, which will be introduced in subsequent subsections.

12.6.2 The Game Pong

This section introduces the Pong (Fig. 12.1(d)), which is the simplest game among all Atari games.

This game has a left racket and a right racket, where the left racket is controlled by some built-in logic of the game, and the right racket is controlled by the player. Our RL agent will control the right racket. There is also a ball in the middle, which may not be shown in the first few frames of each round. After few frames, the ball will move starting from the center of the screen. The ball will be bounced back when it encounters a racket. If a ball moves out of the screen across the left boundary or the right boundary, the round ends. There are also two digits in the upper part of the screen to show the number of rounds both parties have won in this episode.

In this game, each episode has multiple rounds. For each round, the player will be rewarded +1 if it wins the round, or it will be rewarded −1 if it loses the round. When a round finishes, the next round begins, until the player has wined or lost 21 rounds, or the maximum number of steps of the episode has been reached. Therefore, the reward in each episode ranges from −21 to 21. A smart agent can obtain average episode reward around 21.

The observation space of the game without ram is `Box(0, 255, (210, 160, 3), uint8)`. The action space is `Discrete(6)`. The action candidates are $\{0, 1, 2, 3, 4, 5\}$, where the action 0 and 1 mean do nothing, and action 2 and 4 try to move the pad upward (decrease the X value), and the action 3 and 5 try to move the pad downward (increase the X value).

Code 12.1 shows a closed-form solution for the environment PongNoFrameskip-v4. This solution first determines the X axis (i.e. horizontal axis) of the right racket and the ball by comparing with the colors. The RGB color of the right racket is (92, 186, 92), while the RGB color of the ball is (236, 236, 236). After getting the colors, compare the X values between the right racket and the ball. If the X value of the right racket is smaller than the ball, the agent use action 3 to increase the X value of the right racket. If the X value of the right racket is smaller than the ball, the agent use action 4 to decrease the X value of the right racket.

Code 12.1 Closed-form solution of `PongNoFrameskip-v4`.

PongNoFrameskip-v4_ClosedForm.ipynb

```
class ClosedFormAgent:
    def __init__(self, _):
        pass

    def reset(self, mode=None):
        pass

    def step(self, observation, reward, terminated):
        racket = np.where((observation[34:193, :, 0] == 92).any(
                axis=1))[0].mean()
        ball = np.where((observation[34:193, :, 0] == 236).any(
                axis=1))[0].mean()
        return 2 + int(racket < ball)

    def close(self):
        pass


agent = ClosedFormAgent(env)
```

12.6.3 Wrapper Class of Atari Environment

The wrapper class `AtariPreprocessing` and the wrapper class `FrameStack` are wrapper classes especially implemented for Atari games. Their functionalities include:

- Choose initial actions randomly: Randomly choose actions in the first 30 steps of the environment. This is because many Atari games, like Pong, do not really start the game at the beginning. It also prevents the agent from remembering some certain patterns at the episode start.
- Each step covers 4 frames: This limits the frequency that the agent changes the action in order to have a fair comparison with human.
- Rescale screen: Resize the screen to reduce computation, and remove the insignificant part. After this step, the screen size becomes (84, 84).
- Gray scale: Convert the colorful images to gray images.

Online Contents

Advanced readers can check the explanation of the class `gym.wrapper.AtariPreprocessing` and the class `gym.wrapper.FrameStack` in GitHub repo of this book.

Code 12.2 shows how to obtain an environment object using these two wrapper classes.

Code 12.2 Wrapped environment class.
`PongNoFrameskip-v4_CategoricalDQN_tf.ipynb`

```
1  env = gym.make('PongNoFrameskip-v4')
2  env = FrameStack(AtariPreprocessing(env), num_stack=4)
3  for key in vars(env):
4      logging.info('%s: %s', key, vars(env)[key])
5  for key in vars(env.spec):
6      logging.info('%s: %s', key, vars(env.spec)[key])
```

12.6.4 Use Categorical DQN to Solve Pong

This section implements Categorical DQN. Codes 12.3 and 12.4 show the codes that use TensorFlow and PyTorch respectively.

Code 12.3   Categorical DQN agent (TensorFlow version).

PongNoFrameskip-v4_CategoricalDQN_tf.ipynb

```
1   class CategoricalDQNAgent:
2       def __init__(self, env):
3           self.action_n = env.action_space.n
4           self.gamma = 0.99
5           self.epsilon = 1. # exploration
6
7           self.replayer = DQNReplayer(capacity=100000)
8
9           atom_count = 51
10          self.atom_min = -10.
11          self.atom_max = 10.
12          self.atom_difference = (self.atom_max - self.atom_min) / \
13                  (atom_count - 1)
14          self.atom_tensor = tf.linspace(self.atom_min, self.atom_max,
15                  atom_count)
16
17          self.evaluate_net = self.build_net(self.action_n, atom_count)
18          self.target_net = models.clone_model(self.evaluate_net)
19
20      def build_net(self, action_n, atom_count):
21          net = keras.Sequential([
22                  keras.layers.Permute((2, 3, 1), input_shape=(4, 84, 84)),
23                  layers.Conv2D(32, kernel_size=8, strides=4,
24                  activation=nn.relu),
25                  layers.Conv2D(64, kernel_size=4, strides=2,
26                  activation=nn.relu),
27                  layers.Conv2D(64, kernel_size=3, strides=1,
28                  activation=nn.relu),
29                  layers.Flatten(),
30                  layers.Dense(512, activation=nn.relu),
31                  layers.Dense(action_n * atom_count),
32                  layers.Reshape((action_n, atom_count)), layers.Softmax()])
33          optimizer = optimizers.Adam(0.0001)
34          net.compile(loss=losses.mse, optimizer=optimizer)
35          return net
36
37      def reset(self, mode=None):
38          self.mode = mode
39          if mode == 'train':
40              self.trajectory = []
41
42      def step(self, observation, reward, terminated):
43          state_tensor = tf.convert_to_tensor(np.array(observation)[np.newaxis],
44                  dtype=tf.float32)
45          prob_tensor = self.evaluate_net(state_tensor)
46          q_component_tensor = prob_tensor * self.atom_tensor
47          q_tensor = tf.reduce_mean(q_component_tensor, axis=2)
48          action_tensor = tf.math.argmax(q_tensor, axis=1)
49          actions = action_tensor.numpy()
50          action = actions[0]
51          if self.mode == 'train':
52              if np.random.rand() < self.epsilon:
53                  action = np.random.randint(0, self.action_n)
54
55              self.trajectory += [observation, reward, terminated, action]
56              if len(self.trajectory) >= 8:
57                  state, _, _, act, next_state, reward, terminated, _ = \
58                          self.trajectory[-8:]
59                  self.replayer.store(state, act, reward, next_state, terminated)
60              if self.replayer.count >= 1024 and self.replayer.count % 10 == 0:
61                  self.learn()
62          return action
```

```
63
64    def close(self):
65        pass
66
67    def update_net(self, target_net, evaluate_net, learning_rate=0.005):
68        average_weights = [(1. - learning_rate) * t + learning_rate * e for
69                t, e in zip(target_net.get_weights(),
70                evaluate_net.get_weights())]
71        target_net.set_weights(average_weights)
72
73    def learn(self):
74        # replay
75        batch_size = 32
76        states, actions, rewards, next_states, terminateds = \
77                self.replayer.sample(batch_size)
78        state_tensor = tf.convert_to_tensor(states, dtype=tf.float32)
79        reward_tensor = tf.convert_to_tensor(rewards[:, np.newaxis],
80                dtype=tf.float32)
81        terminated_tensor = tf.convert_to_tensor(terminateds[:, np.newaxis],
82                dtype=tf.float32)
83        next_state_tensor = tf.convert_to_tensor(next_states, dtype=tf.float32)
84
85        # compute target
86        next_prob_tensor = self.target_net(next_state_tensor)
87        next_q_tensor = tf.reduce_sum(next_prob_tensor * self.atom_tensor,
88                axis=2)
89        next_action_tensor = tf.math.argmax(next_q_tensor, axis=1)
90        next_actions = next_action_tensor.numpy()
91        indices = [[idx, next_action] for idx, next_action in
92                enumerate(next_actions)]
93        next_dist_tensor = tf.gather_nd(next_prob_tensor, indices)
94        next_dist_tensor = tf.reshape(next_dist_tensor,
95                shape=(batch_size, 1, -1))
96
97        # project
98        target_tensor = reward_tensor + self.gamma * tf.reshape(
99                self.atom_tensor, (1, -1)) * (1. - terminated_tensor)
100               # broadcast
101       clipped_target_tensor = tf.clip_by_value(target_tensor,
102               self.atom_min, self.atom_max)
103       projection_tensor = tf.clip_by_value(1. - tf.math.abs(
104               clipped_target_tensor[:, np.newaxis, ...]
105               - tf.reshape(self.atom_tensor, shape=(1, -1, 1)))
106               / self.atom_difference, 0, 1)
107       projected_tensor = tf.reduce_sum(projection_tensor * next_dist_tensor,
108               axis=-1)
109
110       with tf.GradientTape() as tape:
111           all_q_prob_tensor = self.evaluate_net(state_tensor)
112           indices = [[idx, action] for idx, action in enumerate(actions)]
113           q_prob_tensor = tf.gather_nd(all_q_prob_tensor, indices)
114
115           cross_entropy_tensor = -tf.reduce_sum(tf.math.xlogy(
116                   projected_tensor, q_prob_tensor + 1e-8))
117           loss_tensor = tf.reduce_mean(cross_entropy_tensor)
118       grads = tape.gradient(loss_tensor, self.evaluate_net.variables)
119       self.evaluate_net.optimizer.apply_gradients(
120               zip(grads, self.evaluate_net.variables))
121
122       self.update_net(self.target_net, self.evaluate_net)
123
124       self.epsilon = max(self.epsilon - 1e-5, 0.05)
125
126
127 agent = CategoricalDQNAgent(env)
```

Code 12.4   Categorical DQN agent (PyTorch version).

`PongNoFrameskip-v4_CategoricalDQN_torch.ipynb`

```python
class CategoricalDQNAgent:
    def __init__(self, env):
        self.action_n = env.action_space.n
        self.gamma = 0.99
        self.epsilon = 1.  # exploration

        self.replayer = DQNReplayer(capacity=100000)

        self.atom_count = 51
        self.atom_min = -10.
        self.atom_max = 10.
        self.atom_difference = (self.atom_max - self.atom_min) \
                / (self.atom_count - 1)
        self.atom_tensor = torch.linspace(self.atom_min, self.atom_max,
                self.atom_count)

        self.evaluate_net = nn.Sequential(
                nn.Conv2d(4, 32, kernel_size=8, stride=4), nn.ReLU(),
                nn.Conv2d(32, 64, kernel_size=4, stride=2), nn.ReLU(),
                nn.Conv2d(64, 64, kernel_size=3, stride=1), nn.ReLU(),
                nn.Flatten(),
                nn.Linear(3136, 512), nn.ReLU(inplace=True),
                nn.Linear(512, self.action_n * self.atom_count))
        self.target_net = copy.deepcopy(self.evaluate_net)
        self.optimizer = optim.Adam(self.evaluate_net.parameters(), lr=0.0001)

    def reset(self, mode=None):
        self.mode = mode
        if mode == 'train':
            self.trajectory = []

    def step(self, observation, reward, terminated):
        state_tensor = torch.as_tensor(observation,
                dtype=torch.float).unsqueeze(0)
        logit_tensor = self.evaluate_net(state_tensor).view(-1, self.action_n,
                self.atom_count)
        prob_tensor = logit_tensor.softmax(dim=-1)
        q_component_tensor = prob_tensor * self.atom_tensor
        q_tensor = q_component_tensor.mean(2)
        action_tensor = q_tensor.argmax(dim=1)
        actions = action_tensor.detach().numpy()
        action = actions[0]
        if self.mode == 'train':
            if np.random.rand() < self.epsilon:
                action = np.random.randint(0, self.action_n)

            self.trajectory += [observation, reward, terminated, action]
            if len(self.trajectory) >= 8:
                state, _, _, act, next_state, reward, terminated, _ = \
                        self.trajectory[-8:]
                self.replayer.store(state, act, reward, next_state, terminated)
            if self.replayer.count >= 1024 and self.replayer.count % 10 == 0:
                self.learn()
        return action

    def close(self):
        pass

    def update_net(self, target_net, evaluate_net, learning_rate=0.005):
        for target_param, evaluate_param in zip(
                target_net.parameters(), evaluate_net.parameters()):
            target_param.data.copy_(learning_rate * evaluate_param.data
```

```
63                    + (1 - learning_rate) * target_param.data)
64
65    def learn(self):
66        # replay
67        batch_size = 32
68        states, actions, rewards, next_states, terminateds = \
69                self.replayer.sample(batch_size)
70        state_tensor = torch.as_tensor(states, dtype=torch.float)
71        reward_tensor = torch.as_tensor(rewards, dtype=torch.float)
72        terminated_tensor = torch.as_tensor(terminateds, dtype=torch.float)
73        next_state_tensor = torch.as_tensor(next_states, dtype=torch.float)
74
75        # compute target
76        next_logit_tensor = self.target_net(next_state_tensor).view(-1,
77                self.action_n, self.atom_count)
78        next_prob_tensor = next_logit_tensor.softmax(dim=-1)
79        next_q_tensor = (next_prob_tensor * self.atom_tensor).sum(2)
80        next_action_tensor = next_q_tensor.argmax(dim=1)
81        next_actions = next_action_tensor.detach().numpy()
82        next_dist_tensor = next_prob_tensor[np.arange(batch_size),
83                next_actions, :].unsqueeze(1)
84
85        # project
86        target_tensor = reward_tensor.reshape(batch_size, 1) + self.gamma \
87                * self.atom_tensor.repeat(batch_size, 1) \
88                * (1. - terminated_tensor).reshape(-1, 1)
89        clipped_target_tensor = target_tensor.clamp(self.atom_min,
90                self.atom_max)
91        projection_tensor = (1. - (clipped_target_tensor.unsqueeze(1)
92                - self.atom_tensor.view(1, -1, 1)).abs()
93                / self.atom_difference).clamp(0, 1)
94        projected_tensor = (projection_tensor * next_dist_tensor).sum(-1)
95
96        logit_tensor = self.evaluate_net(state_tensor).view(-1, self.action_n,
97                self.atom_count)
98        all_q_prob_tensor = logit_tensor.softmax(dim=-1)
99        q_prob_tensor = all_q_prob_tensor[range(batch_size), actions, :]
100
101        cross_entropy_tensor = -torch.xlogy(projected_tensor, q_prob_tensor
102                + 1e-8).sum(1)
103        loss_tensor = cross_entropy_tensor.mean()
104        self.optimizer.zero_grad()
105        loss_tensor.backward()
106        self.optimizer.step()
107
108        self.update_net(self.target_net, self.evaluate_net)
109
110        self.epsilon = max(self.epsilon - 1e-5, 0.05)
111
112
113 agent = CategoricalDQNAgent(env)
```

The observation of the environment is the stack of multiple images. For the case when the network input is images, we usually use Convolutional Neural Network (CNN). Codes 12.3 and 12.4 use the neural networks in Fig. 12.2. The convolution part uses three ReLU activated convolution layers, while the fully connected part uses two ReLU activated linear layers. Since the action space is finite, we only input states to the network, while the actions are used to select among the outputs, which are the probability estimates of all actions.

The size of support size of categorical distribution is set to 51. In fact, 5 supporting values suffice for the game Pong. Here we use 51 for being consistent with the original paper.

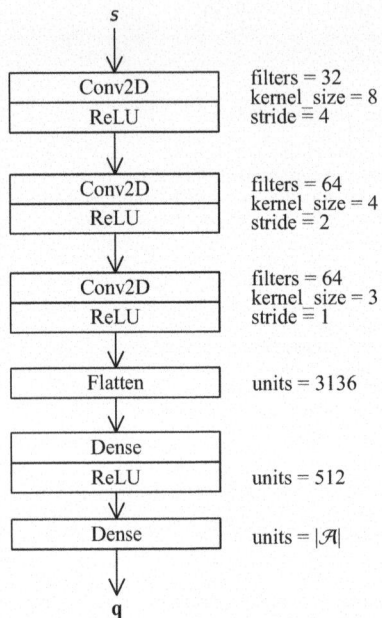

Fig. 12.2   Neural network for Categorical DQN.

Exploration uses ε-greedy, where the parameter ε decreases during the training. Interaction between this agent and the environment once again uses Code 1.3.

12.6.5 Use QR-DQN to Solve Pong

Codes 12.5 and 12.6 implement QR-DQN agent. It uses the same neural network in Fig. 12.2. Once again, since the action space is finite, only states are used as the input of the network, while the actions are used to select among outputs, which are the quantile values of all actions. Neither TensorFlow nor PyTorch has implemented the QR Huber loss, so we need to implement it by ourselves.

Code 12.5   QR-DQN agent (TensorFlow version).
PongNoFrameskip-v4_QRDQN_tf.ipynb

```
1  class QRDQNAgent:
2      def __init__(self, env):
3          self.action_n = env.action_space.n
4          self.gamma = 0.99
5          self.epsilon = 1.
6
7          self.replayer = DQNReplayer(capacity=100000)
8
9          quantile_count = 64
```

```
10          self.cumprob_tensor = tf.range(1 / (2 * quantile_count),
11                  1, 1 / quantile_count)[np.newaxis, :, np.newaxis]
12
13          self.evaluate_net = self.build_net(self.action_n, quantile_count)
14          self.target_net = models.clone_model(self.evaluate_net)
15
16      def build_net(self, action_n, quantile_count):
17          net = keras.Sequential([
18                  keras.layers.Permute((2, 3, 1), input_shape=(4, 84, 84)),
19                  layers.Conv2D(32, kernel_size=8, strides=4,
20                  activation=nn.relu),
21                  layers.Conv2D(64, kernel_size=4, strides=2,
22                  activation=nn.relu),
23                  layers.Conv2D(64, kernel_size=3, strides=1,
24                  activation=nn.relu),
25                  layers.Flatten(),
26                  layers.Dense(512, activation=nn.relu),
27                  layers.Dense(action_n * quantile_count),
28                  layers.Reshape((action_n, quantile_count))])
29          optimizer = optimizers.Adam(0.0001)
30          net.compile(optimizer=optimizer)
31          return net
32
33      def reset(self, mode=None):
34          self.mode = mode
35          if mode == 'train':
36              self.trajectory = []
37
38      def step(self, observation, reward, terminated):
39          state_tensor = tf.convert_to_tensor(np.array(observation)[np.newaxis],
40                  dtype=tf.float32)
41          q_component_tensor = self.evaluate_net(state_tensor)
42          q_tensor = tf.reduce_mean(q_component_tensor, axis=2)
43          action_tensor = tf.math.argmax(q_tensor, axis=1)
44          actions = action_tensor.numpy()
45          action = actions[0]
46          if self.mode == 'train':
47              if np.random.rand() < self.epsilon:
48                  action = np.random.randint(0, self.action_n)
49
50              self.trajectory += [observation, reward, terminated, action]
51              if len(self.trajectory) >= 8:
52                  state, _, _, act, next_state, reward, terminated, _ = \
53                          self.trajectory[-8:]
54                  self.replayer.store(state, act, reward, next_state, terminated)
55              if self.replayer.count >= 1024 and self.replayer.count % 10 == 0:
56                  self.learn()
57          return action
58
59      def close(self):
60          pass
61
62      def update_net(self, target_net, evaluate_net, learning_rate=0.005):
63          average_weights = [(1. - learning_rate) * t + learning_rate * e for
64                  t, e in zip(target_net.get_weights(),
65                  evaluate_net.get_weights())]
66          target_net.set_weights(average_weights)
67
68      def learn(self):
69          # replay
70          batch_size = 32
71          states, actions, rewards, next_states, terminateds = \
72                  self.replayer.sample(batch_size)
73          state_tensor = tf.convert_to_tensor(states, dtype=tf.float32)
74          reward_tensor = tf.convert_to_tensor(rewards[:, np.newaxis],
75                  dtype=tf.float32)
76          terminated_tensor = tf.convert_to_tensor(terminateds[:, np.newaxis],
```

```
77              dtype=tf.float32)
78          next_state_tensor = tf.convert_to_tensor(next_states, dtype=tf.float32)
79
80          # compute target
81          next_q_component_tensor = self.evaluate_net(next_state_tensor)
82          next_q_tensor = tf.reduce_mean(next_q_component_tensor, axis=2)
83          next_action_tensor = tf.math.argmax(next_q_tensor, axis=1)
84          next_actions = next_action_tensor.numpy()
85          all_next_q_quantile_tensor = self.target_net(next_state_tensor)
86          indices = [[idx, next_action] for idx, next_action in
87                  enumerate(next_actions)]
88          next_q_quantile_tensor = tf.gather_nd(all_next_q_quantile_tensor,
89                  indices)
90          target_quantile_tensor = reward_tensor + self.gamma \
91                  * next_q_quantile_tensor * (1. - terminated_tensor)
92
93          with tf.GradientTape() as tape:
94              all_q_quantile_tensor = self.evaluate_net(state_tensor)
95              indices = [[idx, action] for idx, action in enumerate(actions)]
96              q_quantile_tensor = tf.gather_nd(all_q_quantile_tensor, indices)
97
98              target_quantile_tensor = target_quantile_tensor[:, np.newaxis, :]
99              q_quantile_tensor = q_quantile_tensor[:, :, np.newaxis]
100             td_error_tensor = target_quantile_tensor - q_quantile_tensor
101             abs_td_error_tensor = tf.math.abs(td_error_tensor)
102             hubor_delta = 1.
103             hubor_loss_tensor = tf.where(abs_td_error_tensor < hubor_delta,
104                     0.5 * tf.square(td_error_tensor),
105                     hubor_delta * (abs_td_error_tensor - 0.5 * hubor_delta))
106             comparison_tensor = tf.cast(td_error_tensor < 0, dtype=tf.float32)
107             quantile_regression_tensor = tf.math.abs(self.cumprob_tensor -
108                     comparison_tensor)
109             quantile_huber_loss_tensor = tf.reduce_mean(tf.reduce_sum(
110                     hubor_loss_tensor * quantile_regression_tensor, axis=-1),
111                     axis=1)
112             loss_tensor = tf.reduce_mean(quantile_huber_loss_tensor)
113         grads = tape.gradient(loss_tensor, self.evaluate_net.variables)
114         self.evaluate_net.optimizer.apply_gradients(
115                 zip(grads, self.evaluate_net.variables))
116
117         self.update_net(self.target_net, self.evaluate_net)
118
119         self.epsilon = max(self.epsilon - 1e-5, 0.05)
120
121
122 agent = QRDQNAgent(env)
```

Code 12.6   QR-DQN agent (PyTorch version).
PongNoFrameskip-v4_QRDQN_torch.ipynb

```
1  class QRDQNAgent:
2      def __init__(self, env):
3          self.action_n = env.action_space.n
4          self.gamma = 0.99
5          self.epsilon = 1.
6
7          self.replayer = DQNReplayer(capacity=100000)
8
9          self.quantile_count = 64
10         self.cumprob_tensor = torch.arange(1 / (2 * self.quantile_count),
11                 1, 1 / self.quantile_count).view(1, -1, 1)
12
13         self.evaluate_net = nn.Sequential(
14                 nn.Conv2d(4, 32, kernel_size=8, stride=4), nn.ReLU(),
15                 nn.Conv2d(32, 64, kernel_size=4, stride=2), nn.ReLU(),
```

```python
                nn.Conv2d(64, 64, kernel_size=3, stride=1), nn.ReLU(),
                nn.Flatten(),
                nn.Linear(in_features=3136, out_features=512), nn.ReLU(),
                nn.Linear(in_features=512,
                    out_features=self.action_n * self.quantile_count))
        self.target_net = copy.deepcopy(self.evaluate_net)
        self.optimizer = optim.Adam(self.evaluate_net.parameters(), lr=0.0001)

        self.loss = nn.SmoothL1Loss(reduction="none")

    def reset(self, mode=None):
        self.mode = mode
        if mode == 'train':
            self.trajectory = []

    def step(self, observation, reward, terminated):
        state_tensor = torch.as_tensor(observation,
                dtype=torch.float).unsqueeze(0)
        q_component_tensor = self.evaluate_net(state_tensor).view(-1,
                self.action_n, self.quantile_count)
        q_tensor = q_component_tensor.mean(2)
        action_tensor = q_tensor.argmax(dim=1)
        actions = action_tensor.detach().numpy()
        action = actions[0]
        if self.mode == 'train':
            if np.random.rand() < self.epsilon:
                action = np.random.randint(0, self.action_n)

            self.trajectory += [observation, reward, terminated, action]
            if len(self.trajectory) >= 8:
                state, _, _, act, next_state, reward, terminated, _ = \
                        self.trajectory[-8:]
                self.replayer.store(state, act, reward, next_state, terminated)
            if self.replayer.count >= 1024 and self.replayer.count % 10 == 0:
                self.learn()
        return action

    def close(self):
        pass

    def update_net(self, target_net, evaluate_net, learning_rate=0.005):
        for target_param, evaluate_param in zip(
                target_net.parameters(), evaluate_net.parameters()):
            target_param.data.copy_(learning_rate * evaluate_param.data
                    + (1 - learning_rate) * target_param.data)

    def learn(self):
        # replay
        batch_size = 32
        states, actions, rewards, next_states, terminateds = \
                self.replayer.sample(batch_size)
        state_tensor = torch.as_tensor(states, dtype=torch.float)
        reward_tensor = torch.as_tensor(rewards, dtype=torch.float)
        terminated_tensor = torch.as_tensor(terminateds, dtype=torch.float)
        next_state_tensor = torch.as_tensor(next_states, dtype=torch.float)

        # compute target
        next_q_component_tensor = self.evaluate_net(next_state_tensor).view(
                -1, self.action_n, self.quantile_count)
        next_q_tensor = next_q_component_tensor.mean(2)
        next_action_tensor = next_q_tensor.argmax(dim=1)
        next_actions = next_action_tensor.detach().numpy()
        all_next_q_quantile_tensor = self.target_net(next_state_tensor
                ).view(-1, self.action_n, self.quantile_count)
        next_q_quantile_tensor = all_next_q_quantile_tensor[
                range(batch_size), next_actions, :]
        target_quantile_tensor = reward_tensor.reshape(batch_size, 1) \
```

```
83          + self.gamma * next_q_quantile_tensor \
84          * (1. - terminated_tensor).reshape(-1, 1)
85
86     all_q_quantile_tensor = self.evaluate_net(state_tensor).view(-1,
87              self.action_n, self.quantile_count)
88     q_quantile_tensor = all_q_quantile_tensor[range(batch_size), actions,
89          :]
90
91     target_quantile_tensor = target_quantile_tensor.unsqueeze(1)
92     q_quantile_tensor = q_quantile_tensor.unsqueeze(2)
93     hubor_loss_tensor = self.loss(target_quantile_tensor,
94          q_quantile_tensor)
95     comparison_tensor = (target_quantile_tensor
96          < q_quantile_tensor).detach().float()
97     quantile_regression_tensor = (self.cumprob_tensor
98          - comparison_tensor).abs()
99     quantile_huber_loss_tensor = (hubor_loss_tensor
100         * quantile_regression_tensor).sum(-1).mean(1)
101    loss_tensor = quantile_huber_loss_tensor.mean()
102    self.optimizer.zero_grad()
103    loss_tensor.backward()
104    self.optimizer.step()
105
106    self.update_net(self.target_net, self.evaluate_net)
107
108    self.epsilon = max(self.epsilon - 1e-5, 0.05)
109
110
111 agent = QRDQNAgent(env)
```

Interaction between this agent and the environment once again uses Code 1.3.

12.6.6 Use IQN to Solve Pong

The network structure of IQN differs from those of Categorical DQN and QR-DQN. In Algo. 12.4, the inputs of the quantile network are state–action pair (s, a) and cumulative probability ω. For the same reason in Categorical DQN and QR-DQN, since the action space of Pong is finite, we do not use action as input, but instead we use action to select among network outputs. Additionally, we not only need to input state s to the network, but also need to input the cumulative probability ω to the vector $\cos(\pi\omega\iota)$, where $\iota = (1, 2, \ldots, k)^\top$, k is the length of the vector, and $\cos()$ is element-wise cosine function. Then we pass the vector to the fully connected network and get the embedded features. Then we conduct element-wise multiplications on CNN features and embedded features to combine these two parts. The combined results are further passed to another fully connected network to get the final quantile outputs. The structure and codes of the network are shown in Fig. 12.3, Codes 12.7 and 12.8.

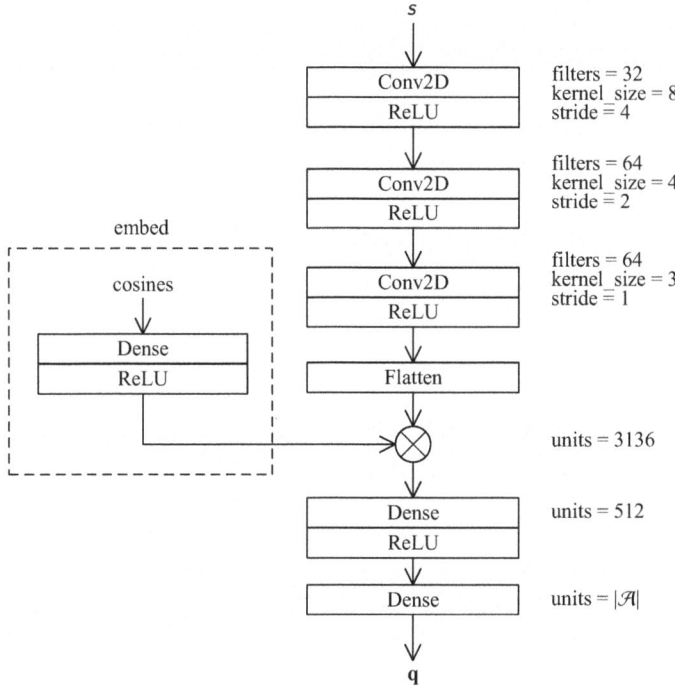

Fig. 12.3 Neural network for IQN.

Code 12.7   Quantile network (TensorFlow version).

PongNoFrameskip-v4_IQN_tf.ipynb

```
class Net(keras.Model):
    def __init__(self, action_n, sample_count, cosine_count):
        super().__init__()
        self.cosine_count = cosine_count
        self.conv = keras.Sequential([
                keras.layers.Permute((2, 3, 1), input_shape=(4, 84, 84)),
                layers.Conv2D(32, kernel_size=8, strides=4,
                activation=nn.relu),
                layers.Conv2D(64, kernel_size=4, strides=2,
                activation=nn.relu),
                layers.Conv2D(64, kernel_size=3, strides=1,
                activation=nn.relu),
                layers.Reshape((1, 3136))])
        self.emb = keras.Sequential([
                layers.Dense(3136, activation=nn.relu,
                input_shape=(sample_count, cosine_count))])
        self.fc = keras.Sequential([
                layers.Dense(512, activation=nn.relu),
                layers.Dense(action_n),
                layers.Permute((2, 1))])

    def call(self, input_tensor, cumprob_tensor):
```

```
23        logit_tensor = self.conv(input_tensor)
24        index_tensor = tf.range(1, self.cosine_count + 1, dtype=tf.float32)[
25                np.newaxis, np.newaxis, :]
26        cosine_tensor = tf.math.cos(index_tensor * np.pi * cumprob_tensor)
27        emb_tensor = self.emb(cosine_tensor)
28        prod_tensor = logit_tensor * emb_tensor
29        output_tensor = self.fc(prod_tensor)
30        return output_tensor
```

Code 12.8   Quantile network (PyTorch version).

PongNoFrameskip-v4_IQN_torch.ipynb

```
1  class Net(nn.Module):
2      def __init__(self, action_n, sample_count, cosine_count=64):
3          super().__init__()
4          self.sample_count = sample_count
5          self.cosine_count = cosine_count
6          self.conv = nn.Sequential(
7                  nn.Conv2d(4, 32, kernel_size=8, stride=4), nn.ReLU(),
8                  nn.Conv2d(32, 64, kernel_size=4, stride=2), nn.ReLU(),
9                  nn.Conv2d(64, 64, kernel_size=3, stride=1), nn.ReLU(),
10                 nn.Flatten())
11         self.emb = nn.Sequential(
12                 nn.Linear(in_features=64, out_features=3136), nn.ReLU())
13         self.fc = nn.Sequential(
14                 nn.Linear(in_features=3136, out_features=512), nn.ReLU(),
15                 nn.Linear(in_features=512, out_features=action_n))
16
17     def forward(self, input_tensor, cumprob_tensor):
18         batch_size = input_tensor.size(0)
19         logit_tensor = self.conv(input_tensor).unsqueeze(1)
20         index_tensor = torch.arange(start=1, end=self.cosine_count + 1).view(1,
21                 1, self.cosine_count)
22         cosine_tensor = torch.cos(index_tensor * np.pi * cumprob_tensor)
23         emb_tensor = self.emb(cosine_tensor)
24         prod_tensor = logit_tensor * emb_tensor
25         output_tensor = self.fc(prod_tensor).transpose(1, 2)
26         return output_tensor
```

Codes 12.9 and 12.10 implement the IQN agent. Each time we estimate the action value, we sample 8 cumulated probability values.

Code 12.9   IQN agent (TensorFlow version).

PongNoFrameskip-v4_IQN_tf.ipynb

```
1  class IQNAgent:
2      def __init__(self, env):
3          self.action_n = env.action_space.n
4          self.gamma = 0.99
5          self.epsilon = 1.
6
7          self.replayer = DQNReplayer(capacity=100000)
8
9          self.sample_count = 8
10         self.evaluate_net = self.build_net(action_n=self.action_n,
11                 sample_count=self.sample_count)
12         self.target_net = self.build_net(action_n=self.action_n,
13                 sample_count=self.sample_count)
14
15     def build_net(self, action_n, sample_count, cosine_count=64):
16         net = Net(action_n, sample_count, cosine_count)
17         loss = losses.Huber(reduction="none")
18         optimizer = optimizers.Adam(0.0001)
```

```
19          net.compile(loss=loss, optimizer=optimizer)
20          return net
21
22      def reset(self, mode=None):
23          self.mode = mode
24          if mode == 'train':
25              self.trajectory = []
26
27      def step(self, observation, reward, terminated):
28          state_tensor = tf.convert_to_tensor(np.array(observation)[np.newaxis],
29                  dtype=tf.float32)
30          prob_tensor = tf.random.uniform((1, self.sample_count, 1))
31          q_component_tensor = self.evaluate_net(state_tensor, prob_tensor)
32          q_tensor = tf.reduce_mean(q_component_tensor, axis=2)
33          action_tensor = tf.math.argmax(q_tensor, axis=1)
34          actions = action_tensor.numpy()
35          action = actions[0]
36          if self.mode == 'train':
37              if np.random.rand() < self.epsilon:
38                  action = np.random.randint(0, self.action_n)
39              self.trajectory += [observation, reward, terminated, action]
40              if len(self.trajectory) >= 8:
41                  state, _, _, act, next_state, reward, terminated, _ = \
42                          self.trajectory[-8:]
43                  self.replayer.store(state, act, reward, next_state, terminated)
44                  if self.replayer.count >= 1024 and self.replayer.count % 10 == 0:
45                      self.learn()
46          return action
47
48      def close(self):
49          pass
50
51      def update_net(self, target_net, evaluate_net, learning_rate=0.005):
52          average_weights = [(1. - learning_rate) * t + learning_rate * e for
53                  t, e in zip(target_net.get_weights(),
54                  evaluate_net.get_weights())]
55          target_net.set_weights(average_weights)
56
57      def learn(self):
58          # replay
59          batch_size = 32
60          states, actions, rewards, next_states, terminateds = \
61                  self.replayer.sample(batch_size)
62          state_tensor = tf.convert_to_tensor(states, dtype=tf.float32)
63          reward_tensor = tf.convert_to_tensor(rewards[:, np.newaxis],
64                  dtype=tf.float32)
65          terminated_tensor = tf.convert_to_tensor(terminateds[:, np.newaxis],
66                  dtype=tf.float32)
67          next_state_tensor = tf.convert_to_tensor(next_states, dtype=tf.float32)
68
69          # calculate target
70          next_cumprob_tensor = tf.random.uniform((batch_size, self.sample_count,
71                  1))
72          next_q_component_tensor = self.evaluate_net(next_state_tensor,
73                  next_cumprob_tensor)
74          next_q_tensor = tf.reduce_mean(next_q_component_tensor, axis=2)
75          next_action_tensor = tf.math.argmax(next_q_tensor, axis=1)
76          next_actions = next_action_tensor.numpy()
77          next_cumprob_tensor = tf.random.uniform((batch_size, self.sample_count,
78                  1))
79          all_next_q_quantile_tensor = self.target_net(next_state_tensor,
80                  next_cumprob_tensor)
81          indices = [[idx, next_action] for idx, next_action in
82                  enumerate(next_actions)]
83          next_q_quantile_tensor = tf.gather_nd(all_next_q_quantile_tensor,
84                  indices)
85          target_quantile_tensor = reward_tensor + self.gamma \
```

```
86              * next_q_quantile_tensor * (1. - terminated_tensor)
87
88        with tf.GradientTape() as tape:
89            cumprob_tensor = tf.random.uniform((batch_size, self.sample_count,
90                    1))
91            all_q_quantile_tensor = self.evaluate_net(state_tensor,
92                    cumprob_tensor)
93            indices = [[idx, action] for idx, action in enumerate(actions)]
94            q_quantile_tensor = tf.gather_nd(all_q_quantile_tensor, indices)
95            target_quantile_tensor = target_quantile_tensor[:, np.newaxis, :]
96            q_quantile_tensor = q_quantile_tensor[:, :, np.newaxis]
97            td_error_tensor = target_quantile_tensor - q_quantile_tensor
98            abs_td_error_tensor = tf.math.abs(td_error_tensor)
99            hubor_delta = 1.
100           hubor_loss_tensor = tf.where(abs_td_error_tensor < hubor_delta,
101                   0.5 * tf.square(td_error_tensor),
102                   hubor_delta * (abs_td_error_tensor - 0.5 * hubor_delta))
103           comparison_tensor = tf.cast(td_error_tensor < 0, dtype=tf.float32)
104           quantile_regression_tensor = tf.math.abs(cumprob_tensor -
105                   comparison_tensor)
106           quantile_huber_loss_tensor = tf.reduce_mean(tf.reduce_sum(
107                   hubor_loss_tensor * quantile_regression_tensor, axis=-1),
108                   axis=1)
109           loss_tensor = tf.reduce_mean(quantile_huber_loss_tensor)
110       grads = tape.gradient(loss_tensor, self.evaluate_net.variables)
111       self.evaluate_net.optimizer.apply_gradients(
112               zip(grads, self.evaluate_net.variables))
113
114       self.update_net(self.target_net, self.evaluate_net)
115
116       self.epsilon = max(self.epsilon - 1e-5, 0.05)
117
118
119 agent = IQNAgent(env)
```

Code 12.10   IQN agent (PyTorch version).
PongNoFrameskip-v4_IQN_torch.ipynb

```
1  class IQNAgent:
2      def __init__(self, env):
3          self.action_n = env.action_space.n
4          self.gamma = 0.99
5          self.epsilon = 1.
6
7          self.replayer = DQNReplayer(capacity=100000)
8
9          self.sample_count = 8
10         self.evaluate_net = Net(action_n=self.action_n,
11                 sample_count=self.sample_count)
12         self.target_net = copy.deepcopy(self.evaluate_net)
13         self.optimizer = optim.Adam(self.evaluate_net.parameters(), lr=0.0001)
14         self.loss = nn.SmoothL1Loss(reduction="none")
15
16     def reset(self, mode=None):
17         self.mode = mode
18         if mode == 'train':
19             self.trajectory = []
20
21     def step(self, observation, reward, terminated):
22         state_tensor = torch.as_tensor(observation,
23                 dtype=torch.float).unsqueeze(0)
24         cumprod_tensor = torch.rand(1, self.sample_count, 1)
25         q_component_tensor = self.evaluate_net(state_tensor, cumprod_tensor)
26         q_tensor = q_component_tensor.mean(2)
27         action_tensor = q_tensor.argmax(dim=1)
```

```
28          actions = action_tensor.detach().numpy()
29          action = actions[0]
30          if self.mode == 'train':
31              if np.random.rand() < self.epsilon:
32                  action = np.random.randint(0, self.action_n)
33
34              self.trajectory += [observation, reward, terminated, action]
35              if len(self.trajectory) >= 8:
36                  state, _, _, act, next_state, reward, terminated, _ = \
37                          self.trajectory[-8:]
38                  self.replayer.store(state, act, reward, next_state, terminated)
39                  if self.replayer.count >= 1024 and self.replayer.count % 10 == 0:
40                      self.learn()
41          return action
42
43      def close(self):
44          pass
45
46      def update_net(self, target_net, evaluate_net, learning_rate=0.005):
47          for target_param, evaluate_param in zip(
48                  target_net.parameters(), evaluate_net.parameters()):
49              target_param.data.copy_(learning_rate * evaluate_param.data
50                      + (1 - learning_rate) * target_param.data)
51
52      def learn(self):
53          # replay
54          batch_size = 32
55          states, actions, rewards, next_states, terminateds = \
56                  self.replayer.sample(batch_size)
57          state_tensor = torch.as_tensor(states, dtype=torch.float)
58          reward_tensor = torch.as_tensor(rewards, dtype=torch.float)
59          terminated_tensor = torch.as_tensor(terminateds, dtype=torch.float)
60          next_state_tensor = torch.as_tensor(next_states, dtype=torch.float)
61
62          # calculate target
63          next_cumprob_tensor = torch.rand(batch_size, self.sample_count, 1)
64          next_q_component_tensor = self.evaluate_net(next_state_tensor,
65                  next_cumprob_tensor)
66          next_q_tensor = next_q_component_tensor.mean(2)
67          next_action_tensor = next_q_tensor.argmax(dim=1)
68          next_actions = next_action_tensor.detach().numpy()
69          next_cumprob_tensor = torch.rand(batch_size, self.sample_count, 1)
70          all_next_q_quantile_tensor = self.target_net(next_state_tensor,
71                  next_cumprob_tensor)
72          next_q_quantile_tensor = all_next_q_quantile_tensor[
73                  range(batch_size), next_actions, :]
74          target_quantile_tensor = reward_tensor.reshape(batch_size, 1) \
75                  + self.gamma * next_q_quantile_tensor \
76                  * (1. - terminated_tensor).reshape(-1, 1)
77
78          cumprob_tensor = torch.rand(batch_size, self.sample_count, 1)
79          all_q_quantile_tensor = self.evaluate_net(state_tensor, cumprob_tensor)
80          q_quantile_tensor = all_q_quantile_tensor[range(batch_size), actions,
81                  :]
82          target_quantile_tensor = target_quantile_tensor.unsqueeze(1)
83          q_quantile_tensor = q_quantile_tensor.unsqueeze(2)
84          hubor_loss_tensor = self.loss(target_quantile_tensor,
85                  q_quantile_tensor)
86          comparison_tensor = (target_quantile_tensor <
87                  q_quantile_tensor).detach().float()
88          quantile_regression_tensor = (cumprob_tensor - comparison_tensor).abs()
89          quantile_huber_loss_tensor = (hubor_loss_tensor *
90                  quantile_regression_tensor).sum(-1).mean(1)
91          loss_tensor = quantile_huber_loss_tensor.mean()
92          self.optimizer.zero_grad()
93          loss_tensor.backward()
94          self.optimizer.step()
```

```
95
96        self.update_net(self.target_net, self.evaluate_net)
97
98        self.epsilon = max(self.epsilon - 1e-5, 0.05)
99
100
101   agent = IQNAgent(env)
```

Interaction between this agent and the environment once again uses Code 1.3.

12.7 Summary

- Distributional RL considers the distribution of values, especially action values.
- Distributional RL can maximize utility, including von Neumann Morgenstern (VNM) utility and Yarri utility. VNM utility function u is applied to the action values, while distortion function of the Yarri utility $\beta : [0, 1] \rightarrow [0, 1]$ is applied to cumulated probability values.
- Categorical DQN uses neural networks to output the probability mass function of categorical distribution to approximate the probability mass function or probability distribution function. The training of Categorical DQN tries to minimize cross-entropy loss.
- Quantile Regression Deep Q Network (QR-DQN) and Implicit Quantile Network (IQN) use neural networks to approximate the quantile functions of action values. Their training tries to minimize QR Huber loss. QR-DQN pre-define some cumulated probability values evenly among the interval $[0, 1]$, while IQN algorithm samples randomly among the interval $[0, 1]$.
- We can stack multiple images to track the movement of the images. When observation is images, we can use CNN to extract features.

12.8 Exercises

12.8.1 Multiple Choices

12.1 Which of the following type of RL algorithms is most consistent with distributional RL algorithms: (   )

A. Optimal value RL algorithms.
B. Policy-gradient RL algorithms.
C. Actor–critic RL algorithms.

12.2 Consider a continuous random variable X. Its PDF is p and quantile function is ϕ. Then its expectation satisfies: (   )

A. $E[X] = E\big[p(X)\big]$.

B. $E[X] = E_{\Omega \sim \text{uniform}[0,1]}\left[\phi(\Omega)\right]$.

C. $E[X] = E\left[p(X)\right]$ and $E[X] = E_{\Omega \sim \text{uniform}[0,1]}\left[\phi(\Omega)\right]$.

12.3 Consider a continuous random variable X. Its quantile function is ϕ. Given the Yarri distorted function $\beta : [0,1] \rightarrow [0,1]$, the distorted expectation of X can be expressed as: (   )

A. $E^{\langle\beta\rangle}[X] = \int_0^1 \phi(\omega)\beta(\omega)\,d\omega$.

B. $E^{\langle\beta\rangle}[X] = \int_0^1 \phi(\omega)\,d\beta(\omega)$.

C. $E^{\langle\beta\rangle}[X] = \int_0^1 \phi(\omega)\beta^{-1}(\omega)\,d\omega$.

12.4 On distributional RL algorithms, choose the correct one: (   )

A. Categorical DQN and QR-DQN try to minimize QR Huber loss.

B. Categorical DQN and IQN try to minimize QR Huber loss.

C. QR-DQN and IQN try to minimize QR Huber loss.

12.5 On distributional RL algorithms, choose the correct one: (   )

A. Categorical DQN randomly samples multiple cumulated probability values for decision.

B. QR-DQN randomly samples multiple cumulated probability values for decision.

C. IQN randomly samples multiple cumulated probability values for decision.

12.6 On Categorical DQN, when the support set of the categorical distribution is in the form of $q^{(i)} = q^{(0)} + i\Delta q$ $(i \in \mathcal{I})$, the project ratio from $u^{(j)} = r + \gamma q^{(j)}$ $(j \in \mathcal{I})$ to $q^{(i)}$ $(i \in \mathcal{I})$ is: (   )

A. $\text{clip}\left(\frac{u^{(j)} - u^{(i)}}{\Delta q}, 0, 1\right)$

B. $1 - \text{clip}\left(\frac{u^{(j)} - u^{(i)}}{\Delta q}, 0, 1\right)$

C. $1 - \text{clip}\left(\frac{\text{clip}\left(u^{(j)}, q^{(0)}, q^{(|\mathcal{I}|-1)}\right) - u^{(i)}}{\Delta q}, 0, 1\right)$

12.8.2 Programming

12.7 Use an algorithm in this chapter to play Atari game `BreakoutNoFrameskip-v4`.

12.8.3 Mock Interview

12.8 What are the advantages of introducing probability distribution in distributional RL algorithms?

12.9 What RL algorithms can maximize utility or minimize statistics risks? Why can those algorithms do that?

Chapter 13
Minimize Regret

This chapter covers

- online RL
- regret
- Multi-Arm Bandit (MAB)
- ε-greedy algorithm
- Upper Confidence Bound (UCB) algorithm
- UCB1 algorithm
- Bayesian UCB algorithm
- Thompson sampling algorithm
- Upper Confidence Bound Value Iteration (UCBVI) algorithm
- tradeoff between exploration and exploitation
- customizing a Gym environment

13.1 Regret

RL adapts the concept of regret in general online machine learning. First, let us review this concept in general machine learning.

Interdisciplinary Reference 13.1
Machine Learning: Online Learning and Regret

Regret is usually the most important performance metric in online ML tasks, including classification tasks, regression tasks, and RL tasks.

Naïve ML tasks have access to all data in the beginning. Such tasks can also be called offline tasks. The most important performance metric of offline tasks is usually called the loss. For classification tasks, the loss can be accuracy. For regression tasks, the loss can be mean squared error. For RL tasks, the loss can be average episode costs or equivalently the negative of average episode rewards. We can leverage all these data to minimize the loss. Let ℓ_{min} be the minimal possible loss. Besides, we also consider the metrics such as convergence rate and sample complexity. Large convergence rate and little sample complexity mean that we can reach the optimal loss faster. However, even an algorithm has large convergence rate and little sample complexity, the loss during the training can be very large. Specially, an algorithm may only have little loss when it convergences to the optimal, but its loss can always be large before it convergences to the optimal.

In contrast to offline tasks, online tasks can access more data with more time. We can use a positive integer k to denote the time when new data are available. The new data at time k can be denoted as $\mathcal{D}^{(k)}$. In the beginning, $k = 0$, no data are available. Before time k, the available data are $\bigcup_{\kappa=1}^{k-1} \mathcal{D}^{(\kappa)}$. After time k, available data are $\bigcup_{\kappa=1}^{k} \mathcal{D}^{(\kappa)}$.

Online learning typically co-exists with multi-task learning where the tasks vary through time. Since the task is changing, each task lasts for limited time. Therefore, we want the task performing well during the whole training process. Another typical scenario is combination of a pre-train task and an online task: First there is a pre-training task, which uses pre-training data to train, and we do not care about the loss during pre-training. However, the training results will be migrated to real-world use scenario, which is not identical to the pre-training environment. We may care about the losses in the real-use use scenario. For example, we can first pre-train an RL agent in a simulator, and the failure in the simulator is cheap. The pre-training agent then is applied to the real-world task, and the cost in the real world can be costly. The later task is also an online learning task.

The concept of regret is introduced to quantify the performance during the training process. The definition of regret for an ML algorithm is as follows: When first κ data are available, the algorithm can get a classifier, regressor, or an RL policy, which leads to loss ℓ_κ. Accordingly, we can define the regret of the first k data $(k = 1, 2, 3, \dots)$ as

$$\text{regret}_{\le k} \stackrel{\text{def}}{=} \sum_{\kappa=1}^{k} (\ell_\kappa - \ell_{min}).$$

The intuition of regret is, if we had known the optimal classifier, regressor, or policy, how much room did we have to do better.

We want to minimize the regret.

Episodic RL tasks usually use episode rewards to define regret. Consider an episode task, let g_{π_*} be the largest average episode reward for this task. For an RL algorithm that attains average episode reward g_{π_κ} at k-th episode, its regret among the first k episodes ($k = 1, 2, \ldots$) is defined as

$$\text{regret}_{\leq k} \overset{\text{def}}{=} \sum_{\kappa=1}^{k} \left(g_{\pi_*} - g_{\pi_\kappa} \right).$$

It's very difficult, usually impossible to calculate the accurate regret for an algorithm. We usually only calculate its asymptotic value.

13.2 MAB: Multi-Arm Bandit

Among researches on regret, Multi-Arm Bandit is the most famous, and most known task. This section we will learn how to solve this task.

13.2.1 MAB Problem

The problem of **Multi-Arm Bandit** (MAB) is as follows: In a casino, a slot machine has $|\mathcal{A}|$ arms. Each time a gambler can only play one arm. After a play, the gambler will be rewarded, and distribution of the reward relates to which arm is played. The gambler wants to maximize the expectation of rewards, or minimize the regret in repeated plays.

MAB problem can be modeled as an episodic RL task. Each episode has only $T = 1$ step. Its state space is a singleton $S = \{s\}$, and its action space $\mathcal{A}$ is usually assumed to be a finite set. The episode reward G equals the only reward R. Let the action value $q(a) \overset{\text{def}}{=} E[R|A = a] = \sum_r r \Pr[r|A = a]$ denote the expectation of reward R when the armed a ($a \in \mathcal{A}$) is played. The conditional distribution $\Pr[R|A = a]$ and conditional expectation $E[R|A = a]$ are unknown beforehand. In the beginning, the agent knows nothing about the dynamics of the environment, so it has to try randomly. After the agent has more interaction with the environment, the agent knows more about the environment, so it can be smarter. If the agent can interact with the environment infinite times, it can find the optimal policy, which always plays the arm $a_* = \max_a q(a)$.

Although MAB is much simpler than a general MDP, it needs to consider the trade-off between exploration and exploitation, which is a core issue of RL. On the one hand, the agent needs to exploit and tend to choose the arm with the largest action value estimate; on the other hand, the agent needs to explore and try those arms that do not have the largest action value estimates for the time being.

In some tasks, MAB problem may have extra assumptions. For example,

- Bernoulli-reward MAB: This is an MAB whose reward space is a discrete space $\{0, 1\}$. The distribution of the reward R when the action $a \in \mathcal{A}$ is chosen obeys $R \sim \text{bernoulli}(q(a))$, where $q(a) \in [0, 1]$ is an unknown parameter that relies on the action.
- Gaussian-reward MAB: This is an MAB whose reward space is a continuous space $(-\infty, +\infty)$. The distribution of the reward R when the action $a \in \mathcal{A}$ is chosen obeys $R \sim \text{normal}\big(q(a), \sigma^2(a)\big)$, where $q(a)$ and $\sigma^2(a)$ are parameters that rely on the action. $q(a)$ is unknown beforehand, while $\sigma^2(a)$ can be either known or unknown.

13.2.2 ε-Greedy Algorithm

Starting from this section, we try to solve MAB.

Section 4.1.3 introduced ε-greedy algorithm, which is an algorithm that trades-off exploitation and exploration. This algorithm can be used to solve MAB (see Algo. 13.1).

Algorithm 13.1 ε-greedy.

Parameter: exploration probability ε (may vary in different iterations).

1. (Initialize) Initialize the action value estimates $q(a) \leftarrow$ arbitrary value ($a \in \mathcal{A}$). If we use incremental implementation to update values, initialize the sum of weights $c(a) \leftarrow 0$ ($a \in \mathcal{A}$).
2. (MC update) For every episode:

    2.1. (Decide and sample) Use the ε-greedy policy derived from the action value estimate q to determine action A (especially, pick a random action with probability ε, and pick the action $\arg\max_a q(a)$ with probability $1 - \varepsilon$). Execute the action A and observe reward R.

    2.2. (Update action value estimate) Update $q(A)$ to reduce $[R - q(A)]^2$. (For incremental implementation, set $c(A) \leftarrow c(A) + 1$, $q(A) \leftarrow q(A) + \frac{1}{c(A)}[R - q(A)]$.)

Unfortunately, ε-greedy algorithm can not minimize regret. We will introduce algorithms that can attain smaller asymptotic regret in the subsequent subsections.

13.2.3 UCB: Upper Confidence Bound

This section introduces UCB. We will first learn the general UCB algorithm, then learn the UCB1 algorithm, which is specially designed for MAB whose rewards are within the range [0, 1]. We will also calculate the regret of UCB1.

The intuition of **Upper Confidence Bound** (**UCB**) is as follows (Lai, 1985): We provide an estimate of reward $u(a)$ for each arm such that $q(a) < u(a)$ holds with large probability. The estimate $u(a)$ is called UCB. It is usually of the form

$$u(a) = \hat{q}(a) + b(a), \quad a \in \mathcal{A},$$

where $\hat{q}(a)$ is the estimate of action value, and $b(a)$ is the **bonus** value.

We need to designate the bonus value strategically. In the beginning, the agent knows nothing about the reward, so it needs very large bonus that is close to positive infinity, which leads to random selection of arms. This random selection corresponds to exploration. During training, the agent interacts more with the environment, and we can estimate action values with increasing accuracy, and the bonus can be much smaller. This corresponds to fewer exploration and more exploitation. The choice of bonus value is critical to the performance.

UCB algorithm is shown in Algo. 13.2.

Algorithm 13.2   UCB (including UCB1).

---

1. Initialize:
   - **1.1.** (Initialize counters) $c(a) \leftarrow 0$ ($a \in \mathcal{A}$).
   - **1.2.** (Initialize action value estimates) $q(a) \leftarrow$ arbitrary value ($a \in \mathcal{A}$).
   - **1.3.** (Initialize policy) Set the optimal action estimate $A \leftarrow$ arbitrary value.

2. (MC update) For every episode:
   - **2.1.** (Execute) Execute the action A, and observe the reward R.
   - **2.2.** (Update counter) $c(A) \leftarrow c(A) + 1$.
   - **2.3.** (Update action value estimate) $q(A) \leftarrow q(A) + \frac{1}{c(A)}[R - q(A)]$.
   - **2.4.** (Update policy) For each action $a \in \mathcal{A}$, calculate the bonus $b(a)$. (For example, UCB1 algorithm for Bernoulli-reward MAB uses $b(a) \leftarrow \sqrt{\frac{2\ln c(\mathcal{A})}{c(a)}}$, where $c(\mathcal{A}) \leftarrow \sum_a c(a)$.) Calculate UCB $u(a) \leftarrow q(a) + b(a)$. The optimal action estimate $A \leftarrow \arg\max_a u(a)$.

MABs with different additional properties require different ways to choose bonus. MAB with reward space $\mathbb{R} \subseteq [0, 1]$ can choose the bonus as

$$b(a) = \sqrt{\frac{2\ln c(\mathcal{A})}{c(a)}},$$

where

$$c(\mathcal{A}) = \sum_{a \in \mathcal{A}} c(a).$$

Such UCB algorithm is called **UCB1** algorithm (Auer, 2002). The regret of UCB1 algorithm is $\text{regret}_{\leq k} = O\left(|\mathcal{A}| \ln k\right)$.

The remaining part of this section will prove this regret. This proof is one of the easiest among all proofs for regret, and it uses a routine on regret calculation. In order to bound the regret of an algorithm, we usually consider an event with large probability. The final regret can be partitioned into two parts: the regret contributed when the event does not happen, and the regret contributed when the event happens. From the view of trade-off between exploration and exploitation, when the event does not happen, the agent explores. When the agent explores, each episode may have large regret. But we can use some concentration inequality to bound the probability that the event does not happen, so that the total exploration regret is bounded. When the event happens, the agent exploits, and the episode regret is small, and the total exploitation regret is bounded, too.

Interdisciplinary Reference 13.2
Probability Theory: Hoeffding's Inequality

Let X_i ($0 \leq i < c$) denote the independent random variable within the range $[x_{i,\min}, x_{i,\max}]$. Let $\bar{X} = \frac{1}{c} \sum_{i=0}^{c-1} X_i$. Then we have

$$\Pr\left[\bar{X} - E[\bar{X}] \geq \varepsilon\right] \leq \exp\left(-\frac{2c^2\varepsilon^2}{\sum_{i=0}^{c-1} \left(x_{i,\max} - x_{i,\min}\right)^2}\right)$$

$$\Pr\left[\bar{X} - E[\bar{X}] \leq -\varepsilon\right] \leq \exp\left(-\frac{2c^2\varepsilon^2}{\sum_{i=0}^{c-1} \left(x_{i,\max} - x_{i,\min}\right)^2}\right)$$

for any $\varepsilon > 0$.

Especially, if X_i is within $[0, 1]$ ($0 \leq i < c$), we have

$$\Pr\left[\bar{X} - E[\bar{X}] \geq \varepsilon\right] \leq \exp\left(-2c\varepsilon^2\right)$$

$$\Pr\left[\bar{X} - E[\bar{X}] \leq -\varepsilon\right] \leq \exp\left(-2c\varepsilon^2\right)$$

for any $\varepsilon > 0$.

Hoeffding's inequality can lead to the following propositions:

- Let $\tilde{q}_c(a)$ denote the action value estimate of the action a when the first c reward samples are given. For any positive integer $\kappa > 0$, any action $a \in \mathcal{A}$, and any positive integer $c_a > 0$, we have

$$\Pr\left[q(a) + \sqrt{\frac{2\ln\kappa}{c_a}} \le \tilde{q}_{c_a}(a)\right] \le \frac{1}{\kappa^4},$$

$$\Pr\left[\tilde{q}_{c_a}(a) + \sqrt{\frac{2\ln\kappa}{c_a}} \le q(a)\right] \le \frac{1}{\kappa^4}.$$

(Proof: Obviously, $\mathrm{E}\left[\tilde{q}_{c_a}(a)\right] = q(a)$. Plugging $\varepsilon = \sqrt{\frac{2\ln\kappa}{c_a}} > 0$ into Hoeffding's inequality leads to

$$\Pr\left[\tilde{q}_{c_a}(a) - q(a) \ge \sqrt{\frac{2\ln\kappa}{c_a}}\right] \le \exp\left(-2c_a\left(\sqrt{\frac{2\ln\kappa}{c_a}}\right)^2\right),$$

$$\Pr\left[\tilde{q}_{c_a}(a) - q(a) \le -\sqrt{\frac{2\ln\kappa}{c_a}}\right] \le \exp\left(-2c_a\left(\sqrt{\frac{2\ln\kappa}{c_a}}\right)^2\right).$$

Simplifying the above inequalities completes the proof.)

• Let a_* denote the optimal action. For any positive integer $\kappa > 0$ and positive integer $c_* > 0$, we have

$$\Pr\left[\tilde{q}_{c_*}(a_*) + \sqrt{\frac{2\ln\kappa}{c_*}} \le q(a_*)\right] \le \frac{1}{\kappa^4}.$$

(Proof: This can be obtained by assigning a with a_* and c_a with c_* in the second part of the former proposition.)

Let us continue the proof. Now we conduct some preparations.

For any positive integer $\kappa > 0$ and an arbitrary action $a \in \mathcal{A}$, we can define $\underline{c}_\kappa(a) \overset{\text{def}}{=} \frac{8\ln\kappa}{\left(q(a_*)-q(a)\right)^2}$. For any positive integer $c_a > \underline{c}_\kappa(a)$, we have $q(a_*) > q(a) + \sqrt{\frac{2\ln\kappa}{c_a}} + \sqrt{\frac{2\ln\kappa}{c_a}}$, so $\Pr\left[q(a_*) \le q(a) + \sqrt{\frac{2\ln\kappa}{c_a}} + \sqrt{\frac{2\ln\kappa}{c_a}}\right] = 0$.

For any positive integer $\kappa > 0$, an arbitrary action $a \in \mathcal{A}$, any positive integer $c_* > 0$, and any positive integer $c_* > \underline{c}_\kappa(a)$, if the following inequality holds

$$\tilde{q}_{c_*}(a_*) + \sqrt{\frac{2\ln\kappa}{c_*}} \le \tilde{q}_{c_a}(a) + \sqrt{\frac{2\ln\kappa}{c_a}},$$

at least one of the following three inequalities holds: (1) $\tilde{q}_{c_*}(a_*) + \sqrt{\frac{2\ln\kappa}{c_*}} \le q(a_*)$; (2) $q(a_*) \le q(a) + \sqrt{\frac{2\ln\kappa}{c_a}} + \sqrt{\frac{2\ln\kappa}{c_a}}$; (3) $q(a) + \sqrt{\frac{2\ln\kappa}{c_a}} \le \tilde{q}_{c_a}(a)$. (This can be proved by contradiction. If none of the three inequalities hold, we have

$$\tilde{q}_{c_*}(a_*) + \sqrt{\frac{2\ln \kappa}{c_*}} > q(a_*) > q(a) + \sqrt{\frac{2\ln \kappa}{c_a}} + \sqrt{\frac{2\ln \kappa}{c_a}} > \tilde{q}_{c_a}(a) + \sqrt{\frac{2\ln \kappa}{c_a}},$$

which leads to contradiction.) Therefore,

$$\Pr\left[\tilde{q}_{c_*}(a_*) + \sqrt{\frac{2\ln \kappa}{c_*}} \leq \tilde{q}_{c_a}(a) + \sqrt{\frac{2\ln \kappa}{c_a}}\right]$$

$$\leq \Pr\left[\tilde{q}_{c_*}(a_*) + \sqrt{\frac{2\ln \kappa}{c_*}} \leq q(a_*)\right] + \Pr\left[q(a_*) \leq q(a) + \sqrt{\frac{2\ln \kappa}{c_a}} + \sqrt{\frac{2\ln \kappa}{c_a}}\right]$$

$$+ \Pr\left[q(a) + \sqrt{\frac{2\ln \kappa}{c_a}} \leq \tilde{q}_{c_a}(a)\right]$$

$$\leq \frac{1}{\kappa^4} + 0 + \frac{1}{\kappa^4}$$

$$= \frac{2}{\kappa^4}.$$

Furthermore, for any positive integer $\kappa > 0$ and an arbitrary action $a \in \mathcal{A}$, let $c_\kappa(a)$ denote the number of times that action a appears in the first κ episodes. Then

$$\Pr\left[\tilde{q}_{c_\kappa(a_*)}(a_*) + \sqrt{\frac{2\ln \kappa}{c_\kappa(a_*)}} \leq \tilde{q}_{c_\kappa(a)}(a) + \sqrt{\frac{2\ln \kappa}{c_\kappa(a)}}, c_\kappa(a) > \underline{c}_\kappa(a)\right]$$

$$\leq \Pr\left[\min_{1 \leq c_* \leq \kappa} \tilde{q}_{c_*}(a_*) + \sqrt{\frac{2\ln \kappa}{c_*}} \leq \max_{\underline{c}_\kappa(a) \leq c_* \leq \kappa} \tilde{q}_{c_a}(a) + \sqrt{\frac{2\ln \kappa}{c_a}}\right]$$

$$\leq \sum_{c_*=1}^{\kappa} \sum_{c_*=\underline{c}_\kappa(a)}^{\kappa} \Pr\left[\tilde{q}_{c_*}(a_*) + \sqrt{\frac{2\ln \kappa}{c_*}} \leq \tilde{q}_{c_a}(a) + \sqrt{\frac{2\ln \kappa}{c_a}}\right]$$

$$\leq \kappa \cdot \kappa \cdot \frac{2}{\kappa^4}$$

$$= \frac{2}{\kappa^2}$$

When $A_\kappa = a$, $\tilde{q}_{c_\kappa(a_*)}(a_*) + \sqrt{\frac{2\ln \kappa}{c_\kappa(a_*)}} \leq \tilde{q}_{c_\kappa(a)}(a) + \sqrt{\frac{2\ln \kappa}{c_\kappa(a)}}$. Therefore,

$$\Pr\left[A_\kappa = a, c_\kappa(a) > \underline{c}_\kappa(a)\right]$$

$$\leq \Pr\left[\tilde{q}_{c_\kappa(a_*)}(a_*) + \sqrt{\frac{2\ln \kappa}{c_\kappa(a_*)}} \leq \tilde{q}_{c_\kappa(a)}(a) + \sqrt{\frac{2\ln \kappa}{c_\kappa(a)}}, c_\kappa(a) > \underline{c}_\kappa(a)\right]$$

$$\leq \frac{2}{\kappa^2}.$$

We continue finding the regret of UCB1. For an arbitrary positive integer $k > 0$ and any action $a \neq a_*$, we can partition the expectation of counting into two parts:

$$E\left[c_k(a)\right]$$

$$= \sum_{\kappa=1}^{k} \Pr\left[A_\kappa = a\right]$$

$$= \sum_{\kappa=1}^{k} \Pr\left[A_\kappa = a, c_\kappa(a) \leq \underline{c}_k(a)\right] + \sum_{\kappa=1}^{k} \Pr\left[A_\kappa = a, c_\kappa(a) > \underline{c}_k(a)\right]$$

$$\leq \underline{c}_k(a) + \sum_{\kappa=1}^{k} \frac{2}{\kappa^2}$$

$$\leq \frac{8 \ln k}{\left(q(a_*) - q(a)\right)^2} + \frac{\pi^2}{3}.$$

Therefore, the regret

$$\text{regret}_{\leq k}$$

$$= \sum_{\kappa=1}^{k} \left(q(a_*) - q(a_\kappa)\right)$$

$$= \sum_{a \in \mathcal{A}: a \neq a_*} \left(q(a_*) - q(a)\right) E\left[c_k(a)\right]$$

$$\leq \sum_{a \in \mathcal{A}: a \neq a_*} \left(q(a_*) - q(a)\right) \left(\frac{8 \ln k}{\left(q(a_*) - q(a)\right)^2} + \frac{\pi^2}{3}\right)$$

$$= \sum_{a \in \mathcal{A}: a \neq a_*} \frac{8}{q(a_*) - q(a)} \ln k + \frac{\pi^2}{3} \sum_{a \in \mathcal{A}} \left(q(a_*) - q(a)\right)$$

$$\leq \frac{8}{\min_{a \in \mathcal{A}: a \neq a_*} \left(q(a_*) - q(a)\right)} |\mathcal{A}| \ln k + \frac{\pi^2}{3} \sum_{a \in \mathcal{A}} \left(q(a_*) - q(a)\right).$$

So we prove that the regret equals $O\left(|\mathcal{A}| \ln k\right)$.

This asymptotic regret contains a constant coefficient $8/\min_{a \in \mathcal{A}: a \neq a_*} \left(q(a_*) - q(a)\right)$. If the optimal action value is close to the second optimal action value, the coefficient may be very large.

13.2.4 Bayesian UCB

We have known that UCB algorithm needs to design $u(a) = \hat{q}(a) + b(a)$ so that the probability of $u(a) > q(a)$ is sufficiently small. Bayesian UCB is a way to determine UCB using Bayesian method.

The intuition of **Bayesian Upper Confidence Bound (Bayesian UCB)** algorithm is as follows: Assume that the reward obeys a distribution parameterized by $\theta(a)$ when the arm $a \in \mathcal{A}$ is played. We can determine the action value from distribution and its parameter. If the parameter $\theta(a)$ itself obeys a distribution, the action value obeys some distribution determined by the parameter $\theta(a)$ too. Therefore, I can choose a value, such as $\mu(\theta(a)) + 3\sigma(\theta(a))$ where $\mu(\theta(a))$ and $\sigma(\theta(a))$ are the mean and standard deviation of the action value respectively, such that the probability that the real action value is greater than my value is small. With more and more available data, the standard deviation of action value is getting smaller, and the value I pick will become increasingly accurate.

Bayesian UCB requires agent to determine the form of likelihood distribution a priori, and we usually set the distribution of parameters as the conjugate distribution in order to make the form of parameter distribution consistent throughout all iterations. Some examples of conjugate distributions are shown in the sequel.

Interdisciplinary Reference 13.3
Probability Theory: Conjugate Distribution

In Bayesian probability theory, consider the likelihood $p(x|\theta)$, where θ is the distribution parameter and x is the data. If its prior distribution and posterior distribution share the same form, such form will be called **conjugate distribution** of the likelihood distribution.

Beta distribution is the conjugate distribution of the binomial distribution. Specifically speaking, consider the likelihood distribution $X \sim \text{binomial}(n, \theta)$, whose PMF function is $p(x|\theta) = \binom{n}{k}\theta^x(1-\theta)^{n-x}$. Let the prior distribution be $\Theta \sim \text{beta}(\alpha, \beta)$, whose PDF function is $p(\theta) = \frac{\Gamma(\alpha+\beta)}{\Gamma(\alpha)\Gamma(\beta)}\theta^{\alpha-1}(1-\theta)^{\beta-1}$. Then the posterior distribution becomes $\Theta \sim \text{beta}(\alpha + x, \beta + n - x)$, whose PDF function is $p(\theta|x) = \frac{\Gamma(\alpha+\beta+n)}{\Gamma(\alpha+x)\Gamma(\beta+n-x)}\theta^{\alpha+x-1}(1-\theta)^{\beta+n-x-1}$. The prior distribution and the posterior distribution share the form of beta distribution.

Bernoulli distribution is a special case of the binomial distribution, so beta distribution is the conjugate distribution of Bernoulli distribution too. Specifically, consider the likelihood $X \sim \text{bernoulli}(\theta)$, whose PMF is $p(x|\theta) = \theta^x(1-\theta)^{n-x}$. Let the prior distribution be $\Theta \sim \text{beta}(\alpha, \beta)$, whose PDF is $p(\theta) = \frac{\Gamma(\alpha+\beta)}{\Gamma(\alpha)\Gamma(\beta)}$. Then the posterior distribution becomes $\Theta \sim \text{beta}(\alpha + x, \beta + 1 - x)$, whose PDF is

$p(\theta|x) = \frac{\Gamma(\alpha+\beta+n)}{\Gamma(\alpha+x)\Gamma(\beta+n-x)}\theta^{\alpha+x-1}(1-\theta)^{\beta-x}$. The prior distribution and the posterior distribution shares the same form of beta distribution.

Gaussian distribution is the conjugate distribution of Gaussian distribution itself. Specifically, consider the likelihood $\text{normal}\left(\theta, \sigma^2_{\text{likelihood}}\right)$, whose PDF

is $p(x|\theta) = \frac{1}{\sqrt{2\pi\sigma^2_{\text{likelihood}}}}\exp\left(-\frac{(x-\theta)^2}{2\sigma^2_{\text{likelihood}}}\right)$. Let the prior distribution be

$\Theta \sim \text{normal}\left(\mu_{\text{prior}}, \sigma^2_{\text{prior}}\right)$, and then the posterior distribution becomes $\Theta \sim$

$\text{normal}\left(\frac{\frac{\mu_{\text{prior}}}{\sigma^2_{\text{prior}}}+\frac{x}{\sigma^2_{\text{likelihood}}}}{\frac{1}{\sigma^2_{\text{prior}}}+\frac{1}{\sigma^2_{\text{likelihood}}}}, \frac{1}{\frac{1}{\sigma^2_{\text{prior}}}+\frac{1}{\sigma^2_{\text{likelihood}}}}\right)$. The prior distribution and the posterior

distribution share the same form of Gaussian distribution.

Section 13.2.1 mentioned two special MABs: Bernoulli-reward MAB and Gaussian-reward MAB. Bayesian UCB algorithm uses the following setting in these two MABs.

- Bernoulli-reward MAB: When the action $a \in \mathcal{A}$ is chosen, the reward R obeys the Bernoulli distribution $R \sim \text{bernoulli}(q(a))$, where $q(a)$ is an unknown parameter. Since beta distribution is the conjugate distribution of Bernoulli distribution, Bernoulli-reward MAB uses beta distribution as the prior distribution.

- Gaussian-reward MAB: When the action $a \in \mathcal{A}$ is chosen, the reward R obeys the Gaussian distribution $R \sim \text{normal}\left(q(a), \sigma^2_{\text{likelihood}}\right)$, where $q(a)$ is an unknown parameter, and $\sigma^2_{\text{likelihood}}$ is a pre-determined parameter. Since Gaussian distribution is the conjugate distribution of Gaussian distribution, so Gaussian-reward MAB uses Gaussian distribution as the prior distribution.

Bayesian UCB is shown in Algo. 13.3, which explicitly remarks for Bernoulli-reward MAB and Gaussian-reward MAB.

Algorithm 13.3   Bayesian UCB.

Parameters: reward likelihood, such as Bernoulli reward and Gaussian reward.

1. Initialize:

   1.1. (Initialize distribution parameters) For Bernoulli-reward MAB, the prior distribution is the beta distribution $\text{beta}(\alpha(a), \beta(a))$, so we initialize the parameter $\alpha(a) \leftarrow 1$, $\beta(a) \leftarrow 1$. For Gaussian-reward MAB, the prior distribution is the Gaussian distribution $\text{normal}\left(\mu(a), \sigma^2(a)\right)$, so we initialize the parameter $\mu(a) \leftarrow 0$, $\sigma(a) \leftarrow 1$.

1.2. (Initialize policy) Set the optimal action estimate $A \leftarrow$ arbitrary value.

2. (MC update) For every episode:

    2.1. (Execute) Execute the action A, and observe the reward R.

    2.2. (Update distribution parameters) For Beta-reward MAB: $\alpha(A) \leftarrow \alpha(A) + R$ and $\beta(A) \leftarrow \beta(A) + 1 - R$. For Gaussian-reward MAB with known $\sigma_{\text{likelihood}}^2$, $\mu(A) \leftarrow \dfrac{\frac{\mu(A)}{\sigma^2(A)} + \frac{R}{\sigma_{\text{likelihood}}^2}}{\frac{1}{\sigma^2(A)} + \frac{1}{\sigma_{\text{likelihood}}^2}}$, $\sigma(A) \leftarrow \dfrac{1}{\sqrt{\frac{1}{\sigma^2(A)} + \frac{1}{\sigma_{\text{likelihood}}^2}}}$.

    2.3. (Update policy) For each action $a \in \mathcal{A}$, calculate the mean $\mu(a)$ and the standard deviation $\sigma(a)$ of the distributions respectively. (For beta distribution $\mu(a) \leftarrow \frac{\alpha(a)}{\alpha(a)+\beta(a)}$, $\sigma(a) \leftarrow \frac{1}{\alpha(a)+\beta(a)}\sqrt{\frac{\alpha(a)\beta(a)}{\alpha(a)+\beta(a)+1}}$. For Gaussian distribution, the mean and the standard deviation are parameters.) Calculate UCB $u(a) \leftarrow \mu(a) + 3\sigma(a)$. The optimal action estimate $A \leftarrow \arg\max_a u(a)$.

Since Bayesian UCB introduces the conjugate prior, this algorithm actually tries to minimize the regret over the prior distribution.

13.2.5 Thompson Sampling

Thompson sampling algorithm is similar to Bayesian UCB algorithm. But Thompson sampling algorithm does not obtain UCB from distribution parameters. Instead, it uses the distribution to generate samples, and then uses the samples to make decision. Such sampling can act as probability matching.

Algorithm 13.4 shows the Thompson sampling algorithm. This algorithm only differs from Bayesian UCB (Algo. 13.3) during decision in Step 2.3.

Algorithm 13.4   Thompson Sampling.

Parameters: reward likelihood, such as Bernoulli reward and Gaussian reward.

1. Initialize:

    1.1. (Initialize distribution parameters) For Bernoulli-reward MAB, the prior distribution is the beta distribution $\text{beta}(\alpha(a), \beta(a))$, so we initialize the parameter $\alpha(a) \leftarrow 1$, $\beta(a) \leftarrow 1$. For Gaussian-reward MAB, the prior distribution is the Gaussian distribution $\text{normal}(\mu(a), \sigma^2(a))$, so we initialize the parameter $\mu(a) \leftarrow 0$, $\sigma(a) \leftarrow 1$.

1.2. (Initialize policy) Set the optimal action estimate $A \leftarrow$ arbitrary value.

2. (MC update) For every episode:

2.1. (Execute) Execute the action A, and observe the reward R.

2.2. (Update distribution parameters) For Beta-reward MAB: $\alpha(A) \leftarrow \alpha(A) + R$ and $\beta(A) \leftarrow \beta(A) + 1 - R$. For Gaussian-reward MAB with known $\sigma_{\text{likelihood}}^2$, $\mu(A) \leftarrow \dfrac{\frac{\mu(A)}{\sigma^2(A)} + \frac{R}{\sigma_{\text{likelihood}}^2}}{\frac{1}{\sigma^2(A)} + \frac{1}{\sigma_{\text{likelihood}}^2}}$, $\sigma(A) \leftarrow \dfrac{1}{\sqrt{\frac{1}{\sigma^2(A)} + \frac{1}{\sigma_{\text{likelihood}}^2}}}$.

2.3. (Update policy) For each action $a \in \mathcal{A}$, sample according to the distribution and obtain a sample $q(a)$ ($a \in \mathcal{A}$). Set the optimal action estimate $A \leftarrow \arg\max_a q(a)$.

Thompson sampling algorithm and Bayesian UCB algorithm share the same asymptotic regret.

13.3 UCBVI: Upper Confidence Bound Value Iteration

The MAB in the previous section is a special type of MDP such that it has only one state and one step. This section considers the general finite MDP.

Upper Confidence Bound Value Iteration (UCBVI) is an algorithm to minimize regret of finite MDP (Azar, 2017). The idea of UCBVI is: Maintaining UCB for both action values and state values such that the probability of true action values less than the action value upper bound is tiny, and the probability of true state values less than the state value upper bound is tiny, too. When the mathematical model of dynamics model is unknown, this algorithm uses the visitation counting to estimate the dynamics. Therefore, this algorithm is model-based. Algorithm 13.5 shows this algorithm. The algorithms use the upper bound on the number of steps (denoted as $t_{\max}$), the upper bound on the return of an episode (denoted as $g_{\max}$), and the number of episodes k. The bonus can be in the form of

$$ b(S_t, A_t) \leftarrow 2g_{\max}\sqrt{\frac{\ln|\mathcal{S}\|\mathcal{A}|k^2 t_{\max}^2}{c(S_t, A_t)}}. $$

Algorithm 13.5   UCBVI.

Parameters: range of episode return $[0, g_{\max}]$, upper bound of steps in an episode $t_{\max}$, episode number k, and other parameters.

1. Initialize:

1.1. (Initialize counting and transition probability estimate) Initialize the counting $c(s, a, s') \leftarrow 0$ ($s \in S, a \in \mathcal{A}, s' \in S^+$), $c(s, a) \leftarrow 0$ ($s \in S, a \in \mathcal{A}$).
Initialize the transition probability estimate $p(s'|s, a) \leftarrow \frac{1}{|S^+|}$ ($s \in S, a \in \mathcal{A}, s' \in S^+$).

1.2. (Initialize UCB) Initialize action-value UCB $u^{(q)}(s, a) \leftarrow g_{\max}$ ($s \in S, a \in \mathcal{A}$), and initialize state-value UCB $u^{(v)}(s) \leftarrow g_{\max}$ ($s \in S$).

2. (Value iteration) Iterate k episodes:

2.1. (Sample) Use the policy derived from action-value UCB $u^{(q)}$ to generate a trajectory $S_0, A_0, R_1, S_1, \ldots, S_{T-1}, A_{T-1}, R_T, S_T$, where $T \leq t_{\max}$.

2.2. For each $t \leftarrow T - 1, T - 2, \ldots, 0$:

2.2.1. (Update counting and transition probability estimate) Update counting $c(S_t, A_t, S_{t+1}) \leftarrow c(S_t, A_t, S_{t+1}) + 1$, $c(S_t, A_t) \leftarrow c(S_t, A_t) + 1$, and update transition probability estimate $p(s'|S_t, A_t) \leftarrow \frac{c(S_t, A_t, s')}{c(S_t, A_t)}$ ($s' \in S^+$) for the whole trajectory.

2.2.2. (Update action-value UCB) $u^{(q)}(S_t, A_t) \leftarrow R_{t+1} + b(S_t, A_t) + \sum_{s'} p(s'|S_t, A_t) u^{(v)}(s')$ (the bonus can be $b(S_t, A_t) \leftarrow 2g_{\max} \sqrt{\frac{\ln |S||\mathcal{A}|k^2 t_{\max}^2}{c(S_t, A_t)}}$). $u^{(q)}(S_t, A_t) \leftarrow \min \left\{ u^{(q)}(S_t, A_t), g_{\max} \right\}$.

2.2.3. (Update state-value UCB) $u^{(v)}(S_t) \leftarrow \max_a u^{(q)}(S_t, a)$.

Interdisciplinary Reference 13.4
Asymptotic Complexity: $\tilde{O}$ Notation

$\tilde{O}$ notation (read as soft-O notation): $f(x) = \tilde{O}(g(x))$ means that there exists $n \in (0, +\infty)$ such that $f(x) = O(g(x) \ln^n g(x))$.

Algorithm 13.5 with the bonus $b(s, a) \leftarrow 2g_{\max} \sqrt{\frac{\ln |S||\mathcal{A}|k^2 t_{\max}^2}{c(s, a)}}$ can attain regret $\tilde{O}\left(g_{\max} |S| \sqrt{|\mathcal{A}|k t_{\max}}\right)$. Its proof is omitted since it is too complex to address here, but the main idea is to partition the regret to two parts: the exploration part and exploitation part. The exploration part has large regret, but its probability is small, so its contribution to the expected regret is bounded; the exploitation part has large probability, but the regret is small, so its contribution to the expected regret is bounded too.

13.4 Case Study: Bernoulli-Reward MAB

This section will implement and solve the Bernoulli-reward MAB environment.

13.4.1 Create Custom Environment

The environments in previous chapters are either provided by the Gym itself, or provided by a third-party library. Different from those, this section will implement our own environment from the scratch. We will implement the interface of Gym for the environment, so that the environment can be used in the same way as other environments.

From the usage of Gym, we know what we need to implement a custom environment.

- Since we need to use the function gym.make() to get an environment env, which is an object of gym.Env, our custom environment needs to be a class that extends gym.Env, and somehow registered in the library Gym so that it can be accessed by gym.make().
- For an environment object env, we can get its observation space by using env.observation_space and its action space by using env.action_space. Therefore, we need to overwrite the constructor of the custom environment class so that it constructs self.observation_space and self.action_space.
- The core of a Gym environment is the logic of environment model. Especially, env.reset() initializes the environment, and env.step() drives to the next step(). Our custom environment of course needs to implement both of them. Besides, we also need to implement env.render() and env.close() correctly to visualize and release resources respectively.

From the above analysis, we know that, in order to implement a custom environment, we need to extend the class gym.Env. We need to overwrite the constructor and construct the member observation_space and action_space, overwrite the reset() and step() to implement the model, overwrite render() to visualize the environment, and overwrite close() to release resources. Then we register the class into Gym.

Code 13.1 shows the codes for the environment class BernoulliMABEnv. The environment that is compatible with Gym should be derived from gym.Env. We need to define the state space and action space in the constructor, and we need to implement the member step(). Additionally, we can optionally implement the member reset(), close(), and render(). Here we only implement reset(), and forsake close() and render().

Code 13.1   The environment class `BernoulliMABEnv`.

`BernoulliMABEnv-v0_demo.ipynb`

```python
class BernoulliMABEnv(gym.Env):
    """ Multi-Armed Bandit (MAB) with Bernoulli rewards """

    def __init__(self, n=10, means=None):
        super(BernoulliMABEnv, self).__init__()
        self.observation_space = spaces.Box(low=0, high=0, shape=(0,),
                dtype=float)
        self.action_space = spaces.Discrete(n)
        self.means = means or self.np_random.random(n)

    def reset(self, *, seed=None, options=None):
        super().reset(seed=seed)
        return np.empty(0, dtype=float), {}

    def step(self, action):
        mean = self.means[action]
        reward = self.np_random.binomial(1, mean)
        observation = np.empty(0, dtype=float)
        return observation, reward, True, False, {}
```

After we implement the class `BernoulliMABEnv`, we need to register it into Gym so that we can access it via `gym.make()` later. Registration needs to use the function `gym.envs.registration.register()`. This function has keyword parameters such as `id` and `entry_point`. The parameter `id` is of type `str`, meaning the environment ID we call `gym.make()`. The parameter `entry_point` points to the environment class. Code 13.2 registers the environment class `BernoulliMABEnv` as the task `BernoulliMABEnv-v0`.

Code 13.2   Register the environment class `BernoulliMABEnv` into Gym.

`BernoulliMABEnv-v0_demo.ipynb`

```python
from gym.envs.registration import register
register(id='BernoulliMABEnv-v0', entry_point=BernoulliMABEnv)
```

Now we have successfully implemented this environment.

13.4.2 ε-Greedy Solver

This section uses ε-greedy policy to solve this environment. The codes for agent are shown in Code 13.3.

Code 13.3 ε-greedy policy agent.

`BernoulliMABEnv-v0_demo.ipynb`

```python
class EpsilonGreedyAgent:
    def __init__(self, env):
        self.epsilon = 0.1
        self.action_n = env.action_space.n
        self.counts = np.zeros(self.action_n, dtype=float)
        self.qs = np.zeros(self.action_n, dtype=float)

    def reset(self, mode=None):
        self.mode = mode

    def step(self, observation, reward, terminated):
        if np.random.rand() < self.epsilon:
            action = np.random.randint(self.action_n)
        else:
            action = self.qs.argmax()
        if self.mode == 'train':
            if terminated:
                self.reward = reward  # save reward
            else:
                self.action = action  # save action
        return action

    def close(self):
        if self.mode == 'train':
            self.counts[self.action] += 1
            self.qs[self.action] += (self.reward - self.qs[self.action]) / \
                    self.counts[self.action]
```

Interaction between this agent and the environment once again uses Code 1.3.

We can get one regret sample during one training process. If we want to evaluate the average regret, we need to average over multiple regret samples from multiple training processed. Code 13.4 trains 100 agents to evaluate the average regret.

Code 13.4   Evaluate average regret.

`BernoulliMABEnv-v0_demo.ipynb`

```python
trial_regrets = []
for trial in range(100):
    # create a new agent for each trial — change agent here
    agent = EpsilonGreedyAgent(env)

    # train
    episode_rewards = []
    for episode in range(1000):
        episode_reward, elapsed_steps = play_episode(env, agent, seed=episode,
                mode='train')
        episode_rewards.append(episode_reward)
    regrets = env.means.max() - np.array(episode_rewards)
    trial_regret = regrets.sum()
    trial_regrets.append(trial_regret)

    # test
    episode_rewards = []
    for episode in range(100):
        episode_reward, elapsed_steps = play_episode(env, agent)
        episode_rewards.append(episode_reward)
    logging.info(
            'trial %d: average episode reward = %.2f ± %.2f, regret = %.2f',
            trial, np.mean(episode_rewards), np.std(episode_rewards),
```

```
24              trial_regret)
25
26   logging.info('average regret = %.2f ± %.2f',
27           np.mean(trial_regrets), np.std(trial_regrets))
```

13.4.3 UCB1 Solver

Since the reward of Bernoulli-reward MAB is within the range $[0, 1]$, we can use UCB1 to solve the problem. Code 13.5 shows the UCB1 agent.

Code 13.5   UCB1 agent.
`BernoulliMABEnv-v0_demo.ipynb`

```
1   class UCB1Agent:
2       def __init__(self, env):
3           self.action_n = env.action_space.n
4           self.counts = np.zeros(self.action_n, dtype=float)
5           self.qs = np.zeros(self.action_n, dtype=float)
6
7       def reset(self, mode=None):
8           self.mode = mode
9
10      def step(self, observation, reward, terminated):
11          total_count = max(self.counts.sum(), 1)  # lower bounded by 1
12          sqrts = np.sqrt(2 * np.log(total_count) / self.counts.clip(min=0.01))
13          ucbs = self.qs + sqrts
14          action = ucbs.argmax()
15          if self.mode == 'train':
16              if terminated:
17                  self.reward = reward  # save reward
18              else:
19                  self.action = action  # save action
20          return action
21
22      def close(self):
23          if self.mode == 'train':
24              self.counts[self.action] += 1
25              self.qs[self.action] += (self.reward - self.qs[self.action]) / \
26                      self.counts[self.action]
```

Interaction between this agent and the environment once again uses Code 1.3. We can also evaluate the average regret by replacing the agent in Code 13.4.

13.4.4 Bayesian UCB Solver

Code 13.6 shows the implementation of Bayesian UCB agent. Bernoulli-reward MAB chooses beta distribution as the prior distribution. The agent maintains the parameters of beta distribution. In the beginning, the prior distribution is beta$(1, 1)$, which is equivalent to the uniform distribution uniform $[0, 1]$.

Code 13.6 Bayesian UCB agent.

`BernoulliMABEnv-v0_demo.ipynb`

```python
class BayesianUCBAgent:
    def __init__(self, env):
        self.action_n = env.action_space.n
        self.alphas = np.ones(self.action_n, dtype=float)
        self.betas = np.ones(self.action_n, dtype=float)

    def reset(self, mode=None):
        self.mode = mode

    def step(self, observation, reward, terminated):
        means = stats.beta.mean(self.alphas, self.betas)
        stds = stats.beta.std(self.alphas, self.betas)
        ucbs = means + 3 * stds
        action = ucbs.argmax()
        if self.mode == 'train':
            if terminated:
                self.reward = reward  # save reward
            else:
                self.action = action  # save action
        return action

    def close(self):
        if self.mode == 'train':
            self.alphas[self.action] += self.reward
            self.betas[self.action] += (1. - self.reward)
```

Interaction between this agent and the environment once again uses Code 1.3. We can also evaluate the average regret by replacing the agent in Code 13.4.

13.4.5 Thompson Sampling Solver

Code 13.7 implements the Thompson sampling algorithm. Similarly, the initial prior distribution is set to beta$(1, 1)$, which is equivalent to the uniform distribution uniform $[0, 1]$.

Code 13.7 Thompson sampling agent.

`BernoulliMABEnv-v0_demo.ipynb`

```python
class ThompsonSamplingAgent:
    def __init__(self, env):
        self.action_n = env.action_space.n
        self.alphas = np.ones(self.action_n, dtype=float)
        self.betas = np.ones(self.action_n, dtype=float)

    def reset(self, mode=None):
        self.mode = mode

    def step(self, observation, reward, terminated):
        samples = [np.random.beta(max(alpha, 1e-6), max(beta, 1e-6))
                for alpha, beta in zip(self.alphas, self.betas)]
        action = np.argmax(samples)
        if self.mode == 'train':
            if terminated:
                self.reward = reward  # save reward
```

```
17        else:
18            self.action = action  # save action
19        return action
20
21    def close(self):
22        if self.mode == 'train':
23            self.alphas[self.action] += self.reward
24            self.betas[self.action] += (1. - self.reward)
```

Interaction between this agent and the environment once again uses Code 1.3. We can also evaluate the average regret by replacing the agent in Code 13.4.

13.5 Summary

- Regret is an important performance metric for online RL. Its definition is

$$\text{regret}_{\leq k} \stackrel{\text{def}}{=} \sum_{\kappa=1}^{k} \left(g_{\pi_*} - g_{\pi_\kappa} \right).$$

- Multi-Arm Bandit (MAB) is a type of single-step singular-state RL task. Its special cases include Bernoulli-reward MAB and Gaussian-reward MAB.
- The algorithms to minimize regrets for MABs include Upper Confidence Bound (UCB), Bayesian UCB, and Thompson sampling.
- The form of UCB is

$$\text{action value UCB:} \quad u^{(q)}(s, a) = \hat{q}(s, a) + b(s, a), \quad s \in \mathcal{S}, a \in \mathcal{A}$$

$$\text{state value UCB:} \quad u^{(v)}(s) = \max_a u^{(q)}(s, a), \quad s \in \mathcal{S},$$

where b is the bonus.
- The algorithms to minimize regrets for finite MDPs include Upper Confidence Bound Value Iteration (UCBVI) algorithm, which is a model-based algorithm.
- A usual way to get regret bound: Consider an event that is very likely to happen. When the event is not happening, the agent is exploring. We can use concentration inequality to prove that the total exploration probability is small, so the total exploration regret is bounded. When the event is happening, the agent is exploiting. We can use concentration inequality to prove that the total exploitation regret is bounded, too.
- Implement a customized environment for Gym: Derive from the class gym.Env, override interface functions such as reset() and step(), and register the environment class using the function gym.envs.registration.register().

13.6 Exercises

13.6.1 Multiple Choices

13.1 Which of the following performance metrics is particularly interested in online RL tasks: (   )

  A. Regret.
  B. Convergence speed.
  C. Sample complexity.

13.2 On regret, choose the correct one: (   )

  A. Regret is an important performance metric for online learning tasks.
  B. Regret is an important performance metric for offline learning tasks.
  C. Regret is an important performance metric for both online learning tasks and offline learning tasks.

13.3 On UCB, choose the correct one: (   )

  A. UCB algorithm can be only used in tasks that have bounded rewards.
  B. UCB1 algorithm can be only used in tasks that have bounded rewards.
  C. Bayesian UCB algorithm can be only used in tasks that have bounded rewards.

13.4 Which of the following algorithms use the Bayesian method: (   )

  A. ε-greedy algorithm.
  B. UCB algorithm.
  C. Bayesian UCB algorithm.

13.5 On MAB, choose the correct one: (   )

  A. The reward distribution by playing every arm is i.i.d. distributed.
  B. The reward of Bernoulli-reward MAB is always within the range $[0, 1]$.
  C. When using Bayesian UCB algorithm to solve the Bernoulli-reward MAB task, we usually assume the prior distribution is Bernoulli distribution.

13.6 On UCBVI on finite MDP, choose the correct one: (   )

  A. UCBVI algorithm needs to know the dynamic of the environment.
  B. UCBVI algorithm is a model-based algorithm.
  C. UCBVI algorithm can guarantee the regret $O\left(g_{max}|\mathcal{S}|\sqrt{|\mathcal{A}|kt_{max}}\right)$ for finite MDP.

13.6.2 Programming

13.7 Implement an environment for Gaussian-reward MAB, and solve the environment. (Reference codes can be found in `GaussianMABEnv_demo.ipynb`)

13.6.3 Mock Interview

13.8 What is regret? Why online RL cares about regret?

13.9 What is MAB? Why do we need to consider MAB?

13.10 What tasks can UCB1 solve? What tasks can UCBVI solve? What are similarities and differences between the two?

Chapter 14
Tree Search

This chapter covers

- tree search
- exhaustive search
- heuristic search
- Monte Carlo Tree Search (MCTS) algorithm
- Upper Confidence bounds applied to Tree (UCT) algorithm
- AlphaGo algorithm
- AlphaGo Zero algorithm
- AlphaZero algorithm
- MuZero algorithm

Now we have seen many algorithms that leverage environment models to solve MDP, such as LP in Chap. 2, and DP in Chap. 3. Both LP and DP need the environment model as inputs. If the environment model is not an input, we can also estimate the environment model, such as the UCB algorithm and UCBVI algorithm in the previous chapter. This chapter will take a step further, and consider how to conduct model-based RL for a general MDP.

Based on the timing of planning, model-based RL algorithms can be classified into the following two categories:

- Planning beforehand, a.k.a. background planning: Agents prepare the policy on all possible states before they interact with the environment. For example, linear programming method, and dynamic programming method are background planning.

© The Author(s), under exclusive license to Springer Nature Singapore Pte Ltd. 2024

Z. Xiao, *Reinforcement Learning*, https://doi.org/10.1007/978-981-19-4933-3_14

• Planning on decision: Agents plan while they interact with the environment. Usually, an agent plans for the current state only when it observes something during the interaction.

This chapter will focus on Monte Carlo Tree Search, which is a planning-on-decision algorithm.

14.1 MCTS: Monte Carlo Tree Search

For a planning-on-decision algorithm, the agent starts to plan for a state only when it arrives at the state. We can draw all subsequent states starting from this state as a tree, henceforth called **search tree**. As Fig. 14.1(a), the root of the tree is the current state, denoted as s_{root}. Every layer in the tree represents the states that the agent can arrive at by a single interaction with the environment. If the task is a sequential task, the tree may have infinite depth. The nodes in a layer represent all states that the agent can arrive. If there are infinite states in the state space, there may be infinite nodes in a layer. **Tree search** is the process of searching starting from the root note s_{root} to its subsequent reachable states.

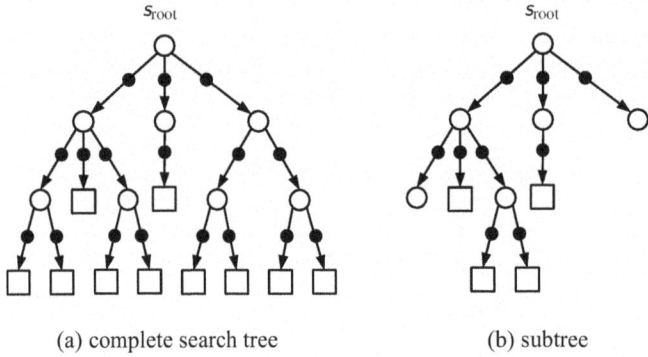

(a) complete search tree   (b) subtree

Fig. 14.1   Search tree.

Exhaustive search is the simplest tree search algorithm. It considers all subsequent reachable states all the way to the end of the episode. It searches thoroughly, and the result is more trustful. However, it does not suitable for the task where state space and/or action space are very large. Even for MDP with medium size, the computation is not small.

Heuristic search is a way to resolve the issue that exhaustive search meets when the space is too large. Heuristic search uses a heuristic rule to select action rather than enumerate all possible actions, and so it may not search the entire tree. This de facto **prunes** the search tree. For example, Fig. 14.1(b) is the pruned tree of

Fig. 14.1(a), so the number of nodes in Fig. 14.1(b) is smaller than the number of nodes in Fig. 14.1(a).

Now we introduce a heuristic search algorithm: **Monte Carlo Tree Search** (**MCTS**). Algorithm 14.1 shows its steps. After the agent reaches a new state s_{root}, Steps 1 and 2 build the search tree whose root is s_{root}. Specifically, Step 1 initializes the tree as a tree with only one root s_{root}, and Step 2 extends the search tree gradually so that the search tree contains more nodes. The tree search in Step 2 consists of the following steps (Fig. 14.2):

- Select: Starting from the root node s_{root}, sample a trajectory within the search tree, until a leaf node s_{leaf} that is just outside the subtree.
- Expand: Initialize the action of the new leaf node s_{leaf} so that the search in the future can use these actions to find subsequent nodes.
- Evaluate: Estimate the action value after the leaf node s_{leaf}.
- Backup, a.k.a. backpropagate: Starting from the leaf node s_{leaf} backward to the root node s_{root}, update the action value estimate of each state–action pair.

After the search tree is constructed, Step 3 uses this search tree to decide.

Algorithm 14.1   MCTS.

Inputs: current state s_{root}, dynamics p or dynamics network $p(\phi)$, and prediction network $f(\theta) = (\pi(\theta), v(\theta))$.
Output: estimate of the optimal policy of the current state $\pi(\cdot|s_{\text{root}})$.
Parameters: parameters of the variant of PUCT, parameters that limit the searching times, parameters of experience replay, and parameter for final decision $\kappa > 0$.

1. (Initialize search tree) Initial the tree as a tree consisting of only one root node s_{root}.
2. (Construct search tree) Tree search multiple times to refine the search tree:

   2.1. (Select) Starting from the root node to a leaf node, use the dynamics and the variant of PUCT to choose action. Record the trajectory of selection $s_{\text{root}}, a_{\text{root}}, \ldots, s_{\text{leaf}}$ for the backup in the sequel. If this search is the first search, there is only one node in the tree, so the trajectory has only one state, and $s_{\text{leaf}} \leftarrow s_{\text{root}}$.

   2.2. (Expand and evaluate) Initialize the action value estimate $q(s_{\text{leaf}}, a) \leftarrow 0$ ($a \in \mathcal{A}(s_{\text{leaf}})$), counting $c(s_{\text{leaf}}, a) \leftarrow 0$ ($a \in \mathcal{A}(s_{\text{leaf}})$), and use the prediction net to calculate the a priori selection probability and state value estimate $(\pi_{\text{PUCT}}(\cdot|s_{\text{leaf}}), v(s_{\text{leaf}})) \leftarrow f(s_{\text{leaf}}; \theta)$. Select the action with the largest a priori selection probability $a_{\text{leaf}} \leftarrow \arg\max_{a \in \mathcal{A}(s_{\text{leaf}})} \pi_{\text{PUCT}}(a|s_{\text{leaf}})$.

   2.3. (Backup) Starting from the leaf node s_{leaf} (included) to the root node s_{root}, use increment implementation to update the state–action pair:

$$g \leftarrow \begin{cases} r + \gamma g, & s \neq s_{\text{leaf}} \\ v(s_{\text{leaf}}), & s = s_{\text{leaf}}, \end{cases} c(s,a) \leftarrow c(s,a)+1, q(s,a) \leftarrow q(s,a)+$$
$$\frac{1}{c(s,a)}\left[g - q(s,a)\right].$$

3. (Decide using the tree search) Output the policy $\pi(a|s_{\text{root}}) \leftarrow \frac{\left[c(s_{\text{root}},a)\right]^{\kappa}}{\sum_{a' \in \mathcal{A}(s_{\text{root}})}\left[c(s_{\text{root}},a')\right]^{\kappa}}$ $(a \in \mathcal{A}(s_{\text{root}}))$.

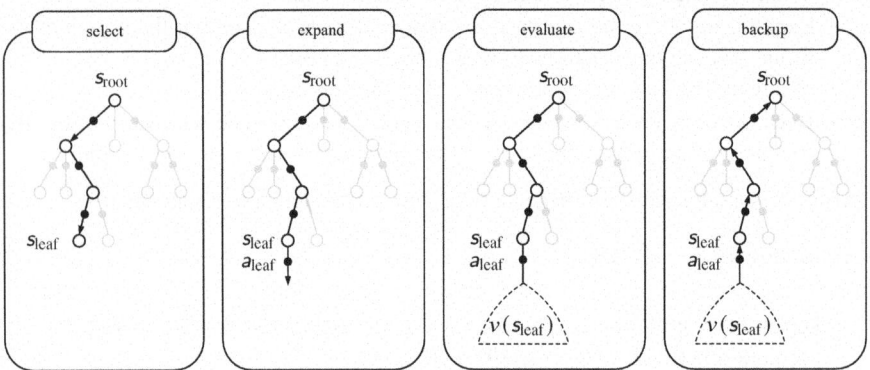

Fig. 14.2   Steps of MCTS.

This algorithm is complex. We will explain it further in the next subsections.

14.1.1 Select

From this subsection, we will talk about the steps of MCTS. This subsection introduces the first step: selection.

Selection in MCTS is as follows: According to a policy, starting from the root node s_{root}, decide and execute, until reaching a leaf node s_{leaf}. The leaf node here should be understood as follows: Before the selection, node s_{leaf} was outside of the search tree. During this search, the selection step selects an action of a node within the search tree, which directs to the state s_{leaf}. Then, the node is included in the search tree, and becomes a leaf node of the search tree. The policy that selection uses is called **selection policy**, a.k.a. **tree policy**.

The optimal selection policy among all possible selection policies is the optimal policy of DTMDP. However, we do not know the optimal policy. MCTS will start from a dummy initial policy and gradually improve it, so that it becomes smarter.

Upper Confidence bounds applied to Trees (UCT) is the most commonly used selection algorithm. Recall that, Sect. 13.2 introduced how to use UCB on MAB tasks for the tradeoff of exploitation and exploration. MAB is a simple MDP that has only one initial state and only one step. UCT applies UCB into trees in the following way: When reaching a state node s, we need to select an action from its action space $\mathcal{A}(s)$. For every state–action pair (s, a), let $c(s, a)$ denote its visitation count and $q(s, a)$ denote its action value estimate. Using the visitation count and action value estimate, we can calculate the UCB for each state–action pair. For example, if we use UCB1, the UCB becomes $u(s, a) = q(s, a) + b(s, a)$, where the bonus $b(s, a) = \lambda_{\text{UCT}}\sqrt{\frac{\ln c(s,a)}{c(s)}}$, and $c(s) = \sum_a c(s, a)$, $\lambda_{\text{UCT}} = \sqrt{2}$.

For a particular node s, if it's visited for the first time, all state–action pairs (s, a) have not been visited yet, so $c(s, a) = 0$ and $q(s, a)$ keep its initial value (usually, $q(s, a) = 0$). In this case, all upper confidence bound $u(s, a)$ is the same for all actions, so an action is randomly picked. With more visitation of the node s, the visitation count and action value estimate for the state–action pair (s, a) $(a \in \mathcal{A}(s))$ will change, and the upper confidence bound will change, so the selection policy will change. Changes of counting and action value estimates will happen in the backup step. We will explain this in Sect. 14.1.3. We hope that the selection policy will be the same as the optimal policy of MDP after the end of training.

We just noticed that, if a state s is visited for the first time, the visitation counting and the action value estimates of all state–action pairs can all be zero, and the selected action is uniformly distributed. If we can give a proper prediction of selection according to the instinct of state, we can make the selection smarter. The **variant of Predict-UCT (variant of PUCT)** adapts this idea: It uses some prediction to change the probability of selection. The variant of PUCT somehow estimates the selection probability $\pi_{\text{PUCT}}(a|s)$ for each state–action pair (s, a), and then uses this probability to calculate the bonus:

$$b(s, a) = \lambda_{\text{PUCT}}\pi_{\text{PUCT}}(a|s)\frac{\sqrt{c(s)}}{1 + c(s, a)},$$

where $c(s) = \sum_{a \in \mathcal{A}(s)} c(s, a)$, and λ_{PUCT} is a parameter, and then it calculates the UCB as $u(s, a) = q(s, a) + b(s, a)$. Here, the selection probability prediction $\pi_{\text{PUCT}}(a|s)$ is called a priori selection probability, and it will be determined in the expansion step.

The selection step needs the dynamics of the environment to determine the next state and the reward after each action execution. If the dynamics p is available as inputs, we can directly use it. Otherwise, we need to estimate the dynamics. A possible way is to use a neural network to maintain the estimate of dynamics, in the form of $(s', r) = p(s, a; \phi)$, where ϕ is the parameter of the neural network. Such neural network is called **dynamics network**. Its input is state–action pair, and outputs are next state and reward. The training of and dynamics network, which is independent of the tree search, will be explained in Sect. 14.1.5.

14.1.2 Expand and Evaluate

This subsection explains the expansion and evaluation in MCTS.

The intuition of **expansion** is as follows: The selection step ends up at a leaf node s_{leaf}. We hope that, in future selections, we can search further from this leaf node s_{leaf} to more nodes. Therefore, we need to maintain the values that enable the selection from the leaf node s_{leaf}. According to the expression of the variant of PUCT, we can know that the variant of PUCT needs the following information:

- The visitation count $c(s, a)$ for each state–action pair (s, a);
- The action value estimate $q(s, a)$ for each state–action pair (s, a);
- The a priori selection probability $\pi_{\text{PUCT}}(a|s)$ for each state–action pair (s, a).

The expansion step initializes all three pieces of information for the leaf node s_{leaf}. We can initialize the visitation counts and the action value estimates as 0, and use a neural network to initialize the a priori selection probability. The neural network to generate the a priori selection probability is called policy network, denoted as $\pi_{\text{PUCT}}(s; \boldsymbol{\theta}_{\text{PUCT}})$, where $\boldsymbol{\theta}_{\text{PUCT}}$ is the network parameter. The design of this network is the same as the design of the policy network in Sect. 7.1: The direct output is usually in the form of an action preference vector when the action space is discrete, and the direct output is usually in the form of parameters of continuous distribution when the action space is continuous. We can also select an action a_{leaf} that maximizes the a priori selection probability.

The intuition of **evaluation** is as follows: Recall the exhaustive search algorithm, each time the agent reaches a state–action pair (s, a), it will continue finding the next state until it reaches the leaf node that represents terminal states, and then update the action value estimate and visitation counting of the states and their actions backwardly starting from the leaf node to the root node. (The update is actually at the backup step, which will be explained in the next subsection.) As contrast, MCTS prunes the tree and does not search until the terminal state. Therefore, after MCTS gets the new leaf node s_{leaf} and an action a_{leaf}, it needs some other method to estimate the action value of the state–action pair. There are many ways to estimate the values, one of them is to use a neural network. The input of the neural network can be the state s_{leaf}. Theoretically speaking, the input of the neural network can be the state–action pair $(s_{\text{leaf}}, a_{\text{leaf}})$, but the action a_{leaf} is determined by s_{leaf} through a neural network, so essentially the input is s_{leaf}. This neural network is called state value network, denoted as $v(s_{\text{leaf}}; \mathbf{w})$, where $\mathbf{w}$ is the parameter of state value network. There are other evaluation methods, such as "rollout" in the AlphaGo algorithm (in Sect. 14.2.4). We ignore other methods here.

The policy network $\pi_{\text{PUCT}}(s_{\text{leaf}}; \boldsymbol{\theta}_{\text{PUCT}})$ and the state value network $v(s_{\text{leaf}}; \mathbf{w})$ share the same input s_{leaf}. Therefore, they can use a common part to fetch the same feature. The combined implementation of the two networks is called **prediction network**, denoted as $\big(\pi_{\text{PUCT}}(\cdot|s_{\text{leaf}}), v(s_{\text{leaf}})\big) = f(s_{\text{leaf}}; \boldsymbol{\theta})$, where

- the parameters of the prediction network are $\boldsymbol{\theta}$;
- the input of the prediction network is the leaf node s_{leaf};

- the outputs of the prediction network are the a priori selection probability $\pi_{\text{PUCT}}(a|s_{\text{leaf}})$ $(a \in \mathcal{A}(s_{\text{leaf}}))$ and the state value estimate $v(s_{\text{leaf}})$.

With the prediction network, the expansion and evaluation can work together: We can input the leaf node s_{leaf} to the prediction network $f(\boldsymbol{\theta})$ to get the output $\left(\pi_{\text{PUCT}}(\cdot|s_{\text{leaf}}), v(s_{\text{leaf}})\right)$. Then expand the leaf node using

$$
\begin{aligned}
c(s_{\text{leaf}}, a) &= 0, & a &\in \mathcal{A}(s_{\text{leaf}}) \\
q(s_{\text{leaf}}, a) &= 0, & a &\in \mathcal{A}(s_{\text{leaf}}) \\
\pi_{\text{PUCT}}(a|s_{\text{leaf}}) &= \pi_{\text{PUCT}}(a|s_{\text{leaf}}; \boldsymbol{\theta}), & a &\in \mathcal{A}(s_{\text{leaf}}),
\end{aligned}
$$

and use $v(s_{\text{leaf}})$ as the evaluation result for the next backup step.

Training the parameters of the prediction network, which is independent of tree search, will be explained in Sect. 14.1.5.

14.1.3 Backup

The selection policy needs the visitation count, action value estimate, and a priori selection probability. These values are initialized at the expansion step and updated at the backup step. This subsection explains the last step of MCTS: backup.

Backup is the process that updates action value estimates and visitation counting for each state–action pair. We do not need to update the a priori selection probability. This process is called "backup" since the order of the update is backward from the leaf node s_{leaf} (included) to the root node s_{root} for easy calculation of the return. Specifically speaking, the previous evaluation step has already got the estimate of action value of the state–action pair $(s_{\text{leaf}}, a_{\text{leaf}})$ as $v(s_{\text{leaf}})$, which is exactly a return sample of $(s_{\text{leaf}}, a_{\text{leaf}})$. Then backward from the leaf node, we use the form $g \leftarrow r + \gamma g$ to calculate the return sample g for all state–action pairs (s, a) in the selection trajectory.

Upon obtaining the return sample g for every state–action pair (s, a), we can use increment implementation to update the action value estimate $q(s, a)$ and visitation frequency $c(s, a)$ for the state–action pair (s, a).

The backup step can be mathematically expressed as

$$
\begin{aligned}
g &\leftarrow \begin{cases} r + \gamma g, & s \neq s_{\text{leaf}} \\ v(s_{\text{leaf}}), & s = s_{\text{leaf}}, \end{cases} \\
c(s, a) &\leftarrow c(s, a) + 1, \\
q(s, a) &\leftarrow q(s, a) + \frac{1}{c(s, a)} \left[g - q(s, a) \right].
\end{aligned}
$$

Executing MCTS repeatedly will update the visitation count for each node, especially the root node.

14.1.4 Decide

We can use the search tree to decide the policy to interact with the real environment. A usual policy is using the visitation count to determine the policy at the root node s_{root} as

$$\pi(a|s_{\text{root}}) = \frac{\left[c(s_{\text{root}}, a)\right]^{\kappa}}{\sum_{a' \in \mathcal{A}(s_{\text{root}})} \left[c(s_{\text{root}}, a')\right]^{\kappa}}, \quad a \in \mathcal{A}(s_{\text{root}})$$

where κ is the parameter for the tradeoff between exploration and exploitation.

> **❗ Note**
>
> The decision only uses the visitation count, rather than using the selection policy. The selection policy is only used at the selection step of MCTS. Table 14.1 compares the selection policy and decision policy.

Table 14.1 Compare selection policy and decision policy.

policy	when to use	popular implementation
selection policy	selection step in MCTS, used for deciding within the search tree	variant of PUCT $\pi_{\text{select}}(s) = \arg\max_a \left[q(s, a) + \lambda_{\text{PUCT}}\, \pi_{\text{PUCT}}(a|s) \frac{\sqrt{c(s)}}{1 + c(s,a)} \right].$
decision policy	interact with the real environment	proportional to the power of visitation count $\pi(a|s) = \frac{\left[c(s,a)\right]^{\kappa}}{\sum_{a'}\left[c(s,a')\right]^{\kappa}}, \quad a \in \mathcal{A}(s).$

14.1.5 Train Networks in MCTS

MCTS uses the prediction network. For the tasks where the dynamics are not included in the inputs, MCTS also uses dynamics network. Table 14.2 summarizes the neural networks in MCTS. This subsection introduces how to train the parameters of these networks.

Training prediction can be implemented by minimizing the cross-entropy loss of the policy and ℓ_2 loss of the state value estimate. Specifically speaking, consider a state S in a real trajectory. Let $\Pi(\cdot|S)$ denote its decision policy and let G denote its return sample. Feeding the state S into the prediction network, we can get the a priori selection probability $\pi_{\text{PUCT}}(\cdot|S)$ and the state value estimate $v(S)$. In this sample, the cross-entropy loss of the policy is $-\sum_{a \in \mathcal{A}(S)} \Pi(a|S) \log \pi_{\text{PUCT}}(a|S)$, and the ℓ_2

Table 14.2 **Neural networks in MCTS.**

neural networks	inputs	outputs	when to use	remark
prediction network $f(s; \theta)$	state s	a priori selection probability $\pi_{\text{PUCT}}(\cdot\|s)$ and state value estimate $v(s)$	expansion and evaluation	
dynamics network $p(s, a; \varphi)$	state–action pair (s, a)	next state s' and reward r	selection	only need when the dynamics of the environment is unknown.

loss of state value is $\left(G - v(S)\right)^2$. Besides minimizing these two losses, we usually penalize the parameters, such as using ℓ_2 penalty to minimize $\|\theta\|_2^2$. So the sample loss becomes

$$- \sum_{a \in \mathcal{A}(S)} \Pi(a|S) \log \pi_{\text{PUCT}}(a|S) + \left(G - v(S)\right)^2 + \lambda_2 \|\theta\|_2^2,$$

where λ_2 is a parameter.

Algorithm 14.2 shows the MCTS with training prediction network. The inputs of algorithm include the dynamics model of environment, so it does not need a dynamics network. It can be viewed as the AlphaZero algorithm for single-agent tasks. In the beginning, the agent initializes the parameters of prediction network θ, and then interacts with the environment. During the interaction, each time the agent reaches a new state, it builds a new search tree, and then decides using the built tree. The experience of interaction with the environment is saved in the form of (S, Π, G), and the saved experiences will be used to train the prediction network.

Algorithm 14.2 AlphaZero.

Input: dynamics p.
Output: optimal policy estimate.
Parameters: parameters of MCTS, parameters for training the prediction network, and parameters of experience replay.

1. Initialize:

   1.1. (Initialize parameters) Initialize the parameters of prediction network θ.
   1.2. (Initialize experience storage) $\mathcal{D} \leftarrow \varnothing$.

2. For each episode:

2.1. (Collect experiences) Select an initial state, and use MCTS to select action until the end of the episode or meeting a predefined breaking condition. Save each state S in the trajectory, the decision policy Π of the state, and the return G in the form of (S, Π, G) into storage.
2.2. (Use experiences) Do the following once or multiple times:
  2.2.1. (Replay) Sample a batch of experiences $\mathcal{B}$ from the storage $\mathcal{D}$. Each entry is in the form of (S, Π, G).
  2.2.2. (Update prediction network parameter) Update θ to minimize

$$\frac{1}{|\mathcal{B}|} \sum_{(S,\Pi,G)\in\mathcal{B}} \left[-\sum_{a\in\mathcal{A}(S)} \Pi(a|S) \log \pi_{\text{PUCT}}(a|S) + (G - v(S))^2 \right] + \lambda_2 \|\theta\|_2^2.$$

MCTS is named as the tree search with MC, because it uses unbiased estimates when updating the state value network, which is an MC update.

The dynamics network is needed when the dynamics is not available as inputs. We can train the dynamics networks by comparing the next state prediction and the real next state generated by the environment. For example, in a real trajectory, the state–action pair (S, A) is followed by the next state S' and reward R. The dynamics network outputs the probability of the next state $p(S'|S, A)$ and reward $r(S, A)$, and then the cross-entropy sample loss of the next state part is $-\ln p(S'|S, A)$ and ℓ_2 sample loss of the reward part is $(R - r(S, A))^2$. There are also other probabilities. For example, if the state space $S = \mathbb{R}^n$ and the output of the dynamics network can be a definite value, we can use ℓ_2 loss between the predicted next state and the real next state.

Algorithm 14.3 shows the usage of MCTS in tasks without known dynamics. Since the algorithm inputs do not contain the dynamics, the agent can only interact with the environment. Therefore, dynamics network is needed. In order to train the dynamics network, it needs to store the transition (S, A, R, S', D'). This algorithm can be viewed as the MuZero algorithm in single-agent tasks.

Algorithm 14.3 MuZero.

Input: environment (for interaction only).
Output: optimal policy estimate.
Parameters: parameters of MCTS, parameters for training the prediction network and dynamics network, and parameters of experience replay.

1. Initialize:

  1.1. (Initialize parameters) Initialize the parameters of prediction network θ and dynamics network ϕ.
  1.2. (Initialize experience storage) $\mathcal{D} \leftarrow \varnothing$.

2. For each episode:

2.1. (Collect experiences) Select an initial state, and use MCTS to select action until the end of the episode or meeting a predefined breaking condition. Save each state S in the trajectory, and the decision policy Π of the state, the follow-up action A, the next state S', reward R, terminal state indicator D', and the return G in the form of $(S, \Pi, A, R, S', D', G)$ into storage.

2.2. (Use experiences) Do the following once or multiple times:

    2.2.1. (Replay) Sample a batch of experiences $\mathcal{B}$ from the storage $\mathcal{D}$. Each entry is in the form of $(S, \Pi, A, R, S', D', G)$.

    2.2.2. (Update prediction network parameter) Update θ using (S, Π, G).

    2.2.3. (Update dynamics network parameter) Update ϕ using (S, A, R, S', D').

The advantages and disadvantages of MCTS are as follows:

- Advantage: MCTS uses heuristic search to invest computation resources to the most important branches, and can stop at any time to adjust for varied resource limitation.
- Disadvantage: Each time the agent meets a new state, it needs to construct a new tree. The decision policy is wise only when many MCTS can be conducted, so it is computation intensive. This has become the main issue of MCTS.

Since MCTS uses massive computation resources, MCTS is usually deployed in multiple processors and is computed in parallel. In order to better utilize the computation resources in the parallel computation scenario, MCTS is usually implemented as **Asynchronous Policy and Value MCTS (APV-MCTS)**. Here, the "asynchronous" means that the training of neural networks and tree search can be conducted asynchronously in different devices. The device for training neural networks does not need to care whether tree search is ongoing, since it can always sample from experience storage and update the parameters of neural networks accordingly. The device for tree search does not need to care whether neural network training is ongoing, since it can always use the latest neural network parameters to search. Since the computation of tree search is much larger than the computation of neural networks, there usually are multiple devices for tree search, and only one device for training neural networks.

14.2 Application in Board Game

The application of MCTS in board games such as the game of Go has been widely applauded by the general public. This section will introduce the characteristics of board games, and how to apply MCTS in the board games.

14.2.1 Board Games

Board games are games that place or move pieces on a board. There are varied kinds of board games: Some board games involve two players, such as chess, gobang, Go, shogi; while some board games involve more players, such as Chinese checkers, four-player Luzhanqi. Some board games have stochastic, or players do not have complete information, such as aeroplane chess depending on the result of dice rolling; while some board games are deterministic and players have complete information, such as Go and chess. In some board games such as Go, gobang, reversi, and Chinese checker, all pieces are of the same kind; in some board games such as chess, there are different kinds of pieces that have distinct behaviors.

This section considers a specific type of two-player zero-sum deterministic sequential board games. The two-player zero-sum deterministic sequential board games are the board games with the following properties:

- Two-player: There are two players. So the task is a two-agent task.
- **Zero-sum**: For each episode, the sum of the episode reward of the two players is zero. The outcome can be either a player wins over another player, or the game ends with a tie. The outcome can be mathematically designated as follows: Let there be no discount (i.e. discount factor $\gamma = 1$). The winner gets episode reward $+1$. The loser gets episode reward -1. Both players get episode reward 0 if the game ends with a tie. Therefore, the total episode rewards of the two players are always 0. Using this property, the algorithm for single-agent tasks can be adapted to two-agent tasks.
- Sequential: Two players take turns to play. The player who plays first is called the black player, while the player who players last is called the white player.
- Deterministic: Both players have full current information, and there is no stochastic in the environment.

Among many such games, this section focuses on games with the following additional characteristic:

- Square grid board: The board is a square grid board.
- Placement game: Each time, a player places a piece in a vacant position.
- Homogenous pieces: There is only one type of pieces for each player.

Table 14.3 lists some games. They have different rules and difficulties, but they all meet the aforementioned characteristics.

Interdisciplinary Reference 14.1
Board Game: Tic-Tac-Toe and Gomoku

Tic-Tac-Toe and freestyle Gomoku are alignment games. Academically speaking, both of them are special cases of (m, n, k) games. In a (m, n, k) game, two players take turns to place one piece on an $m \times n$ board. A play wins when they successfully place $\geq k$ pieces in a direction, either vertically, horizontally, or diagonally. The

Table 14.3 Some two-player zero-sum deterministic sequential board games.

game	type	board shape	maximum possible steps	reference average steps	has been solved or not	complexity categories of the generalized version
Tic-Tac-Toe	alignment game	3×3	9	9	Yes (tie)	PSPACE-complete
freestyle Gomoku	alignment game	15×15	225	30	Yes (black wins)	PSPACE-completely
Reversi	counting game	8×8	60	58	No	PSPACE-complete
Go	territory game	19×1	$+\infty$	150	No	EXPTIME-complete

game ends up with a tie if neither player wins when the board is full. Tic-Tac-Toe is exactly the $(3, 3, 3)$ game. The freestyle Gomoku is exactly the $(15, 15, 5)$ game.

According to the "m,n,k-game" entry in Wikipedia, some (m, n, k) games have been solved. When both players play optimally, the results of the games are as follows:

- $k = 1$ and $k = 2$: Black wins except that $(1, 1, 2)$ and $(2, 1, 2)$ apparently lead to a tie.
- $k = 3$: Tic-Tac-Toe $(3, 3, 3)$ ends in a tie. $\min \{m, n\} < 3$ ends in a tie, too. Black wins in other cases. The game ends in a tie for all $k \geq 3$ and $k > \min m, n$.
- $k = 4$: $(5, 5, 4)$ and $(6, 6, 5)$ end in a tie. Black wins for $(6, 5, 4)$. For $(m, 4, 4)$, black wins for $m \geq 30$, and tie for $m \leq 8$.
- $k = 5$: On $(m, m, 5)$, tie when $m = 6, 7, 8$, and black wins for the freestyle Gomoku ($m = 15$).
- $k = 6, 7, 8$: $k = 8$ ties for infinite board size, but unknown for finite board size. $k = 6$ and $k = 7$ unknown for infinite board size. $(9, 6, 6)$ and $(7, 7, 6)$ end in a tie.
- $k \geq 9$ ends in a tie.

Interdisciplinary Reference 14.2
Board Game: Reversi

Reversi, or Othello, is a count game on a 8×8 board. At the beginning of an episode, there are four pieces (a.k.a. disks) on the board. Two pieces are black and two pieces are white. Then the black and the white take turns to place their piece, while the new piece should reverse some opponent's pieces to themselves by putting them between the new piece and an old own piece in a line. The game ends when no players can play. A player wins when it has more pieces at the end of the game.

For example, an episode starts with 4 pieces in Fig. 14.3(a). The numbers on the left and top of the board are used to represent the positions of the grid. Now the black player can place a stone in one position among $(2, 4)$, $(3, 5)$, $(4, 2)$, and $(5, 3)$ and reverse a white piece. Now the black places at $(2, 4)$, which turns the white piece at $(3, 4)$ into a black piece. Now it is white's turn. The white can choose a place among $(2, 3)$, $(2, 5)$, and $(4, 5)$, all of which can turn one black piece into white. If the white plays $(2, 5)$, it will lead to Fig. 14.3(c).

If the reversi is extended into an $n \times n$ board, it can be proved that the problem is PSPACE-complete. If two players play optimally, white wins when $n = 4$ or 6. $n = 8$ is not completely solved yet, but people generally believe that the game should end in a tie.

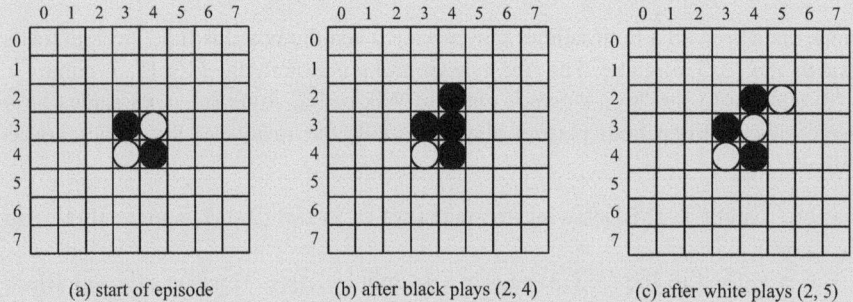

(a) start of episode   (b) after black plays (2, 4)   (c) after white plays (2, 5)

Fig. 14.3   First two steps of the reversi opening "Chimney".

Interdisciplinary Reference 14.3
Board Game: Go

The game of Go is a territory game on the 19×19 board. The rule of Go is as follows:

- Link four-way: A piece (a.k.a. stone) in a board can link to the piece with the same color in the four directions on the board, but it can not link diagonally.
- Survive with liberties: A liberty of a piece is an empty place that links to the current piece. If a piece does not have liberties, they die and should be removed from the board.
- Take turns to place: The black player and the white player take turns to place pieces. Each time they place one piece on the board. But the place should meet some criteria. One of the most famous criteria is komi: If a player just places a piece to kill a piece of another player, the other player can not immediately place a piece at the killed place to kill back, unless the placing can kill > 1 pieces. There may be other rules: For example, some rules do not allow suicide,

meaning that a player can not place piece that dies immediately if it can not kill any pieces of the opponent.

- Win on calculation: There are many variants of rules to determine the winner. One among them is the Chinese rule, which determines the winner according to the number of places each play occupies. If the number of places black occupies minus the number of places white occupies exceeds a pre-designated number, say 3.72, black wins; otherwise, white wins.

Go is believed to be one of the most complex board games, much more complex than Tic-Tac-Toe, Gomoku, and Reversi. The primary reason is Go has the largest board. Besides, the following characteristics of Go also increase the complexity:

- For (m, n, k) game and reversi, the maximum number of steps in an episode is bounded by a predefined integer. That is because each step will occupy one position on the board so that the number of steps in an episode is bounded by the number of positions on the board, which is finite and pre-defined. Consequently, if we generalize the size of the board as a parameter, the complexity of this problem is PSPACE-complete. However, this is not true for the game of Go. In Go, the pieces can be removed from the board when they are killed, so the number of steps in an episode can be infinite. Consequently, if we generalize the size of the board as a parameter, the complexity of this problem is probably PSPACE-hard rather than PSPACE-complete.
- For (m, n, k) game and reversi, the information of the current board situation and the current player together fully determine the state of the game. However, this is not true for the game of Go. There is the rule of Komi in the game of Go, which relies on recent information of killed position to determine whether a position is valid or not.
- For (m, n, k) game and reversi, changing the colors of all pieces, that is, changing all white pieces into black and changing all black pieces into white, an end game that white wins will become an end game where black wins. However, this is not always true for the game of Go. As a territory game, the rule of Go usually designates a threshold such as 3.75, and the black player wins only if the occupation of the black player minus the occupation of the white player exceeds the threshold. they need to occupy more positions than the white players. If the black player only occupies very few more positions than the white (say, 1), the winner is the white player. If we change all colors of pieces in such end game, the white player occupies more positions than the black, the winner is still the white player. From this sense, Go has some asymmetric on the players.

Since the game of Go is much more complex than Tic-Tac-Toe, Gomoku, and reversi, developing an agent for Go requires much more resources than developing agents for Tic-Tac-Toe, Gomoku, and reversi.

Combinatorial game theory studies the solution of the above games.

Interdisciplinary Reference 14.4
Combinatorial Game Theory: Game Tree

In combinatorial game theory, a **game tree** is a tree that is used to represent the states all players can meet. Fig. 14.4 shows the game tree of Tic-Tac-Toe. Initially, the board is empty, and it is black's turn. The black player may have lots of different choices, each of which may lead to different next states. In different next states, the white player has different actions to choose, and each action will lead to different subsequent actions of the black player. Each node in the tree represents a state, which contains both the information of board status and the player to play.

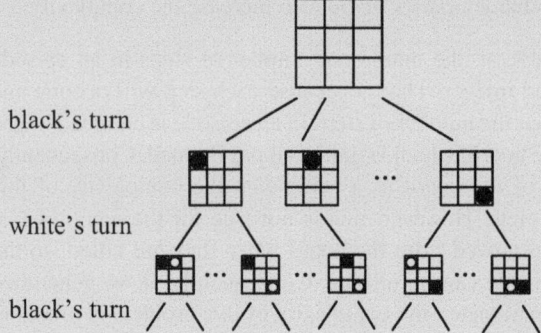

Fig. 14.4   Game tree of Tic-Tac-Toe.

Interdisciplinary Reference 14.5
Combinatorial Game Theory: Maximin and Minimax

Maximin and **minimax** are strategies for two-player zero-sum games. Maximin is with regard to reward, while minimax is with regard to cost. Their idea is as follows: For the current player, no matter which action the current player chooses, the opponent will always try to maximize the opponent's reward or equivalently minimize the opponent's cost in the next step. Since the game is zero-sum, the current player will always get the minimal reward, or equivalently maximal cost. The current player should take this into account when they make a decision: the current player should maximize the minimal reward, or minimize the maximum cost.

We can search exhaustively for small-scale games. Exhaustive search starts from the current state to find all possible states until the leaf nodes that represent the end of game. With such tree, the player can choose the optimal action by applying maximin strategy in a backward direction. For example, Fig. 14.5 shows an example tree of Tic-Tac-Toe. The state in depth 0 will lead to multiple states in depth 1. If the black player plays $(1, 0)$ (the leftmost branch), it can win no matter how the white player reacts. Therefore, the black player can find a winning policy using tree search and maximum strategy.

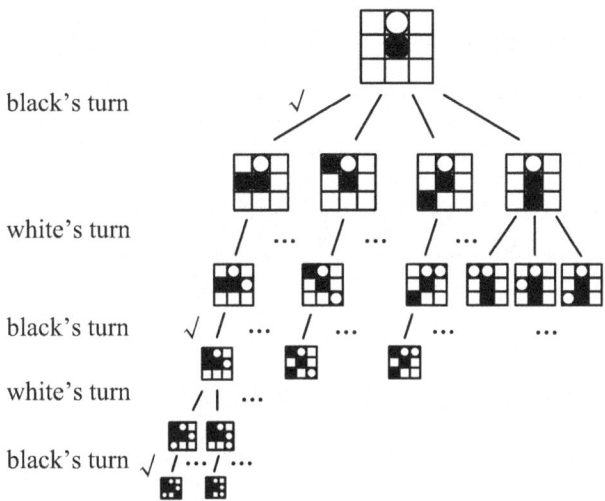

Fig. 14.5   **Maximin decision of Tic-Tac-Toe.**

On Tic-Tac-Toe, the state space has at most $3^9 = 19683$ states. In fact, if we exclude invalid states carefully, such as removing the board with five blacks but no white, or the board with both 3 blacks in a row and 3 whites in a row, the number of states becomes 5478. The number of trajectories is at most $9! = 362880$. If we exclude the invalid trajectories carefully, since some trajectories may end within ≤ 8 steps, the number of trajectories is 255168. In such scale, we can exhaustively search Tic-Tac-Toe. However, for Gomoku, reversi, and Go, the number of states and trajectories are too large, so we can not search exhaustively.

14.2.2 Self-Play

The previous section mentioned that we may not be able to exhaustively search the board game with large state space and trajectory space. This section will use MCTS to solve those tasks.

We have learned the MCTS for single-agent tasks. But a two-player zero-sum game has two players. In order to resolve this gap, **self-play** is proposed to apply the single-agent algorithm to two-player zero-sum tasks. The idea of self-player is as follows: Two players in the game use the same rules and the same parameters to conduct MCTS. Fig. 14.6 shows the MCTS in the board game. During the game, the current player, either black or white, starts from its current state, and uses the latest prediction networks (and optionally dynamics network) to MCTS, and uses the built search tree to decide.

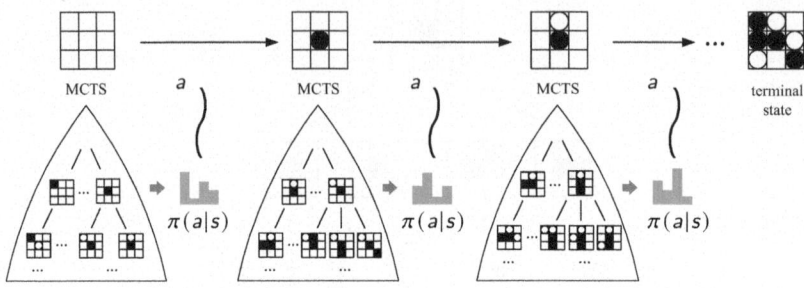

Fig. 14.6 MCTS with self-play.

Algorithm 14.1 can also be used to show the MCTS with self-play. Remarkably, we should notice the following differences when applying Algo. 14.1 into the board games this section interests:

- In Step 2.1: During the selection step of MCTS, we should always select on behalf of the interest of the player of the state node. The action value estimate is also with regard to the player of the state node.
- In Step 2.3: Note that the board game sets the discount factor $\gamma = 1$, and there are non-zero rewards only when reaching the terminal states. Additionally, the sum of rewards of the two players is zero. Therefore, the update of return sample should be

$$ g \leftarrow \begin{cases} -g, & s \neq s_{\text{leaf}} \\ v(s_{\text{leaf}}), & s = s_{\text{leaf}}. \end{cases} $$

Algorithms 14.2 and 14.3 can also be used to show the updating of neural networks when MCTS with self-play is used. Remarkably, we should save the experience of both players, and we should use the experience of both players when training the networks.

We have known that the MCTS is very resource intensive for single-agent tasks, since it needs to construct a new search tree each time the agent meets a new state. This is still true when MCTS is applied to the two-player board games. If we can have some simple ways to extend every existing experience to multiple experiences, we can obviously accelerate the preparation of training data. Here we introduce a method to improve the sample efficiency using the symmetric of board games.

Some board games have opportunities to re-use experience using symmetric. For example, for the selected board games in this section, we may rotate the board and corresponding policy in multiple ways, including horizontal flipping, vertical flipping, and 90° rotation, which may extend a transition to 8 different transitions

(See Table 14.4). For (m, n, k) games and reversi, we may also change the colors of all pieces, from black to white and from white to black. All these extensions should be based on the understanding of the symmetry of the board games.

Table 14.4   8 equivalent boards.

	rotate	rotate after transpose
rotate 0° counter-clockwise	board	np.transpose(board)
rotate 90° counter-clockwise	np.rot90(board)	np.flipud(board)
rotate 180° counter-clockwise	np.rot90(board, k=2)	np.rot90(np.flipud(board))
rotate 270° counter-clockwise	np.rot90(board, k=3)	np.fliplr(board)

14.2.3 Neural Networks for Board Games

Table 14.2 in Sect. 14.1.5 summarizes the neural networks in MCTS. This section introduces how to design the structure of neural networks for the selected board games.

First, let us consider the inputs of the networks. We can construct the inputs of the networks using the concept of canonical board. For the board game with board shape $n \times n$, the canonical board is a matrix $\left(b_{i,j} : 0 \le i, j < n\right)$ of the shape (n, n). Every element in the matrix is an element in $\{+1, 0, -1\}$, where

- $b_{i,j} = 0$ means that the position (i, j) is vacant;
- $b_{i,j} = +1$ means that there is a piece of the current player at the position (i, j);
- $b_{i,j} = -1$ means that there is a piece of the opponent at the position (i, j).

For example, consider the board in Fig. 14.7(a), if it's black's turn, the canonical board is Fig. 14.7(a). If it's white's turn, the canonical board is Fig. 14.7(b). For Tic-Tac-Toe, Gomoku, and reversi, we only need one canonical board as the inputs of neural networks, since this canonical board suffices for determining the a priori selection probability and state value estimate. The game of Go has the rule of ko and komi, so it needs some additional canonical boards to represent the current player and the status of ko.

Different games have different board sizes. For games such as Gomoku and Go, the size of board is not small. We usually use residual networks to extract

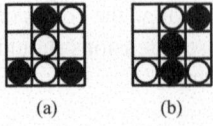

(a)   (b)

Fig. 14.7   Reverse the color of all pieces on the board.

features from boards. Figure 14.8 shows a possible structure of the prediction network for the game of Go. The probability network and the value network share some common part, which consists of some convolution layers and ResNet layers. Both probability network and value network have their independent part, such as independent convolution layers. The output of probability network is activated by softmax to ensure the output is in the format of probability. The output of value network is activated by tanh to ensure the range of output is within $[-1, 1]$.

Interdisciplinary Reference 14.6
Deep Learning: Residual Network

Residual Network (ResNet) is a CNN structure proposed in (Kaiming, 2016). Figure 14.9 shows its common structure, which consists of convolution layers, batch normalization layers, and activation layers. Additionally, there is a shortcut from the input. The batch normalization and shortcut can help avoid the vanishing gradients problem and accuracy saturation problem.

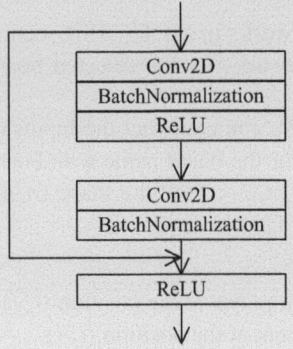

Fig. 14.9   Residual network.
This picture is adapted from (Kaiming, 2016).

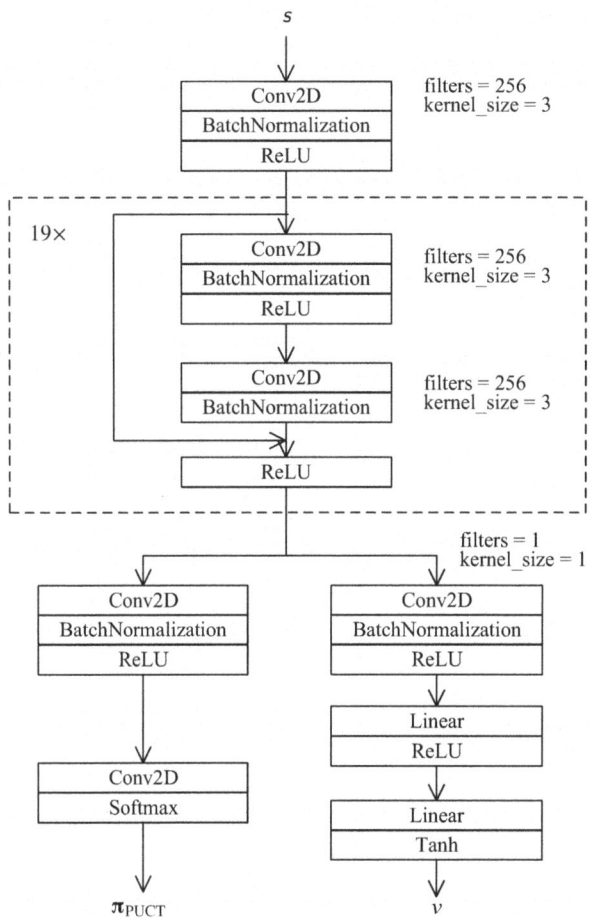

Fig. 14.8 Example structure prediction network for the game of Go.

14.2.4 From AlphaGo to MuZero

The DRL algorithms that apply MCTS on two-player zero-sum board games include
AlphaGo, AlphaGo Zero, AlphaZero, and MuZero (Table 14.5). All these algorithms
use APV-MCTS with the variant of PUCT, self-play, and ResNet. They all have their
own characteristic, too. AlphaGo is the first DRL algorithm of this kind, and it was
designed only for the game of Go. Subsequent algorithms fixed the design deficits
in AlphaGo, and extended its use case. Algorithms 14.2 and 14.3 in Sect. 14.1 can
be regarded as the AlphaZero algorithm and MuZero algorithm. This subsection
introduces and compares these four algorithms.

Table 14.5   Some DRL algorithms that apply MCTS to board games.

algorithm	tasks	inputs	neural networks	resource for Go	paper
AlphaGo	Go	environment dynamics (i.e. rules of the games), interaction history between exports and knowledge	policy network, state value network, behavior cloning network	(distributed version) self-play: 1202 CPUs, neural network training: 176 GPUs	(Silver, 2016)
AlphaGo Zero	Go	environment dynamics (i.e. rules of the games)	prediction network	self-play: 64 GPUs, neural network training: 4 first-generation TPUs	(Silver, 2017)
AlphaZero	Go, chess, and shogi	environment dynamics (i.e. rules of the games)	prediction network	self-play: 5000 first-generation TPUs, neural network training: 16 second-generation TPUs	(Silver, 2018)
MuZero	Go, chess, shogi, and Atari games	environment (interaction only)	prediction network, dynamics network, representation network	self-play: 1000 third-generation TPUs, neural network training: 16 third-generation TPUs	(Schrittwieser, 2020)

Let us discuss the most famous **AlphaGo** algorithm. Before AlphaGo, there are already applications that use MCTS to solve Go. AlphaGo introduced deep learning into such setting, and created an agent that was much more powerful than ever. AlphaGo was specifically designed for the game of Go, and its inputs include the mathematical models of Go.

AlphaGo maintains three neural networks: policy network, state value network, and behavior cloning network. AlphaGo did not combine the policy network and state value network into a single prediction network, which leads to some performance degradation. AlphaGo also uses the behavior cloning algorithm in imitation learning, so it needs the interaction history of expert policies as inputs, and maintains a neural network called behavior cloning network. It uses supervised learning to train this behavior cloning network. We will talk about imitation learning and behavior cloning in Chap. 16.

AlphaGo also designs some features that were specifically designed for the game of go, such as the feature to denote the number of liberties for every piece, and features to denote whether there are "ladder captures" (a terminology in the game of Go).

AlphaGo also uses "rollout" in the evaluation step. **Rollout**, a.k.a. playout, is as follows: In order to understand the return after the leaf node, a policy is used to sample many trajectories, starting from the leaf node to the terminal states. Then the trajectories are used to estimate the state value of the leaf node. The evaluation containing the rollout can also be called simulation. The policy for rollout, a.k.a. **rollout policy**, needs to be called massively, so it is simplified with some loss of accuracy. Since rollout requires massive computation resources, the follow-up algorithm discards it.

AlphaGo Zero is an improvement of AlphaGo. It discards some unnecessary designs in AlphaGo, which simplifies the implementation and achieves better performance. The improvement of AlphaGo Zero includes:

- Remove the imitation learning, so the experience history of expert policy and the behavior cloning network is no longer needed.
- Remove rollout. Removing this inefficient computation can improve the performance with the same computation resources.
- Combine the policy network and state value network into prediction network, which can improve the sample efficiency.

The paper of AlphaGo Zero used TPUs for the first time. According to the paper, when training the AlphaGo Zero algorithm, self-play uses 64 GPUs, and training neural networks uses 4 first-generation TPUs. Additional resources are used to evaluate the performance of the agent during training.

AlphaZero extends the usage from Go to other board games, such as chess and shogi. Therefore, its feature design is more generalized. Compared to the game of Go, chess and shogi have their own characteristic:

- Go has only one type of pieces, but chess and shogi have many other kinds of shogi. For chess and shogi, each type of pieces has its board as inputs. It is like one-hot coding on the type of pieces.
- Rotating the board of Go 90° still leads to a valid Go board, but this is not true for chess and shogi. Therefore, the sample extension based on rotation can be used for Go but can not be used for chess and shogi.
- The Go game can rarely end in a tie. In most cases, a player either wins or loses. Therefore, the agent tries to maximize the winning probability. Chess and shogi more commonly end in a tie, so the agent tries to maximize the expectation.

In the AlphaZero paper, in order to train a Go agent, self-play uses 5000 first-generation TPUs, and neural network training uses 16 second-generation TPUs.

MuZero further extends so that the mathematical model of environment dynamics is no longer needed. Since dynamics model is no longer known beforehand, we need to use dynamics network to estimate dynamics.

In MuZero paper, in order to train a Go agent, self-play uses 1000 third-generation TPUs, and neural network training uses 16 third-generation TPUs.

Additionally, MuZero paper also considers the partially observable environment. MuZero introduces the representation function h maps the observation o into a hidden state $s \in S$, where S is called embedding space, which is very similar to state space. This representation function is implemented as a neural network. The paper uses MuZero with representation network to train agents for Atari video games.

14.3 Case Study: Tic-Tac-Toe

This section implements an agent that uses MCTS with self-play to play a board game.

MCTS needs massive computation resources, so it is generally not suitable for a common PC. Therefore, this section considers a tiny board game: Tic-Tac-Toe. We also somehow simplify the MCTS so that the training is feasible in a common PC.

14.3.1 boardgame2: Board Game Environment

The latest version of Gym does not include any built-in board game environments. This section will develop boardgame2, a boardgame package that extends Gym. The codes of boardgame2 can be downloaded from:

```
https://github.com/ZhiqingXiao/boardgame2/
```

Please put the downloaded codes in the executing directory to run this package.

The package boardgame2 has the following Python files:

- boardgame2/__init__.py: This file registers all environments into Gym.
- boardgame2/env.py: This file implements the base class BoardGameEnv for two-player zero-sum game, as well as some constants and utility functions.
- boardgame2/kinarow.py: This file implements the class KInARowEnv for the k-in-a-row game.
- Other files: Those files implement other boardgame environments such as the class ReversiEnv for reversi, unit tests, and the version.

The implementation of boardgame2 uses the following concepts:

- player: It is an int value within $\{-1, +1\}$, where $+1$ is the black player (defined as the constant boardgame2.BLACK), and -1 is the white player (defined as the constant boardgame2.WHITE). Here we assign the two players as $+1$ and -1 respectively so that it will be easier to reverse a board.
- board: It's a np.array object, whose elements are int values among $\{-1, 0, +1\}$. 0 means the position is vacant, and $+1$ and -1 mean that the position has a piece of corresponding player.
- winner: It is either an int value among $\{-1, 0, +1\}$ or None. For a game whose winner is clear, the winner is a value among $\{-1, 0, +1\}$, where -1 means the white player wins, 0 means the game ends in a tie, and $+1$ means the black player wins. If the winner of the game is still unknown, the winner is None.

- location: It is a np.array object of the shape (2,). It is the indices of the cross point in the board.
- valid: It is a np.array' object, whose elements are int values among {0, 1}. 0 means that the player can not place a piece in the position, while 1 means the player can place a piece there.

With the above concepts, boardgame2 defines the state and action as follows:

- State: It is defined as a tuple consisting of board and the next player player.
- Action: It is a position like a location. However, it can also be the constant env.PASS or the constant env.RESIGN, where env.PASS means that the player wants to skip this step, and env.RESIGN means that the player wants to resign.

Now we have determined the state space and the action space. We can implement the environment class. boardgame2 uses the class BoardGameEnv as the base class of all environment classes.

Code 14.1 implements the constructor of the class. The parameters of the constructor include:

- The parameter board_shape is to control the shape of the board. Its type is either int or tuple[int, int]. If its type is int, the board is a squared board with length of a side length board_shape; if its type is tuple[int, int], the board is a rectangular board with width board_shape[0] and height board_shape[1].
- The parameter illegal_action_model is to control how will the environment involve after it is input an illegal action. Its type is str. The default value 'resign' means that inputting an illegal action is equivalent to resigning.
- The parameter render_characters to control the visualization. It's a str of length 3. The three characters are the character for idle space, for the player black, and for the player white respectively.
- The parameter allow_pass to control whether to allow a player to skip a step. Its type is bool. The default value True means allowing the skip.

The constructor also defines the observation space and the action space. The observation space is of type spaces.Tuple with two components. The first component is the board situation, and the second component is the current player to play. For some board games (such as Tic-Tac-Toe), such observation can fully determine the state. The action space is of the type spaces.Box with two dimensions. The action is the coordination in the 2-D board.

Code 14.1 The constructor of the class BoardGameEnv.
https://github.com/ZhiqingXiao/boardgame2/blob/master/boardgame2/env.py

```
def __init__(self, board_shape: int | tuple[int, int],
        illegal_action_mode: str='resign',
        render_characters: str='+ox', allow_pass: bool=True):
    self.allow_pass = allow_pass

    if illegal_action_mode == 'resign':
        self.illegal_equivalent_action = self.RESIGN
```

```
8    elif illegal_action_mode == 'pass':
9        self.illegal_equivalent_action = self.PASS
10   else:
11       raise ValueError()
12
13   self.render_characters = {player : render_characters[player] for player \
14           in [EMPTY, BLACK, WHITE]}
15
16   if isinstance(board_shape, int):
17       board_shape = (board_shape, board_shape)
18   self.board = np.zeros(board_shape)
19
20   observation_spaces = [
21           spaces.Box(low=-1, high=1, shape=board_shape, dtype=np.int8),
22           spaces.Box(low=-1, high=1, shape=(), dtype=np.int8)]
23   self.observation_space = spaces.Tuple(observation_spaces)
24   self.action_space = spaces.Box(low=-np.ones((2,)),
25           high=np.array(board_shape)-1, dtype=np.int8)
```

The boardgame2 also needs to provide APIs for the dynamics of the environment, so the algorithm that requires the environment dynamics can work. Recap in those MCTS algorithms, the algorithms need to know whether a state s is a terminated state. Furthermore, the algorithms need to know who is the winner if the game is terminated, and what is the action space $\mathcal{A}(s)$ and the next state after each action $a \in \mathcal{A}(s)$ if the game is not terminated. Table 14.1 lists these APIs for getting this information.

Table 14.6   APIs for environment dynamics.

member function	description	parameters	return values
get_winner()	determine whether a state is terminal and who is the winner if it is terminal.	state (but the information of the current player is not used)	winner (can be None)
get_valid()	get all possible actions of a state.	state	the board to show vaild actions valid
next_step()	get the next state after a state–action pair.	state–action pair	next state, reward, terminated, and info (a dict)

Next, we consider the implementation of these APIs.

Code 14.2 provides the functions that are related to determine whether an action is valid. The functions include:

- is_valid(state, action) -> bool: Given a state s and an action a, check whether the action a belongs to $\mathcal{A}(s)$. For most of board games including Tic-Tac-Toe, we only need to check whether the index is within a certain range and whether the position is vacant. The former check is implemented by the auxiliary function boardgame2.is_index(board, action).
- has_valid(state) -> bool: Given a state s, check whether $\mathcal{A}(s)$ is empty.

- get_valid(state): Given a state s, get $\mathcal{A}(s)$. The type of return value valid is np.array, and its elements indicate that whether the positions are valid actions.

The function has_valid() and the function get_valid() are implemented by repeatedly calling the function is_valid().

Code 14.2   The member function is_valid(), has_valid(), and get_valid() in the class BoardGameEnv.

https://github.com/ZhiqingXiao/boardgame2/blob/master/boardgame2/env.py

```
def is_valid(self, state, action) -> bool:
    board, _ = state
    if not is_index(board, action):
        return False
    x, y = action
    return board[x, y] == EMPTY

def get_valid(self, state):
    board, _ = state
    valid = np.zeros_like(board, dtype=np.int8)
    for x in range(board.shape[0]):
        for y in range(board.shape[1]):
            valid[x, y] = self.is_valid(state, np.array([x, y]))
    return valid

def has_valid(self, state) -> bool:
    board = state[0]
    for x in range(board.shape[0]):
        for y in range(board.shape[1]):
            if self.is_valid(state, np.array([x, y])):
                return True
    return False
```

Code 14.3 implements the function get_winner(), which is used to determine the end of the game. The class KInARowEnv, which is used for k-in-a-row game such as Tic-Tac-Toe, overwrites the function of the base class BoardGameEnv. For a k-in-a-row game, it checks whether many pieces are in the same line. The parameter of get_winner() is the observation state. It has two components for board and player, but only the first component is used in this function. The return value winner can be None or an int value. As stated previously, if the game ends, it is an int value that indicates who wins the game.

Code 14.3   The member function get_winner() in the class KInARowEnv.

https://github.com/ZhiqingXiao/boardgame2/blob/master/boardgame2/kinarow.py

```
def get_winner(self, state):
    board, _ = state
    for player in [BLACK, WHITE]:
        for x in range(board.shape[0]):
            for y in range(board.shape[1]):
                for dx, dy in [(1, -1), (1, 0), (1, 1), (0, 1)]:
                        # 8 directions here
                    xx, yy = x, y
                    for count in itertools.count():
                        if not is_index(board, (xx, yy)) or \
                                board[xx, yy] != player:
                            break
```

```
13                          xx, yy = xx + dx, yy + dy
14                      if count >= self.target_length:
15                          return player
16          for player in [BLACK, WHITE]:
17              if self.has_valid((board, player)):
18                  return None
19          return 0
```

Code 14.4 implements the function next_step() to obtain the next state, and its auxiliary function get_next_step(). The function get_next_step() returns the state of the current player after it places its stone. It will raise error if action does not correspond to a valid location in the board. The function next_step() wraps accordingly. Its parameters are still the state and the action, but its return values are the next observation next_state, the reward reward, the indicator for termination terminated, and extra information for valid action info. This calls next_step() for the next state of the opponent. If the opponent does not have any valid actions, the opponent must use the action PASS, and proceed to next step further.

Code 14.4 The member function next_step() and get_next_state() in the class BoardGameEnv.
https://github.com/ZhiqingXiao/boardgame2/blob/master/boardgame2/env.py

```
1   def get_next_state(self, state, action):
2       board, player = state
3       x, y = action
4       if self.is_valid(state, action):
5           board = copy.deepcopy(board)
6           board[x, y] = player
7       return board, -player
8
9   def next_step(self, state, action):
10      if not self.is_valid(state, action):
11          action = self.illegal_equivalent_action
12      if np.array_equal(action, self.RESIGN):
13          return state, -state[1], True, {}
14      while True:
15          state = self.get_next_state(state, action)
16          winner = self.get_winner(state)
17          if winner is not None:
18              return state, winner, True, {}
19          if self.has_valid(state):
20              break
21          action = self.PASS
22      return state, 0., False, {}
```

In order to access the environment class using Gym APIs, Code 14.5 overwrites the function reset(), the function step(), and the function render(). The function reset() initializes an empty board, and determines the player who moves first. The function step() calls the environment dynamics function next_step(). The function render() has a parameter mode, which can be either 'ansi' or 'human'. When mode is 'ansi', the function returns a string; when mode is 'human', the string is printed to standard output. The possible parameter values are stored in the member metadata. The core logic of formatting a board to a string is implemented in the function boardgame2.strfboard().

Code 14.5 **The member function** `reset()`, `step()`, **and** `render()` **in the class** **BoardGameEnv**.

`https://github.com/ZhiqingXiao/boardgame2/blob/master/boardgame2/`
`env.py`

```python
def reset(self, *, seed=None, options=None):
    super().reset(seed=seed)
    self.board = np.zeros_like(self.board, dtype=np.int8)
    self.player = BLACK
    return self.board, self.player

def step(self, action):
    state = (self.board, self.player)
    next_state, reward, terminated, info = self.next_step(state, action)
    self.board, self.player = next_state
    return next_state, reward, terminated, truncated, info

metadata = {"render_modes": ["ansi", "human"]}

def render(self):
    mode = self.render_mode
    outfile = StringIO() if mode == 'ansi' else sys.stdout
    s = strfboard(self.board, self.render_characters)
    outfile.write(s)
    if mode != 'human':
        return outfile
```

Now we have known how Tic-Tac-Toe is implemented in the package boardgame2. boardgame2 also implements other board games, such as Reversi. All full implementation can be found in the GitHub repo.

14.3.2 Exhaustive Search

This section uses exhaustive search to solve the Tic-Tac-Toe task. The scale of the Tic-Tac-Toe task is tiny, so exhaustive search is feasible.

Code 14.6 implements the exhaustive search agent. Its member `learn()` implements the learning process. It first initializes the member `self.winner`, which stores the winner of each state. Since the state is of type `tuple`, it can not be directly used as keys of a `dict`. Therefore, we use its string representation as keys; additionally, its value is the winner. In the beginning, `self.winner` is an empty `dict`, meaning that we have not decided the winners for any states yet. During the searching, we will get to know the winner of a state, then the corresponding information will be saved in `self.winner`. Next, we initialize the member `self.policy`, which stores the action of each state. Its keys are the string representations of states, while its values are actions of those states. In the beginning, `self.policy` is an empty `dict`, meaning that we have not decided the actions for any states yet. During the searching, we will know the action for some state, and then the string representation of the state will be saved as a key, and corresponding action will be saved as a value. The core of learning process is to call the member search() recursively to conduct tree search. The member search() has a parameter state, presenting the state to search. It first judges that whether this state has been searched or not. If it has been searched, it will be in self.winner, and then we can directly use the result without further searching. If this state has not been searched yet, we can first check whether this state is a terminal state. If the state is a terminal state, save the winner information to self.winner. If the state is not a terminal state, we need to find out all available valid actions, and get their next states, and search recursively. After obtaining the results for those recursive search, it combines the searching results to determine the winner and action of the state, and save them in self.winner and self.policy respectively.

Code 14.6   Exhaustive search agent.

TicTacToe-v0_ExhaustiveSearch.ipynb

```python
class ExhaustiveSearchAgent:
    def __init__(self, env):
        self.env = env
        self.learn()

    def learn(self):
        self.winner = {}  # str(state) -> player
        self.policy = {}  # str(state) -> action
        init_state = np.zeros_like(env.board, dtype=np.int8), BLACK
        self.search(init_state)

    def search(self, state):  # Tree Search: recursive implementation
        s = str(state)
        if s not in self.winner:  # the node has not been calculated
            self.winner[s] = self.env.get_winner(state)
            if self.winner[s] is None:  # do not have immediate winner
                # try all next valid actions
                valid = self.env.get_valid(state)
                winner_actions = {}
                for x in range(valid.shape[0]):
                    for y in range(valid.shape[1]):
                        if valid[x, y]:
                            action = np.array([x, y])
                            next_state = self.env.get_next_state(state, action)
                            winner = self.search(next_state)
                            winner_actions[winner] = action

                # choose the best action
                _, player = state
                for winner in [player, EMPTY, -player]:
                    if winner in winner_actions:
                        action = winner_actions[winner]
                        self.policy[s] = action
                        self.winner[s] = winner
                        break
        return self.winner[s]

    def reset(self, mode=None):
        pass

    def step(self, observation, reward, terminated):
        s = str(observation)
        action = self.policy[s]
        return action

    def close(self):
        pass
```

```
49
50   agent = ExhaustiveSearchAgent(env=env)
```

Code 14.7 implements the self-play. It is similar to Code 1.3. Its display part prints out the board. Its return values are winner and elapsed_steps.

Code 14.7   Self-play.

TicTacToe-v0_ExhaustiveSearch.ipynb

```
1    def play_boardgame2_episode(env, agent, mode=None, verbose=False):
2        observation, _ = env.reset()
3        winner = 0
4        terminated, truncated = False, False
5        agent.reset(mode=mode)
6        elapsed_steps = 0
7        while True:
8            if verbose:
9                board, player = observation
10               print(boardgame2.strfboard(board))
11           action = agent.step(observation, winner, terminated)
12           if verbose:
13               logging.info('step %d: player %d, action %s', elapsed_steps,
14                            player, action)
15           observation, winner, terminated, truncated, _ = env.step(action)
16           if terminated or truncated:
17               if verbose:
18                   board, _ = observation
19                   print(boardgame2.strfboard(board))
20               break
21           elapsed_steps += 1
22       agent.close()
23       return winner, elapsed_steps
24
25
26   winner, elapsed_steps = play_boardgame2_episode(env, agent, mode='test',
27           verbose=True)
28   logging.info('test episode: winner = %d, steps = %d', winner, elapsed_steps)
```

14.3.3 Heuristic Search

This section uses AlphaZero algorithm to solve the Tic-Tac-Toe task.

Code 14.8 implements the experience replayer. Each entry of experience has four components. The first two components player and board together store the state; the third component prob stores the policy, and the last component winner stores the reward.

Code 14.8   Replay buffer of AlphaZero agent.

TicTacToe-v0_AlphaZero_tf.ipynb

```
1    class AlphaZeroReplayer:
2        def __init__(self):
3            self.fields = ['player', 'board', 'prob', 'winner']
4            self.memory = pd.DataFrame(columns=self.fields)
5
6        def store(self, df):
7            self.memory = pd.concat([self.memory, df[self.fields]],
8                    ignore_index=True)
```

```
9
10   def sample(self, size):
11       indices = np.random.choice(self.memory.shape[0], size=size)
12       return (np.stack(self.memory.loc[indices, field]) for field in
13               self.fields)
```

Since the dynamic is known, we only need prediction network. Codes 14.9 and 14.10 implement the prediction network. The network structure is simplified from Fig. 14.8. The common part has 1 convolution layer and 2 ResNet layers. The probability part has 2 convolution layers. The value part also has one additional convolution layer and one fully-connected layer.

Code 14.9 Network of AlphaZero agent (TensorFlow version).
TicTacToe-v0_AlphaZero_tf.ipynb

```
1  class AlphaZeroNet(keras.Model):
2      def __init__(self, input_shape, regularizer=regularizers.l2(1e-4)):
3          super().__init__()
4
5          # common net
6          self.input_net = keras.Sequential([
7                  layers.Reshape(input_shape + (1,)),
8                  layers.Conv2D(256, kernel_size=3, padding='same',
9                  kernel_regularizer=regularizer, bias_regularizer=regularizer),
10                 layers.BatchNormalization(), layers.ReLU()])
11         self.residual_nets = [keras.Sequential([
12                 layers.Conv2D(256, kernel_size=3, padding='same',
13                 kernel_regularizer=regularizer, bias_regularizer=regularizer),
14                 layers.BatchNormalization()]) for _ in range(2)]
15
16         # probability net
17         self.prob_net = keras.Sequential([
18                 layers.Conv2D(256, kernel_size=3, padding='same',
19                 kernel_regularizer=regularizer, bias_regularizer=regularizer),
20                 layers.BatchNormalization(), layers.ReLU(),
21                 layers.Conv2D(1, kernel_size=3, padding='same',
22                 kernel_regularizer=regularizer, bias_regularizer=regularizer),
23                 layers.Flatten(), layers.Softmax(),
24                 layers.Reshape(input_shape)])
25
26         # value net
27         self.value_net = keras.Sequential([
28                 layers.Conv2D(1, kernel_size=3, padding='same',
29                 kernel_regularizer=regularizer, bias_regularizer=regularizer),
30                 layers.BatchNormalization(), layers.ReLU(),
31                 layers.Flatten(),
32                 layers.Dense(1, activation=nn.tanh,
33                 kernel_regularizer=regularizer, bias_regularizer=regularizer)
34                 ])
35
36     def call(self, board_tensor):
37         # common net
38         x = self.input_net(board_tensor)
39         for i_net, residual_net in enumerate(self.residual_nets):
40             y = residual_net(x)
41             if i_net == len(self.residual_nets) - 1:
42                 y = y + x
43             x = nn.relu(y)
44         common_feature_tensor = x
45
46         # probability net
47         prob_tensor = self.prob_net(common_feature_tensor)
48
49         # value net
```

```
50        v_tensor = self.value_net(common_feature_tensor)
51
52        return prob_tensor, v_tensor
```

Code 14.10   Network of AlphaZero agent (PyTorch version).

TicTacToe-v0_AlphaZero_torch.ipynb

```
1   class AlphaZeroNet(nn.Module):
2       def __init__(self, input_shape):
3           super().__init__()
4
5           self.input_shape = input_shape
6
7           # common net
8           self.input_net = nn.Sequential(
9                   nn.Conv2d(1, 256, kernel_size=3, padding="same"),
10                  nn.BatchNorm2d(256), nn.ReLU())
11          self.residual_nets = [nn.Sequential(
12                  nn.Conv2d(256, 256, kernel_size=3, padding="same"),
13                  nn.BatchNorm2d(256)) for _ in range(2)]
14
15          # probability net
16          self.prob_net = nn.Sequential(
17                  nn.Conv2d(256, 256, kernel_size=3, padding="same"),
18                  nn.BatchNorm2d(256), nn.ReLU(),
19                  nn.Conv2d(256, 1, kernel_size=3, padding="same"))
20
21          # value net
22          self.value_net0 = nn.Sequential(
23                  nn.Conv2d(256, 1, kernel_size=3, padding="same"),
24                  nn.BatchNorm2d(1), nn.ReLU())
25          self.value_net1 = nn.Sequential(
26                  nn.Linear(np.prod(input_shape), 1), nn.Tanh())
27
28      def forward(self, board_tensor):
29          # common net
30          input_tensor = board_tensor.view(-1, 1, *self.input_shape)
31          x = self.input_net(input_tensor)
32          for i_net, residual_net in enumerate(self.residual_nets):
33              y = residual_net(x)
34              if i_net == len(self.residual_nets) - 1:
35                  y = y + x
36              x = torch.clamp(y, 0)
37          common_feature_tensor = x
38
39          # probability net
40          logit_tensor = self.prob_net(common_feature_tensor)
41          logit_flatten_tensor = logit_tensor.view(-1)
42          prob_flatten_tensor = functional.softmax(logit_flatten_tensor, dim=-1)
43          prob_tensor = prob_flatten_tensor.view(-1, *self.input_shape)
44
45          # value net
46          v_feature_tensor = self.value_net0(common_feature_tensor)
47          v_flatten_tensor = v_feature_tensor.view(-1, np.prod(self.input_shape))
48          v_tensor = self.value_net1(v_flatten_tensor)
49
50          return prob_tensor, v_tensor
```

Codes 14.11 and 14.12 implement the agent class AlphaZeroAgent. The TensorFlow version defines a function categorical_crossentropy_2d() inside the agent class and this function is used together with losses.MSE as the two losses of the prediction network. The PyTorch version uses the loss BCELoss and MSELoss directly, and it does not need to define customized loss function. In the beginning of every episode, the member function `reset_mcts()` is called to initialize the search tree. The initialization includes the initialization of action value estimate `self.q` and the counting of action visitation `self.count`. Both members are of type `dict` with default values that are board-shape `np.array` whose all elements are zero. The member `step()` calls `self.search()` recursively to search and the search updates `self.count`. Then the decision policy uses the formula $\pi(a|s) = \frac{c(s,a)}{\sum_{a' \in \mathcal{A}(s)} c(s,a')}$ ($a \in \mathcal{A}(s)$), which is exactly the form of decision policy in Sect. 14.1.4 with $\kappa = 1$.

Code 14.11   AlphaZero agent (TensorFlow version).
`TicTacToe-v0_AlphaZero_tf.ipynb`

```python
class AlphaZeroAgent:
    def __init__(self, env):
        self.env = env

        self.replayer = AlphaZeroReplayer()

        self.board = np.zeros_like(env.board)
        self.net = self.build_net()

        self.reset_mcts()

    def build_net(self, learning_rate=0.001):
        net = AlphaZeroNet(input_shape=self.board.shape)

        def categorical_crossentropy_2d(y_true, y_pred):
            labels = tf.reshape(y_true, [-1, self.board.size])
            preds = tf.reshape(y_pred, [-1, self.board.size])
            return losses.categorical_crossentropy(labels, preds)

        loss = [categorical_crossentropy_2d, losses.MSE]
        optimizer = optimizers.Adam(learning_rate)
        net.compile(loss=loss, optimizer=optimizer)
        return net

    def reset_mcts(self):
        def zero_board_factory():  # for construct default_dict
            return np.zeros_like(self.board, dtype=float)
        self.q = collections.defaultdict(zero_board_factory)
                # q estimates: board —> board
        self.count = collections.defaultdict(zero_board_factory)
                # q count visitation: board —> board
        self.policy = {}  # policy: board —> board
        self.valid = {}  # valid position: board —> board
        self.winner = {}  # winner: board —> None or int

    def reset(self, mode):
        self.mode = mode
        if mode == "train":
            self.trajectory = []

    def step(self, observation, winner, _):
        board, player = observation
        canonical_board = player * board
        s = boardgame2.strfboard(canonical_board)
        while self.count[s].sum() < 200:  # conduct MCTS 200 times
            self.search(canonical_board, prior_noise=True)
        prob = self.count[s] / self.count[s].sum()

        # sample
        location_index = np.random.choice(prob.size, p=prob.reshape(-1))
        action = np.unravel_index(location_index, prob.shape)
```

```
53          if self.mode == 'train':
54              self.trajectory += [player, board, prob, winner]
55          return action
56
57      def close(self):
58          if self.mode == 'train':
59              self.save_trajectory_to_replayer()
60              if len(self.replayer.memory) >= 1000:
61                  for batch in range(2):  # learn multiple times
62                      self.learn()
63                  self.replayer = AlphaZeroReplayer()
64                          # reset replayer after the agent changes itself
65                  self.reset_mcts()
66
67      def save_trajectory_to_replayer(self):
68          df = pd.DataFrame(
69                  np.array(self.trajectory, dtype=object).reshape(-1, 4),
70                  columns=['player', 'board', 'prob', 'winner'], dtype=object)
71          winner = self.trajectory[-1]
72          df['winner'] = winner
73          self.replayer.store(df)
74
75      def search(self, board, prior_noise=False):  # MCTS
76          s = boardgame2.strfboard(board)
77
78          if s not in self.winner:
79              self.winner[s] = self.env.get_winner((board, BLACK))
80          if self.winner[s] is not None:  # if there is a winner
81              return self.winner[s]
82
83          if s not in self.policy:  # leaf that has not calculate the policy
84              boards = board[np.newaxis].astype(float)
85              pis, vs = self.net.predict(boards, verbose=0)
86              pi, v = pis[0], vs[0]
87              valid = self.env.get_valid((board, BLACK))
88              masked_pi = pi * valid
89              total_masked_pi = np.sum(masked_pi)
90              if total_masked_pi <= 0:
91                  # all valid actions do not have probabilities. rarely occur
92                  masked_pi = valid  # workaround
93                  total_masked_pi = np.sum(masked_pi)
94              self.policy[s] = masked_pi / total_masked_pi
95              self.valid[s] = valid
96              return v
97
98          # calculate PUCT
99          count_sum = self.count[s].sum()
100         c_init = 1.25
101         c_base = 19652.
102         coef = (c_init + np.log1p((1 + count_sum) / c_base)) * \
103                 math.sqrt(count_sum) / (1. + self.count[s])
104         if prior_noise:
105             alpha = 1. / self.valid[s].sum()
106             noise = np.random.gamma(alpha, 1., board.shape)
107             noise *= self.valid[s]
108             noise /= noise.sum()
109             prior_exploration_fraction = 0.25
110             prior = (1. - prior_exploration_fraction) * self.policy[s] \
111                     + prior_exploration_fraction * noise
112         else:
113             prior = self.policy[s]
114         ub = np.where(self.valid[s], self.q[s] + coef * prior, np.nan)
115         location_index = np.nanargmax(ub)
116         location = np.unravel_index(location_index, board.shape)
117
118         (next_board, next_player), _, _, _ = self.env.next_step(
119                 (board, BLACK), np.array(location))
```

```
120        next_canonical_board = next_player * next_board
121        next_v = self.search(next_canonical_board)   # recursive
122        v = next_player * next_v
123
124        self.count[s][location] += 1
125        self.q[s][location] += (v - self.q[s][location]) / \
126                self.count[s][location]
127        return v
128
129    def learn(self):
130        players, boards, probs, winners = self.replayer.sample(64)
131        canonical_boards = (players[:, np.newaxis, np.newaxis] * boards).astype
                (
132                float)
133        vs = (players * winners)[:, np.newaxis].astype(float)
134        self.net.fit(canonical_boards, [probs, vs], verbose=0)
135
136
137 agent = AlphaZeroAgent(env=env)
```

Code 14.12 AlphaZero agent (PyTorch version).
TicTacToe-v0_AlphaZero_torch.ipynb

```
1 class AlphaZeroAgent:
2    def __init__(self, env):
3        self.env = env
4        self.board = np.zeros_like(env.board)
5        self.reset_mcts()
6
7        self.replayer = AlphaZeroReplayer()
8
9        self.net = AlphaZeroNet(input_shape=self.board.shape)
10        self.prob_loss = nn.BCELoss()
11        self.v_loss = nn.MSELoss()
12        self.optimizer = optim.Adam(self.net.parameters(), 1e-3,
13                weight_decay=1e-4)
14
15    def reset_mcts(self):
16        def zero_board_factory():   # for construct default_dict
17            return np.zeros_like(self.board, dtype=float)
18        self.q = collections.defaultdict(zero_board_factory)
19                # q estimates: board -> board
20        self.count = collections.defaultdict(zero_board_factory)
21                # q count visitation: board -> board
22        self.policy = {}   # policy: board -> board
23        self.valid = {}   # valid position: board -> board
24        self.winner = {}   # winner: board -> None or int
25
26    def reset(self, mode):
27        self.mode = mode
28        if mode == "train":
29            self.trajectory = []
30
31    def step(self, observation, winner, _):
32        board, player = observation
33        canonical_board = player * board
34        s = boardgame2.strfboard(canonical_board)
35        while self.count[s].sum() < 200:   # conduct MCTS 200 times
36            self.search(canonical_board, prior_noise=True)
37        prob = self.count[s] / self.count[s].sum()
38
39        # sample
40        location_index = np.random.choice(prob.size, p=prob.reshape(-1))
41        action = np.unravel_index(location_index, prob.shape)
42
```

```
43          if self.mode == 'train':
44              self.trajectory += [player, board, prob, winner]
45          return action
46
47      def close(self):
48          if self.mode == 'train':
49              self.save_trajectory_to_replayer()
50              if len(self.replayer.memory) >= 1000:
51                  for batch in range(2):  # learn multiple times
52                      self.learn()
53                  self.replayer = AlphaZeroReplayer()
54                      # reset replayer after the agent changes itself
55              self.reset_mcts()
56
57      def save_trajectory_to_replayer(self):
58          df = pd.DataFrame(
59                  np.array(self.trajectory, dtype=object).reshape(-1, 4),
60                  columns=['player', 'board', 'prob', 'winner'], dtype=object)
61          winner = self.trajectory[-1]
62          df['winner'] = winner
63          self.replayer.store(df)
64
65      def search(self, board, prior_noise=False):  # MCTS
66          s = boardgame2.strfboard(board)
67
68          if s not in self.winner:
69              self.winner[s] = self.env.get_winner((board, BLACK))
70          if self.winner[s] is not None:  # if there is a winner
71              return self.winner[s]
72
73          if s not in self.policy:  # leaf that has not calculate the policy
74              board_tensor = torch.as_tensor(board, dtype=torch.float).view(1, 1,
75                      *self.board.shape)
76              pi_tensor, v_tensor = self.net(board_tensor)
77              pi = pi_tensor.detach().numpy()[0]
78              v = v_tensor.detach().numpy()[0]
79              valid = self.env.get_valid((board, BLACK))
80              masked_pi = pi * valid
81              total_masked_pi = np.sum(masked_pi)
82              if total_masked_pi <= 0:
83                  # all valid actions do not have probabilities. rarely occur
84                  masked_pi = valid  # workaround
85                  total_masked_pi = np.sum(masked_pi)
86              self.policy[s] = masked_pi / total_masked_pi
87              self.valid[s] = valid
88              return v
89
90          # calculate PUCT
91          count_sum = self.count[s].sum()
92          c_init = 1.25
93          c_base = 19652.
94          coef = (c_init + np.log1p((1 + count_sum) / c_base)) * \
95                  math.sqrt(count_sum) / (1. + self.count[s])
96          if prior_noise:
97              alpha = 1. / self.valid[s].sum()
98              noise = np.random.gamma(alpha, 1., board.shape)
99              noise *= self.valid[s]
100             noise /= noise.sum()
101             prior_exploration_fraction=0.25
102             prior = (1. - prior_exploration_fraction) * self.policy[s] \
103                     + prior_exploration_fraction * noise
104         else:
105             prior = self.policy[s]
106         ub = np.where(self.valid[s], self.q[s] + coef * prior, np.nan)
107         location_index = np.nanargmax(ub)
108         location = np.unravel_index(location_index, board.shape)
109
```

```
110        (next_board, next_player), _, _, _ = self.env.next_step(
111            (board, BLACK), np.array(location))
112        next_canonical_board = next_player * next_board
113        next_v = self.search(next_canonical_board)  # recursive
114        v = next_player * next_v
115
116        self.count[s][location] += 1
117        self.q[s][location] += (v - self.q[s][location]) / \
118            self.count[s][location]
119        return v
120
121    def learn(self):
122        players, boards, probs, winners = self.replayer.sample(64)
123        canonical_boards = players[:, np.newaxis, np.newaxis] * boards
124        targets = (players * winners)[:, np.newaxis]
125
126        target_prob_tensor = torch.as_tensor(probs, dtype=torch.float)
127        canonical_board_tensor = torch.as_tensor(canonical_boards,
128            dtype=torch.float)
129        target_tensor = torch.as_tensor(targets, dtype=torch.float)
130
131        prob_tensor, v_tensor = self.net(canonical_board_tensor)
132
133        flatten_target_prob_tensor = target_prob_tensor.view(-1, self.board.
               size)
134        flatten_prob_tensor = prob_tensor.view(-1, self.board.size)
135        prob_loss_tensor = self.prob_loss(flatten_prob_tensor,
136            flatten_target_prob_tensor)
137        v_loss_tensor = self.v_loss(v_tensor, target_tensor)
138        loss_tensor = prob_loss_tensor + v_loss_tensor
139        self.optimizer.zero_grad()
140        loss_tensor.backward()
141        self.optimizer.step()
142
143
144 agent = AlphaZeroAgent(env=env)
```

14.4 Summary

- Monte Carlo Tree Search (MCTS) is a heuristic search algorithm that maintains a subtree of the complete search tree.
- The steps of MCTS include selection, expansion, evaluation, and backup.
- The selection of MCTS usually uses a variant of Predictor-Upper Confidence Bounds applied to Trees (variant of PUCT), whose bonus is

$$b(s, a) = \lambda_{\text{PUCT}} \pi_{\text{PUCT}}(a|s) \frac{\sqrt{c(s)}}{1 + c(s, a)}, \quad s \in \mathcal{S}, a \in \mathcal{A}(s).$$

It requires the information of visitation count $c(s, a)$, action value estimate $q(s, a)$, and the a priori selection probability $\pi_{\text{PUCT}}(a|s)$ for every state–action pair.
- The selection of MCTS relies on the mathematical model of environment dynamics, if the inputs do not have dynamics model, we can maintain a dynamics network to estimate the dynamics. The dynamics network is in the

form of

$$(s', r) = p(s, a; \varphi).$$

We use the experience in the form of (S, A, R, S', D') to train the dynamics network.

- The expansion and evaluation of MCTS uses prediction network. The prediction network is in the form of

$$(\pi_{\mathrm{PUCT}}, v) = f(s; \boldsymbol{\theta}).$$

We use the experience in the form of (S, Π, G) to train the prediction network.

- Asynchronous Policy Value MCTS (APV-MCTS) searches and trains neural networks asynchronously.
- We can use MCTS with self-play to train agents for two-player zero-sum board games.
- ResNet is usually used in the agents for board games.
- MCTS algorithms for the game of Go include AlphaGo, AlphaGo Zero, AlphaZero, and MuZero. Therein, AlphaGo, AlphaGo Zero, and AlphaZero use dynamics model as inputs, while MuZero does not need dynamics model as inputs.

14.5 Exercises

14.5.1 Multiple Choices

14.1 Which of the following RL algorithms are model-free algorithms: (   )

A. Dynamic Programming.
B. Policy Gradient.
C. MCTS.

14.2 Among the algorithms to find optimal policy for MDP, which of the following algorithm does not require the mathematical model of the environment as the algorithm input: (   )

A. Dynamic Programming.
B. Linear Programming.
C. MuZero algorithm.

14.3 MCTS algorithm uses: (   )

A. Exploration policy and exploitation policy.
B. Searching policy and expansion policy.
C. Agent policy and environment policy.

14.4 MCTS algorithm consists of the following steps: (   )

A. Selection, expansion, evaluation, and backup.

B. Simulation, evaluation, selection, and backup.

C. Selection, evaluation, backup, and bootstrap.

14.5 On MCTS with variant of PUCT, the node in the tree should maintain: (   )

A. State value estimate, visitation count, a priori selection probability.

B. Action value estimate, visitation count, a priori selection probability.

C. State value estimate, action value estimate, visitation count, selection probability.

14.6 On MCTS, which of the following is correct: (   )

A. MCTS can only be used in single-agent tasks, but can not be used in multi-agent tasks.

B. MCTS can not be used in single-agent tasks, but can be used in multi-agent tasks.

C. MCTS can be used in both single-agent tasks and multi-agent tasks.

14.5.2 Programming

14.7 Use AlphaZero and MuZero to solve Connect4 game. You need to implement the Connect4 by yourself.

14.5.3 Mock Interview

14.8 Can you enumerate some model-based RL algorithms? Among the algorithms, which ones are deep RL algorithms?

14.9 What are the shortcomings of MCTS?

14.10 What is AlphaGo algorithm? Why AlphaGo is powerful? What algorithms further extend AlphaGo?

Chapter 15
More Agent–Environment Interfaces

This chapter covers

- average-reward
- differential value
- Continuous-Time MDP (CTMDP)
- non-homogenous MDP
- Semi-MDP (SMDP)
- Partially Observable MDP (POMDP)
- belief
- belief MDP
- belief value
- α-vector
- Point-Based Value Iteration (PBVI) algorithm

Mathematical models are always simplification of real tasks. If we do not simplify the real tasks, the tasks will be too complex to solve; if we simplify the real tasks too much, we may miss the key aspects in the tasks, so that the solutions to the simplified tasks do not work in the real tasks. Discounted DTMDP is a popular and effective model, so this book chooses it. However, there are other models. This chapter will address some of them.

© The Author(s), under exclusive license to Springer Nature Singapore Pte Ltd. 2024
Z. Xiao, *Reinforcement Learning*, https://doi.org/10.1007/978-981-19-4933-3_15

15.1 Average Reward DTMDP

Since Chap. 2, we work on discounted DTMDP. Specifically, we defined the discounted return, defined the concept of "values" accordingly, and introduced some algorithms to maximize the expectation of discounted returns, such as the LP method, numeric iterative algorithms, MC updates, and TD updates. In fact, besides the expectation of discounted return, there are other ways to define the performance of an episode. This section will consider some of them, especially the average return.

15.1.1 Average Reward

For a DTMDP with defined initial probability, dynamics, and policy, we can use the following ways to derive episode rewards from single-step rewards:

- Given a positive integer h, the total reward expectation with horizon h is defined as

$$g_\pi^{[h]} \stackrel{\text{def}}{=} \mathrm{E}_\pi \left[\sum_{\tau=1}^{h} R_\tau \right].$$

- Given a discount factor $\gamma \in (0, 1]$, the expectation of discounted return is defined as

$$\bar{g}_\pi^{(\gamma)} \stackrel{\text{def}}{=} \mathrm{E}_\pi \left[\sum_{\tau=1}^{+\infty} \gamma^\tau R_\tau \right]$$

or

$$g_\pi^{(\gamma)} \stackrel{\text{def}}{=} \mathrm{E}_\pi \left[\sum_{\tau=1}^{+\infty} \gamma^{\tau-1} R_\tau \right].$$

These two versions differ in γ. We always used the second one in the previous chapters throughout the book.

- **Average reward**: The most common definition is

$$\bar{r}_\pi \stackrel{\text{def}}{=} \lim_{h \to +\infty} \mathrm{E}_\pi \left[\frac{1}{h} \sum_{\tau=1}^{h} R_\tau \right].$$

There are other definitions, such as the upper limit average reward $\limsup_{h \to +\infty} \mathrm{E}_\pi \left[\frac{1}{h} \sum_{\tau=1}^{h} R_\tau \right]$, lower limit average reward $\liminf_{h \to +\infty} \mathrm{E}_\pi \left[\frac{1}{h} \sum_{\tau=1}^{h} R_\tau \right]$, and limited discounted average reward $\lim_{\gamma \to 1^-} \lim_{h \to +\infty} \mathrm{E}_\pi \left[\frac{1}{h} \sum_{\tau=1}^{h} \gamma^{\tau-1} R_\tau \right]$. These variants are usually more general than the most common definition. This section adopts the most common definition for simplicity.

Average reward is usually used in sequential tasks. This metric does not depend on a discount factor, and it assumes the rewards over different time are equal. Compared to the discounted return expectation with discount factor $\gamma = 1$, average reward is more likely to converge.

For episodic tasks, we can convert them to sequential tasks by concatenating all episodes sequentially. Applying average rewards in such cases can help consider length of episodes.

We can define values according to each episode reward metric.

- Values with finite horizon h:

$$\text{state value} \quad v_\pi^{[h]}(s) \overset{\text{def}}{=} \mathrm{E}_\pi\left[\sum_{\tau=1}^{h} R_{t+\tau} \middle| S_t = s\right], \qquad s \in \mathcal{S}$$

$$\text{action value} \quad q_\pi^{[h]}(s, a) \overset{\text{def}}{=} \mathrm{E}_\pi\left[\sum_{\tau=1}^{h} R_{t+\tau} \middle| S_t = s, A_t = a\right], \quad s \in \mathcal{S}, a \in \mathcal{A}.$$

- Values with discount factor $\gamma \in (0, 1]$:

$$\text{state value} \quad \bar{v}_\pi^{(\gamma)}(s) \overset{\text{def}}{=} \mathrm{E}_\pi\left[\sum_{\tau=1}^{+\infty} \gamma^\tau R_{t+\tau} \middle| S_t = s\right], \qquad s \in \mathcal{S}$$

$$\text{action value} \quad \bar{q}_\pi^{(\gamma)}(s, a) \overset{\text{def}}{=} \mathrm{E}_\pi\left[\sum_{\tau=1}^{+\infty} \gamma^\tau R_{t+\tau} \middle| S_t = s, A_t = a\right], \quad s \in \mathcal{S}, a \in \mathcal{A},$$

or

$$\text{state value} \quad v_\pi^{(\gamma)}(s) \overset{\text{def}}{=} \mathrm{E}_\pi\left[\sum_{\tau=1}^{+\infty} \gamma^{\tau-1} R_{t+\tau} \middle| S_t = s\right], \qquad s \in \mathcal{S}$$

$$\text{action value} \quad q_\pi^{(\gamma)}(s, a) \overset{\text{def}}{=} \mathrm{E}_\pi\left[\sum_{\tau=1}^{+\infty} \gamma^{\tau-1} R_{t+\tau} \middle| S_t = s, A_t = a\right], \quad s \in \mathcal{S}, a \in \mathcal{A}.$$

- Average-reward values:

$$\text{state value} \quad \bar{v}_\pi(s) \overset{\text{def}}{=} \lim_{h \to +\infty} \mathrm{E}_\pi\left[\frac{1}{h}\sum_{\tau=1}^{h} R_{t+\tau} \middle| S_t = s\right], \qquad s \in \mathcal{S}$$

$$\text{action value} \quad \bar{q}_\pi(s, a) \overset{\text{def}}{=} \lim_{h \to +\infty} \mathrm{E}_\pi\left[\frac{1}{h}\sum_{\tau=1}^{h} R_{t+\tau} \middle| S_t = s, A_t = a\right], \quad s \in \mathcal{S}, a \in \mathcal{A}.$$

We can also define visitation frequency with each performance metric.

- Visitation frequency with finite horizon h:

$$\eta_\pi^{[h]}(s) \overset{\text{def}}{=} \mathbb{E}_\pi \left[\sum_{\tau=0}^{h-1} 1_{[S_t=s]} \right]$$

$$= \sum_{\tau=0}^{h-1} \Pr_\pi[S_t = s], \qquad\qquad s \in S$$

$$\rho_\pi^{[h]}(s, a) \overset{\text{def}}{=} \mathbb{E}_\pi \left[\sum_{\tau=0}^{h-1} 1_{[S_t=s,A_t=a]} \right]$$

$$= \sum_{\tau=0}^{h-1} \Pr_\pi[S_t = s, A_t = a], \quad s \in S, a \in \mathcal{A}.$$

- Visitation frequency with discount factor $\gamma \in (0, 1]$:

$$\eta_\pi^{(\gamma)}(s) \overset{\text{def}}{=} \mathbb{E}_\pi \left[\sum_{\tau=0}^{+\infty} \gamma^\tau 1_{[S_t=s]} \right]$$

$$= \sum_{\tau=0}^{+\infty} \gamma^\tau \Pr_\pi[S_t = s], \qquad\qquad s \in S$$

$$\rho_\pi^{(\gamma)}(s, a) \overset{\text{def}}{=} \mathbb{E}_\pi \left[\sum_{\tau=0}^{+\infty} \gamma^\tau 1_{[S_t=s,A_t=a]} \right]$$

$$= \sum_{\tau=0}^{+\infty} \gamma^\tau \Pr_\pi[S_t = s, A_t = a], \quad s \in S, a \in \mathcal{A}.$$

There are other definitions, such as multiplying $(1 - \gamma)$ on the above definitions.
- Average-reward visitation frequency:

$$\bar{\eta}_\pi(s) \overset{\text{def}}{=} \lim_{h \to +\infty} \mathbb{E}_\pi \left[\frac{1}{h} \sum_{\tau=1}^{h} 1_{[S_t=s]} \right]$$

$$= \lim_{h \to +\infty} \frac{1}{h} \sum_{\tau=1}^{h} \Pr_\pi[S_t = s], \qquad\qquad s \in S$$

$$\bar{\rho}_\pi(s, a) \overset{\text{def}}{=} \lim_{h \to +\infty} \mathbb{E}_\pi \left[\frac{1}{h} \sum_{\tau=1}^{h} 1_{[S_t=s,A_t=a]} \right]$$

$$= \lim_{h \to +\infty} \frac{1}{h} \sum_{\tau=1}^{h} \Pr_\pi[S_t = s, A_t = a], \quad s \in S, a \in \mathcal{A}.$$

In these definitions, it does not matter whether the summation range starts from either 0 or 1, or ends with $h - 1$ or h.

The three versions of values and visitation frequencies have the following relationship:

state value

$$\bar{v}_\pi(s) = \lim_{h \to +\infty} \frac{1}{h} v_\pi^{[h]}(s) \quad = \lim_{\gamma \to 1^-} (1 - \gamma) v_\pi^{(\gamma)}(s), \qquad s \in S$$

action value

$$\bar{q}_\pi(s, a) = \lim_{h \to +\infty} \frac{1}{h} q_\pi^{[h]}(s, a) = \lim_{\gamma \to 1^-} (1 - \gamma) q_\pi^{(\gamma)}(s, a), \quad s \in S, a \in \mathcal{A}$$

state distribution

$$\bar{\eta}_\pi(s) = \lim_{h \to +\infty} \frac{1}{h} \eta_\pi^{[h]}(s) \quad = \lim_{\gamma \to 1^-} (1 - \gamma) \eta_\pi^{(\gamma)}(s), \qquad s \in S$$

state–action distribution

$$\bar{\rho}_\pi(s, a) = \lim_{h \to +\infty} \frac{1}{h} \rho_\pi^{[h]}(s, a) = \lim_{\gamma \to 1^-} (1 - \gamma) \rho_\pi^{(\gamma)}(s, a), \quad s \in S, a \in \mathcal{A}.$$

(Proof: In order to prove all the four results, we only need to prove the following relationship

$$\lim_{h \to +\infty} E_\pi \left[\frac{1}{h} \sum_{\tau=1}^{h} X_\tau \right] = \lim_{h \to +\infty} \frac{1}{h} E_\pi \left[\sum_{\tau=1}^{h} X_\tau \right] = \lim_{\gamma \to 1^-} (1 - \gamma) E_\pi \left[\sum_{\tau=1}^{+\infty} \gamma^{\tau-1} X_\tau \right]$$

holds for an arbitrary discrete-time stochastic process $X_0, X_1, \ldots$. We assume that the first equality always holds. The second equality holds because

$$\lim_{h \to +\infty} \frac{1}{h} E_\pi \left[\sum_{\tau=1}^{h} X_\tau \right]$$

$$= \lim_{h \to +\infty} \frac{\lim_{\gamma \to 1^-} E_\pi \left[\sum_{\tau=1}^{h} \gamma^{\tau-1} X_\tau \right]}{\lim_{\gamma \to 1^-} \sum_{\tau=1}^{h} \gamma^{\tau-1}}$$

$$= \lim_{\gamma \to 1^-} \frac{\lim_{h \to +\infty} E_\pi \left[\sum_{\tau=1}^{h} \gamma^{\tau-1} X_\tau \right]}{\lim_{h \to +\infty} \sum_{\tau=1}^{h} \gamma^{\tau-1}}$$

$$= \lim_{\gamma \to 1^-} \frac{\lim_{h \to +\infty} E_\pi \left[\sum_{\tau=1}^{h} \gamma^{\tau-1} X_\tau \right]}{\frac{1}{1-\gamma}}$$

$$= \lim_{\gamma \to 1^-} (1 - \gamma) E_\pi \left[\sum_{\tau=1}^{h} \gamma^{\tau-1} X_\tau \right].$$

The proof completes.)

When this book considered the discounted DTMDP, we derived the relationship to use state values to back up action values from the iterative relationship of the discounted return $G_t = R_{t+1} + \gamma G_{t+1}$, got the Bellman equation, and then discovered several algorithms to solve DTMDP such as the LP method and VI algorithms. However, average-reward does not use discounted return and does not have such an iterative relationship, and we do not have easy ways to use average-reward state value to back up average-reward action values. We will discuss how to find the optimal average-reward values starting from the next subsection.

15.1.2 Differential Values

In order to solve average-reward DTMDP, we introduce a new DTMDP: **differential MDP**. Differential MDP is a new discounted DTMDP derived from the original average-reward DTMDP and its average reward $\bar{r}_\pi$. Differential MDP uses tilde "~" over the letters, including:

- **differential reward**:
$$\tilde{R}_t \overset{\text{def}}{=} R_t - \bar{r}_\pi,$$

  which subtracts the average reward $\bar{r}_\pi$ from the original reward R_t.
- **differential return**:
$$\tilde{G}_t \overset{\text{def}}{=} \sum_{\tau=1}^{+\infty} \tilde{R}_{t+\tau},$$

  which is in the form of discounted return with the discount factor $\gamma = 1$.
- expectation of differential return as the total episode reward:

$$\tilde{g}_\pi \overset{\text{def}}{=} \mathrm{E}_\pi \left[\tilde{G}_0 \right].$$

Other quantities need to be adjusted accordingly, such as the dynamics $\tilde{p}(s', \tilde{r}|s, a)$.

Since differential MDP is a special case of discounted DTMDP, the definition of values and their relationship in Chap. 2 all hold.

The values derived from the differential return are called **differential values**.

- Differential state values:

$$\tilde{v}_\pi(s) \overset{\text{def}}{=} \mathrm{E}_\pi \left[\tilde{G}_t \middle| S_t = s \right], \quad s \in \mathcal{S}.$$

- Differential action values:

$$\tilde{q}_\pi(s, a) \overset{\text{def}}{=} \mathrm{E}_\pi \left[\tilde{G}_t \middle| S_t = s, A_t = a \right], \quad s \in \mathcal{S}, a \in \mathcal{A}.$$

The relationship among differential values includes:

- Use the differential action value at time t to back up the differential state value at time t:

$$\tilde{v}_\pi(s) = \sum_a \pi(a|s)\tilde{q}_\pi(s,a), \quad s \in \mathcal{S}.$$

- Use the differential state value at time $t+1$ to back up the differential action value at time t:

$$\tilde{q} = \tilde{r}(s,a) + \sum_{s'} p(s'|s,a)\tilde{v}(s')$$

$$= \sum_{s',\tilde{r}} \tilde{p}(s',\tilde{r}|s,a)[\tilde{r} + \tilde{v}_\pi(s')]$$

$$= \sum_{s',r} \tilde{p}(s',r|s,a)[r - \tilde{r} + \tilde{v}_\pi(s')], \quad s \in \mathcal{S}, a \in \mathcal{A}.$$

- Use the differential state value at time $t+1$ to back up the differential state value at time t:

$$\tilde{v}(s) = \sum_a \pi(a|s)\left[\tilde{r}(s,a) + \sum_{s'} p(s'|s,a)\tilde{v}_\pi(s')\right]$$

$$= \sum_a \pi(a|s)\sum_{s'} p(s'|s,a)[\tilde{r}(s,a,s') - \tilde{r}_\pi + \tilde{v}_\pi(s')], \quad s \in \mathcal{S}.$$

- Use the differential action value at time $t+1$ to back up the differential action value at time t:

$$\tilde{q}(s,a) = \sum_{s',\tilde{r}} \tilde{p}(s',\tilde{r}|s,a)\left[\tilde{r} + \sum_{a'} \pi(a'|s')\tilde{q}_\pi(s',a')\right]$$

$$= \sum_{s',r} \tilde{p}(s',r|s,a)\sum_{a'} \pi(a'|s')[r - \tilde{r} + \tilde{q}_\pi(s',a')], \quad s \in \mathcal{S}, a \in \mathcal{A}.$$

The relationship among differential values, finite-horizon values, and discounted values is as follows:

$$\tilde{v}_\pi(s') - \tilde{v}_\pi(s'')$$

$$= \lim_{h\to+\infty}\left[v_\pi^{[h]}(s') - v_\pi^{[h]}(s'')\right]$$

$$= \lim_{\gamma\leftarrow 1^-}\left[v_\pi^{(\gamma)}(s') - v_\pi^{(\gamma)}(s'')\right], \qquad s', s'' \in \mathcal{S}$$

$$\tilde{q}_\pi(s',a') - \tilde{q}_\pi(s'',a'')$$

$$= \lim_{h\to+\infty}\left[q_\pi^{[h]}(s',a') - q_\pi^{[h]}(s'',a'')\right]$$

$$= \lim_{\gamma\leftarrow 1^-}\left[q_\pi^{(\gamma)}(s',a') - q_\pi^{(\gamma)}(s'',a'')\right], \quad s', s'' \in \mathcal{S}, a', a'' \in \mathcal{A}.$$

This relationship tells us that the differential values can differentiate the value difference between two states or two state–action pairs. That is the reason why we call them "differential".

After getting differential values, we can use differential values to derive average-reward values and average reward.

The way to use differential values to calculate average-reward values is:

$$\tilde{v}_\pi(s) = r_\pi(s) + \sum_{s'} p_\pi(s'|s)\left[\tilde{v}_\pi(s') - \tilde{v}_\pi(s)\right], \qquad s \in S$$

$$\tilde{q}_\pi(s,a) = r(s,a) + \sum_{s',a'} p_\pi(s',a'|s,a)\left[\tilde{q}_\pi(s',a') - \tilde{q}_\pi(s,a)\right], \quad s \in S, a \in \mathcal{A}.$$

(Proof: We first consider the relationship of the state value. Considering an ordinary discounted values, the relationship that uses state values to back up state values is:

$$v_\pi^{(\gamma)}(s) = r_\pi(s) + \gamma \sum_{s'} p_\pi(s'|s)v_\pi^{(\gamma)}(s'), \quad s \in S.$$

Now we subtract both sides of the equation by $\gamma v_\pi^{(\gamma)}(s)$, and we get

$$(1-\gamma)v_\pi^{(\gamma)}(s) = r_\pi(s) + \gamma \sum_{s'} p_\pi(s'|s)\left[v_\pi^{(\gamma)}(s') - v_\pi^{(\gamma)}(s)\right], \quad s \in S.$$

Taking the limit $\gamma \to 1^-$, with $\tilde{v}_\pi(s) = \lim_{\gamma \to 1^-}(1-\gamma)v_\pi^{(\gamma)}(s)$ $(s \in S)$ and $\tilde{v}_\pi(s') - \tilde{v}_\pi(s) = \lim_{\gamma \to 1^-}\left[v_\pi^{(\gamma)}(s') - v_\pi^{(\gamma)}(s)\right]$ $(s, s' \in S)$ into consideration, we have

$$\tilde{v}_\pi(s) = r_\pi(s) + \sum_{s'} p_\pi(s'|s)\left[\tilde{v}_\pi(s') - \tilde{v}_\pi(s)\right], \quad s \in S.$$

Next, we consider the relationship of the action values. The relationship that uses discounted action values to back up discounted action values in a discounted DTMDP is expressed by

$$q_\pi^{(\gamma)}(s,a) = r(s,a) + \gamma \sum_{s',a'} p_\pi(s',a'|s,a)q_\pi^{(\gamma)}(s',a'), \quad s \in S, a \in \mathcal{A}.$$

Subtracting both sides by $\gamma q_\pi^{(\gamma)}(s,a)$ and taking the limit $\gamma \to 1^-$ finishes the proof.)

The way to use differential values to calculate average reward is:

$$\bar{r}_\pi = \sum_a \pi(a|s) \sum_{s'} p(s'|s,a)\left[r(s,a,s') - \tilde{v}_\pi(s) + \tilde{v}_\pi(s')\right], \quad s \in S,$$

$$\bar{r}_\pi = \sum_{s',r} p(s',r|s,a) \sum_{a'} \pi(a'|s')\left[r - \tilde{q}_\pi(s,a) + \tilde{q}_\pi(s',a')\right], \quad s \in S, a \in \mathcal{A}.$$

(Proof: Applying the relationship that uses differential state values at time $t + 1$ to back up the differential state values at time t

$$\tilde{v}_\pi(s) = \sum_a \pi(a|s) \sum_{s'} p(s'|s, a) \left[r(s, a, s') - \bar{r}_\pi + \tilde{v}_\pi(s') \right], \quad s \in S,$$

and the relationship that uses differential action values at time $t + 1$ to back up the differential action value at time t

$$\tilde{q}_\pi(s, a) = \sum_{s',r} p(s', r|s, a) \sum_{a'} \pi(a'|s') \left[r - \bar{r}_\pi + \tilde{q}_\pi(s', a') \right], \quad s \in S, a \in \mathcal{A}$$

can get the results.) The above relationship can also be written as

$$\bar{r}_\pi = \mathrm{E}_\pi \left[R_{t+1} + \tilde{v}_\pi(S_{t+1}) - \tilde{v}_\pi(S_t) \right],$$
$$\bar{r}_\pi = \mathrm{E}_\pi \left[R_{t+1} + \tilde{q}_\pi(S_{t+1}, A_{t+1}) - \tilde{q}_\pi(S_t, A_t) \right].$$

When the environment dynamic is unknown, we can use stochastic approximation that leverages the above relationship to estimate the average reward. Specifically, we can gather samples of $R_{t+1} + \tilde{v}_\pi(S_{t+1}) - \tilde{v}_\pi(S_t)$ and $R_{t+1} + \tilde{q}_\pi(S_{t+1}, A_{t+1}) - \tilde{q}_\pi(S_t, A_t)$ to estimate the average reward $\bar{r}_\pi$.

At the end of this section, let us see a property of average-reward state values. This property relates to properties of MDP.

Interdisciplinary Reference 15.1
Stochastic Process: Properties of Markov Process

Consider two states of an MDP $s', s'' \in S$, if there exists time τ such that the transition probability from state s' to state s'' satisfies $p^{[\tau]}(s''|s') > 0$, we say that state s'' is accessible from state s'.

Consider two states of an MDP $s', s'' \in S$, if state s' is accessible from state s'', and state s'' is accessible from state s', we say that state s' and state s'' communicate.

Consider a state of an MDP $s \in S$, if there exists time $\tau > 0$ such that the transition probability from state s to itself satisfies $p^{[\tau]}(s|s)$, we say the state s is recurrent.

For an MDP, if its all recurrent states communicate with each other, the MDP is unichain. Otherwise, it is multichain.

All states in the same chain share the same average-reward state values $\bar{v}_\pi(s)$. That is because, we can reach a recurrent state from any state in the chain in finite steps, which can be negligible in the long-run average. If an MDP is unichain, the average-reward state values of all states equal the average reward.

15.1.3 Optimal Policy

Different definitions of values will further lead to different definitions of optimal values.

- Given the positive integer h, we have finite-horizon optimal values

$$\text{state value} \quad v_*^{[h]}(s) \overset{\text{def}}{=} \sup_{\pi} v_\pi^{[h]}(s), \qquad s \in \mathcal{S}$$

$$\text{action value} \quad q_*^{[h]}(s, a) \overset{\text{def}}{=} \sup_{\pi} q_\pi^{[h]}(s, a), \quad s \in \mathcal{S}, a \in \mathcal{A}.$$

- Given the discount factor $\gamma \in (0, 1]$, we have discounted average optimal values

$$\text{state value} \quad \bar{v}_*^{(\gamma)}(s) \overset{\text{def}}{=} \sup_{\pi} \bar{v}_\pi^{(\gamma)}(s), \qquad s \in \mathcal{S}$$

$$\text{action value} \quad \bar{q}_*^{(\gamma)}(s, a) \overset{\text{def}}{=} \sup_{\pi} \bar{q}_\pi^{(\gamma)}(s, a), \quad s \in \mathcal{S}, a \in \mathcal{A},$$

or

$$\text{state value} \quad v_*^{(\gamma)}(s) \overset{\text{def}}{=} \sup_{\pi} v_\pi^{(\gamma)}(s), \qquad s \in \mathcal{S}$$

$$\text{action value} \quad q_*^{(\gamma)}(s, a) \overset{\text{def}}{=} \sup_{\pi} q_\pi^{(\gamma)}(s, a), \quad s \in \mathcal{S}, a \in \mathcal{A}.$$

- Average reward optimal values

$$\text{state value} \quad \bar{v}_*(s) \overset{\text{def}}{=} \sup_{\pi} v_\pi(s), \qquad s \in \mathcal{S}$$

$$\text{action value} \quad \bar{q}_*(s, a) \overset{\text{def}}{=} \sup_{\pi} q_\pi(s, a), \quad s \in \mathcal{S}, a \in \mathcal{A}.$$

A policy is ε-optimal when its values are smaller than the optimal values no more than ε. A policy is optimal when its values equal the optimal values. For simplicity, we assume the optimal policy always exists.

Like the discounted DTMDP, we can use varied ways to find the optimal policy for average-reward DTMDP, such as LP, VI, and TD update. This section shows some algorithms without proof.

LP method: The prime problem is

$$\underset{\substack{\bar{v}(s):s\in\mathcal{S} \\ \tilde{v}(s):s\in\mathcal{S}}}{\text{minimize}} \quad \sum_{s} c(s)\bar{v}(s)$$

$$\text{s.t.} \quad \bar{v}(s) \geq \sum_{s'} p(s'|s, a)\bar{v}(s'), \qquad\qquad s \in \mathcal{S}, a \in \mathcal{A}$$

$$\bar{v}(s) \geq r(s, a) + \sum_{s'} p(s'|s, a)\tilde{v}(s') - \tilde{v}(s), \quad s \in \mathcal{S}, a \in \mathcal{A},$$

where $c(s) > 0$ $(s \in S)$. Solving this problem results in optimal average-reward state values and optimal differential state values. The dual problem is

$$\underset{\substack{\bar{\rho}(s,a):s\in S,a\in \mathcal{A} \\ \rho(s,a):s\in S,a\in \mathcal{A}}}{\text{maximize}} \sum_{s,a} r(s,a)\bar{\rho}(s,a)$$

$$\text{s.t.} \quad \sum_{a'}\bar{\rho}(s',a') - \sum_{s,a}p(s'|s,a)\bar{\rho}(s,a) = 0, \quad s' \in S$$

$$\sum_{a'}\bar{\rho}(s',a') + \sum_{a'}\rho(s',a')$$

$$- \sum_{s,a}p(s'|s,a)\rho(s,a) = c(s'), \quad s' \in S$$

$$\bar{\rho}(s,a) \geq 0, \qquad\qquad\qquad s \in S, a \in \mathcal{A}$$

$$\rho(s,a) \geq 0, \qquad\qquad\qquad s \in S, a \in \mathcal{A}.$$

For a unichain MDP, all states share the same average-reward state value. Therefore, the prime problem is simplified to

$$\underset{\substack{\bar{r}\in \mathbb{R} \\ \tilde{v}(s):s\in S}}{\text{minimize}} \quad \bar{r}$$

$$\text{s.t.} \quad \bar{r} \geq r(s,a) + \sum_{s'}p(s'|s,a)\tilde{v}(s') - \tilde{v}(s), \quad s \in S, a \in \mathcal{A}.$$

The resulting $\bar{r}$ is the optimal average reward, and $\tilde{v}$ is the optimal differential state values. The dual problem is:

$$\underset{\rho(s,a):s\in S,a\in \mathcal{A}}{\text{maximize}} \sum_{s,a} r(s,a)\rho(s,a)$$

$$\text{s.t.} \quad \sum_{a'}\rho(s',a') - \sum_{s,a}p(s'|s,a)\rho(s,a) = 0, \quad s' \in S$$

$$\sum_{a'}\rho(s',a') = 1, \qquad\qquad\qquad s' \in S$$

$$\rho(s,a) \geq 0, \qquad\qquad\qquad s \in S, a \in \mathcal{A}.$$

Relative VI: Since differential values can differentiate states, we can fix a state $s_{\text{fix}} \in S$, and consider the relativeness:

$$\tilde{v}_{k+1}(s) \leftarrow \max_{a}\left[r(s,a) + \sum_{s'}p(s'|s,a)\tilde{v}(s')\right]$$

$$- \max_{a}\left[r(s_{\text{fix}},a) + \sum_{s'}p(s'|s_{\text{fix}},a)\tilde{v}(s')\right].$$

TD updates: Algorithm 15.1 shows the differential semi-gradient algorithm to evaluate action values and the differential SARSA algorithm to find the optimal policy. Algorithm 15.2 shows the differential expected SARSA algorithm and differential Q learning algorithm. These algorithms need to learn both differential action values and average reward. The differential action values are denoted in the form of $\tilde{q}(S, A; \mathbf{w})$ in the algorithms, where the parameters $\mathbf{w}$ need to be updated during learning. Average reward is denoted as $\bar{R}$ in the algorithms. We have noticed that the average reward can be estimated from the samples of $R_{t+1} + (1 - D_{t+1})\tilde{q}_\pi(S_{t+1}, A_{t+1}) - \tilde{q}_\pi(S_t, A_t)$. If we use incremental implementation to estimate it, the update can be formulated as

$$\bar{R} \leftarrow \bar{R} + \alpha^{(r)}\left[R + \left(1 - D'\right)\tilde{q}(S', A'; \mathbf{w}) - \tilde{q}(S, A; \mathbf{w}) - \bar{R}\right].$$

Additionally, notice the differential TD is $\tilde{\varDelta} \stackrel{\text{def}}{=} \tilde{U} - \tilde{q}(S, A; \mathbf{w}) = \bar{R} + (1 - D')\tilde{q}(S', A'; \mathbf{w}) - \tilde{q}(S, A; \mathbf{w})$, so the above update formula can be written as

$$\bar{R} \leftarrow \bar{R} + \alpha^{(r)}\tilde{\varDelta}.$$

Algorithm 15.1   Semi-gradient descent policy evaluation to estimate action values or SARSA policy optimization.

Parameters: optimizers (include learning rates $\alpha^{(\mathbf{w})}$ and $\alpha^{(r)}$), and the parameters to control the number of episodes and the number of steps in each episode.

1. Initialize parameters:

   1.1. (Initialize differential action value parameters) $\mathbf{w} \leftarrow$ arbitrary values.
   1.2. (Initialize average reward estimate) $\bar{R} \leftarrow$ arbitrary value.

2. For each episode:

   2.1. (Initialize state–action pair) Choose the initial state S.
   For policy evaluation, use the policy $\pi(\cdot|S)$ to determine the action A.
   For policy optimization, use the policy derived from current optimal action value estimates $\tilde{q}(S, \cdot; \mathbf{w})$ (such as ε-greedy policy) to determine the action A.
   2.2. Loop until the episode ends:
       2.2.1. (Sample) Execute the action A, and then observe the reward R, the next state S', and the indicator of episode end D'.
       2.2.2. (Decide) For policy evaluation, use the policy $\pi(\cdot|S')$ to determine the action A'. For policy optimization, use the policy derived from the current optimal action value estimates $\tilde{q}(S', \cdot; \mathbf{w})$ (for example, ε-greedy policy) to determine the action A'. (The action can be arbitrarily chosen if $D' = 1$.)
       2.2.3. (Calculate differential reward) $\tilde{R} \leftarrow R - \bar{R}$.

2.2.4. (Calculate differential TD return) $\tilde{U} \leftarrow \tilde{R} + (1 - D')\tilde{q}(S', A'; \mathbf{w})$.

2.2.5. (Calculate differential TD error) $\tilde{\Delta} \leftarrow \tilde{U} - \tilde{q}(S, A; \mathbf{w})$.

2.2.6. (Update differential action value parameters) Update the parameter $\mathbf{w}$ to reduce $\left[\tilde{U} - \tilde{q}(S, A; \mathbf{w})\right]^2$ (For example, $\mathbf{w} \leftarrow \mathbf{w} + \alpha^{(\mathbf{w})}\tilde{\Delta}\nabla\tilde{q}(S, A; \mathbf{w})$).
Note, you should not re-calculate $\tilde{U}$, and you should not calculate the gradient of $\tilde{U}$ with respect to $\mathbf{w}$.

2.2.7. (Update average reward estimate) Update $\tilde{R}$ to reduce $\left[R + (1 - D')\tilde{q}(S', A'; \mathbf{w}) - \tilde{q}(S, A; \mathbf{w}) - \bar{R}\right]^2$ (For example, $\bar{R} \leftarrow \bar{R} + \alpha^{(r)}\tilde{\Delta}$).

2.2.8. $S \leftarrow S', A \leftarrow A'$.

Algorithm 15.2 Semi-gradient descent differential expected SARSA policy optimization, or differential Q learning.

Parameters: optimizers (include learning rates $\alpha^{(\mathbf{w})}$ and $\alpha^{(r)}$), and the parameters to control the number of episodes and the number of steps in each episode.

1. Initialize parameters:

    1.1. (Initialize differential action value parameters) $\mathbf{w} \leftarrow$ arbitrary values.

    1.2. (Initialize average reward estimate) $\bar{R} \leftarrow$ arbitrary value.

2. For each episode:

    2.1. (Initialize state) Choose the initial state S.

    2.2. Loop until the episode ends:

        2.2.1. (Decide) For policy evaluation, use the policy $\pi(\cdot|S)$ to determine the action A. For policy optimization, use the policy derived from the current optimal action value estimates $\tilde{q}(S, \cdot; \mathbf{w})$ (for example, ε-greedy policy) to determine the action A.

        2.2.2. (Sample) Execute the action A, and then observe the reward R, the next state S', and the indicator of episode end D'.

        2.2.3. (Calculate differential reward) $\tilde{R} \leftarrow R - \bar{R}$.

        2.2.4. (Calculate differential return) For expected SARSA, set $\tilde{U} \leftarrow \tilde{R} + (1 - D')\sum_a \pi(a|S'; \mathbf{w})\tilde{q}(S', a; \mathbf{w})$, where $\pi(\cdot|S'; \mathbf{w})$ is the policy derived from $\tilde{q}(S', \cdot; \mathbf{w})$ (for example, ε-greedy policy). For Q learning, set $\tilde{U} \leftarrow \tilde{R} + (1 - D')\max_a \tilde{q}(S', a; \mathbf{w})$.

        2.2.5. (Calculate differential TD error) $\tilde{\Delta} \leftarrow \tilde{U} - \tilde{q}(S, A; \mathbf{w})$.

2.2.6. (Update differential action value parameters) Update the parameter $\mathbf{w}$ to reduce $\left[\tilde{U} - \tilde{q}(S, A; \mathbf{w})\right]$ (For example, $\mathbf{w} \leftarrow \mathbf{w} + \alpha^{(\mathbf{w})} \tilde{\Delta} \nabla \tilde{q}(S, A; \mathbf{w})$).

Note, you should not re-calculate $\tilde{U}$, and you should not calculate the gradient of $\tilde{U}$ with respect to $\mathbf{w}$.

2.2.7. (Update average reward estimate) Update $\bar{R}$ to reduce $\left[R - \tilde{q}(S, A; \mathbf{w}) + \tilde{q}(S', A'; \mathbf{w}) - \bar{R}\right]^2$ (For example, $\bar{R} \leftarrow \bar{R} + \alpha^{(r)} \tilde{\Delta}$).

2.2.8. $S \leftarrow S'$.

15.2 CTMDP: Continuous-Time MDP

Previous contents in this book considered discrete time index. However, the time index can be not discrete. For example, the time index set can be the set of real numbers or its continuous subset. An MDP with such time index is called **Continuous-Time MDP (CTMDP)**. This section considers CTMDP.

Interdisciplinary Reference 15.2
Stochastic Process: Continuous-Time Markov Process

Following Interdisciplinary Reference 2.1 in Sect. 2.1.1, we discuss the property of CTMDP.

For simplicity, we always assume the transition probability satisfies the following condition:

$$\lim_{\tau \to 0} p^{[\tau]}(s'|s) = p^{[0]}(s'|s), \quad s, s' \in \mathcal{S},$$

which can be written as

$$\lim_{\tau \to 0} \mathbf{P}^{[\tau]} = \mathbf{P}^{[0]}.$$

Consider a $|\mathcal{S}| \times |\mathcal{S}|$ matrix $\mathbf{Q} = (q(s'|s) : s \in \mathcal{S}, s' \in \mathcal{S})$. If it satisfies

$$-\infty \leq q(s|s) \leq 0, \qquad s \in \mathcal{S},$$
$$-\infty < q(s'|s) < +\infty, \quad s, s' \in \mathcal{S},$$
$$\sum_{s'} q(s'|s) \leq 0, \qquad s \in \mathcal{S},$$

the matrix is called a Q matrix. If a Q matrix satisfies

$$\sum_{s'} q(s'|s) = 0, \quad s \in \mathcal{S},$$

the matrix is called a conservative Q matrix.

Define the transition rate matrix as

$$\mathbf{Q} \overset{\text{def}}{=} \lim_{\tau \to 0^+} \frac{1}{\tau} \left(\mathbf{P}^{[\tau]} - \mathbf{I} \right),$$

where $\mathbf{P}^{[\tau]}$ is the transition probability matrix, $\mathbf{I}$ is the identity matrix, and the limit is element-wise. We can prove that the transition rate matrix is a Q matrix. If the state space is finite, the transition rate matrix is conservative. For simplicity, we always assume the transition rate matrix is conservative.

Excluding actions from CTMDP will lead to Continuous-Time Markov Reward Process (CTMRP). Excluding rewards from CTMRP will lead to CTMP. Therefore, the transition in CTMDP can be quantified by transition rate in the form of $q(s'|s, a)$ ($s \in \mathcal{S}, a \in \mathcal{A}, s' \in \mathcal{S}$).

All of state space, action space, and cumulative reward space can be either discrete or continuous. The changing of these values through time can be gradual, or there may be a jump sometimes.

Similar to DTMDP, CTMDP also has discounted return expectation and average reward.

- Discounted return expectation: The discounted return is the integral of reward rate over time. For episodic tasks, the discounted return can be written as $G_t \overset{\text{def}}{=} \int_t^T \gamma^{\tau-t} \mathrm{d}R_\tau + \gamma^{T-t} G_T$, where $G_T = F_g(T, X_T)$ is a random variable related to the terminal state. The reward rate $\frac{\mathrm{d}R_t}{\mathrm{d}t}$ can depend on both S_t and A_t. Since there may be jumps on the discounted return, the reward date $\frac{\mathrm{d}R_t}{\mathrm{d}t}$ can be not in $(-\infty, +\infty)$. For sequential tasks, the discounted return can be written as $G_t \overset{\text{def}}{=} \int_0^{+\infty} \gamma^\tau \mathrm{d}R_{t+\tau}$. We can combine the two notations as

$$G_t \overset{\text{def}}{=} \int_0^{+\infty} \gamma^\tau \mathrm{d}R_{t+\tau}.$$

The expectation of the discounted return is:

$$\bar{g}_\pi^{(\gamma)} \overset{\text{def}}{=} \mathrm{E}_\pi \left[\int_0^{+\infty} \gamma^\tau \mathrm{d}R_\tau \right].$$

- Average reward: The average reward can be written as

$$\bar{r}_\pi \overset{\text{def}}{=} \lim_{h \to +\infty} \mathrm{E}_\pi \left[\frac{1}{h} \int_0^h \mathrm{d}R_\tau \right].$$

There are other definitions, such as upper limit average reward, lower limit average reward, and limited discounted average reward.

Derived from the definition of discounted return expectation and average rewards, we can define values.

- Discounted values with discount factor $\gamma \in (0, 1]$:

state value $\quad \bar{v}_\pi^{(\gamma)}(s) \stackrel{\text{def}}{=} \mathrm{E}_\pi\left[\int_0^{+\infty} \gamma^\tau \mathrm{d}R_{t+\tau} \middle| S_t = s\right], \qquad\qquad s \in \mathcal{S}$

action value $\quad \bar{q}_\pi^{(\gamma)}(s, a) \stackrel{\text{def}}{=} \mathrm{E}_\pi\left[\int_0^{+\infty} \gamma^\tau \mathrm{d}R_{t+\tau} \middle| S_t = s, A_t = a\right], \quad s \in \mathcal{S}, a \in \mathcal{A}.$

- Average-reward values:

state value $\quad \bar{v}_\pi(s) \stackrel{\text{def}}{=} \lim_{h\to+\infty} \mathrm{E}_\pi\left[\frac{1}{h}\int_0^h \gamma^\tau \mathrm{d}R_{t+\tau} \middle| S_t = s\right], \quad s \in \mathcal{S}$

action value $\quad \bar{q}_\pi(s, a) \stackrel{\text{def}}{=} \lim_{h\to+\infty} \mathrm{E}_\pi\left[\frac{1}{h}\int_0^h \gamma^\tau \mathrm{d}R_{t+\tau} \middle| S_t = s, A_t = a\right],$

$$s \in \mathcal{S}, a \in \mathcal{A}.$$

We can further define optimal values:

- Discounted optimal values with discount factor $\gamma \in (0, 1]$:

state value $\quad \bar{v}_*^{(\gamma)}(s) \stackrel{\text{def}}{=} \sup_\pi \bar{v}_\pi^{(\gamma)}(s), \qquad s \in \mathcal{S}$

action value $\quad \bar{q}_*^{(\gamma)}(s, a) \stackrel{\text{def}}{=} \sup_\pi \bar{q}_\pi^{(\gamma)}(s, a), \quad s \in \mathcal{S}, a \in \mathcal{A}.$

- Average-reward optimal values:

state value $\quad \bar{v}_*(s) \stackrel{\text{def}}{=} \sup_\pi \bar{v}_\pi(s), \qquad s \in \mathcal{S}$

action value $\quad \bar{q}_*(s, a) \stackrel{\text{def}}{=} \sup_\pi \bar{q}_\pi(s, a), \quad s \in \mathcal{S}, a \in \mathcal{A}.$

A policy is an ε-optimal policy if its values differ from optimal values by $\leq \varepsilon$. A policy is an optimal policy if its values equal the optimal values.

The analysis of continuous-time is more complex than discrete-time, but the form of results can be similar. Remarkably, the definition of the transition rate matrix $q(s'|s, a)$ differs between $s = s'$ and $s \neq s'$ in a constant 1.

For example, consider the LP method on an average-reward CTMDP whose reward rate is deterministic and denoted as $r(s, a)$. The primary problem is:

$$\operatorname*{minimize}_{\substack{\bar{v}(s):s\in\mathcal{S} \\ \bar{v}(s):s\in\mathcal{S}}} \quad \sum_s c(s)\bar{v}(s)$$

$$\text{s.t.} \qquad 0 \geq \sum_{s'} q(s'|s, a)\bar{v}(s'), \qquad\qquad s \in \mathcal{S}, a \in \mathcal{A}$$

$$\bar{v}(s) \geq r(s, a) + \sum_{s'} q(s'|s, a)\bar{v}(s'), \quad s \in \mathcal{S}, a \in \mathcal{A}$$

where $c(s) > 0$ $(s \in S)$, and the results are optimal average-reward state values and optimal differential state values. The dual problem is:

$$\underset{\substack{\bar{\rho}(s,a):s\in S,a\in\mathcal{A}\\\rho(s,a):s\in S,a\in\mathcal{A}}}{\text{maximize}} \quad \sum_{s,a} r(s,a)\bar{\rho}(s,a)$$

$$\text{s.t.} \quad \sum_{s,a} q(s'|s,a)\bar{\rho}(s,a) = 0, \qquad\qquad s' \in S$$

$$\sum_{a'}\bar{\rho}(s',a') - \sum_{s,a}q(s'|s,a)\rho(s,a) = c(s'), \quad s' \in S$$

$$\bar{\rho}(s,a) \geq 0, \qquad\qquad\qquad s \in S, a \in \mathcal{A}$$

$$\rho(s,a) \geq 0, \qquad\qquad\qquad s \in S, a \in \mathcal{A}.$$

For a unichain MDP, all states share the same average-reward state values. Therefore, the primary problem degrades to

$$\underset{\substack{\bar{r}\in\mathbb{R}\\\bar{v}(s):s\in S}}{\text{minimize}} \quad \bar{r}$$

$$\text{s.t.} \quad \bar{r} \geq r(s,a) + \sum_{s'}q(s'|s,a)\bar{v}(s'), \quad s \in S, a \in \mathcal{A}.$$

And the dual problem degrades to

$$\underset{\rho(s,a):s\in S,a\in\mathcal{A}}{\text{maximize}} \quad \sum_{s,a}r(s,a)\rho(s,a)$$

$$\text{s.t.} \quad \sum_{s,a}q(s'|s,a)\rho(s,a) = 0, \quad s' \in S$$

$$\sum_{a'}\rho(s',a') = 1, \qquad\quad s' \in S$$

$$\rho(s,a) \geq 0, \qquad\qquad s \in S, a \in \mathcal{A}.$$

Using the similarity of the relationship between values, the relationship of CTMDP can be represented into that of DTMDP. This is called uniformization technique. For example, given an average-time CTMDP, the dynamic $\breve{p}$, reward $\breve{r}$, and discount factor $\breve{\gamma}$ of the discounted DTMDP after the uniformization are:

$$\breve{r} = \frac{r(s,a)}{1 + q_{max}}, \quad s \in S, a \in \mathcal{A}$$

$$\breve{p}(s'|s,a) = \frac{q(s'|s,a)}{1 + q_{max}}, \quad s \in S, a \in \mathcal{A}, s' \in S$$

$$\breve{\gamma} = \frac{q_{max}}{1 + q_{max}},$$

where $q_{max} = \sup_{s,a} |q(s|s,a)|$.

15.3 Non-Homogenous MDP

MDP can be either homogenous or non-homogenous. We will discuss non-homogenous MDP in this section.

15.3.1 Representation of Non-Stationary States

This subsection considers an alternative way to represent states in an agent–environment interface.

Since Chap. 1, we discussed the agent–environment interface, where we denote the state at time t as $S_t \in S$. The state S_t may or may not contain the information of the time index t. If state S_t contains the information of time t, the state can also be written as $S_t = (t, X_t)$, where $X_t \in \mathcal{X}$ contains all information except the time. Furthermore, we can partition the state space S into $S = \bigcup_{t \in \mathcal{T}} \{t\} \times \mathcal{X}_t$, where $\mathcal{T}$ is the time index, and $\mathcal{X}_t$ is the state space at time $t \in \mathcal{T}$. $\{t\} \times \mathcal{X}_t$ contains the pair (t, x_t) where $x_t \in \mathcal{X}_t$. The state space with the terminal state is denoted as $\mathcal{X}_t^+$. If we can reach the terminal state at time t, $\mathcal{X}_t^+ = \mathcal{X}_t \cup \{x_{\text{end}}\}$; otherwise, $\mathcal{X}_t^+ = \mathcal{X}_t$.

If the dynamic of an MDP is time-invariant, the MDP is homogenous. If an MDP is homogenous, the state X_t can function as $S_t = (t, X_t)$, or equivalently the state S_t does not need to contain the time information. In this case, it is unnecessary to partition S_t into $S_t = (t, X_t)$. Even when the MDP is non-homogenous, we can still put the time information into states as $S_t = (t, X_t)$, so that the non-homogenous MDP is converted to a homogenous MDP.

Example 15.1 (Hamilton–Jacobi–Bellman Equation) Consider a non-homogenous CTMDP with given dynamics. The discounted optimal values $v_{*,i}$ satisfy the Hamilton–Jacobi–Bellman (HJB) equations (Bellman, 1957):

$$\dot{v}_{*,t}(X_t) + \max_a \left[\nabla v_{*,t}(X_t)\dot{X}_t + \frac{dR_t}{dt} \right] = 0,$$

where $\dot{v}_{*,t}(\cdot)$ is the partial derivatives of $v_{*,t}(\cdot)$ over the subscript t, and $\nabla v_{*,t}(\cdot)$ is the partial derivatives of $v_{*,t}(\cdot)$ with respect to the parameter within parenthesis. (Proof: Plugging in the following Taylor expansion

$$v_{*,t+dt}(X_{t+dt}) = v_{*,t}(X_t) + \dot{v}_{*,t}(X_t) + \nabla v_{*,t}(X_t) + o(dt)$$

$$\int_t^{t+dt} \gamma^{\tau-t} dR_\tau = dR_t + o(dt)$$

to the relationship of the optimal values

$$v_{*,t}(X_t) = \max_a \left[v_{*,t+dt}(X_{t+dt}) + \int_t^{t+dt} \gamma^{\tau-t} dR_\tau \right]$$

will lead to

$$v_{*,t}(X_t) = \max_a \left[v_{*,t}(X_t) + \dot{v}_{*,t}(X_t)dt + \nabla v_{*,t}(X_t)dX_t + o(dt) + dR_t + o(dt) \right].$$

Simplification will lead to

$$\dot{v}_{*,t}(X_t) + \max_a \left[\nabla v_{*,t}(X_t) \frac{dX_t}{dt} + \frac{dR_t}{dt} + \frac{o(dt)}{dt} \right] = 0.$$

Applying the limit $dt \to 0^+$ completes the proof.)

15.3.2 Bounded Time Index

Consider an MDP using the state form X_t. If its time index is bounded, the MDP must be non-homogenous. For example, if the time index set is $\mathcal{T} = \{0, 1, \ldots, t_{max}\}$, the state space $\mathcal{X}_t$ has many states when $t < t_{max}$, but only has the terminal state x_{end} when $t = t_{max}$. Therefore, the MDP is not homogenous. Meanwhile, we can combine t and X_t as the state $S_t = (t, X_t)$ of new homogenous MDP. The state space of the new homogenous MDP is

$$\mathcal{S} = \{ (t, x) : t \in \mathcal{T}, x \in \mathcal{X}_t \},$$

which has more elements than each individual $\mathcal{X}_t$. Therefore, the form of X_t may be better than the form of S_t for such tasks.

When the time index set is bound, we can define return as

$$G_t \overset{\text{def}}{=} \sum_{\tau \in \mathcal{T}: \tau > t} R_\tau.$$

For continuous-time index, the summation here should be understood as integral. This return can be interpreted as the discounted return with discount factor $\gamma = 1$. The performance of this policy can be evaluated by the total reward expectation

$$\bar{g}_{0,\pi} \overset{\text{def}}{=} E_\pi[G_0].$$

Accordingly, we can define values as

$$\text{state value} \quad v_{t,\pi}(x) \overset{\text{def}}{=} E_\pi \left[\sum_{\tau > 0: t+\tau \in \mathcal{T}} R_{t+\tau} \middle| X_t = x \right], \quad t \in \mathcal{T}, x \in \mathcal{X}_t$$

$$\text{action value} \quad q_{t,\pi}(x, a) \overset{\text{def}}{=} E_\pi \left[\sum_{\tau > 0: t+\tau \in \mathcal{T}} R_{t+\tau} \middle| X_t = x, A_t = a \right],$$

$$t \in \mathcal{T}, x \in \mathcal{X}_t, a \in \mathcal{A}_t.$$

We can further define the optimal values as

$$\text{state value} \quad v_{t,*}(x) \overset{\text{def}}{=} \sup_{\pi} v_{t,\pi}(x), \qquad t \in \mathcal{T}, x \in \mathcal{X}_t$$

$$\text{action value} \quad q_{t,*}(x, a) \overset{\text{def}}{=} \sup_{\pi} q_{t,\pi}(x, a), \quad t \in \mathcal{T}, x \in \mathcal{X}_t, a \in \mathcal{A}_t.$$

Example 15.2 (Dynamic Programming in Finite-horizontal Finite MDP) Chapter 3 introduces the DP method to estimate the optimal values of homogenous DTMDP. We can also apply DP on a non-homogenous DTMDP with the time index $\mathcal{T} = \{0, 1, \ldots, t_{\max}\}$. As shown in Algo. 15.3, we can calculate the exact optimal values in the reversed order $t = t_{\max}, t_{\max} - 1, \ldots, 0$. If we used the DP on homogenous MDP in Sect. 3.3, the state space would be much larger, and the computation would be more intensive.

Algorithm 15.3   Model-based VI for fixed-horizontal episode.

Inputs: The number of steps in an episode $t_{\max}$, and the dynamics p_t ($0 \leq t < t_{\max}$).
Outputs: Optimal state value estimate v_t ($0 \leq t < t_{\max}$), and optimal policy estimate π_t ($0 \leq t < t_{\max}$).

1. (Initialize) $v_{t_{\max}}(x_{\text{end}}) \leftarrow 0$.
2. (Iterate) For $t \leftarrow t_{\max} - 1, t_{\max} - 2, \ldots, 1, 0$:

   2.1. (Update optimal action value estimates) $q_t(x, a) \leftarrow r_t(x, a) + \gamma \sum_{x' \in \mathcal{X}_{t+1}} p_t(x'|x, a) v_{t+1}(x')$ ($x \in \mathcal{X}_t, a \in \mathcal{A}_t(x)$).
   2.2. (Update optimal policy estimates) $\pi_t(x) \leftarrow \arg\max_{a \in \mathcal{A}_t(x)} q_t(x, a)$ ($x \in \mathcal{X}_t$).
   2.3. (Update optimal state value estimates) $v_t(x) \leftarrow \max_{a \in \mathcal{A}_t(x)} q_t(x, a)$ ($x \in \mathcal{X}_t$).

15.3.3 Unbounded Time Index

When the time index set is unbounded, we can define the following time-variant metrics:

- Expectation of discounted return with discount factor $\gamma \in (0, 1]$:

$$\bar{g}_{t,\pi}^{(\gamma)} \overset{\text{def}}{=} E_\pi \left[\sum_{\tau > 0} \gamma^\tau R_{t+\tau} \right], \quad t \in \mathcal{T}.$$

There is also an alternative definition for discrete-time, whose value differs in γ.
- Average reward

$$\bar{r}_{t,\pi} \overset{\text{def}}{=} \lim_{h \to +\infty} \mathrm{E}_\pi \left[\frac{1}{h} \sum_{0 < \tau \le h} R_{t+\tau} \right], \quad t \in \mathcal{T}.$$

There are other versions of average reward, which are omitted here.

These two metrics are suitable for both DTMDP and CTMDP. For CTMDP, the summation is in fact integral. Generally, we need to maximize the metrics at $t = 0$.
From these two metrics, we can further define values.

- Discounted values:

$$\text{state value} \quad \bar{v}_{t,\pi}^{(\gamma)}(x) \overset{\text{def}}{=} \mathrm{E}_\pi \left[\sum_{\tau > 0} \gamma^\tau R_{t+\tau} \Big| X_t = x \right],$$
$$t \in \mathcal{T}, x \in \mathcal{X}_t$$

$$\text{action value} \quad \bar{q}_{t,\pi}^{(\gamma)}(x, a) \overset{\text{def}}{=} \mathrm{E}_\pi \left[\sum_{\tau > 0} \gamma^\tau R_{t+\tau} \Big| X_t = x, A_t = a \right],$$
$$t \in \mathcal{T}, x \in \mathcal{X}_t, a \in \mathcal{A}_t.$$

- Average-reward values:

$$\text{state value} \quad \bar{v}_{t,\pi}(x) \overset{\text{def}}{=} \mathrm{E}_\pi \left[\frac{1}{h} \sum_{\tau > 0} R_{t+\tau} \Big| X_t = x \right],$$
$$t \in \mathcal{T}, x \in \mathcal{X}_t$$

$$\text{action value} \quad \bar{q}_{t,\pi}(x, a) \overset{\text{def}}{=} \mathrm{E}_\pi \left[\frac{1}{h} \sum_{\tau > 0} R_{t+\tau} \Big| X_t = x, A_t = a \right],$$
$$t \in \mathcal{T}, x \in \mathcal{X}_t, a \in \mathcal{A}_t.$$

We can further define optimal values.

- Discounted optimal values:

$$\text{state value} \quad \bar{v}_{t,*}^{(\gamma)}(x) \overset{\text{def}}{=} \sup_\pi \bar{v}_{t,\pi}^{(\gamma)}(x), \qquad t \in \mathcal{T}, x \in \mathcal{X}_t$$

$$\text{action value} \quad \bar{q}_{t,*}^{(\gamma)}(x, a) \overset{\text{def}}{=} \sup_\pi \bar{q}_{t,\pi}^{(\gamma)}(x, a), \quad t \in \mathcal{T}, x \in \mathcal{X}_t, a \in \mathcal{A}_t.$$

- Average-reward optimal values:

$$\text{state value} \quad \bar{v}_{t,*}(x) \overset{\text{def}}{=} \sup_\pi \bar{v}_{t,\pi}(x), \qquad t \in \mathcal{T}, x \in \mathcal{X}_t$$

$$\text{action value} \quad \bar{q}_{t,*}(x, a) \overset{\text{def}}{=} \sup_\pi \bar{q}_{t,\pi}(x, a), \quad t \in \mathcal{T}, x \in \mathcal{X}_t, a \in \mathcal{A}_t.$$

15.4 SMDP: Semi-MDP

This section introduces **Semi-Markov Decision Process (SMDP)**.

15.4.1 SMDP and its Values

Interdisciplinary Reference 15.3
Stochastic Process: Semi-Markov Process

In a **Semi-Markov Process (SMP)**, the state only switches on some random time. Let T_i be the i-th switch time ($i \in \mathbb{N}$), the state after the switch is $\hat{S}_i$. The initial switch time is $T_0 = 0$, and the initial state is $\hat{S}_0$. The difference between two switching times is called **sojourn time**, denoted as $\mathcal{T}_i = T_{i+1} - T_i$, which is also random. At time t, the state of SMP is

$$S_t = \hat{S}_i, \quad T_i \le t < T_{i+1}.$$

For an SMP, either Discrete-Time SMP (DTSMP) or Continuous-Time SMP (CTSMP), since the switch is a discrete event, the stochastic process $\left(\hat{S}_i : i \in \mathbb{N}\right)$ is a DTMP. The trajectory of the DTMP derived from the SMP can be written as

$$\hat{S}_0, \hat{S}_1, \dots.$$

The trajectory of an SMP, either DTSMP or CTSMP, can be presented by the trajectory of SMP with sojourn time:

$$\hat{S}_0, \mathcal{T}_0, \hat{S}_1, \mathcal{T}_1, \dots.$$

! Note

Since the Greek letter τ and the English letter t share the same upper case T, here we use bigger $\mathcal{T}$ to indicate the random variable of the sojourn time.

Section 2.1.1 told that, discarding actions in an MDP leads to an MRP, and discarding rewards in an MRP leads to an MP. Similarly, discarding actions in an SMDP leads to Semi-MRP (SMRP), and discarding rewards in an SMRP leads to Semi-Markov Process.

For an SMDP, if its time index set is the integer set or its subset, the SMDP is a Discrete-Time SMDP (DTSMDP); if its time index set is the real number set or its continuous subset, the SMDP is a Continuous-Time SMDP (CTSMDP).

Similar to the case of SMP, sampling at the switching time of SMDP, we can get a DTMDP. SMDP only decides at switching time, the resulting action A_{T_i} is exactly the action $\hat{A}_i$ at the DTMDP. SMDP can have rewards during the whole timeline, including non-switching time. Therefore, the reward of the corresponding DTMDP, $\hat{R}_{i+1}$, should include all rewards during the time $(T_i, T_{i+1}]$. If the SMDP is discounted, $\hat{R}_{i+1}$ should be the discounted total reward during this period.

> **⚠ Note**
>
> For discounted SMDP, the reward of the corresponding DTMDP depends on the discount factor.

The DTMDP derived from SMDP can be written as

$$\hat{S}_0, \hat{A}_0, \hat{R}_1, \hat{S}_1, \hat{A}_1, \hat{R}_2, \ldots.$$

The trajectory of SMDP, either DTSMDP or CTSMDP, can be presented by the trajectory of DTMDP with sojourn time:

$$\hat{S}_0, \hat{A}_0, \mathcal{T}_0, \hat{R}_1, \hat{S}_1, \hat{A}_1, \mathcal{T}_1, \hat{R}_2, \ldots.$$

The DTMDP with sojourn time can be obtained from the initial state distribution $p_{S_0}(s)$ and the dynamic with sojourn time

$$\hat{p}(\tau, s', r | s, a) \stackrel{\text{def}}{=} \Pr \left[\mathcal{T}_i = \tau, \hat{S}_{i+1} = s', \hat{R}_{i+1} = r \middle| \hat{S}_i = s, \hat{A}_i = a \right],$$

$$s \in \mathcal{S}, a \in \mathcal{A}, r \in \mathcal{R}, \tau \in \mathcal{T}, s' \in \mathcal{S}^+.$$

Example 15.3 (Atari) For the Atari games whose id does not have "`Deterministic`", each step may skip $\mathcal{T}$ frames, where $\mathcal{T}$ is a random number within $\{2, 3, 4\}$. This is a DTSMDP.

We can get the following derived quantities from the dynamics with sojourn time:

- Sojourn time expectation given the state–action pair

$$\tau(s, a) \stackrel{\text{def}}{=} \mathrm{E} \left[\mathcal{T}_i \middle| \hat{S}_i = s, \hat{A}_i = a \right], \quad s \in \mathcal{S}, a \in \mathcal{A}.$$

- Single-step reward expectation given the state–action pair

$$\hat{r}(s, a) \stackrel{\text{def}}{=} \mathrm{E} \left[\hat{R}_{i+1} \middle| \hat{S}_i = s, \hat{A}_i = a \right], \quad s \in \mathcal{S}, a \in \mathcal{A}.$$

The long-term reward of an SMDP can also be quantified by discounted return expectation or average reward.

- Discounted return expectation with the discount factor $\gamma \in (0, 1]$:

$$g_\pi^{(\gamma)} \overset{\text{def}}{=} \left[\sum_{\tau>0} \gamma^\tau R_\tau \right] = \mathrm{E}_\pi \left[\sum_{i=1}^{+\infty} \gamma^{T_i} \hat{R}_i \right].$$

For DTSMDP, the definition may differ in γ. However, it is not important, so this section will always use this definition. For CTSMDP, summation over $\tau > 0$ is in fact integral.

- Average-reward values:

$$\bar{r}_\pi \overset{\text{def}}{=} \lim_{h\to+\infty} \mathrm{E}_\pi \left[\frac{1}{h} \sum_{0<\tau\le h} R_\tau \right] = \lim_{h\to+\infty} \mathrm{E}_\pi \left[\frac{1}{T_h} \sum_{i=1}^{h} \hat{R}_i \right].$$

Values derived from them:

- Discounted values with discount factor $\gamma \in (0, 1]$:

    state value

$$v_\pi^{(\gamma)}(s) \overset{\text{def}}{=} \mathrm{E}_\pi \left[\sum_{\tau>0} \gamma^\tau R_{T_i+\tau} \middle| S_{T_i} = s \right]$$

$$= \mathrm{E}_\pi \left[\sum_{\iota>1}^{+\infty} \gamma^{T_{i+\iota}-T_i} \hat{R}_{i+\iota} \middle| \hat{S}_i = s \right], \qquad s \in S,$$

    action value

$$q_\pi^{(\gamma)}(s, a) \overset{\text{def}}{=} \mathrm{E}_\pi \left[\sum_{\tau>0} \gamma^\tau R_{T_i+\tau} \middle| S_{T_i} = s, A_{T_i} = a \right]$$

$$= \mathrm{E}_\pi \left[\sum_{\iota>1}^{+\infty} \gamma^{T_{i+\iota}-T_i} \hat{R}_{i+\iota} \middle| \hat{S}_i = s, \hat{A}_i = a \right], \quad s \in S, a \in \mathcal{A}.$$

- Average reward:

    state value

$$\bar{v}_\pi(s) \overset{\text{def}}{=} \lim_{h\to+\infty} \mathrm{E}_\pi \left[\frac{1}{h} \sum_{0<\tau\le h} R_{t+\tau} \middle| S_t = s \right]$$

$$= \lim_{h\to+\infty} \mathrm{E}_\pi \left[\frac{1}{T_{i+h} - T_i} \sum_{\iota=1}^{h} \hat{R}_{i+\iota} \middle| \hat{S}_i = s \right], \qquad s \in S,$$

    action value

$$\bar{q}_\pi(s,a) \stackrel{\text{def}}{=} \lim_{h \to +\infty} \mathrm{E}_\pi \left[\frac{1}{h} \sum_{0 < \tau \leq h} R_{t+\tau} \middle| S_t = s, A_t = a \right]$$

$$= \lim_{h \to +\infty} \mathrm{E}_\pi \left[\frac{1}{T_{i+h} - T_i} \sum_{t=1}^{h} \hat{R}_{i+t} \middle| \hat{S}_i = s, \hat{A}_i = a \right], \quad s \in \mathcal{S}, a \in \mathcal{A}.$$

15.4.2 Find Optimal Policy

This section considers finding the optimal policy in the setting of discounted SMDP. Average-reward SMDP is more complex, which may involve differential SMDP, so it is omitted here.

For discounted SMDP, its discounted return satisfies the following iterative formula:

$$G_{T_i} = \hat{R}_{i+1} + \gamma^{\mathcal{T}_i} G_{T_{i+1}}.$$

Compared to the formula $G_t = R_{t+1} + \gamma G_{t+1}$ that shows the relationship of return discounted return for DTMDP, the discount factor has a power $\mathcal{T}_i$, which is a random variable.

Given the policy π, the relationships among discounted values include:

- Use action values at time t to back up state values at time t:

$$v_\pi(s) = \sum_a \pi(a|s) q_\pi(s,a), \quad s \in \mathcal{S}.$$

- Use state values at time $t+1$ to back up action values at time t:

$$q_\pi(s,a) = \hat{r}(s,a) + \sum_{s',r} \hat{p}(s', \tau | s, a) v_\pi(s')$$

$$= \sum_{s', \hat{r}, \tau} \hat{p}(s', \hat{r}, \tau | s, a) [\hat{r} + \gamma^\tau v_\pi(s')], \quad s \in \mathcal{S}, a \in \mathcal{A}.$$

Similarly, we can define discounted optimal values, and the discounted optimal values have the following relationship:

- Use the optimal action value at time t to back up the optimal state values at time t:

$$v_*(s) = \sum_a \pi(a|s) q_*(s,a), \quad s \in \mathcal{S}.$$

- Use the optimal state value at time $t+1$ to back up the optimal action values at time t:

$$q_*(s,a) = \hat{r}(s,a) + \sum_{s',\tau} \gamma^\tau \hat{p}(s', \tau | s, a) v_*(s')$$

$$= \sum_{s', \hat{r}, \tau} \hat{p}(s', \hat{r}, \tau | s, a) \left[\hat{r} + \gamma^{\tau} v_{*}(s') \right], \quad s \in \mathcal{S}, a \in \mathcal{A}.$$

Using these relationships, we can design algorithms such LP, VI, and TD update.

For example, in SARSA algorithm and Q learning algorithm, we can introduce the sojourn time into the TD target $\hat{U}_i$:

- The target of SARSA becomes $\hat{U}_i = \hat{R}_{i+1} + \gamma^{T_i} \left(1 - \hat{D}_{i+1} \right) q \left(\hat{S}_{i+1}, \hat{A}_{i+1} \right)$.
- The target of Q learning becomes $\hat{U}_i = \hat{R}_{i+1} + \gamma^{T_i} \left(1 - \hat{D}_{i+1} \right) \max_a q \left(\hat{S}_{i+1}, a \right)$.

Then use the SARSA algorithm or Q learning algorithm for DTMDP to learn action values.

15.4.3 HRL: Hierarchical Reinforcement Learning

Hierarchical Reinforcement Learning is an RL method based on SMDP.

The idea of **Hierarchical Reinforcement Learning (HRL)** is as follows: In order to solve a task with a complex goal, we may decompose the goal into many subgoals. Then we try to reach these subgoals in an order. For example, consider a task that controls a robot arm to put ice cream into a running refrigerator. We can separate this goal into many subgoals: We can first open the door of the refrigerator, put the ice cream in, and then close the door of the refrigerator. These three subgoals can also be viewed as high-level action, a.k.a. option. We can view this RL task hierarchically: The high-level RL task chooses among subgoals, i.e. first open the door, second put the ice cream in, and lastly close the door; the low-level RL tasks decide how to use the raw actions to reach the subgoals.

The advantages of HRL include:

- Both high-level tasks and low-level tasks are simpler than the original tasks. Therefore, the complexity of the tasks is reduced.
- The agent can know what subgoals it is pursuing when it is interacting with the environment, which can make the policy more interpretable.

The disadvantages of HRL include:

- Partitioning a task into multiple levels may put additional constraints on the policy space. Therefore, the optimal policy of the original task may be no longer in the policy space of the hierarchy policy space.
- The design of subgoals may be not optimal. For example, sometimes some subgoals can be considered jointly, but high-level decision can only consider one subgoal at a time.

The high-level decision process of HRL is theoretically SMDP. The reason is, after a high-level action is executed, the next time to make a high-level decision is a random variable.

15.5 POMDP: Partially Observable Markov Decision Process

Section 1.3 told us that an agent in the agent–environment interface can observe observations O_t. For MDP, we can recover the state S_t from the observation O_t. If the observation O_t does not include all information about the state S_t, we say the environment is partially observed. For a partially observed decision process, if its state process is a Markov Process, we say the process is a **Partially Observable MDP (POMDP)**.

Interdisciplinary Reference 15.4
Stochastic Process: Hidden Markov Model

In a **Hidden Markov Model (HMM)**, there is a Markov Process $\{S_t : t \in \mathcal{T}\}$, but this MP can not be directly observable. There is an observable process $\{O_t : t \in \mathcal{T}\}$. The conditional probability of observation O_t given the state S_t is called the emission probability or output probability, denoted as

$$o(o|s') \stackrel{\text{def}}{=} \Pr[O_t = o|S_t = s'], \quad s' \in \mathcal{S}, o \in O.$$

15.5.1 DTPOMDP: Discrete-Time POMDP

This section considers Discrete-Time POMDP (DTPOMDP). The trajectory of the environment in a DTPOMDP has the form $R_0, S_0, O_0, A_0, R_1, S_1, O_1, A_1, R_2, \ldots$. Due to the Markovian, we have

$$\Pr[R_{t+1}, S_{t+1}|R_0, S_0, O_0, A_0, \ldots, R_t, S_t, O_t, A_t] = \Pr[R_{t+1}, S_{t+1}|S_t, A_t]$$
$$\Pr[O_{t+1}|R_0, S_0, O_0, A_0, \ldots, R_t, S_t, O_t, A_t, R_{t+1}, S_{t+1}] = \Pr[O_{t+1}|A_t, S_{t+1}]$$

for $t \geq 0$. A DTPOMDP can be specified as follows: Let the reward space be $\mathcal{R}$. Let the state space be $\mathcal{S}$. Let the observation space be O. Let the action space be $\mathcal{A}$. Then define

- initial distribution, including the initial reward–state distribution

$$p_{S_0,R_0}(s, r) \stackrel{\text{def}}{=} \Pr[S_0 = s, R_0 = r], \quad s \in \mathcal{S}, r \in \mathcal{R},$$

and initial emission distribution

$$o_0(o|s') \stackrel{\text{def}}{=} \Pr[O_0 = o|S_0 = s'], \quad s' \in \mathcal{S}, o \in O.$$

- dynamics, including the transition probability from state–action pair to next reward–state pair

$$p(s', r|s, a) \overset{\text{def}}{=} \Pr\left[S_{t+1} = s', R_{t+1} = r | S_t = s, A_t = a\right],$$
$$s \in S, a \in \mathcal{A}, r \in \mathcal{R}, s' \in S^+$$

and the transition probability from action–next-state to next observation

$$o(o|a, s') \overset{\text{def}}{=} \Pr\left[O_{t+1} = o | A_t = a, S_{t+1} = s'\right], \quad a \in \mathcal{A}, s' \in S, o \in O.$$

In many tasks, the initial reward R_0 is always 0. In such cases, we can ignore R_0 and exclude it out of the trajectory. Consequentially, the initial reward–state distribution degrades to the initial state distribution $p_{S_0}(s)$ ($s \in S$). In some tasks, the initial observation O_0 are trivial (For example, the agent does not observe before the first action, or the initial observation is meaningless), we can ignore O_0 and exclude O_0 out of trajectory. In such case, we do not need the initial observation probability.

Example 15.4 (Tiger) The task "Tiger" is one of the most well-known POMDP tasks. The task is as follows: There are two doors. there are treasures behind one of the doors, while there is a fierce tiger behind another door. The two doors have equal probability to have either tiger or treasures. The agent can choose between three actions: open the left door, open the right door, or listen. If the agent chooses to open a door, it gets reward +10 if the door with treasures is opened, and the agent gets reward -100 if the door with tiger is opened, and the episode ends. If the agent chooses to listen, it has probability 85% to hear tiger's roar from the door with tiger, and probability 15% to hear tiger's roar from the door with treasures. The reward of the listening step is -1, and the episode continues so the agent can make another choice. This task can be modeled as a POMDP, which involves the following spaces:

- The reward space $\mathcal{R} = \{0, -1, +10, -100\}$, where 0 is only used for the initial reward R_0. If R_0 is discarded, the reward space degrades to $\mathcal{R} \in \{-1, +10, -100\}$.
- The state space $S = \{s_{\text{left}}, s_{\text{right}}\}$, and the state space with the terminal state is $S^+ = \{s_{\text{left}}, s_{\text{right}}, s_{\text{end}}\}$.
- The action space $\mathcal{A} = \{a_{\text{left}}, a_{\text{right}}, a_{\text{listen}}\}$.
- The observation space $O = \{o_{\text{init}}, o_{\text{left}}, o_{\text{right}}\}$, where o_{init} is only used for the initial observation O_0. If O_0 is discarded, the observation space is $O = \{o_{\text{left}}, o_{\text{right}}\}$.

Tables 15.1 and 15.2 show the initial probability, while Tables 15.3 and 15.4 show the dynamics.

15.5.2 Belief

In POMDP, the agent can not directly observe states. It can only infer states from observations. If the agent can also observe rewards during interaction, the observed

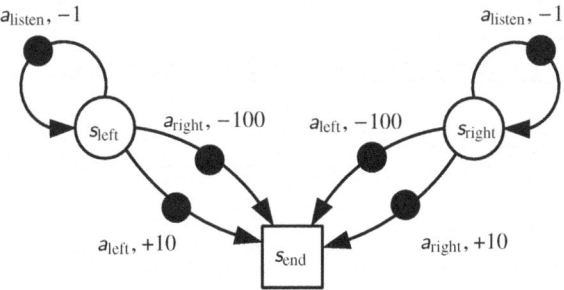

Fig. 15.1 MDP of the task "Tiger".
Observations are not illustrated in the figure.

Table 15.1 Initial probability in the task "Tiger". We can delete the reward row since the initial reward is trivial.

r	s	$p_{S_0,R_0}(s,r)$
0	s_{left}	0.5
0	s_{right}	0.5
others		0

Table 15.2 Initial emission probability in the task "Tiger". We can delete this table since the initial observation is trivial.

s'	o	$o_0(o\|s')$
s_{left}	o_{init}	1
s_{right}	o_{init}	1
others		0

Table 15.3 Dynamics in the task "Tiger". We can delete the reward column if rewards are not considered.

s	a	r	s'	$p(s',r\|s,a)$
s_{left}	a_{left}	+10	s_{end}	1
s_{left}	a_{right}	−100	s_{end}	1
s_{left}	a_{listen}	−1	s_{left}	1
s_{right}	a_{left}	−100	s_{end}	1
s_{right}	a_{right}	+10	s_{end}	1
s_{right}	a_{listen}	−1	s_{right}	1
others				0

Table 15.4   Observation probability in the task "Tiger".

a	s'	o	$o(o\|a, s')$
a_{listen}	s_{left}	o_{left}	0.85
a_{listen}	s_{left}	o_{right}	0.15
a_{listen}	s_{right}	o_{left}	0.15
a_{listen}	s_{right}	o_{right}	0.85
others			0

rewards are also treated as a form of observations and can be used to infer states. In some tasks, the agent can not obtain rewards in real-time (for example, the rewards can obtain only at the end of the episode), so the agent can not use reward information to guess states.

Belief $B_t \in \mathcal{B}$ is used to present the guess of S_t at time t, where $\mathcal{B}$ is called **belief space**. After the concept of belief is introduced, the trajectory of DTPOMDP maintained by the agent can be written as

$$R_0, O_0, B_0, A_0, R_1, O_1, B_1, A_1, R_2, \ldots,$$

where R_0 and O_0 can be omitted if they are trivial. For episodic tasks, the belief becomes terminal belief b_{end} when the state reaches the terminal state s_{end}. The belief space with the terminal belief is denoted as $\mathcal{B}^+ = \mathcal{B} \cup \{b_{\text{end}}\}$.

Figure 15.2 illustrates the trajectories maintain by the environment and the agent. There are three regions in this figure:

- the region only maintained by the environment, including the state S_t;
- the region only maintained by agents, including the belief B_t;
- the region maintained by both the environment and agents, including the observation O_t, the action A_t, and the reward R_t. Note that the reward is not available to agents during interactions, so agents can not use rewards to update belief.

Belief can be quantified by the conditional probability distribution over the state space $\mathcal{S}$. Let $b_t : \mathcal{S} \to \mathbb{R}$ be the conditional probability over the state space $\mathcal{S}$ at time t:

$$b_t(s) \stackrel{\text{def}}{=} \Pr[S_t = s|O_0, A_0, \ldots, O_{t-1}, A_{t-1}, R_t, O_t], \quad s \in \mathcal{S}.$$

Vector representation for this is $\mathbf{b}_t = \left(b_t(s) : s \in \mathcal{S}\right)^\top$. Besides, the terminal belief for episodic tasks is still $b_{\text{end}} \in \mathcal{B}^+\backslash\mathcal{B}$, which is an abstract element, not a vector.

! Note

The belief has some representation besides the condition probabilities over states. We will see an example of other representations in the latter part of this chapter. In

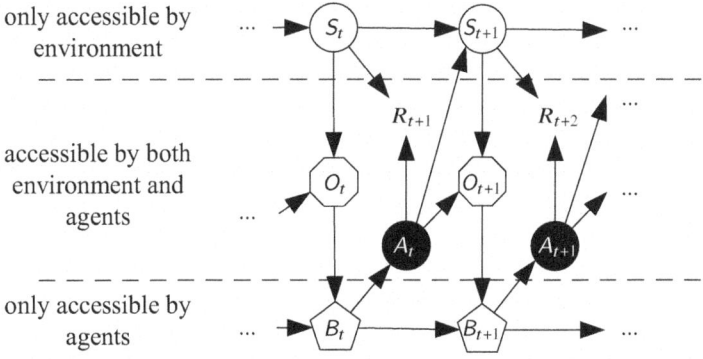

Fig. 15.2   Trajectories maintained by the environment and the agent.
"→" means probabilistic dependency.

this chapter, the belief in the form of conditional probability distribution is written in the serif typeface.

After we define the belief as the conditional probability distribution, we can further introduce the probability conditioned on belief and action, denoted as ω:

- The transition probability from belief–action pair to next observation–state pair:

$$\omega(s', o|b, a) \stackrel{\text{def}}{=} \Pr\left[S_{t+1} = s', O_{t+1} = o | B_t = b, A_t = a\right],$$
$$b \in \mathcal{B}, a \in \mathcal{A}, s' \in \mathcal{S}^+, o \in \mathcal{O}.$$

We can prove that

$$\omega(s', o|b, a)$$
$$\stackrel{\text{def}}{=} \Pr\left[S_{t+1} = s', O_{t+1} = o | B_t = b, A_t = a\right]$$
$$= \Pr\left[O_{t+1} = o | B_t = b, A_t = a, S_{t+1} = s'\right] \Pr\left[S_{t+1} = s' | B_t = b, A_t = a\right]$$
$$= \Pr\left[O_{t+1} = o | A_t = a, S_{t+1} = s'\right]$$
$$\qquad \sum_s \Pr\left[S_t = s, S_{t+1} = s' | B_t = b, A_t = a\right]$$
$$= \Pr\left[O_{t+1} = o | A_t = a, S_{t+1} = s'\right]$$
$$\qquad \sum_s \Pr\left[S_{t+1} = s' | S_t = s, B_t = b, A_t = a\right] \Pr\left[S_t = s | B_t = b, A_t = a\right]$$
$$= o(o|a, s') \sum_s p(s'|s, a) b(s), \qquad b \in \mathcal{B}, a \in \mathcal{A}, s' \in \mathcal{S}^+, o \in \mathcal{O}.$$

- The emission probability from belief–action pair to next observation:

$$\omega(o|b, a) \stackrel{\text{def}}{=} \Pr\left[O_{t+1} = o | B_t = b, A_t = a\right], \qquad b \in \mathcal{B}, a \in \mathcal{A}, o \in O.$$

We can prove that

$$
\begin{aligned}
\omega(o|b, a) &\stackrel{\text{def}}{=} \Pr\left[O_{t+1} = o | B_t = b, A_t = a\right] \\
&= \sum_{s'} \Pr\left[S_{t+1} = s', O_{t+1} = o | B_t = b, A_t = a\right] \\
&= \sum_{s'} \omega(s', o|b, a), \qquad b \in \mathcal{B}, a \in \mathcal{A}, o \in O.
\end{aligned}
$$

We can know that the next belief B_{t+1} is fully determined by the belief B_t, action A_t, and next observation O_{t+1}. Therefore, we can define the belief update operator $\mathfrak{u} : \mathcal{B} \times \mathcal{A} \times O \to \mathcal{B}$ as $b' = \mathfrak{u}(b, a, o)$, where

$$
\begin{aligned}
\mathfrak{u}&(b, a, o)(s') \\
&= \Pr\left[S_{t+1} = s' | B_t = b, A_t = a, O_{t+1} = o\right] \\
&= \frac{\Pr\left[S_{t+1} = s', O_{t+1} = o | B_t = b, A_t = a\right]}{\Pr\left[O_{t+1} = o | B_t = b, A_t = a\right]} \\
&= \frac{\omega(o|a, s')\sum_s p(s'|s, a)b(s)}{\sum_{s''} \omega(o|a, s'')\sum_s p(s''|s, a)b(s)}, \qquad b \in \mathcal{B}, a \in \mathcal{A}, o \in O, s' \in S^+.
\end{aligned}
$$

Example 15.5 (Tiger) We consider the task "Tiger", where the state space $S = \{s_{\text{left}}, s_{\text{right}}\}$ has two elements, so the belief $b(s)$ ($s \in S$) can be denoted as a vector of length 2, where the two elements are the conditional probability of s_{left} and s_{right} respectively. We can write the belief as $\left(b_{\text{left}}, b_{\text{right}}\right)^{\mathrm{T}}$ for short when there is no confusion. We can get Tables 15.5, 15.6, and the belief update operator (Table 15.7) from the dynamics.

Table 15.5 Conditional probability $\omega(s', o|b, a)$ in the task "Tiger".

b	a	s'	o	$\omega(s', o|b, a)$
$\left(b_{\text{left}}, b_{\text{right}}\right)^{\mathrm{T}}$	a_{listen}	s_{left}	o_{left}	$0.85 b_{\text{left}}$
$\left(b_{\text{left}}, b_{\text{right}}\right)^{\mathrm{T}}$	a_{listen}	s_{left}	o_{right}	$0.15 b_{\text{right}}$
$\left(b_{\text{left}}, b_{\text{right}}\right)^{\mathrm{T}}$	a_{listen}	s_{right}	o_{left}	$0.15 b_{\text{left}}$
$\left(b_{\text{left}}, b_{\text{right}}\right)^{\mathrm{T}}$	a_{listen}	s_{right}	o_{right}	$0.85 b_{\text{right}}$
others				0

Table 15.6   Conditional probability $\omega(o|b, a)$ in the task "Tiger".

b	a	o	$\omega(o|b, a)$
$\left(b_{\text{left}}, b_{\text{right}}\right)^{\top}$	a_{listen}	o_{left}	$0.85b_{\text{left}} + 0.15b_{\text{right}}$
$\left(b_{\text{left}}, b_{\text{right}}\right)^{\top}$	a_{listen}	o_{right}	$0.15b_{\text{left}} + 0.85b_{\text{right}}$
	others		0

Table 15.7   Belief updates in the task "Tiger".

b	a	o	$\mathrm{u}(b, a, o)$
$\begin{bmatrix} b_{\text{left}} \\ b_{\text{right}} \end{bmatrix}$	a_{listen}	o_{left}	$\frac{1}{0.85b_{\text{left}}+0.15b_{\text{right}}}\begin{bmatrix} 0.85b_{\text{left}} \\ 0.15b_{\text{right}} \end{bmatrix}$
$\begin{bmatrix} b_{\text{left}} \\ b_{\text{right}} \end{bmatrix}$	a_{listen}	o_{right}	$\frac{1}{0.15b_{\text{left}}+0.85b_{\text{right}}}\begin{bmatrix} 0.15b_{\text{left}} \\ 0.85b_{\text{right}} \end{bmatrix}$

15.5.3 Belief MDP

Consider the trajectory maintained by the agent. We can treat the beliefs as states of the agent, and than the decision process maintained by the agent is an MDP with beliefs as states. This MDP is called **belief MDP**. The states in a belief MDP are called belief states.

We can derive the initial probability and the dynamic of belief MDP from the initial probability and the dynamics of the original POMDP:

- Initial belief state distribution $p_{B_0}(b)$: This is a single-point distribution on the support set b_0:

$$
\begin{aligned}
b_0(s) &\stackrel{\text{def}}{=} \Pr\left[S_0 = s | O_0 = o\right] \\
&= \frac{\Pr\left[S_0 = s, O_0 = o\right]}{\Pr\left[O_0 = o\right]} \\
&= \frac{\Pr\left[S_0 = s, O_0 = o\right]}{\sum_{s''}\Pr\left[S_0 = s'', O_0 = o\right]} \\
&= \frac{\Pr\left[O_0 = o | S_0 = s\right]\Pr\left[S_0 = s\right]}{\sum_{s''}\Pr\left[O_0 = o | S_0 = s''\right]\Pr\left[S_0 = s''\right]} \\
&= \frac{o_0(o|s)p_{S_0}(s)}{\sum_{s''}o_0(o|s'')p_{S_0}(s'')}.
\end{aligned}
$$

If the initial observation O_0 is trivial, the initial state distribution b_0 equals the distribution of the initial state S_0. Otherwise, we can get the initial information from O_0.

- Dynamics: Use the transition probability

$$\Pr\left[B_{t+1} = b' \middle| B_t = b, A_t = a\right]$$

$$= \sum_{o \in O} \Pr\left[B_{t+1} = b', O_{t+1} = o \middle| B_t = b, A_t = a\right]$$

$$= \sum_{o \in O} \Pr\left[B_{t+1} = b' \middle| B_t = b, A_t = a, O_{t+1} = o\right]$$

$$\Pr\left[O_{t+1} = o \middle| B_t = b, A_t = a\right]$$

$$= \sum_{o \in O} 1_{\left[b' = \mathrm{u}(b,a,o)\right]} \omega(o|b, a)$$

$$= \sum_{o \in O : b' = \mathrm{u}(b,a,o)} \omega(o|b, a), \quad b \in \mathcal{B}, a \in \mathcal{A}, b' \in \mathcal{B}^+.$$

This can be also written as

$$\Pr\left[B_{t+1} = b' \middle| B_t = b, A_t = a\right]$$

$$= \sum_{o \in O : b' = \mathrm{u}(b,a,o)} \sum_{s'} o(o|a, s') \sum_{s} p(s'|s, a) b(s),$$

$$b \in \mathcal{B}, a \in \mathcal{A}, b' \in \mathcal{B}^+.$$

We also can get quantities such as

$$r(b, a) \overset{\text{def}}{=} \mathrm{E}_\pi[R_{t+1}|B_t = b, A_t = a] = \sum_{s} r(s, a) b(s), \quad b \in \mathcal{B}, a \in \mathcal{A}.$$

Example 15.6 (Tiger) The task "Tiger" has Table 15.8.

Table 15.8 $r(b, a)$ **in the task "Tiger".**

b	a	$r(b, a)$
$\left(b_{\text{left}}, b_{\text{right}}\right)^\top$	a_{left}	$+10 b_{\text{left}} - 100 b_{\text{right}}$
$\left(b_{\text{left}}, b_{\text{right}}\right)^\top$	a_{right}	$-100 b_{\text{left}} + 10 b_{\text{right}}$
$\left(b_{\text{left}}, b_{\text{right}}\right)^\top$	a_{listen}	-1

It can be proved that, if the agent has observed c_{left} observation o_{left} and c_{right} observation o_{right} from start of an episode to a particular time in the episode, the belief is $\frac{1}{0.85^{\Delta c} + 0.15^{\Delta c}} \begin{pmatrix} 0.85^{\Delta c} \\ 0.15^{\Delta c} \end{pmatrix}$, where $\Delta c = c_{\text{left}} - c_{\text{right}}$. (Proof: We can use mathematical induction. In the beginning, $c_{\text{left}} = c_{\text{right}} = 0$. The initial belief $(0.5, 0.5)^\top$ satisfies the assumption. Now we assume that the belief becomes $\begin{pmatrix} b_{\text{left}} \\ b_{\text{right}} \end{pmatrix} = \frac{1}{0.85^{\Delta c} + 0.15^{\Delta c}} \begin{pmatrix} 0.85^{\Delta c} \\ 0.15^{\Delta c} \end{pmatrix}$ at a step. If the next observation is o_{left}, the belief is updated to

$$\frac{1}{0.85b_{\text{left}} + 0.15b_{\text{right}}} \begin{pmatrix} 0.85b_{\text{left}} \\ 0.15b_{\text{right}} \end{pmatrix}$$

$$= \frac{1}{0.85 \times 0.85^{\triangle c} + 0.15 \times 0.15^{\triangle c}} \begin{pmatrix} 0.85 \times 0.85^{\triangle c} \\ 0.15 \times 0.15^{\triangle c} \end{pmatrix}$$

$$= \frac{1}{0.85^{\triangle c+1} + 0.15^{\triangle c+1}} \begin{pmatrix} 0.85^{\triangle c+1} \\ 0.15^{\triangle c+1} \end{pmatrix}$$

which satisfies the assumption. If the next observation is o_{right}, the belief is updated to

$$\frac{1}{0.85b_{\text{left}} + 0.15b_{\text{right}}} \begin{pmatrix} 0.85b_{\text{left}} \\ 0.15b_{\text{right}} \end{pmatrix}$$

$$= \frac{1}{0.15 \times 0.85^{\triangle c} + 0.85 \times 0.15^{\triangle c}} \begin{pmatrix} 0.15 \times 0.85^{\triangle c} \\ 0.85 \times 0.15^{\triangle c} \end{pmatrix}$$

$$= \frac{1}{0.85^{\triangle c-1} + 0.15^{\triangle c-1}} \begin{pmatrix} 0.85^{\triangle c-1} \\ 0.15^{\triangle c-1} \end{pmatrix}$$

which satisfies the assumption, too. The proof completes.)

In the task "Tiger", the guess of agent on states can be fully specified by $\triangle c$. Therefore, we can also use $\triangle c$ to quantify beliefs. Using this way, the belief is an integer, and the belief space $\mathcal{B}$ is the integer set. Such belief MDP is illustrated in Fig. 15.3.

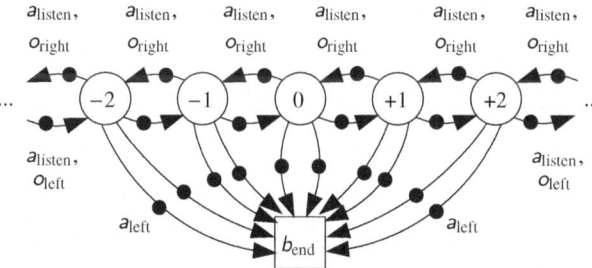

Fig. 15.3 Belief MDP of the task "Tiger".
The number in the circle of belief state is $\triangle c$. Rewards are not depicted in the figure.

In belief MDP, the agent chooses action according to the belief. Therefore, the policy of the agent is $\pi : \mathcal{B} \times \mathcal{A} \to \mathbb{R}$:

$$\pi(a|b) \overset{\text{def}}{=} \Pr[A_t = a|B_t = b], \quad b \in \mathcal{B}, a \in \mathcal{A}.$$

15.5.4 Belief Values

Belief MDP has its own state values and action values. Taking the setting of discounted returns as an example, given a policy $\pi : \mathcal{B} \rightarrow \mathcal{A}$, we can define discounted return belief values as follows:

- Belief state values $v_\pi(b)$:

$$v_\pi(b) \overset{\text{def}}{=} \mathrm{E}_\pi[G_t|B_t = b], \quad b \in \mathcal{B}.$$

- Belief action values $q_\pi(b, a)$:

$$q_\pi(b, a) \overset{\text{def}}{=} \mathrm{E}_\pi[G_t|B_t = b], \quad b \in \mathcal{B}, a \in \mathcal{A}.$$

Since the policy π is also the policy of the original POMDP, the belief state values and belief action values here can be also viewed as the belief state values and belief action values of the original POMDP.

The relationship between belief values includes:

- Use the belief action values at time t to back up the belief state values at time t:

$$v_\pi(b) = \sum_a \pi(a|b) q_\pi(b, a), \quad b \in \mathcal{B}.$$

(Proof:

$$
\begin{aligned}
v_\pi(b) &= \mathrm{E}_\pi[G_t|B_t = b] \\
&= \sum_g g \Pr[G_t = g|B_t = b] \\
&= \sum_g g \sum_a \Pr[G_t = g, A_t = a|B_t = b] \\
&= \sum_g g \sum_a \Pr[A_t = a|B_t = b] \Pr[G_t = g|B_t = b, A_t = a] \\
&= \sum_a \Pr[A_t = a|B_t = b] \sum_g g \Pr[G_t = g|B_t = b, A_t = a] \\
&= \sum_a \Pr[A_t = a|B_t = b] \mathrm{E}_\pi[G_t|B_t = b, A_t = a] \\
&= \sum_a \pi(a|b) q_\pi(b, a).
\end{aligned}
$$

The proof completes.)

- Use the belief state values at time $t+1$ to back up the belief action values at time t:

$$q_\pi(b, a) = r(b, a) + \gamma \sum_o \omega(o|b, a) v_\pi(\mathfrak{u}(b, a, o)), \quad b \in \mathcal{B}, a \in \mathcal{A}.$$

(Proof:

$$\mathrm{E}_\pi[G_{t+1}|B_t = b, A_t = a]$$
$$= \sum_g g \Pr[G_{t+1} = g|B_t = b, A_t = a]$$
$$= \sum_g g \sum_{o,b'} \Pr[B_{t+1} = b', O_{t+1} = o, G_{t+1} = g|B_t = b, A_t = a]$$
$$= \sum_g g \sum_{o,b'} \Pr[B_{t+1} = b'|B_t = b, A_t = a, O_{t+1} = o]$$
$$\Pr[O_{t+1} = o|B_t = b, A_t = a]$$
$$\Pr[G_{t+1} = g|B_t = b, A_t = a, B_{t+1} = b', O_{t+1} = o]$$
$$= \sum_{o,b'} \Pr[B_{t+1} = b'|B_t = b, A_t = a, O_{t+1} = o]$$
$$\Pr[O_{t+1} = o|B_t = b, A_t = a]$$
$$\sum_g g \Pr[G_{t+1} = g|B_t = b, A_t = a, B_{t+1} = b', O_{t+1} = o]$$
$$= \sum_{o,b'} 1_{[b'=\mathfrak{u}(b,a,o)]} \Pr[O_{t+1} = o|B_t = b, A_t = a]$$
$$\sum_g g \Pr[G_{t+1} = g|B_t = b, A_t = a, B_{t+1} = b', O_{t+1} = o].$$

Noticing $\Pr[G_{t+1} = g|B_t = b, A_t = a, B_{t+1} = b', O_{t+1} = o] = \Pr[G_{t+1} = g|B_{t+1} = b']$ and the summation is nonzero only at $b' = \mathfrak{u}(b, a, o)$, we have

$$\mathrm{E}_\pi[G_{t+1}|B_t = b, A_t = a]$$
$$= \sum_o \Pr[O_{t+1} = o|B_t = b, A_t = a] \sum_g g \Pr[G_{t+1} = g|B_{t+1} = \mathfrak{u}(b, a, o)]$$
$$= \sum_o \omega(o|b, a) v_\pi(\mathfrak{u}(b, a, o)).$$

Furthermore,

$$q_\pi(b, a) = \mathrm{E}_\pi[R_{t+1}|B_t = b, A_t = a] + \gamma \mathrm{E}_\pi[G_{t+1}|B_t = b, A_t = a]$$
$$= r(b, a) + \gamma \sum_o \omega(o|b, a) v_\pi(\mathfrak{u}(b, a, o)).$$

The proof completes.)

- Use the belief state values at time $t + 1$ to back up the belief state values at time t:

$$v_\pi(b) = \sum_a \pi(a|b)\left[r(b,a) + \gamma\sum_o \omega(o|b,a)v_\pi\big(\mathrm{u}(b,a,o)\big)\right], \quad b \in \mathcal{B}.$$

- Use the belief action values at time $t + 1$ to back up the belief action values at time t:

$$q_\pi(b,a) = r(b,a) + \gamma\sum_o \omega(o|b,a)\sum_{a'} \pi\big(a'|\mathrm{u}(b,a,o)\big)q_\pi\big(\mathrm{u}(b,a,o),a'\big),$$

$$b \in \mathcal{B}, a \in \mathcal{A}.$$

We can further define optimal values and optimal policy.

- Optimal belief state values:

$$v_*(b) \stackrel{\text{def}}{=} \sup_\pi v_\pi(b), \quad b \in \mathcal{B}.$$

- Optimal belief action values:

$$q_*(b,a) \stackrel{\text{def}}{=} \sup_\pi q_\pi(b,a), \quad b \in \mathcal{B}, a \in \mathcal{A}.$$

For simplicity, we always assume that the optimal values exist.

Optimal policy is defined as

$$\pi_*(b) \stackrel{\text{def}}{=} \arg\max_a q_*(b,a), \quad b \in \mathcal{B}.$$

For simplicity, we always assume that the optimal policy exists.

The relationship among optimal values includes:

- Use the optimal action values at time t to back up the optimal state values at time t:

$$v_*(b) = \max_a q_*(b,a), \quad b \in \mathcal{B}.$$

- Use the optimal state values at time $t + 1$ to back up the optimal action values at time t:

$$q_*(b,a) = r(b,a) + \gamma\sum_o \omega(o|b,a)v_*\big(\mathrm{u}(b,a,o)\big), \quad b \in \mathcal{B}, a \in \mathcal{A}.$$

- Use the optimal state values at time $t + 1$ to back up the optimal state values at time t:

$$v_*(b) = \max_a \left[r(b, a) + \gamma \sum_o \omega(o|b, a) v_*\big(\mathfrak{u}(b, a, o)\big) \right], \quad b \in \mathcal{B}.$$

- Use the optimal action values at time $t + 1$ to back up the optimal action values at time t:

$$q_*(b, a) = r(b, a) + \gamma \sum_o \omega(o|b, a) \max_{a'} q_*\big(\mathfrak{u}(b, a, o), a'\big), \quad b \in \mathcal{B}, a \in \mathcal{A}.$$

Example 15.7 (Discounted Tiger) Introducing the discount factor $\gamma = 1$ into the task "Tiger", we can get the optimal belief values and optimal policy of this discounted POMDP using VI (see Table 15.9). VI will be implemented in Sect. 15.6.2.

Table 15.9 Discounted optimal values and optimal policy of the task "Tiger".

belief $\triangle c$	belief b	optimal belief action values $q_*(b, a_{\text{left}})$	optimal belief action values $q_*\big(b, a_{\text{right}}\big)$	optimal belief action values $q_*(b, a_{\text{listen}})$	optimal belief state values $v_*(b)$	optimal policy $\pi_*(b)$
...	...	...	...	...	...	...
-3	$(\approx 0.01, \approx 0.99)$	≈ -99.40	≈ 9.40	≈ 8.58	≈ 9.40	a_{right}
-2	$(\approx 0.03, \approx 0.97)$	≈ -96.68	≈ 6.68	≈ 7.84	≈ 7.84	a_{listen}
-1	$(0.15, 0.85)$	-83.5	-6.5	≈ 6.15	≈ 6.16	a_{listen}
0	$(0.5, 0.5)$	-45	-45	≈ 5.16	≈ 5.16	a_{listen}
1	$(0.85, 0.15)$	-6.5	-83.5	≈ 6.16	≈ 6.16	a_{listen}
2	$(\approx 0.97, \approx 0.03)$	≈ 6.68	≈ -96.68	≈ 7.84	≈ 7.84	a_{listen}
3	$(\approx 0.99, \approx 0.01)$	≈ 9.40	≈ -99.40	≈ 8.58	≈ 9.40	a_{left}
...	...	...	...	...	...	...

Table 15.9 tells us that an agent with the optimal policy will not reach the belief with $|\triangle c| > 3$. Therefore, we can even set the belief space to a set with only 7 elements, each of which represents $\triangle c$ as one of $\{\leq -3, -2, -1, 0, 1, 2, \geq 3\}$. Such an agent is possible to find the optimal policy, too. This is an example of finite belief space.

15.5.5 Belief Values for Finite POMDP

This section discusses the finite POMDP with a fixed number step of episodes. The state is of the form of $x_t \in \mathcal{X}_t$ in Sect. 15.3.1, and the number of steps in an episode is denoted as $t_{\max}$. We will use VI to find optimal values.

The VI algorithm in Sect. 15.3.2 stores value estimates for each state. Unfortunately, the belief space usually has an infinite number of elements, so it is impossible to store value estimates for each belief state. Fortunately, when the state space, action space, and observation are all finite, the optimal belief state values can be represented by a finite number of hyperplanes. This can significantly simplify the representation of optimal belief state values.

Mathematically, at time t, there exists a set $\mathcal{L} = \{\alpha_l : 0 \le l < |\mathcal{L}_t|\}$, where each element is a $|\mathcal{X}_t|$-dim hyperplane, such that the optimal belief state values can be represented as (Sondik, 1971)

$$v_{*,t}(b) = \max_{\alpha \in \mathcal{L}_t} \sum_{x \in \mathcal{X}_t} \alpha(x)b(x), \quad b \in \mathcal{B}_t.$$

Since the hyperplane can be denoted by a vector, each α_l is called α-vector. (Proof: We can use mathematical induction to prove this result. Let $t_{\max}$ be the number of steps in a POMDP. At time $t_{\max}$, we have

$$v_{*,t_{\max}}(b_{\text{end}}) = 0.$$

This can be viewed as the set $\mathcal{L}_{t_{\max}}$ has only a single all-zero vector. So the assumption holds. Now we assume that at time $t + 1$, $v_{*,t+1}$ can be represented as

$$v_{*,t+1}(b) = \max_{\alpha \in \mathcal{L}_{t+1}} \sum_{x \in \mathcal{X}_{t+1}} \alpha(x)b(x), \quad b \in \mathcal{B}_{t+1}$$

where the set $\mathcal{L}_{t+1}$ has $|\mathcal{L}_{t+1}|$ number of α-vectors. Now we try to find a set $\mathcal{L}_t$ such that $v_{*,t}(b)$ can satisfy the assumption. In order to find it, we consider the Bellman equation:

$$v_{*,t}(b) = \max_{a \in \mathcal{A}_t} \left[r_t(b, a) + \sum_{o \in O_t} \omega_t(o|b, a)v_{*,t+1}\big(\mathfrak{u}(b, a, o)\big) \right], \quad b \in \mathcal{B}_t.$$

The outmost operation is $\max_{a \in \mathcal{A}_t}$, so we need to find $\mathcal{L}_t(a)$ for each action $a \in \mathcal{A}_t$, such that $\mathcal{L}_t(a)$ is a subset of $\mathcal{L}_t$ and

$$\mathcal{L}_t = \bigcup_{a \in \mathcal{A}_t} \mathcal{L}_t(a).$$

In order to find $\mathcal{L}_t(a)$, we notice that each element of max operation is a summation of $1+|O_t|$ elements. Therefore, we need to find a set of α-vectors for each component. For example, we can define the part corresponding to $r(b, a)$ as $\mathcal{L}_r(a)$, and the part for each observation $o \in O_t$ is $\mathcal{L}_t(a, o)$. So $\mathcal{L}_t(a)$ can be represented as the summation of their elements, i.e.

$$\mathcal{L}_t(a) = \mathcal{L}_r(a) + \sum_{o \in O_t} \mathcal{L}_t(a, o)$$

$$= \left\{ \alpha_r + \sum_{o \in O_t} \alpha_o : \alpha_r \in \mathcal{L}_r(a), \alpha_o \in \mathcal{L}_t(a, o) \text{ for } \forall o \in O_t \right\}.$$

Then we investigate $\mathcal{L}_r(a)$ and $\mathcal{L}_t(a, o)$. Obviously, the set $\mathcal{L}_r(a)$ has only one element $(r_t(x, a) : x \in \mathcal{X}_t)^\top$, and the set $\mathcal{L}_t(a, o)$ has $|\mathcal{L}_{t+1}|$ elements. The latter is because, for each $\alpha_{t+1} \in \mathcal{L}_{t+1}$, we can construct a α-vector:

$$\left(\gamma \sum_{x'} o_t\left(o|a, x'\right) p_t\left(x'|x, a\right) \alpha_{t+1}\left(x'\right) : x \in \mathcal{X}_t \right)^\top$$

Therefore,

$$\mathcal{L}_t(a, o) = \left\{ \left(\gamma \sum_{x'} o_t\left(o|a, x'\right) p_t\left(x'|x, a\right) \alpha_{t+1}\left(x'\right) : x \in \mathcal{X}_t \right)^\top : \alpha_{t+1} \in \mathcal{L}_{t+1} \right\}.$$

Now we get $\mathcal{L}_r(a)$ and $\mathcal{L}_t(a, o)$. $\mathcal{L}_r(a)$ has only one element, while $\mathcal{L}_t(a, o)$ has $|\mathcal{L}_{t+1}|$ elements. Then we get $\mathcal{L}_t(a)$, which has $|\mathcal{L}_{t+1}|^{|O_t|}$ elements. We can further get $\mathcal{L}_t$, which has $|\mathcal{A}_t|$ elements. We can verify that $\mathcal{L}_t$ satisfies the induction assumption. The proof completes.)

The mathematic induction proof tells us two conclusions:

- We can use the following way to calculate $\mathcal{L}_t$:

$$\mathcal{L}_r(a) \leftarrow \left\{ (r_t(x, a) : x \in \mathcal{X}_t)^\top \right\}, \qquad\qquad\qquad a \in \mathcal{A}_t$$

$$\mathcal{L}_t(a, o) \leftarrow \left\{ \left(\gamma \sum_{x'} o_t\left(o|a, x'\right) p_t\left(x'|x, a\right) \alpha_{t+1}\left(x'\right) : x \in \mathcal{X}_t \right)^\top : \alpha_{t+1} \in \mathcal{L}_{t+1} \right\},$$

$$\qquad\qquad\qquad\qquad\qquad\qquad a \in \mathcal{A}_t, o \in O_t$$

$$\mathcal{L}_t(a) \leftarrow \mathcal{L}_r(a) + \sum_{o \in O_t} \mathcal{L}_t(a, o), \qquad\qquad\qquad a \in \mathcal{A}_t$$

$$\mathcal{L}_t \leftarrow \bigcup_{a \in \mathcal{A}_t} \mathcal{L}_t(a).$$

- The number of hyperplanes satisfies

$$|\mathcal{L}_t| = 1, \qquad\qquad t = t_{\max},$$
$$|\mathcal{L}_t| = |\mathcal{A}_t||\mathcal{L}_{t+1}|^{|O_t|}, \quad t < t_{\max}.$$

Removing the recursion leads to

$$|\mathcal{L}_t| = \prod_{\tau=t}^{t_{\max}-1} |\mathcal{A}_\tau|^{\prod_{\tau'=t}^{\tau-1}|O_{\tau'}|}, \quad t = 0, 1, \ldots, t_{\max} - 1.$$

Especially, if the state space, action space, observation space do not change over time $t < t_{\max}$, we can omit the subscript in the above result, and the number of elements degrades to

$$|\mathcal{L}_t| = |\mathcal{A}|^{\frac{|O|^{t_{max}-t}-1}{|O|-1}}, \quad t = 0, 1, \ldots, t_{max} - 1.$$

The number of hyperplanes is quite large. For most of tasks, it is too complex to calculate all hyperplanes. Therefore, we may have to calculate fewer α-vectors. If we choose to calculate fewer α-vectors, the values are merely lower bounds on the belief state values, rather than the optimal belief state values themselves.

Sample-based algorithms are such type of algorithms. These algorithms sample some beliefs, and only consider values at those belief samples. Specifically, **Point-Based Value Iteration** (**PBVI**) is one of such algorithms (Joelle, 2003). This algorithm first samples a set of belief, say $\mathcal{B}_t$, where $\mathcal{B}_t$ is a finite set of samples, and then estimate an α-vector for every belief in this set. The method of estimation is as follows:

$$\alpha_t(a, o, \alpha_{t+1}) \leftarrow \left(\gamma \sum_{x'} o_t\big(o\big|a, x'\big) p_t\big(x'\big|x, a\big) \alpha_{t+1}\big(x'\big) : x \in X_t \right)^\top,$$

$$a \in \mathcal{A}_t, o \in O_t, \alpha_{t+1} \in \mathcal{L}_{t+1},$$

$$\alpha_t(b, a) \leftarrow \big(r_t(x, a) : x \in X_t \big)^\top + \gamma \sum_{o \in O_t} \underset{\alpha_{t+1} \in \mathcal{L}_{t+1}}{\arg\max} \sum_{x \in X_t} \alpha_t(x; a, o, \alpha_{t+1}) b(x),$$

$$b \in \mathcal{B}_t, a \in \mathcal{A}_t,$$

$$\alpha_t(b) \leftarrow \underset{a \in \mathcal{A}_t}{\arg\max} \sum_{x \in X_t} \alpha_t(x; b, a) b(x), \quad b \in \mathcal{B}_t.$$

Such belief values are the lower bounds of the optimal belief state values.

We can use some other methods to get the upper bound of the optimal belief values. A possible way is as follows: We can modify the observation probability of the POMDP, i.e. $o_t(o|a, x')$, such that we can always recover the state X_t from the observation O_t for all time t. Then the optimal values of the resulting MDP will be not less than the optimal belief values of the original POMDP. This is obvious since the agent can guess the states better using the modified observation, make smarter decisions, and get larger rewards. Therefore, we can find the optimal values for the modified MDP, and then the found optimal values will be the upper bounds of the optimal belief values of the POMDP. Since modified MDP is merely a normal MDP, we use the algorithms in Sect. 15.3.2 to find its optimal values. Therefore, we can get the upper bound on some belief values.

15.5.6 Use Memory

This section uses memory to solve POMDP.

Recapping the task "Pong" in Sect. 12.6, we notice that the information of an individual frame can not determine the momentum of birds and pads. In order to recover the state, the wrapper class `AtariPreprocessing` runs 4 frames each time, and stack 4 frames together into the neural network to get the momentum information. Unfortunately, this solution is only able to process historic observations of finite length, and can not make use of arbitrary length of historic observations to estimate states.

Deep learning, especially its applications in Natural Language Processing, has developed many methods to process sequential information and store them, such as Recurrent Neural Networks and attention. These methods can be combined into existing algorithms for MDP.

For example, **Deep Recurrent Q Network (DRQN)** algorithm combines the Long Short-Term Memory (LSTM) and DQN, where LSTM is used to extract current states from historical observations.

15.6 Case Study: Tiger

This section considers the task "Tiger". The specification of this task has been introduced in Sect. 15.5.

15.6.1 Compare Discounted Return Expectation and Average Reward

The task "Tiger" is an episodic task. When the agent chooses action a_{left} or a_{right}, the episode ends. For such an episode, we usually use return expectation with discount factor $\gamma = 1$ to evaluate the performance of a policy.

In the task "Tiger", the step number in an episode is random. It can be any positive integer. If we take the length of episodes into consideration, and want to get more positive rewards in a shorter time, we may convert the episodic task into a sequential task by concatenating episodic tasks, and consider the average reward of the sequential task.

Code 15.1 implements the environment. The constructor of the class TigerEnv has a parameter episodic. We can deploy an episodic task by setting episodic as True, and deploy a sequential task by setting episodic to False.

Code 15.1 The environment class TigerEnv for the task "Tiger".
Tiger-v0_ClosedForm.ipynb

```
1   class Observation:
2       LEFT, RIGHT, START = range(3)
3
4   class Action:
5       LEFT, RIGHT, LISTEN = range(3)
6
7
8   class TigerEnv(gym.Env):
9
10      def __init__(self, episodic=True):
11          self.action_space = spaces.Discrete(3)
12          self.observation_space = spaces.Discrete(2)
13          self.episodic = episodic
```

```
14
15    def reset(self, *, seed=None, options=None):
16        super().reset(seed=seed)
17        self.state = np.random.choice(2)
18        return Observation.START, {}  # placebo observation
19
20    def step(self, action):
21        if action == Action.LISTEN:
22            if np.random.rand() > 0.85:
23                observation = 1 - self.state
24            else:
25                observation = self.state
26            reward = -1
27            terminated = False
28        else:
29            observation = self.state
30            if action == self.state:
31                reward = 10.
32            else:
33                reward = -100.
34            if self.episodic:
35                terminated = True
36            else:
37                terminated = False
38                observation = self.reset()
39        return observation, reward, terminated, False, {}
```

Code 15.2 registers of environments into Gym. Two versions of environments are registered: One is `TigerEnv-v0`, the episodic version. Another is `TigerEnv200-v0`, which mimics the sequential case by setting the maximum steps of the episode to a very large number 200. We can get the average reward by dividing the total reward by the number of steps.

Code 15.2   Register the environment class `TigerEnv`.

`Tiger-v0_ClosedForm.ipynb`

```
1    from gym.envs.registration import register
2    register(id="Tiger-v0", entry_point=TigerEnv, kwargs={"episodic": True})
3    register(id="Tiger200-v0", entry_point=TigerEnv, kwargs={"episodic": False},
4             max_episode_steps=200)
```

Section 15.5.5 shows the optimal policy that maximizes the return expectation with discount factor $\gamma = 1$. Code 15.3 implements this optimal policy. Using this policy to interact with the environment, the discounted return expectation is about 5, and the average reward is about 1.

Code 15.3   Optimal policy when discount factor $\gamma = 1$.

`Tiger-v0_ClosedForm.ipynb`

```
1    class Agent:
2        def __init__(self, env=None):
3            pass
4
5        def reset(self, mode=None):
6            self.count = 0
7
8        def step(self, observation, reward, terminated):
9            if observation == Observation.LEFT:
10               self.count += 1
11           elif observation == Observation.RIGHT:
12               self.count -= 1
```

```
13        else:  # observation == Observation.START
14            self.count = 0
15
16        if self.count > 2:
17            action = Action.LEFT
18        elif self.count < -2:
19            action = Action.RIGHT
20        else:
21            action = Action.LISTEN
22        return action
23
24    def close(self):
25        pass
26
27
28 agent = Agent(env)
```

15.6.2 Belief MDP

Section 15.5.3 introduces how to construct the belief MDP with finite belief space for the task "Tiger". This section will use VI to find optimal state values for the belief MDP.

Code 15.4 implements the VI algorithm. The codes use a pd.DataFrame object to maintain all results. The pd.DataFrame object is indexed by $\triangle c$, whose values can be $-4, -3, \ldots, 3, 4$. We do not enumerate all integers here, which is impossible either. The results of iterations show that such a choice does not degrade the performance. Then, we calculate $\left(b(s_{\text{left}}), b(s_{\text{right}})\right)^{\mathsf{T}}$, $\left(o(o_{\text{left}}|b, a_{\text{listen}}), o(o_{\text{right}}|b, a_{\text{listen}})\right)^{\mathsf{T}}$, and $\left(r(b, a_{\text{left}}), r(b, a_{\text{right}}), r(b, a_{\text{listen}})\right)^{\mathsf{T}}$ for each $\triangle c$. Then we calculate optimal value estimates iteratively. At last, we use the optimal value estimates to calculate optimal policy estimates.

Code 15.4   Belief VI.

Tiger-v0_Plan_demo.ipynb

```
1  discount = 1.
2
3  df = pd.DataFrame(0., index=range(-4, 5), columns=[])
4  df["h(left)"] = 0.85 ** df.index.to_series()  # preference for S = left
5  df["h(right)"] = 0.15 ** df.index.to_series()  # preference for S = right
6  df["p(left)"] = df["h(left)"] / (df["h(left)"] + df["h(right)"])  # b(left)
7  df["p(right)"] = df["h(right)"] / (df["h(left)"] + df["h(right)"])  # b(right)
8  df["omega(left)"] = 0.85 * df["p(left)"] + 0.15 * df["p(right)"]
9          # omega(left|b, listen)
10 df["omega(right)"] = 0.15 * df["p(left)"] + 0.85 * df["p(right)"]
11         # omega(right|b, listen)
12 df["r(left)"] = 10. * df["p(left)"] - 100. * df["p(right)"]  # r(b, left)
13 df["r(right)"] = -100. * df["p(left)"] + 10. * df["p(right)"]  # r(b, right)
14 df["r(listen)"] = -1.  # r(b, listen)
15
16 df[["q(left)", "q(right)", "q(listen)", "v"]] = 0.  # values
17 for i in range(300):
18     df["q(left)"] = df["r(left)"]
```

```
19    df["q(right)"] = df["r(right)"]
20    df["q(listen)"] = df["r(listen)"] + discount * (
21            df["omega(left)"] * df["v"].shift(-1).fillna(10) +
22            df["omega(right)"] * df["v"].shift(1).fillna(10))
23    df["v"] = df[["q(left)", "q(right)", "q(listen)"]].max(axis=1)
24
25  df["action"] = df[["q(left)", "q(right)", "q(listen)"]].values.argmax(axis=1)
26  df
```

15.6.3 Non-Stationary Belief State Values

Code 15.5 implements the PBVI algorithm. PBVI needs to know the number of steps in the episode beforehand, and here we set the number of steps to $t_{\max} = 10$. It is already very similar to the setting of infinite steps. We first sample 15 points evenly in the belief space, and we use the iteration formula to calculate and obtain the optimal belief state values at each time t.

Code 15.5   PBVI.

Tiger-v0_Plan_demo.ipynb

```
1   class State:
2       LEFT, RIGHT = range(2)   # do not contain the terminate state
3   state_count = 2
4   states = range(state_count)
5
6
7   class Action:
8       LEFT, RIGHT, LISTEN = range(3)
9   action_count = 3
10  actions = range(action_count)
11
12
13  class Observation:
14      LEFT, RIGHT = range(2)
15  observation_count = 2
16  observations = range(observation_count)
17
18
19  # r(S,A): state x action -> reward
20  rewards = np.zeros((state_count, action_count))
21  rewards[State.LEFT, Action.LEFT] = 10.
22  rewards[State.LEFT, Action.RIGHT] = -100.
23  rewards[State.RIGHT, Action.LEFT] = -100.
24  rewards[State.RIGHT, Action.RIGHT] = 10.
25  rewards[:, Action.LISTEN] = -1.
26
27  # p(S'|S,A): state x action x next_state -> probability
28  transitions = np.zeros((state_count, action_count, state_count))
29  transitions[State.LEFT, :, State.LEFT] = 1.
30  transitions[State.RIGHT, :, State.RIGHT] = 1.
31
32  # o(O|A,S'): action x next_state x next_observation -> probability
33  observes = np.zeros((action_count, state_count, observation_count))
34  observes[Action.LISTEN, Action.LEFT, Observation.LEFT] = 0.85
35  observes[Action.LISTEN, Action.LEFT, Observation.RIGHT] = 0.15
36  observes[Action.LISTEN, Action.RIGHT, Observation.LEFT] = 0.15
37  observes[Action.LISTEN, Action.RIGHT, Observation.RIGHT] = 0.85
38
39
```

```python
40   # sample beliefs
41   belief_count = 15
42   beliefs = list(np.array([p, 1-p]) for p in np.linspace(0, 1, belief_count))
43
44   action_alphas = {action: rewards[:, action] for action in actions}
45
46   horizon = 10
47
48   # initialize alpha vectors
49   alphas = [np.zeros(state_count)]
50
51   ss_state_value = {}
52
53   for t in reversed(range(horizon)):
54       logging.info("t = %d", t)
55
56       # Calculate alpha vector for each (action, observation, alpha)
57       action_observation_alpha_alphas = {}
58       for action in actions:
59           for observation in observations:
60               for alpha_idx, alpha in enumerate(alphas):
61                   action_observation_alpha_alphas \
62                       [(action, observation, alpha_idx)] = \
63                       discount * np.dot(transitions[:, action, :], \
64                       observes[action, :, observation] * alpha)
65
66       # Calculate alpha vector for each (belief, action)
67       belief_action_alphas = {}
68       for belief_idx, belief in enumerate(beliefs):
69           for action in actions:
70               belief_action_alphas[(belief_idx, action)] = \
71                   action_alphas[action].copy()
72               def dot_belief(x):
73                   return np.dot(x, belief)
74               for observation in observations:
75                   belief_action_observation_vector = max([
76                       action_observation_alpha_alphas[
77                       (action, observation, alpha_idx)]
78                       for alpha_idx, _ in enumerate(alphas)], key=dot_belief)
79                   belief_action_alphas[(belief_idx, action)] += \
80                       belief_action_observation_vector
81
82       # Calculate alpha vector for each belief
83       belief_alphas = {}
84       for belief_idx, belief in enumerate(beliefs):
85           def dot_belief(x):
86               return np.dot(x, belief)
87           belief_alphas[belief_idx] = max([
88               belief_action_alphas[(belief_idx, action)]
89               for action in actions], key=dot_belief)
90
91       alphas = belief_alphas.values()
92
93       # dump state values for display only
94       df_belief = pd.DataFrame(beliefs, index=range(belief_count),
95                   columns=states)
96       df_alpha = pd.DataFrame(alphas, index=range(belief_count), columns=states)
97       ss_state_value[t] = (df_belief * df_alpha).sum(axis=1)
98
99
100  logging.info("state_value =")
101  pd.DataFrame(ss_state_value)
```

Additionally, the observation space in Code 15.5 and that in Code 15.1 are implemented differently. The class Observation in Code 15.1 consists of the first placebo observation START, but the class Observation in Code 15.5 does not contain this element.

15.7 Summary

- Average reward is defined as

$$\bar{r}_\pi \overset{\text{def}}{=} \lim_{h \to +\infty} \mathrm{E}_\pi \left[\frac{1}{h} \sum_{\tau=1}^{h} R_\tau \right].$$

- Average-reward values are defined as

$$\text{state value} \quad \bar{v}_\pi(s) \overset{\text{def}}{=} \lim_{h \to +\infty} \mathrm{E}_\pi \left[\frac{1}{h} \sum_{\tau=1}^{h} R_{t+\tau} \middle| S_t = s \right], \qquad s \in S$$

$$\text{action value} \quad \bar{q}_\pi(s,a) \overset{\text{def}}{=} \lim_{h \to +\infty} \mathrm{E}_\pi \left[\frac{1}{h} \sum_{\tau=1}^{h} R_{t+\tau} \middle| S_t = s, A_t = a \right], s \in S, a \in \mathcal{A}.$$

- Average-reward visitation frequency

$$\bar{\eta}_\pi(s) \overset{\text{def}}{=} \lim_{h \to +\infty} \mathrm{E}_\pi \left[\frac{1}{h} \sum_{\tau=1}^{h} 1_{[S_\tau = s]} \right]$$

$$= \lim_{h \to +\infty} \frac{1}{h} \sum_{\tau=1}^{h} \Pr_\pi[S_\tau = s], \qquad s \in S$$

$$\bar{\rho}_\pi(s,a) \overset{\text{def}}{=} \lim_{h \to +\infty} \mathrm{E}_\pi \left[\frac{1}{h} \sum_{\tau=1}^{h} 1_{[S_\tau = s, A_\tau = a]} \right]$$

$$= \lim_{h \to +\infty} \frac{1}{h} \sum_{\tau=1}^{h} \Pr_\pi[S_\tau = s, A_\tau = a], \quad s \in S, a \in \mathcal{A}.$$

- Differential MDP uses differential reward

$$\tilde{R}_t \overset{\text{def}}{=} R_t - \bar{r}_\pi,$$

with discount factor $\gamma = 1$.
- Differential values:

$$\text{state value} \quad \tilde{v}_\pi(s) \overset{\text{def}}{=} \lim_{h \to +\infty} \mathrm{E}_\pi \left[\frac{1}{h} \sum_{\tau=1}^{h} \tilde{R}_{t+\tau} \middle| S_t = s \right], \qquad s \in S$$

action value $\tilde{q}_\pi(s, a) \overset{\text{def}}{=} \lim\limits_{h \to +\infty} \mathrm{E}_\pi \left[\dfrac{1}{h} \sum\limits_{\tau=1}^{h} \tilde{R}_{t+\tau} \middle| S_t = s, A_t = a \right], s \in \mathcal{S}, a \in \mathcal{A}.$

- Use differential values to calculate average-reward values:

$$\bar{v}_\pi(s) = r_\pi(s) + \sum_{s'} p_\pi(s'|s) \left(\tilde{v}_\pi(s') - \tilde{v}_\pi(s) \right), \qquad s \in \mathcal{S}$$

$$\bar{q}_\pi(s, a) = r(s, a) + \sum_{s', a'} p_\pi(s', a'|s, a) \left(\tilde{q}_\pi(s', a') - \tilde{q}_\pi(s, a) \right), \quad s \in \mathcal{S}, a \in \mathcal{A}.$$

- LP to find the average-reward optimal state values

$$\operatorname*{minimize}_{\substack{\bar{v}(s):s \in \mathcal{S} \\ \tilde{v}(s):s \in \mathcal{S}}} \quad \sum_s c(s) \bar{v}(s)$$

$$\text{s.t.} \qquad \bar{v}(s) \geq \sum_{s'} p(s'|s, a) \bar{v}(s'), \qquad\qquad s \in \mathcal{S}, a \in \mathcal{A}$$

$$\bar{v}(s) \geq r(s, a) + \sum_{s'} p(s'|s, a) \tilde{v}(s') - \tilde{v}(s), \quad s \in \mathcal{S}, a \in \mathcal{A}.$$

- Relative VI to find the optimal differential state values: Fix $s_{\text{fix}} \in \mathcal{S}$, and use the following equations to update:

$$\tilde{v}_{k+1}(s) \leftarrow \max_a r(s, a) + \sum_{s'} p(s'|s, a) \tilde{v}(s'),$$

$$- \max_a r(s_{\text{fix}}, a) + \sum_{s'} p(s_{\text{fix}}|s, a) \tilde{v}(s'), \quad s \in \mathcal{S}.$$

- TD update algorithm updates the average reward estimate using

$$\bar{r} \leftarrow \bar{r} + \alpha^{(r)} \tilde{\varDelta}.$$

- The transition rate of a CTMDP can be denoted as $q(s'|s, a)$ ($s \in \mathcal{S}, a \in \mathcal{A}, s' \in \mathcal{S}$).
- The state S_t in an MDP may be partitioned to $S_t = (t, X_t)$. The form X_t is usually used in non-homogenous decision processes.
- The sojourn time of SMDP is a random number.
- The environment of DTPOMDP can be mathematically modeled by the initial distribution

$$p_{S_0, R_0}(s, r) \overset{\text{def}}{=} \Pr[S_0 = s, R_0 = r], \quad s \in \mathcal{S}, r \in \mathcal{R},$$

$$o_0(o|s') \overset{\text{def}}{=} \Pr[O_0 = o|S_0 = s'], \quad s' \in \mathcal{S}, o \in O$$

and dynamics

$$p(s', r|s, a) \overset{\text{def}}{=} \Pr\left[S_{t+1} = s', R_{t+1} = r | S_t = s, A_t = a\right],$$
$$s \in S, a \in \mathcal{A}, r \in \mathcal{R}, s' \in S^+$$

$$o(o|a, s') \overset{\text{def}}{=} \Pr\left[O_{t+1} = o | A_t = a, S_{t+1} = s'\right], \quad a \in \mathcal{A}, s' \in S, o \in O.$$

- Belief is the guess of states. Belief is usually quantified by conditional state probability.
- Update of belief:

$$\mathfrak{u}(b, a, o)(s') = \frac{o(o|a, s') \sum_s p(s'|s, a) b(s)}{\sum_{s''} o(o|a, s'') \sum_s p(s''|s, a) b(s)},$$
$$b \in \mathcal{B}, a \in \mathcal{A}, o \in O, s' \in S^+.$$

- The optimal state values of finite POMDP can be represented by

$$v_{*,t}(b) = \max_{\alpha \in \mathcal{L}_t} \sum_{x \in \mathcal{X}_t} \alpha(x), \qquad b \in \mathcal{B}_t,$$

where the set $\mathcal{L}_t = \{\alpha_l : 0 \leq l < |\mathcal{L}_t|\}$ contains α-vectors of dimension $|\mathcal{X}_t|$.
- Point-Based Value Iteration (PBVI) algorithm samples some beliefs and only updates beliefs on those beliefs, so the computation is reduced.

15.8 Exercises

15.8.1 Multiple Choices

15.1 On using the state value at time $t + 1$ to back up the state values at time t in a DTMDP, choose the correct one: (   )

A. Discounted values ($\gamma \in (0, 1)$) satisfy $v_\pi^{(\gamma)}(s) = r_\pi(s) + \sum_{s'} p_\pi(s'|s) v_\pi^{(\gamma)}(s')$ ($s \in S$).
B. Average reward values satisfy $\bar{v}_\pi(s) = r_\pi(s) + \sum_{s'} p_\pi(s'|s) \bar{v}(s')$ ($s \in S$).
C. Differential values satisfy $\tilde{v}_\pi(s) = \tilde{r}(s) + \sum_{s'} p_\pi(s'|s) \tilde{v}(s')$ ($s \in S$).

15.2 On a unichain MDP with given policy π, choose the correct one: (   )

A. All states share the same finite-horizon values $v_\pi^{[h]}(s)$ ($s \in S$).
B. All states share the same discounted values $v_\pi^{(\gamma)}(s)$ ($s \in S$).
C. All states share the same average reward values $\bar{v}_\pi(s)$ ($s \in S$).

15.3 Let $\mathcal{T}$ denote the time index set of an MDP. The MDP must be non-homogenous when $\mathcal{T}$ is: (   )

A. $\{0, 1, \ldots, t_{\max}\}$, where $t_{\max}$ is a positive integer.
B. Natural number set $\mathbb{N}$.

C. $[0, +\infty)$.

15.4 The high-level tasks in HRL can be viewed as: (   )

A. Differential MDP.
B. CTMDP.
C. SMDP.

15.5 Given a POMDP, we can modify the observation so that it becomes a fully-observable MDP. Choose the correct one: (   )

A. The optimal values of POMDP are not less than the optimal values of corresponding MDP.
B. The optimal values of POMDP are equal to the optimal values of corresponding MDP.
C. The optimal values of POMDP are not more than the optimal values of corresponding MDP.

15.8.2 Programming

15.6 The task `GuessingGame-v0` is a DTPOMDP task in the package `gym_toytext`. In this task, the environment has a time-invariant state, which is a number ranging from 0 to 200. In the beginning, the agent observes a trivial value of 0. After that, the agent can guess a number in each step. If the guess number is smaller than the state, the agent observes the value 1; if the guess number equals the state, the agent observes the value 2; if the guess number is greater than the state, the agent observes the value 3. When the guess number differs from the state no more than 1%, the agent gets reward 1 and the episode ends; otherwise, the reward is 0 and the episode continues. Please design a policy to maximize the average reward. What is the belief space? Please implement some codes to calculate the average reward of this policy. (C.f. `GuessingGame-v0_CloseForm.ipynb`)

15.8.3 Mock Interview

15.7 What is the similarity and difference between discounted MDP and average-reward MDP?

15.8 How to convert a non-homogenous MDP to a homogenous MDP?

15.9 What is Semi-MDP?

15.10 Why Deep Recurrent Q Network (DRQN) can help to solve partially observable tasks?

Chapter 16
Learn from Feedback and Imitation Learning

This chapter covers

- Reward Model (RM)
- Inverse Reinforcement Learning (IRL)
- Preference-based Reinforcement Learning (PbRL)
- Reinforcement Learning with Human Feedback (RLHF)
- Generative Pre-trained Transformer (GPT)
- Imitation Learning (IL)
- expert policy
- Behavior Cloning (BC) algorithm
- compounding error
- Generative Adversarial Imitation Learning (GAIL) algorithm

RL learns from reward signals. However, some tasks do not provide reward signals. This chapter will consider applying RL-alike algorithms to solve the tasks without reward signals.

16.1 Learn from Feedback

Training an agent needs some information or data to indicate what kinds of policies are good and what kinds of policies are bad. Besides the reward signals that we have discussed in previous chapters, other data may contain such information. The forms of such data can be as follows:

© The Author(s), under exclusive license to Springer Nature Singapore Pte Ltd. 2024
Z. Xiao, *Reinforcement Learning*, https://doi.org/10.1007/978-981-19-4933-3_16

526 16 Learn from Feedback and Imitation Learning

- Preference data, which compare different states, actions, or trajectories, indicate which state, action, or trajectory is better than others.
- Known good policies, or transition data of interactions between good policies and environments.
- The optimal policies under some conditions. Such data are similar to targets in supervised learning, but may differ in that such optimal actions may not be available for all transitions.
- Modification suggestions under some conditions. Compared to the previous entry, the modification suggestions may not provide the optimal actions. They may just provide better actions.
- Other data that may be used to infer the goodness of policies, such as scores in the form of discrete grading or centesimal grading.

In conventional RL and online RL, all aforementioned forms of data, together with the reward signals in RL, are all feedback data. Learning from feedbacks is an extended concept of RL (Fig. 16.1).

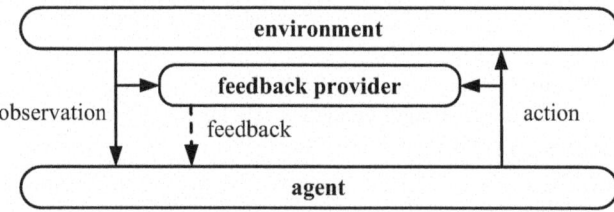

Fig. 16.1 Learning from feedbacks.

Although learning from feedbacks is often called RL from feedbacks, it is not strictly RL when feedback data are not reward signals. If feedback data are optimal actions on all data entries, it is supervised learning. There are cases when a task is neither a supervised learning task nor an RL task, such as imitation learning in Sect. 16.2.

16.1.1 Reward Model

For the tasks that the reward signals are not directly available, we may consider learning a **Reward Model (RM)** or a utility model, and then apply outputs of RM or utility model as reward signals on the RL algorithms. This approach is called **Inverse Reinforcement Learning (IRL)**.

> **! Note**
>
> The concepts of RM and IRL also exist for offline tasks.

RM is the part of environment model that produces reward signals. If an algorithm needs to learn both transition model and RM, it usually learns these two models at the same time.

Learning an RM may encounter the following difficulties:

- Training data can not determine the unique RM. On the one hand, multiple reward models may satisfy the data. On the other hand, data, which may be noisy, may not provide the optimal data, so no reward models can satisfy the data.
- It is computationally intensive to evaluate an RM. Evaluating an RM usually requires applying RL algorithm on this RM to derive a policy for further evaluation, and such RL algorithms may need to be applied repeatedly during the iteration of the RM.

16.1.2 PbRL: Preference-based RL

A task is a **Preference-based RL (PbRL)** task when environment provides preference data rather than reward signals.

Preference data may have the following forms:

- action preference, which compares the goodness among different actions followed by a state;
- state preference, which compares the goodness among different states;
- trajectory preference, which compares the goodness among different trajectories.

Both action preference data and state preference data can be converted to trajectory preference data, so trajectory preference data are the most universal form of preference. Many PbRL algorithms accept trajectory preference data as inputs. A limitation of trajectory preference data is its temporal credit assignment problem, which means that it is difficult to determine which state(s) or action(s) lead to a preference.

PbRL algorithms may have following approaches:

- Learn RM or utility model: This approach first learns an RM, and then applies rewards generated by RM to RL algorithms. PbRL algorithms in this approach are IRL algorithms too. This is the most common approach. Besides learning RMs, we may also consider learning utility models. Utility, which has been introduced in Interdisciplinary References 12.3 and 12.4, can be used in the same way as reward signals for RL algorithms.
- Learn a preference model: Here, the preference model usually means action preference model, which compares among actions under a given state. When

an action preference model has been learned, we can derive a greed policy from
the action preference model.

- Compare and rank policies: This approach directly compares policies or actions,
  and outputs better policies or better actions.
- Bayesian method: This approach assumes the policy is random (denoted
  as Π) and has a distribution. For every policy sample π, generate multiple
  trajectories are generated. This approach further obtains the preference among
  these trajectory samples. (Let T and t denote a random trajectory and a
  trajectory sample respectively. For a particular preference sample $t^{(1)} > t^{(2)}$,
  its likelihood on the condition of a policy sample π can be

$$p\left(t^{(1)} > t^{(2)}\Big|\pi\right) = \Phi\left(\frac{1}{\sqrt{2}\sigma_p}\left(\mathrm{E}_{T\sim\pi}\left[d\left(t^{(1)}, T\right)\right] - \mathrm{E}_{T\sim\pi}\left[d\left(t^{(2)}, T\right)\right]\right)\right)$$

where d is some distance between two trajectories, and Π is the CDF of the
standard normal distribution.) Next, save the preference data into a data set
$\mathcal{D}_<$. Then try to maximize the posterior probability $\Pr[\Pi = \pi|\mathcal{D}_<]$ and get the
optimal policy $\arg\max_\pi \Pr[\Pi = \pi|\mathcal{D}_<]$. One issue of this approach is that there
are usually no known good distances for trajectories. People often have to use
Hamming distance or Euclid distance, but these distances can not reflect the real
difference among trajectories, so the likelihood is not useful.

16.1.3 RLHF: Reinforcement Learning with Human Feedback

It is human to say the final word on judging the goodness of an AI system. The
willingness of human can be reflected by either existing offline data or feedbacks
during training. A task is called **Reinforcement Learning with Human Feedback**
(**RLHF**) when feedbacks are being provided by human during the training (Christina,
2017).

The feedbacks provided by human can be of any forms. For example, people
can provide goodness data for interactions, or provide preferences, or provide
modification suggestions. Furthermore, human can work with other AI: Say, let AI
generate some feedbacks first, and then human process those feedbacks by further
modifications and enrichments, and the enriched data are provided to system as
feedbacks.

One key consideration of designing an RLHF system is to determine the form of
feedbacks. We need to consider the following aspects during the system design:

- Feasibility: We need to consider whether the training can succeed with these
  features; how many feature data are needed, and how completeness should the
  feedback have.

- Cost: We need to consider whether it is difficult to get the feedbacks; how much time, how much money, and how many people are needed to provide the feedbacks.
- Consistency: We need to consider whether different people at different time will provide the same result for the same data entry, whether the feedback will be impacted by the cultural and religious backgrounds of feedback provides, and whether the feedbacks can be impacted by some subjective factors.

The forms of human feedbacks are various. Here are two large categories:

- The human feedbacks can be reward signals or the derivatives of reward signals (such as ranking of rewards). Both the raw reward signals and the derivatives of reward signals have their own advantages. The raw reward signals can be directly used in the training, but different people may provide different, or even contradictory, rewards at different time. It may be easier to provide derivatives of rewards, such as preferences, but those derivatives are not rewards any way, so they can not be directly used in conventional RL algorithms.
- The human feedbacks can be some inputs that make the rewards larger, rather than reward signals or their derivatives. For example, forms of human feedback can be better quantities, texts, categories, and physical actions. The advantage of such feedbacks is, the training data are no longer constrained by data generated by the model. When human provides such feedbacks, we can either let humans provide feedbacks from the scratch, or provide some data so that human can refer to when they provide feedback. Such reference data may make the feedback process easier. However, the reference data may impact the quality of human feedbacks, either positively way or negatively.

Human feedbacks have limitations. It may be time-consuming and costly to get human feedbacks, and human feedbacks may be incorrect and inconsistent. Other limitations include:

- The humans that provide human feedbacks may have bias and limitations. This problem is similar to issues of sampling method in statistics. People who provide the feedbacks may not be the best people. Due to the reasons of availability and costs, developers are likely to choose a human team that has lower costs to provide feedbacks, but such human team may not be the most professional, or have different cultural, religious, and moral backgrounds, and there are even discriminations. Besides, there may be malicious feedback providers who provide adversarial misleading feedbacks, especially when there are no sufficient background checks on the recruiting process of feedback providers.
- The features of feedback providers are not fed into the system. Everyone is unique: they have their own grow-up environment, religious belief, moral standard, educational and professional experience, knowledge background, etc. It is impossible to include all differences into systems. If we ignore the diversity of human in feature dimensions, we may lose some effective information, and degrade the performance. Taking language models for example, we may expect

a language model to communicate in different styles. Some applications want it to communicate in a polite and greasy way, but other applications want it to communicate in a concise and meaningful way. Some applications want it to output creatively, but other applications want it to output reasonably. Some applications want concise outputs, but other applications want detailed outputs. Some applications want to limit the discussion within natural science, but other applications want to take historical and social backgrounds into consideration. Different feedback providers may have different experiences, backgrounds, and communication styles, and each of them may fit for a particular type of application. In this case, features of human are important.

- Human nature can lead to imperfect datasets. For example, a language model may get high rated by flattering and/or avoiding controversial topics, rather than actually solving the problem that needs to be solved, which does not achieve the original intention of the system design.
- Human may not be as smart as machines in some tasks. In some tasks. such as board games including chess and Go, AI can do better than people. In some tasks, human can' t process as much information as data-driven programs. For example, for autonomous driving applications, humans make decisions only based on two-dimensional images and sounds, while AI can process information in three-dimensional space in continuous time.
- In addition, human feedbacks may have other non-technical risks to human feedback, including security risks such as leaks, and regulatory and legal risks.

The following measures can be considered to reduce the negative impacts of human feedbacks.

- For the issue that human feedbacks are costly and time-consuming, we may consider optimize the usage of human resources, and leverage existing AI. For example, we can use an AI to provide some feedback candidates, and human can either pick one from the choices, or write their own feedback only when none feedbacks are satisfactory.
- For the issue that feedbacks are not complete, we can train and evaluate agents in time when feedbacks are being collected, and try to identify issues from behaviors of agents and make adjustments accordingly as early as possible.
- For the issue of feedback quality and adversarial feedbacks, we can audit and check the human feedbacks, such as adding verification samples to check the quality of human feedbacks, or solicit feedbacks from multiple human and compare.
- For the issue of human are not representative, we can apply more strategic method to determine the humans with controlling of human costs. We might use the sampling methods in statistics, such as stratified sampling and cluster sampling, to make the feedback crowd more reasonable.
- For the issue that features of human are not included in the system, we can collect the features of humans, and use those features as part of data features during the training. For example, when training of a large language model, the

professional background of the feedback person (such as lawyers, doctors, etc.) can be recorded and taken into account during training.

- Besides, during the design and implement of the system, we may resort to professional people for their advice to reduce legal and security risks.

16.2 IL: Imitation Learning

Imitation Learning (**IL**) tries to imitate existing good policy from the information of good policy or some experience that the good policy interacts with the environment.

The typical setting of an IL task is as follows: As Fig. 16.2, similar to agent–environment interface, the environment is still driven by the initial state distribution p_{S_0} and the transition probability p, but the environment does not send reward signal R to the agent. The agent does not receive the reward signal from the environment, but it can access to some interaction history between the agent and a policy. The policy that generated the history is a verified good policy, called **expert policy**, denoted as π_E. In the history, there are only observations and actions, but no rewards. The agent does not know the mathematical model of the expert policy. The agent can only leverage the interaction history between the expert policy and the environment, and find an **imitation policy**, hope that the imitation policy can well perform. The imitation policy is usually approximated by a parametric function $\pi(a|s; \boldsymbol{\theta})$ $(s \in \mathcal{S}, a \in \mathcal{A}(s))$, where $\boldsymbol{\theta}$ is the policy parameter, and the function satisfies the constraint $\sum_{a \in \mathcal{A}(s)} \pi(a|s; \boldsymbol{\theta}) = 1$ $(s \in \mathcal{S})$. The forms of the approximation functions are identical to those in Sect. 7.1.1.

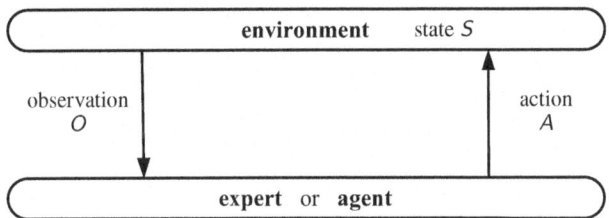

Fig. 16.2 Agent–environment interface of IL.

Strictly speaking, IL differs from RL. The difference between IL and RL is as follows:

- In IL tasks, agents can not receive reward signals directly, they can only obtain the interaction history between expert policy and the environment.
- In RL tasks, agents can directly obtain reward signals for cost signals from the environment.

Section 1.4.1 covers the concept of offline RL and online RL. IL also has offline IL and online RL. Offline IL means that the interaction history between expert policy and the environment is fixed and no more data will be available. Online IL means that the interaction history will be available as a continuous stream, and the agent can continuously learn from new interactions. The performance metrics of these kinds of tasks usually include the sample complexity between expert policy and environment.

Two most popular categories of ILs are: behavior cloning IL and adversarial IL. Behavior cloning IL tries to reduce the KL divergence from imitation policy to expert policy, while adversarial IL tries to reduce the JS divergence from imitation policy to expert policy. In the sequel, we will learn what is f divergence, and why minimizing KL divergence and JS divergence can be used in IL.

16.2.1 f-Divergences and their Properties

This section introduces some theoretical background of IL. We will first review some knowledge of information theory, including the concepts of f-divergence and three special f-divergences, i.e. total variation distance, KL divergence, and JS divergence. Then we prove that the difference of expected returns between two policies is bounded by the total variation distance, KL divergence, and JS divergence between two policies.

Interdisciplinary Reference 16.1
Information Theory: f-Divergence

Consider two probability distributions p and q such that $p \ll q$. Given the convex function $f : (0, +\infty) \to \mathbb{R}$ such that $f(1) = 0$. The f-**divergence** from p to q is defined as

$$d_f(p\|q) \overset{\text{def}}{=} \mathrm{E}_{X\sim q}\left[f\left(\frac{p(X)}{q(X)}\right)\right].$$

- f-divergence with $f_{\text{TV}}(x) \overset{\text{def}}{=} \frac{1}{2}|x-1|$ is **Total Variation distance (TV-distance)**

$$d_{\text{TV}}(p\|q) = \frac{1}{2}\sum_x |p(x) - q(x)|.$$

(Proof: $d_{\text{TV}}(p\|q) \;=\; \mathrm{E}_{X\sim q}\left[f_{\text{TV}}\left(\frac{p(X)}{q(X)}\right)\right] \;=\; \sum_x q(x)\frac{1}{2}\left|\frac{p(x)}{q(x)} - 1\right| \;=\; \frac{1}{2}\sum_x |p(x) - q(x)|.$)

- f-divergence with $f_{\text{KL}}(x) \overset{\text{def}}{=} x\ln x$ is the KL divergence mentioned in Sect. 8.4.1:

$$d_{\text{KL}}(p\|q) = \mathrm{E}_{X\sim p}\left[\ln\left(\frac{p(X)}{q(X)}\right)\right].$$

(Proof: $d_{\mathrm{KL}}(p\|q) \;=\; \mathrm{E}_{X\sim q}\left[f_{\mathrm{KL}}\left(\frac{p(X)}{q(X)}\right)\right] \;=\; \sum_x q(x)\frac{p(x)}{q(x)}\ln\frac{p(x)}{q(x)} \;=\;$
$\sum_x p(x)\ln\frac{p(x)}{q(x)} = \mathrm{E}_{X\sim p}\left[\ln\left(\frac{p(X)}{q(X)}\right)\right].$)

- f-divergence with $f_{\mathrm{JS}} \overset{\text{def}}{=} x\ln x - (x+1)\ln\frac{x+1}{2}$ is **Jensen–Shannon divergence** (**JS divergence**), denoted as $d_{\mathrm{JS}}(p\|q)$.

The properties of f-divergence: We can easily verify that

- $d_{\mathrm{TV}}\left(p\,\middle\|\,\frac{p+q}{2}\right) = d_{\mathrm{TV}}\left(q\,\middle\|\,\frac{p+q}{2}\right) = \frac{1}{2}d_{\mathrm{TV}}(p\|q)$.
- $f_{\mathrm{KL}}\left(p\,\middle\|\,\frac{p+q}{2}\right) = f_{\mathrm{KL}}(p\|p+q) + \ln 2$.

The relationship among different f-divergences include:

- Use KL divergence to represent JS divergence:

$$d_{\mathrm{KL}}(p\|q) = d_{\mathrm{KL}}\left(p\,\middle\|\,\frac{p+q}{2}\right) + d_{\mathrm{KL}}\left(q\,\middle\|\,\frac{p+q}{2}\right)$$
$$= d_{\mathrm{KL}}(p\|p+q) + d_{\mathrm{KL}}(q\|p+q) + \ln 4.$$

(Proof:

$$d_{\mathrm{JS}}(p\|q)$$
$$= \sum_x q(x)\left[\frac{p(x)}{q(x)}\ln\frac{p(x)}{q(x)} - \left(\frac{p(x)}{q(x)}+1\right)\ln\frac{\frac{p(x)}{q(x)}+1}{2}\right]$$
$$= \sum_x\left[p(x)\ln\frac{p(x)}{q(x)} - p(x)\ln\frac{\frac{p(x)+q(x)}{2}}{q(x)} - q(x)\ln\frac{\frac{p(x)+q(x)}{2}}{q(x)}\right]$$
$$= \sum_x p(x)\ln\frac{p(x)}{\frac{p(x)+q(x)}{2}} + \sum_x q(x)\ln\frac{q(x)}{\frac{p(x)+q(x)}{2}}$$
$$= d_{\mathrm{KL}}\left(p\,\middle\|\,\frac{p+q}{2}\right) + d_{\mathrm{KL}}\left(q\,\middle\|\,\frac{p+q}{2}\right).$$

The proof completes.)
- The inequality between TV distance and KL divergence, a.k.a. Pinsker's inequality:

$$d_{\mathrm{TV}}^2(p\|q) \le \frac{1}{2}d_{\mathrm{KL}}(p\|q).$$

(Proof: We first consider the case where both p and q are binomial distributions. Let the binomial distribution p be $p(0) = p_0$ and $p(1) = 1-p_0$. Let the binomial distribution q be $q(0) = q_0$ and $q(1) = 1 - q_0$. We have

$$d_{\text{KL}}(p\|q) - 2d_{\text{TV}}^2(p\|q) = p_0 \ln \frac{p_0}{q_0} + (1 - p_0) \ln \frac{1 - p_0}{1 - q_0} - 2(p_0 - q_0)^2.$$

Now we need to prove that aforementioned is always non-negative. Fix an arbitrary p_0, we calculate its partial derivative with respect to q_0:

$$\frac{\partial}{\partial q_0}\left[d_{\text{KL}}(p\|q) - 2d_{\text{TV}}^2(p\|q)\right] = -\frac{p_0}{q_0} + \frac{1 - p_0}{1 - q_0} + 4(p_0 - q_0)$$

$$= (p_0 - q_0)\left[4 - \frac{1}{q_0(1 - q_0)}\right].$$

Since $q_0(1 - q_0) \in \left[0, \frac{1}{4}\right]$, we have $4 - \frac{1}{q_0(1-q_0)} \leq 0$. Therefore, $d_{\text{KL}}(p\|q) - 2d_{\text{TV}}^2(p\|q)$ reaches its minimum when $q_0 = p_0$, and we can verify the minimum value is 0. Therefore, we prove that $d_{\text{KL}}(p\|q) - 2d_{\text{TV}}^2(p\|q) \geq 0$. For the cases where either p or q is not binomial distributed, we define the event e_0 as $p(X) \leq q(X)$, and define the event e_1 as $p(X) > q(X)$. Define the new distributions p_E and q_E as:

$$p_E(e_0) = \sum_{X:p(X)\leq q(X)} p(X)$$

$$p_E(e_1) = \sum_{X:p(X)>q(X)} p(X)$$

$$q_E(e_0) = \sum_{X:p(X)\leq q(X)} q(X)$$

$$q_E(e_1) = \sum_{X:p(X)>q(X)} q(X).$$

We can verify that

$$d_{\text{TV}}(p\|q) = d_{\text{TV}}(p_E\|q_E)$$
$$d_{\text{KL}}(p\|q) \geq d_{\text{KL}}(p_E\|q_E).$$

Applying the inequality of binomial case, i.e. $2d_{\text{TV}}^2(p_E\|q_E) \leq d_{\text{KL}}(p_E\|q_E)$, completes the proof.)

• The inequality between TV distance and JS divergence:

$$d_{\text{TV}}^2(p\|q) \leq d_{\text{JS}}(p\|q).$$

(Proof: Pinsker's inequality and the relationship between JS divergence and KL divergence together lead to

$$d_{\text{JS}}(p\|q) = d_{\text{KL}}\left(p\left\|\frac{p+q}{2}\right.\right) + d_{\text{KL}}\left(q\left\|\frac{p+q}{2}\right.\right)$$

$$\geq 2d_{\mathrm{TV}}^2\left(p\left\|\frac{p+q}{2}\right.\right) + 2d_{\mathrm{TV}}^2\left(q\left\|\frac{p+q}{2}\right.\right).$$

Then applying the property of TV distance finishes the proof.)

Next, we review the variational representation of the f-divergence. First, let us recap the concept of **convex conjugation**. The convex conjugation of the convex function $f : (0, +\infty) \to \mathbb{R}$, denoted as f^*, is defined as $f^*(y) = \sup_{x \in \mathbb{R}}\left[xy - f(x)\right]$. The properties of convex conjugation: (1) f^* is convex; and (2) $(f^*)^* = f$.

Example 16.1 The convex conjugation of JS divergence is:

$$f_{\mathrm{JS}}^*(y) = -\ln\left(2 - e^y\right).$$

(Proof: $f_{\mathrm{JS}}^*(y) = \sup_y\left[xy - f_{\mathrm{JS}}(x)\right]$. The (x_0, y_0) that nullizes the partial derivative

$$\frac{\partial}{\partial x}(xy - f_{\mathrm{JS}}(x)) = y - \left((1 + \ln x) - \left(1 + \ln\frac{x+1}{2}\right)\right) = y - \left(\ln x - \ln\frac{x+1}{2}\right)$$

satisfies $y_0 = \ln x_0 - \ln\frac{x_0+1}{2}$, or equivalently $\frac{x_0+1}{2} = \frac{1}{2-e^{y_0}}$. Therefore,

$$x_0 y_0 - f_{\mathrm{JS}}(x_0) = x_0\left(\ln x_0 - \ln\frac{x_0+1}{2}\right) - \left(x_0\ln x_0 - (x_0+1)\ln\frac{x_0+1}{2}\right)$$

$$= \ln\frac{x_0+1}{2}$$

$$= -\ln\left(2 - e^{y_0}\right).$$

The proof completes.)

The variational representation of f-divergence is:

$$d_f(p\|q) = \sup_{\psi:\mathcal{X}\to\mathbb{R}} \mathrm{E}_{X\sim p}\left[\psi(X)\right] - \mathrm{E}_{X\sim q}\left[f^*(\psi(X))\right].$$

(Proof:

$$d_f(p\|q)$$

$$= \sum_x q(x)f\left(\frac{p(x)}{q(x)}\right)$$

$$= \sum_x q(x)\sup_{y\in\mathbb{R}}\left(y\frac{p(x)}{q(x)} - f^*(y)\right)$$

$$= \sup_{\psi:\mathcal{X}\to\mathbb{R}}\sum_x q(x)\left(\psi(x)\frac{p(x)}{q(x)} - f^*(\psi(x))\right)$$

$$= \sup_{\psi:\mathcal{X}\to\mathbb{R}}\sum_x p(x)\psi(x) - \sum_x q(x)f^*(\psi(x))$$

$$= \sup_{\psi:\mathcal{X}\to\mathbb{R}} E_{X\sim p}\big[\psi(X)\big] - E_{X\sim q}\big[f^*\big(\psi(X)\big)\big].$$

The proof completes.)

Example 16.2 (Variational Representation of JS Divergence) JS divergence can be represented as

$$d_{JS}\big(p\|q\big) = \sup_{\psi:\mathcal{X}\to\mathbb{R}} E_{X\sim p}\big[\psi(X)\big] + E_{X\sim q}\big[\ln\big(2 - \exp\psi(X)\big)\big].$$

(Proof: Plug in $f_{JS}^*(y) = -\ln\big(2 - e^y\big)$.) Let $\phi(\cdot) \overset{\text{def}}{=} \exp\psi(\cdot)$, and then JS divergence can be further represented as

$$d_{JS}\big(p\|q\big) = \sum_{\phi:\mathcal{X}\to(0,2)} E_{X\sim p}\big[\ln\phi(X)\big] + E_{X\sim q}\big[\ln\big(2 - \phi(X)\big)\big].$$

Now we have reviewed f-divergence and its properties. Next, we apply this knowledge to build the theoretical foundation of IL.

For simplicity, the remaining part of the section only considers MDP with bounded rewards. That is, there exists a positive real number r_{bound} such that $p(s',r|s,a) = 0$ holds when r satisfies $|r| \geq r_{\text{bound}}$.

Fix an MDP with reward bound r_{bound}. For any two policies π' and π'', we have

$$\big|g_{\pi'} - g_{\pi''}\big| \leq 2r_{\text{bound}}d_{TV}\big(\rho_{\pi'}\|\rho_{\pi''}\big).$$

(Proof: Sect. 2.3.5 proved that, for any policy π,

$$g_\pi = \sum_{s,a} r(s,a)\rho_\pi(s,a).$$

So

$$\begin{aligned}
&\big|g_{\pi'} - g_{\pi''}\big| \\
&= \left|\sum_{s,a} r(s,a)\big(\rho_{\pi'}(s,a) - \rho_{\pi''}(s,a)\big)\right| \\
&\leq r_{\text{bound}}\sum_{s,a}\big|\rho_{\pi'}(s,a) - \rho_{\pi''}(s,a)\big| \\
&= 2r_{\text{bound}}d_{TV}\big(\rho_{\pi'}\|\rho_{\pi''}\big).
\end{aligned}$$

The proof completes.) This result shows that, for two policies, if the TV distance between two discounted state–action distribution is small, the difference in return expectation between these two policies is small.

The TV distance of discounted distribution is further restricted by the TV distance of policies. Mathematically speaking, given the MDP and two policies π' and π'', the relationship between the TV distance of discounted distributions and the TV distance of policies includes:

- TV distance of discounted state distribution is bounded by TV distance of policies:

$$d_{\text{TV}}\left(\eta_{\pi'}\big\|\eta_{\pi''}\right) \le \frac{\gamma}{1-\gamma}E_{S\sim\eta_{\pi''}}\left[d_{\text{TV}}\left(\pi'(\cdot|S)\big\|\pi''(\cdot|S)\right)\right].$$

(Proof: For simplicity, we only consider finite MDP here. Recap that Sect. 2.3.2 introduced the vector representation of Bellman expectation equality of discounted state distribution as

$$\boldsymbol{\eta}_\pi = \mathbf{p}_{S_0} + \gamma\mathbf{P}_\pi\boldsymbol{\eta}_\pi,$$

where $\boldsymbol{\eta}_\pi$ is the discounted state visitation frequency vector, $\mathbf{p}_{S_0}$ is the probability of the initial state vector, and $\mathbf{P}$ is the transition matrix. We further have

$$\boldsymbol{\eta}_\pi = \left(\mathbf{I} - \gamma\mathbf{P}_\pi\right)^{-1}\mathbf{p}_{S_0}.$$

So the discounted state distributions of the policies π' and π'', i.e. $\boldsymbol{\eta}_{\pi'}$ and $\boldsymbol{\eta}_{\pi''}$, satisfy

$$
\begin{aligned}
\boldsymbol{\eta}_{\pi'} - \boldsymbol{\eta}_{\pi''} &= \left(\mathbf{I} - \gamma\mathbf{P}_{\pi'}\right)^{-1}\mathbf{p}_{S_0} - \left(\mathbf{I} - \gamma\mathbf{P}_{\pi''}\right)^{-1}\mathbf{p}_{S_0} \\
&= \left[\left(\mathbf{I} - \gamma\mathbf{P}_{\pi'}\right)^{-1} - \left(\mathbf{I} - \gamma\mathbf{P}_{\pi''}\right)^{-1}\right]\mathbf{p}_{S_0} \\
&= \left(\mathbf{I} - \gamma\mathbf{P}_{\pi'}\right)^{-1}\left[\left(\mathbf{I} - \gamma\mathbf{P}_{\pi''}\right) - \left(\mathbf{I} - \gamma\mathbf{P}_{\pi'}\right)\right]\left(\mathbf{I} - \gamma\mathbf{P}_{\pi''}\right)^{-1}\mathbf{p}_{S_0} \\
&= \left(\mathbf{I} - \gamma\mathbf{P}_{\pi'}\right)^{-1}\gamma(\mathbf{P}_{\pi'} - \mathbf{P}_{\pi''})\boldsymbol{\eta}_{\pi''}.
\end{aligned}
$$

Consider the definition of TV metric

$$d_{\text{TV}}\left(\eta_{\pi'}\big\|\eta_{\pi''}\right) = \frac{1}{2}\big\|\boldsymbol{\eta}_{\pi'} - \boldsymbol{\eta}_{\pi''}\big\|_1 = \frac{1}{2}\big\|\left(\mathbf{I} - \gamma\mathbf{P}_{\pi'}\right)^{-1}\gamma(\mathbf{P}_{\pi'} - \mathbf{P}_{\pi''})\boldsymbol{\eta}_{\pi''}\big\|_1,$$

$$E_{S\sim\eta_{\pi''}}\left[d_{\text{TV}}\left(\pi'(\cdot|S)\big\|\pi''(\cdot|S)\right)\right] = \frac{1}{2}\sum_s\eta_{\pi''}(s)\sum_a|\pi'(a|s) - \pi''(a|s)|,$$

and consider

$$\big\|\left(\mathbf{I} - \gamma\mathbf{P}_\pi\right)^{-1}\big\|_1 = \Bigg\|\sum_{t=0}^{+\infty}\gamma^t\mathbf{P}_\pi^t\Bigg\|_1 \le \sum_{t=0}^{+\infty}\gamma^t\|\mathbf{P}_\pi\|_1^t \le \sum_{t=0}^{+\infty}\gamma^t = \frac{1}{1-\gamma}$$

$$\big\|(\mathbf{P}_{\pi'} - \mathbf{P}_{\pi''})\boldsymbol{\eta}_{\pi''}\big\|_1 \le \sum_{s,s'}\eta_{\pi''}(s)\big|p_{\pi'}(s'|s) - p_{\pi''}(s'|s)\big|$$

$$= \sum_{s,s'} \eta_{\pi''}(s) \left| \sum_a p(s'|s,a)(\pi'(a|s) - \pi''(a|s)) \right|$$

$$\leq \sum_{s,s'} \eta_{\pi''}(s) \sum_a p(s'|s,a)|\pi'(a|s) - \pi''(a|s)|$$

$$= \sum_{s,s'} \eta_{\pi''}(s) \sum_a p(s'|s,a)|\pi'(a|s) - \pi''(a|s)|$$

$$= \sum_s \eta_{\pi''}(s) \sum_a |\pi'(a|s) - \pi''(a|s)|,$$

we have

$$d_{\mathrm{TV}}(\eta_{\pi'}\|\eta_{\pi''})$$

$$= \frac{1}{2} \left\| (\mathbf{I} - \gamma \mathbf{P}_{\pi'})^{-1} \gamma (\mathbf{P}_{\pi'} - \mathbf{P}_{\pi''}) \boldsymbol{\eta}_{\pi''} \right\|_1$$

$$\leq \frac{1}{2} \left\| (\mathbf{I} - \gamma \mathbf{P}_{\pi'})^{-1} \right\|_1 \cdot \gamma \cdot \left\| (\mathbf{P}_{\pi'} - \mathbf{P}_{\pi''}) \boldsymbol{\eta}_{\pi''} \right\|_1$$

$$\leq \frac{1}{2} \cdot \frac{1}{1-\gamma} \cdot \gamma \cdot 2 \mathrm{E}_{S \sim \eta_{\pi''}} \left[d_{\mathrm{TV}}(\pi'(\cdot|S)\|\pi''(\cdot|S)) \right]$$

$$= \frac{\gamma}{1-\gamma} \mathrm{E}_{S \sim \eta_{\pi''}} \left[d_{\mathrm{TV}}(\pi'(\cdot|S)\|\pi''(\cdot|S)) \right].$$

The proof completes.)
- TV distance of discounted state–action distributions is bounded by TV distance of the policies:

$$d_{\mathrm{TV}}(\rho_{\pi'}\|\rho_{\pi''}) \leq \frac{1}{1-\gamma} \mathrm{E}_{S \sim \eta_{\pi''}} \left[d_{\mathrm{TV}}(\pi'(\cdot|S)\|\pi''(\cdot|S)) \right].$$

(Proof:

$$d_{\mathrm{TV}}(\rho_{\pi'}\|\rho_{\pi''})$$

$$= \frac{1}{2} \sum_{s,a} |\rho_{\pi'}(s,a) - \rho_{\pi''}(s,a)|$$

$$= \frac{1}{2} \sum_{s,a} |\eta_{\pi'}(s)\pi'(a|s) - \eta_{\pi''}(s)\pi''(a|s)|$$

$$= \frac{1}{2} \sum_{s,a} |(\eta_{\pi'}(s) - \eta_{\pi''}(s))\pi'(a|s) + \eta_{\pi''}(s)(\pi'(a|s) - \pi''(a|s))|$$

$$\leq \frac{1}{2} \sum_{s,a} |(\eta_{\pi'}(s) - \eta_{\pi''}(s))\pi'(a|s)| + \frac{1}{2} \sum_{s,a} |\eta_{\pi''}(s)(\pi'(a|s) - \pi''(a|s))|$$

$$= d_{\mathrm{TV}}(\eta_{\pi'}\|\eta_{\pi''}) + \mathrm{E}_{S \sim \eta_{\pi''}} \left[d_{\mathrm{TV}}(\pi'(\cdot|S)\|\pi''(\cdot|S)) \right].$$

Combining the above inequality and the inequality of discounted state distribution completes the proof.)

These results show that, if the TV distance of two policies is small, the TV distance of two discounted distributions is small, too. (The coefficient is $\frac{\gamma}{1-\gamma}$ or $\frac{1}{1-\gamma}$.) We also knew that, if the TV distance of discounted state–action distribution is small, the difference in return expectation of the two policies is small. Therefore, for two policies, if their TV distance is small, the difference in return expectation is small.

If we further consider the inequality between TV distance and KL divergence (i.e. $d_{\mathrm{TV}}^2(p\|q) \leq \frac{1}{2}d_{\mathrm{KL}}(p\|q)$) and the inequality between TV distance and JS divergence (i.e. $d_{\mathrm{TV}}^2(p\|q) \leq d_{\mathrm{JS}}(p\|q)$), we know that, for two policies, if their KL divergence is small, or JS divergence is small, the difference in return expectation is small. Therefore, minimizing the KL divergence or JS divergence between imitation policy and expert policy can make the return expectation of imitation policy similar to the return expectation of expert policy. This explains the basic idea of IL.

16.2.2 BC: Behavior Cloning

The previous section told us that, minimizing the KL divergence between export policy π_{E} and imitation policy $\pi(\boldsymbol{\theta})$, i.e.

$$\underset{\boldsymbol{\theta}}{\text{minimize}} \quad \mathrm{E}_{S \sim \eta_{\pi_{\mathrm{E}}}}\left[d_{\mathrm{KL}}\big(\pi_{\mathrm{E}}(\cdot|S)\|\pi(\cdot|S;\boldsymbol{\theta}))\big)\right],$$

can realize IL. Since

$$\begin{aligned}
&\mathrm{E}_{S \sim \eta_{\pi_{\mathrm{E}}}}\left[d_{\mathrm{KL}}\big(\pi_{\mathrm{E}}(\cdot|S)\|\pi(\cdot|S;\boldsymbol{\theta}))\big)\right] \\
&= \mathrm{E}_{S \sim \eta_{\pi_{\mathrm{E}}}}\left[\mathrm{E}_{A \sim \pi_{\mathrm{E}}(\cdot|S)}\left[\ln \frac{\pi_{\mathrm{E}}(A|S)}{\pi(A|S;\boldsymbol{\theta})}\right]\right] \\
&= \mathrm{E}_{(S,A) \sim \rho_{\pi_{\mathrm{E}}}}\left[\ln \frac{\pi_{\mathrm{E}}(A|S)}{\pi(A|S;\boldsymbol{\theta})}\right] \\
&= \mathrm{E}_{(S,A) \sim \rho_{\pi_{\mathrm{E}}}}\left[\ln \pi_{\mathrm{E}}(A|S)\right] - \mathrm{E}_{(S,A) \sim \rho_{\pi_{\mathrm{E}}}}\left[\ln \pi(A|S;\boldsymbol{\theta})\right],
\end{aligned}$$

it can also be written as

$$\underset{\boldsymbol{\theta}}{\text{maximize}} \quad \mathrm{E}_{(S,A) \sim \rho_{\pi_{\mathrm{E}}}}\left[\ln \pi(A|S;\boldsymbol{\theta})\right].$$

However, we usually do not have the mathematical formula of the expert policy. We can only use the expert policy history $\mathcal{D}_{\mathrm{E}}$ to estimate the expert policy. Therefore, minimizing the KL divergence from the expert policy to imitation policy can be further converted to a maximum likelihood estimate problem, i.e.

$$\underset{\boldsymbol{\theta}}{\text{maximize}} \quad \sum_{(S,A) \in \mathcal{D}_{\mathrm{E}}} \ln \pi(A|S;\boldsymbol{\theta}),$$

where $(S, A) \in \mathcal{D}_E$ means that the form of expert policy history is the state–action pairs that were generated in the interaction between the expert policy and the environment. This is the idea of **Behavior Cloning (BC)**.

BC algorithms for some common forms of imitation policy are shown below:

- Finite MDP can use a look-up table to maintain the optimal imitation policy estimate. The look-up table is a special form of function approximation anyway. The optimal imitation policy estimate is

$$\pi_*(a|s) = \frac{\sum_{(S,A)\in\mathcal{D}_E} 1_{[S=s,A=a]}}{\sum_{(S,A)\in\mathcal{D}_E} 1_{[S=s]}}, \quad s \in \mathcal{S}, a \in \mathcal{A}(s).$$

In this form, if there exists a state $s \in \mathcal{S}$ such that the denominator is 0, the policy of the state can be set to uniform distribution, i.e. $\pi_*(a|s) = \frac{1}{|\mathcal{A}(s)|}$ $(a \in \mathcal{A}(s))$.

- The tasks with discrete action space can be converted to multi-categorical classification tasks. We can introduce the action preference function $h(s, a; \boldsymbol{\theta})$ $(s \in \mathcal{S}, a \in \mathcal{A}(s))$ so that the optimization problem is converted to

$$\underset{\boldsymbol{\theta}}{\text{maximize}} \quad \sum_{(S,A)\in\mathcal{D}_E} \left(h(S, A; \boldsymbol{\theta}) - \underset{a\in\mathcal{A}(S)}{\text{logsumexp}} \, h(S, a; \boldsymbol{\theta}) \right).$$

- When the action space is continuous, we can limit the form of policy, such as Gaussian policy. Especially, if we restrict the form of policy to Gaussian distribution and fix the standard deviation of the Gaussian distribution, the maximum likelihood estimate problem is converted to a regression problem

$$\underset{\boldsymbol{\theta}}{\text{minimize}} \quad \sum_{(S,A)\in\mathcal{D}_E} \left[A - \mu(S; \boldsymbol{\theta}) \right]^2,$$

where $\mu(S; \boldsymbol{\theta})$ is the mean of Gaussian distribution.

The greatest weakness of BC is compounding error. The **compounding error** can be described as follows. Since the imitation policy is not exactly identical to expert policy, it is inevitable that the imitation policy may act differently from expert policy and reach a different state. If the state is little visited by the expert policy, the imitation policy may not know what to do, and may make some very stupid decision, which deteriorates the performance. As shown in Fig. 16.3(b), the trajectory of the imitation policy may just deviate from the optimal very little at the beginning, but since the expert policy does not tell how to recover from deviation, the imitation policy does not know how to recover after deviation.

BC algorithm can learn from a pre-determined dataset, so it can be used in offline IL tasks.

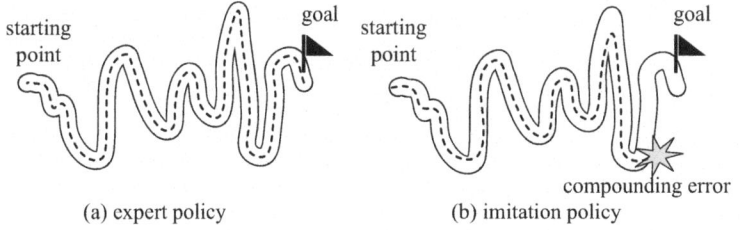

(a) expert policy   (b) imitation policy

Fig. 16.3   Compounding error of imitation policy.

16.2.3 GAIL: Generative Adversarial Imitation Learning

This section introduces another category of imitation learning: adversarial imitation learning. The core idea of adversarial imitation learning is using a discriminative network to make the discounted distribution of imitation policy and discounted distribution of the expert policy similar.

Interdisciplinary Reference 16.2
Machine Learning: Generative Adversarial Network

Generative Adversarial Network (GAN) is a method to generate data using generative network and discriminative network, where the generative network $\pi(\theta)$ maps a random variable $Z \sim p_Z$ to a datum $X_{gen} = \pi(Z; \theta)$, and the **discriminative network** $\phi(\varphi)$, which is a classifier, judges whether a datum is real reference datum $X_{real} \sim p_{real}$ or the generated datum X_{gen} that generated by the generative network. The training process takes turns to train the generative network parameter θ and discriminative network parameter φ to minimax the objective:

$$\underset{\theta}{\text{minimize}}\ \underset{\varphi}{\text{maximize}}\ \mathrm{E}_{X_{gen} \sim p_{gen}(\theta)}\left[\ln \phi\left(X_{gen}; \varphi\right)\right] + \mathrm{E}_{X_{real} \sim p_{real}}\left[\ln\left(1 - \phi(X_{real}; \varphi)\right)\right],$$

where $p_{gen}(\theta)$ is the distribution resulting from passing distribution p_Z through the generative network $\pi(\theta)$.

The training process of GAN can be viewed as optimizing JS divergence. Reason: JS divergence has the property that

$$d_{JS}\left(p_{gen} \| p_{real}\right) = d_{KL}\left(p_{gen} \| p_{gen} + p_{real}\right) + d_{KL}\left(p_{real} \| p_{gen} + p_{real}\right) + \ln 4$$

$$= \mathrm{E}_{X_{gen} \sim p_{gen}}\left[\ln \frac{p_{gen}\left(X_{gen}\right)}{p_{gen}\left(X_{gen}\right) + p_{real}\left(X_{gen}\right)}\right]$$

$$+ \mathrm{E}_{X_{\mathrm{real}} \sim p_{\mathrm{real}}} \left[\ln \frac{p_{\mathrm{real}}(X_{\mathrm{real}})}{p_{\mathrm{gen}}\left(X_{\mathrm{gen}}\right) + p_{\mathrm{real}}\left(X_{\mathrm{gen}}\right)} \right] + \ln 4,$$

which can be written as

$$d_{\mathrm{JS}}\left(p_{\mathrm{gen}} \| p_{\mathrm{real}}\right) = \mathrm{E}_{X_{\mathrm{gen}} \sim p_{\mathrm{gen}}} \left[\ln \phi\left(X_{\mathrm{gen}}\right) \right] + \mathrm{E}_{X_{\mathrm{real}} \sim p_{\mathrm{real}}} \left[\ln \left(1 - \phi(X_{\mathrm{real}})\right) \right] + \ln 4,$$

where $\phi(X) \stackrel{\mathrm{def}}{=} \frac{p_{\mathrm{real}}(X)}{p_{\mathrm{gen}}(X) + p_{\mathrm{real}}(X)}$. This representation of JS divergence only differs from the objective of GAN in a constant $\ln 4$.

Adversarial imitation learning introduces the discriminative network in GAN. The discriminative network tries to output 0 when the experiences of expert policy are inputs, and output 1 when the experiences of imitation policy are inputs. Therefore, we consider the following minimax problem:

$$\underset{\theta}{\mathrm{minimize}} \ \underset{\varphi}{\mathrm{maximize}} \ \mathrm{E}_{(S,A) \sim \rho_{\pi(\theta)}} \left[\ln \phi(S, A; \varphi) \right] + \mathrm{E}_{(S,A) \sim \rho_{\pi_{\mathrm{E}}}} \left[\ln \left(1 - \phi(S, A; \varphi)\right) \right].$$

This form is similar to the variational representation of JS divergence except for a constant difference. Therefore, approximately speaking, adversarial imitation learning is to optimize JS divergence from the discounted state–action distribution of imitation policy $\rho_{\pi(\theta)}$ and the discounted state–action distribution of expert policy $\rho_{\pi_{\mathrm{E}}}$.

The advantage of adversarial imitation learning over BC is: BC only considers the action distribution given each state. For the states that were not visited by expert policy, BC will blindly generate a distribution such that it may lead to another state that was not visited by the expert policy. Contrastingly, adversarial imitation learning considers the whole discounted state–action distribution. If a state is never or seldom visited in the history of expert policy, the agent will let imitation policy avoid such state too. Therefore, adversarial imitation learning is more likely to reduce compounding error.

Generative Adversarial Imitation Learning (GAIL) (Ho, 2016) is the most famous adversarial imitation learning algorithm. This algorithm uses discriminative network to provide reward signal and transfer the IL task to an RL task. Consequently, RL algorithms can be used to update the policy. The original paper combines GAIL with TRPO algorithm, which can be called GAIL-TRPO algorithm. GAIL can also be combined with other RL algorithms. For example, the combination of GAIL and PPO, which is called GAIL-PPO algorithm, is shown in Algo. 16.1. You can compare Algos. 16.1 and 8.6 to understand how discriminative networks provide the reward signal.

Algorithm 16.1 GAIL-PPO.

Inputs: environment (without mathematical model, no reward information), expert policy π_E (without mathematical model) or experience of expert policy $\mathcal{D}_E$.

Output: optimal policy estimate $\pi(\boldsymbol{\theta})$.

Parameters: parameters of PPO algorithm (such as parameter $\varepsilon > 0$ that limits the objective, parameters used for advantage estimation $\lambda \in [0, 1]$, optimizers, discount factor γ), and parameters to train the discriminative network (such as optimizers).

1. Initialize:

   1.1. (Initialize experience of expert policy) If the input is the expert policy rather than the experience of expert policy, use the expert policy π_E to generate the experiences $\mathcal{D}_E$, where the form of each transition is the state–action pair (S, A).

   1.2. (Initialize network parameter) Initialize the policy parameter $\boldsymbol{\theta}$, value parameter $\mathbf{w}$, and discriminative network parameter $\boldsymbol{\varphi}$.

2. (GAIL) Loop:

   2.1. (Initialize imitation experience storage) $\mathcal{D} \leftarrow \varnothing$.

   2.2. (Collect experiences) Do the following once or multiple times:

   2.2.1. (Decide and sample) Use the policy $\pi(\boldsymbol{\theta})$ to generate a trajectory. The trajectory does not have reward information.

   2.2.2. (Calculate reward) Use the discriminative network $\phi(\boldsymbol{\varphi})$ to calculate reward estimate for each state–action pair in the trajectory: $R_t \leftarrow -\ln \phi(S_t, A_t; \boldsymbol{\varphi})$.

   2.2.3. (Calculate old advantage) Use the trajectory to calculate the old advantage, bootstrapping from the value estimates parameterized by $\mathbf{w}$. (For example, $a(S_t, A_t) \leftarrow \sum_{\tau=t}^{T-1} (\gamma\lambda)^{\tau-t} \left[U_{\tau:\tau+1}^{(v)} - v(S_\tau; \mathbf{w}) \right]$.)

   2.2.4. (Store) Save the experience $(S_t, A_t, \pi(A_t|S_t; \boldsymbol{\theta}), a(S_t, A_t; \mathbf{w}), G_t)$ in the storage $\mathcal{D}$.

   2.3. (Use experiences to train the discriminative network) Do the following once or multiple times:

   2.3.1. (Replay) Construct a two-categorical classification dataset in the following way:

   2.3.1.1. Sample a batch of experiences $\mathcal{B}_E$ from the storage $\mathcal{D}_E$. Each entry in the form of (S, A). Assign the classification label $L \leftarrow 0$ for these state–action pairs to indicate that they are experiences of expert policy.

2.3.1.2. Sample a batch of experiences $\mathcal{B}$ from imitation storage $\mathcal{D}$. Each entry in the form of (S, A). Assign the classification label $L \leftarrow 1$ for these state–action pairs to indicate that they are not experiences of expert policy.

2.3.2. (Update discriminative network parameters) Update φ to reduce the cross-entropy loss using the two-categorical classification dataset.

2.4. (Apply imitation experiences to RL) Do the following once or multiple times:

2.4.1. (Replay) Sample a batch of experiences $\mathcal{B}$ from the storage $\mathcal{D}$ (or use the sample results in Step 2.3.1.2). Each entry is in the form of (S, A, Π, A, G).

2.4.2. (Update policy parameter) Update θ to increase $\frac{1}{|\mathcal{B}|} \sum_{(S,A,\Pi,A,G)\in\mathcal{B}} \min \left\{ \frac{\pi(A|S;\theta)}{\Pi} A, A + \varepsilon |A| \right\}$.

2.4.3. (Update value parameter) Update $\mathbf{w}$ to reduce value estimate errors. (For example, minimize $\frac{1}{|\mathcal{B}|} \sum_{(S,A,\Pi,A,G)\in\mathcal{B}} \left[G - v(S; \mathbf{w}) \right]^2$.)

The discriminative network in GAIL can be viewed as a reward model or a utility model. From this sense, GAIL can be viewed as IRL algorithm.

16.3 Application In Training GPT

Generative Pre-trained Transformer (**GPT**) is a series of famous language models, including many versions of GPT such as GPT4, as well as the well-known application ChatGPT. GPT can be viewed as an RL agent. The inputs from users are the observations, and the outputs are the actions. This system does not have existing reward models, but we human can judge whether the outputs of GPT make sense. The human satisfaction can be viewed as implied rewards.

The steps to train GPT are as follows (Fig. 16.4) (Ouyang, 2022):

- Step 1: Use the technique of Natural Language Process (NLP) to generate a very rough language model, which can be treated as a pretrained agent.
- Step 2: Use IL to improve the model. Detailed speaking, this step uses the demonstrated dialogue data, and applies BC algorithm (introduced in Sect. 16.2.2) to improve the agent. After this step, the outputs of agent may make some sense.
- Step 3: Use RLHF to further improve the model. This step consists of two substeps: train an RM, and use RM to conduct RL. Detailed speaking, in order to get an RM, this step first samples more inputs from observation space, and uses agent to provide multiple output actions for each input observation. Let humans

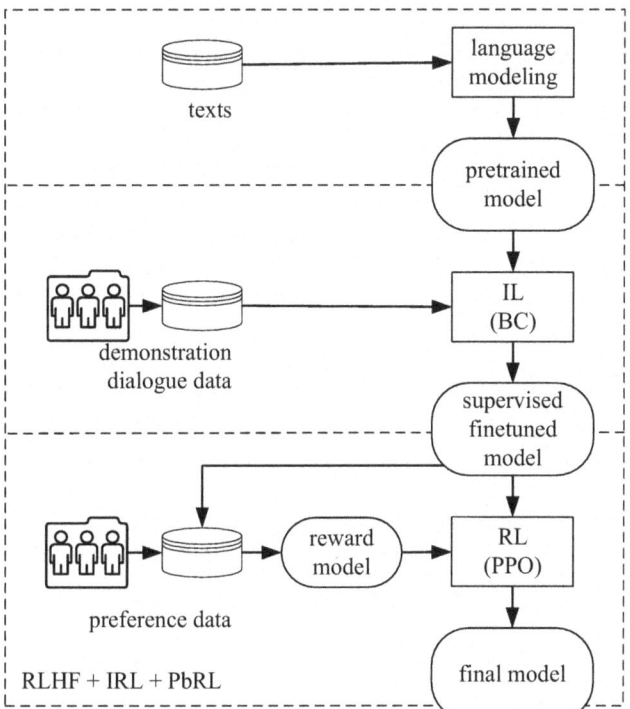

Fig. 16.4 Training GPT.

rank the goodness of these actions, and then use these preference data to learn an RM. Here the preference data are used, because it is easier to rank than to provide an absolute grading, and different humans are more likely to provide a consistent ranking than a consistent grading. After this step, the RM can align to our expectation. Then apply the RM to PPO algorithm (introduced in Sect. 8.3.3) to improve the policy. This step is RLHF since human feedbacks are used, and it is PbRL since feedbacks are preference data. And it is IRL since it learns an RM.

16.4 Case Study: Humanoid

This section considers a task "Humanoid" in PyBullet.

16.4.1 Use PyBullet

Bullet (URL: `https://pybullet.org`) is a physical dynamics open-source library, which realized the motion, collision detection, visualization of 3-d objects. Its Python API is called PyBullet. Many games and movies were developed using Bullet, and many RL researches used Bullet and PyBullet.

This section will introduce the usage of PyBullet, and use the environment provided by PyBullet to train agent.

PyBullet, as a Python extension library, can be installed using the following pip command:

```
pip install --upgrade pybullet
```

The dependent packages will also be installed automatically. (Note: PyBullet <=3.2.5 is not compatible with Gym >=0.26. More details can be found in `https://github.com/bulletphysics/bullet3/issues/4368`.)

After installing PyBullet, we can import it using the following codes:

```
import gym
import pybullet_envs
```

Note that we need to import Gym first, then import PyBullet. When importing PyBullet, we should import `pybullet_envs` rather than `pybullet`. Sequential codes will not use `pybullet_envs` explicitly, but the import statement registers the environments in Gym. Therefore, after the import, we can use `gym.make()` to get the environments in PyBullet, such as

```
env = gym.make('HumanoidBulletEnv-v0')
```

The above statement would fail if we did not import `pybullet_envs`. It can be used only after import `pybullet_envs`.

You may wonder why don't we import PyBullet using `import pybullet`. In fact, this statement can be executed, but it has different usage. PyBullet provides many APIs for users to control and render in a customized way. If we only use Gym-like API for RL training, it is not necessary to import pybullet. If we need to demonstrate the interaction with PyBullet environment, we need to `import pybullet`. If we `import pybullet`, we usually assign an alias p:

```
import pybullet as p
```

Then we demonstrate how to interact with PyBullet and visualize the interaction. Generally speaking, on the one hand, PyBullet environment supports Gym-like API, so we can use `env.reset()`, `env.step()`, and `env.close()` to interact with the environments; on the other hand, PyBullet provides a different set of render APIs to give more freedom to control the rendering, but it is more difficult to use.

Now we show how to interact with the task `HumanoidBulletEnv-v0` and render the interaction.

The task `HumanoidBulletEnv-v0` has a humanoid, and we wish the humanoid can move forward. Using classical method, we can know that its observation space is

Box(-inf, inf, (44,)), the action space is Box(-1, 1, (17,)). The maximum step in each episode is 1000. There is no pre-defined episode reward threshold. The modes of render can be either "human" or "rgb_array".

We use the mode "human" to render. Differ from the classical usage of Gym, we need to call env.render(mode="human") before we call env.reset() to initialize rendering resources, and a new window will pop up to show the motion pictures. We can also use Code 16.1 to adjust the camera. Code 16.1 uses part_name, robot_name = p.getBodyInfo(body_id) to obtain each part, and then find the ID of the torso, and then obtain the position of the torso. Then we adjust the camera so that its distance to the object is 2, the yaw is 0 (i.e. directly facing the object), and the pitch is 20, meaning it slightly looks down at the object.

Code 16.1   Adjust the camera.

HumanoidBulletEnv-v0_ClosedForm_demo.ipynb

```
def adjust_camera():
    distance, yaw, pitch = 2, 0, -20
    num_bodies = p.getNumBodies()
    torse_ids = [body_id for body_id in range(num_bodies) if
            p.getBodyInfo(body_id)[0].decode("ascii") == "torso"]
    torse_id = torse_ids[0]
    position, orientation = p.getBasePositionAndOrientation(torse_id)
    p.resetDebugVisualizerCamera(distance, yaw, pitch, position)
```

Interdisciplinary Reference 16.3
Rotational Kinematics: Principal Axes and Euler's Angles

In rotational kinematics, moving objects or observers can use their facing orientations to determine the principal axis. For moving objects such as walking or running man, car, or plane, we can replace the object as a person who faces the motional direction. For observers such as cameras, we can replace the observer as a person who faces the motional direction. After such replacement, we can define the forward-backward axis as longitudinal axis, the left-right axis as lateral axis, and the up-down axis as vertical axis, as Fig. 16.5. Accordingly, we can further define the set of Euler's angles to determine the rotation position. A set of Euler's angles include yaw, pitch, and roll.

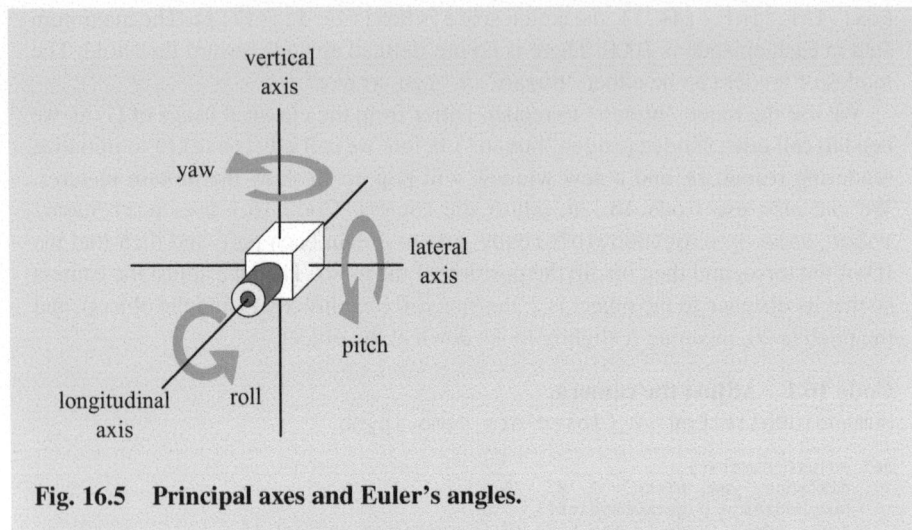

Fig. 16.5   Principal axes and Euler's angles.

Code 16.2 shows an example of interaction and visualization. In order to visualize like an animates, 0.02 second pause is added to every interaction. The GitHub of this book also provides an agent, which maintains a fully connected neural network whose inputs are observations and outputs are actions. This agent performs well: The average episode reward is about 3000, and the humanoid can walk effectively.

Code 16.2   Visualize the interaction with the environment.
HumanoidBulletEnv-v0_ClosedForm_demo.ipynb

```
1  def play_episode(env, agent, seed=None, mode=None, render=False):
2      if render:
3          env.render(mode="human")
4      observation, _ = env.reset(seed=seed)
5      reward, terminated, truncated = 0., False, False
6      agent.reset(mode=mode)
7      episode_reward, elapsed_steps = 0., 0
8      while True:
9          if render:
10             adjust_camera()
11             env.render(mode="human")
12             time.sleep(0.02)
13         action = agent.step(observation, reward, terminated)
14         if terminated or truncated:
15             break
16
17         observation, reward, terminated, truncated, _ = env.step(action)
18         episode_reward += reward
19         elapsed_steps += 1
20     agent.close()
21     return episode_reward, elapsed_steps
```

16.4.2 Use BC to IL

This section implements the BC algorithm. Since the action space of the task is continuous, we can convert the BC algorithm into a conventional regression. In the regression, the features are the states, while the targets are the actions. The objective of training is to minimize the MSE.

The class `SAReplayer` in Code 16.3 stores the state–action pairs that the regression needs, so it is the replayer for BC.

Code 16.3   Experience replayer for state–action pairs.
`HumanoidBulletEnv-v0_BC_tf.ipynb`

```
 1  class SAReplayer:
 2      def __init__(self):
 3          self.fields = ['state', 'action']
 4          self.data = {field: [] for field in self.fields}
 5          self.memory = pd.DataFrame()
 6
 7      def store(self, *args):
 8          for field, arg in zip(self.fields, args):
 9              self.data[field].append(arg)
10
11      def sample(self, size=None):
12          if len(self.memory) < len(self.data[self.fields[0]]):
13              self.memory = pd.DataFrame(self.data, columns=self.fields)
14          if size is None:
15              indices = self.memory.index
16          else:
17              indices = np.random.choice(self.memory.index, size=size)
18          return (np.stack(self.memory.loc[indices, field]) for field in
19                  self.fields)
```

Codes 16.4 and 16.5 implement the BC agents. The constructor of the agent not only accepts an environment object env, but also accepts an agent of expert policy expert_agent. An implementation of expert policy is provided in GitHub repo of this book. Calling the member reset() with parameter mode='expert' can enter the expert mode. In expert mode, the member step() uses expert_agent to decide, and saves the input state and output action in the replayer. At the end of episodes, the saved state–action pairs are used to train the neural network.

Code 16.4   BC agent (TensorFlow version).
`HumanoidBulletEnv-v0_BC_tf.ipynb`

```
 1  class BCAgent:
 2      def __init__(self, env, expert_agent):
 3          self.expert_agent = expert_agent
 4
 5          state_dim = env.observation_space.shape[0]
 6          action_dim = env.action_space.shape[0]
 7
 8          self.net = self.build_net(input_size=state_dim,
 9                  hidden_sizes=[256, 128], output_size=action_dim)
10
11      def build_net(self, input_size=None, hidden_sizes=None, output_size=1,
12                  activation=nn.relu, output_activation=None,
13                  loss=losses.mse, learning_rate=0.001):
14          model = keras.Sequential()
15          for hidden_size in hidden_sizes:
```

```
16              model.add(layers.Dense(units=hidden_size,
17                      activation=activation))
18          model.add(layers.Dense(units=output_size,
19                  activation=output_activation))
20          optimizer = optimizers.Adam(learning_rate)
21          model.compile(optimizer=optimizer, loss=loss)
22          return model
23
24      def reset(self, mode=None):
25          self.mode = mode
26          if self.mode == 'expert':
27              self.expert_agent.reset(mode)
28              self.expert_replayer = SAReplayer()
29
30      def step(self, observation, reward, terminated):
31          if self.mode == 'expert':
32              action = expert_agent.step(observation, reward, terminated)
33              self.expert_replayer.store(observation, action)
34          else:
35              action = self.net(observation[np.newaxis])[0]
36          return action
37
38      def close(self):
39          if self.mode == 'expert':
40              self.expert_agent.close()
41              for _ in range(10):
42                  self.learn()
43
44      def learn(self):
45          states, actions = self.expert_replayer.sample(1024)
46          self.net.fit(states, actions, verbose=0)
47
48
49  agent = BCAgent(env, expert_agent)
```

Code 16.5   BC agent (PyTorch version).

HumanoidBulletEnv-v0_BC_torch.ipynb

```
1   class BCAgent:
2       def __init__(self, env, expert_agent):
3           self.expert_agent = expert_agent
4
5           state_dim = env.observation_space.shape[0]
6           action_dim = env.action_space.shape[0]
7
8           self.net = self.build_net(input_size=state_dim,
9                   hidden_sizes=[256, 128], output_size=action_dim)
10          self.loss = nn.MSELoss()
11          self.optimizer = optim.Adam(self.net.parameters())
12
13      def build_net(self, input_size, hidden_sizes, output_size=1,
14              output_activator=None):
15          layers = []
16          for input_size, output_size in zip(
17                  [input_size,] + hidden_sizes, hidden_sizes + [output_size,]):
18              layers.append(nn.Linear(input_size, output_size))
19              layers.append(nn.ReLU())
20          layers = layers[:-1]
21          if output_activator:
22              layers.append(output_activator)
23          net = nn.Sequential(*layers)
24          return net
25
26      def reset(self, mode=None):
27          self.mode = mode
```

```
28          if self.mode == 'expert':
29              self.expert_agent.reset(mode)
30              self.expert_replayer = SAReplayer()
31
32      def step(self, observation, reward, terminated):
33          if self.mode == 'expert':
34              action = expert_agent.step(observation, reward, terminated)
35              self.expert_replayer.store(observation, action)
36          else:
37              state_tensor = torch.as_tensor(observation, dtype=torch.float
38                      ).unsqueeze(0)
39              action_tensor = self.net(state_tensor)
40              action = action_tensor.detach().numpy()[0]
41          return action
42
43      def close(self):
44          if self.mode == 'expert':
45              self.expert_agent.close()
46              for _ in range(10):
47                  self.learn()
48
49      def learn(self):
50          states, actions = self.expert_replayer.sample(1024)
51          state_tensor = torch.as_tensor(states, dtype=torch.float)
52          action_tensor = torch.as_tensor(actions, dtype=torch.float)
53
54          pred_tensor = self.net(state_tensor)
55          loss_tensor = self.loss(pred_tensor, action_tensor)
56          self.optimizer.zero_grad()
57          loss_tensor.backward()
58          self.optimizer.step()
59
60
61  agent = BCAgent(env, expert_agent)
```

Interaction between this agent and the environment once again uses Code 1.3.

16.4.3 Use GAIL to IL

This section implements GAIL. Codes 16.6 and 16.7 implement the GAIL-PPO algorithm. You may compare them with PPO agents in Codes 8.8 and 8.9. Its constructor saves the agent of expert policy expert_agent, and initializes its corresponding replayer self.expert_replayer alongside the original replayer self.replayer. The replayer self.replayer needs five columns: state, action, ln_prob, advantage, and return, but self.expert_replayer only needs two columns state and action. We use PPOReplayer modified from Code 8.7 to instantiate both replayers to simplify the implementation, while the last three columns of self.expert_replayer are ignored and wasted. We can use reset(mode='expert') to enter the expert mode and use expert_agent to decide. Since IL can not use the reward signal provided by the environment, we should not store reward into trajectory. During the training, we not only need to update the actor network and the critic network, but also need to update discriminator network.

Code 16.6   GAIL-PPO agent (TensorFlow version).

HumanoidBulletEnv-v0_GAILPPO_tf.ipynb

```python
class GAILPPOAgent:
    def __init__(self, env, expert_agent):
        self.expert_agent = expert_agent

        self.expert_replayer = PPOReplayer()
        self.replayer = PPOReplayer()

        self.state_dim = env.observation_space.shape[0]
        self.action_dim = env.action_space.shape[0]
        self.gamma = 0.99
        self.max_kl = 0.01

        self.actor_net = self.build_net(input_size=self.state_dim,
                hidden_sizes=[256, 128], output_size=self.action_dim * 2)
        self.critic_net = self.build_net(input_size=self.state_dim,
                hidden_sizes=[256, 128], output_size=1)
        self.discriminator_net = self.build_net(
                input_size=self.state_dim + self.action_dim,
                hidden_sizes=[256, 128], output_size=1,
                output_activation=nn.sigmoid, loss=losses.binary_crossentropy)

    def build_net(self, input_size=None, hidden_sizes=None, output_size=1,
                activation=nn.relu, output_activation=None,
                loss=losses.mse, learning_rate=0.001):
        model = keras.Sequential()
        for hidden_size in hidden_sizes:
            model.add(layers.Dense(units=hidden_size,
                    activation=activation))
        model.add(layers.Dense(units=output_size,
                activation=output_activation))
        optimizer = optimizers.Adam(learning_rate)
        model.compile(optimizer=optimizer, loss=loss)
        return model

    def get_ln_prob_tensor(self, state_tensor, action_tensor):
        mean_log_std_tensor = self.actor_net(state_tensor)
        mean_tensor, log_std_tensor = tf.split(mean_log_std_tensor,
                2, axis=-1)
        std_tensor = tf.exp(log_std_tensor)
        normal = distributions.Normal(mean_tensor, std_tensor)
        log_prob_tensor = normal.log_prob(action_tensor)
        ln_prob_tensor = tf.reduce_sum(log_prob_tensor, axis=-1)
        return ln_prob_tensor

    def reset(self, mode=None):
        self.mode = mode
        if self.mode == 'expert':
            self.expert_agent.reset(mode)
        if self.mode in ['expert', 'train']:
            self.trajectory = []

    def step(self, observation, reward, terminated):
        if self.mode == 'expert':
            action = expert_agent.step(observation, reward, terminated)
        else:
            mean_ln_stds = self.actor_net.predict(observation[np.newaxis],
                    verbose=0)
            means, ln_stds = np.split(mean_ln_stds, 2, axis=-1)
            if self.mode == 'train':
                stds = np.exp(ln_stds)
                actions = np.random.normal(means, stds)
            else:
```

```
63                     actions = means
64                 action = actions[0]
65             if self.mode in ['train', 'expert']:
66                 self.trajectory += [observation, 0., terminated, action]
67                     # pretend reward is unknown
68             return action
69
70     def close(self):
71         if self.mode == 'expert':
72             self.expert_agent.close()
73         if self.mode in ['train', 'expert']:
74             self.save_trajectory_to_replayer()
75         if self.mode == 'train' and len(self.replayer.memory) >= 2000:
76             self.learn()
77             self.replayer = PPOReplayer()
78                     # reset replayer after the agent changes itself
79
80     def save_trajectory_to_replayer(self):
81         df = pd.DataFrame(
82                 np.array(self.trajectory, dtype=object).reshape(-1, 4),
83                 columns=['state', 'reward', 'terminated', 'action'])
84         if self.mode == 'expert':
85             df['ln_prob'] = float('nan')
86             df['advantage'] = float('nan')
87             df['return'] = float('nan')
88             self.expert_replayer.store(df)
89         elif self.mode == 'train':
90
91             # prepare ln_prob
92             state_tensor = tf.convert_to_tensor(np.stack(df['state']),
93                     dtype=tf.float32)
94             action_tensor = tf.convert_to_tensor(np.stack(df['action']),
95                     dtype=tf.float32)
96             ln_prob_tensor = self.get_ln_prob_tensor(state_tensor,
97                     action_tensor)
98             ln_probs = ln_prob_tensor.numpy()
99             df['ln_prob'] = ln_probs
100
101             # prepare return
102             state_action_tensor = tf.concat([state_tensor, action_tensor],
103                     axis=-1)
104             discrim_tensor = self.discriminator_net(state_action_tensor)
105             reward_tensor = -tf.math.log(discrim_tensor)
106             rewards = reward_tensor.numpy().squeeze()
107             df['reward'] = rewards
108             df['return'] = signal.lfilter([1.,], [1., -self.gamma],
109                     df['reward'][::-1])[::-1]
110
111             # prepare advantage
112             v_tensor = self.critic_net(state_tensor)
113             df['v'] = v_tensor.numpy()
114             df['next_v'] = df['v'].shift(-1).fillna(0.)
115             df['u'] = df['reward'] + self.gamma * df['next_v']
116             df['delta'] = df['u'] - df['v']
117             df['advantage'] = signal.lfilter([1.,], [1., -self.gamma],
118                     df['delta'][::-1])[::-1]
119
120             self.replayer.store(df)
121
122     def learn(self):
123         # replay expert experience
124         expert_states, expert_actions, _, _, _ = self.expert_replayer.sample()
125
126         # replay novel experience
127         states, actions, ln_old_probs, advantages, returns = \
128                 self.replayer.sample()
129         state_tensor = tf.convert_to_tensor(states, dtype=tf.float32)
```

```
130        action_tensor = tf.convert_to_tensor(actions, dtype=tf.float32)
131        ln_old_prob_tensor = tf.convert_to_tensor(ln_old_probs,
132                dtype=tf.float32)
133        advantage_tensor = tf.convert_to_tensor(advantages, dtype=tf.float32)
134
135        # standardize advantage
136        advantage_tensor = (advantage_tensor - tf.reduce_mean(
137                advantage_tensor)) / tf.math.reduce_std(advantage_tensor)
138
139        # update discriminator
140        state_actions = np.concatenate([np.concatenate(
141                [expert_states, expert_actions], axis=-1),
142                np.concatenate([states, actions], axis=-1)], axis=0)
143        expert_batch_size = expert_states.shape[0]
144        batch_size = states.shape[0]
145        labels = np.concatenate([np.zeros(expert_batch_size, dtype=int),
146                np.ones(batch_size, dtype=int)])
147        self.discriminator_net.fit(state_actions, labels, verbose=0)
148
149        # update actor
150        with tf.GradientTape() as tape:
151            ln_pi_tensor = self.get_ln_prob_tensor(state_tensor, action_tensor)
152            surrogate_advantage_tensor = tf.exp(ln_pi_tensor -
153                    ln_old_prob_tensor) * advantage_tensor
154            clip_times_advantage_tensor = 0.1 * surrogate_advantage_tensor
155            max_surrogate_advantage_tensor = advantage_tensor + \
156                    tf.where(advantage_tensor > 0.,
157                    clip_times_advantage_tensor, -clip_times_advantage_tensor)
158            clipped_surrogate_advantage_tensor = tf.minimum(
159                    surrogate_advantage_tensor, max_surrogate_advantage_tensor)
160            loss_tensor = -tf.reduce_mean(clipped_surrogate_advantage_tensor)
161        actor_grads = tape.gradient(loss_tensor, self.actor_net.variables)
162        self.actor_net.optimizer.apply_gradients(
163                zip(actor_grads, self.actor_net.variables))
164
165        # update critic
166        self.critic_net.fit(states, returns, verbose=0)
167
168
169 agent = GAILPPOAgent(env, expert_agent)
```

Code 16.7   GAIL-PPO agent (PyTorch version).

HumanoidBulletEnv-v0_GAILPPO_torch.ipynb

```
1  class GAILPPOAgent:
2      def __init__(self, env, expert_agent):
3          self.expert_agent = expert_agent
4
5          self.expert_replayer = PPOReplayer()
6          self.replayer = PPOReplayer()
7
8          self.state_dim = env.observation_space.shape[0]
9          self.action_dim = env.action_space.shape[0]
10         self.gamma = 0.99
11         self.max_kl = 0.01
12
13         self.actor_net = self.build_net(input_size=self.state_dim,
14                 hidden_sizes=[256, 128], output_size=self.action_dim * 2)
15         self.actor_optimizer = optim.Adam(self.actor_net.parameters(), 0.001)
16         self.critic_net = self.build_net(input_size=self.state_dim,
17                 hidden_sizes=[256, 128], output_size=1)
18         self.critic_loss = nn.MSELoss()
19         self.critic_optimizer = optim.Adam(self.critic_net.parameters())
20         self.discriminator_net = self.build_net(
21                 input_size=self.state_dim + self.action_dim,
```

```python
22                hidden_sizes=[256, 128], output_size=1,
23                output_activator=nn.Sigmoid())
24        self.discriminator_loss = nn.BCELoss()
25        self.discriminator_optimizer = optim.Adam(
26                self.discriminator_net.parameters())
27
28    def build_net(self, input_size, hidden_sizes, output_size=1,
29            output_activator=None):
30        layers = []
31        for input_size, output_size in zip(
32                [input_size,] + hidden_sizes, hidden_sizes + [output_size,]):
33            layers.append(nn.Linear(input_size, output_size))
34            layers.append(nn.ReLU())
35        layers = layers[:-1]
36        if output_activator:
37            layers.append(output_activator)
38        net = nn.Sequential(*layers)
39        return net
40
41    def get_ln_prob_tensor(self, state_tensor, action_tensor):
42        mean_log_std_tensor = self.actor_net(state_tensor)
43        mean_tensor, log_std_tensor = torch.split(mean_log_std_tensor,
44                self.action_dim, dim=-1)
45        std_tensor = torch.exp(log_std_tensor)
46        normal = distributions.Normal(mean_tensor, std_tensor)
47        log_prob_tensor = normal.log_prob(action_tensor)
48        ln_prob_tensor = log_prob_tensor.sum(-1)
49        return ln_prob_tensor
50
51    def reset(self, mode=None):
52        self.mode = mode
53        if self.mode == 'expert':
54            self.expert_agent.reset(mode)
55        if self.mode in ['expert', 'train']:
56            self.trajectory = []
57
58    def step(self, observation, reward, terminated):
59        if self.mode == 'expert':
60            action = expert_agent.step(observation, reward, terminated)
61        else:
62            state_tensor = torch.as_tensor(observation, dtype=torch.float
63                    ).unsqueeze(0)
64            mean_ln_std_tensor = self.actor_net(state_tensor)
65            mean_tensor, ln_std_tensor = torch.split(mean_ln_std_tensor,
66                    self.action_dim, dim=-1)
67            if self.mode == 'train':
68                std_tensor = torch.exp(ln_std_tensor)
69                normal = distributions.Normal(mean_tensor, std_tensor)
70                action_tensor = normal.rsample()
71            else:
72                action_tensor = mean_tensor
73            action = action_tensor.detach().numpy()[0]
74        if self.mode in ['train', 'expert']:
75            self.trajectory += [observation, 0., terminated, action]
76                    # pretend reward is unknown
77        return action
78
79    def close(self):
80        if self.mode == 'expert':
81            self.expert_agent.close()
82        if self.mode in ['train', 'expert']:
83            self.save_trajectory_to_replayer()
84        if self.mode == 'train' and  len(self.replayer.memory) >= 2000:
85            self.learn()
86            self.replayer = PPOReplayer()
87                    # reset replayer after the agent changes itself
88
```

```
89      def save_trajectory_to_replayer(self):
90          df = pd.DataFrame(
91                  np.array(self.trajectory, dtype=object).reshape(-1, 4),
92                  columns=['state', 'reward', 'terminated', 'action'])
93          if self.mode == 'expert':
94              df['ln_prob'] = float('nan')
95              df['advantage'] = float('nan')
96              df['return'] = float('nan')
97              self.expert_replayer.store(df)
98          elif self.mode == 'train':
99
100             # prepare ln_prob
101             state_tensor = torch.as_tensor(df['state'], dtype=torch.float)
102             action_tensor = torch.as_tensor(df['action'], dtype=torch.float)
103             ln_prob_tensor = self.get_ln_prob_tensor(state_tensor,
104                     action_tensor)
105             ln_probs = ln_prob_tensor.detach().numpy()
106             df['ln_prob'] = ln_probs
107
108             # prepare return
109             state_action_tensor = torch.cat([state_tensor, action_tensor],
110                     dim=-1)
111             discrim_tensor = self.discriminator_net(state_action_tensor)
112             reward_tensor = -torch.log(discrim_tensor)
113             rewards = reward_tensor.detach().numpy().squeeze()
114             df['reward'] = rewards
115             df['return'] = signal.lfilter([1.,], [1., -self.gamma],
116                     df['reward'][::-1])[::-1]
117
118             # prepare advantage
119             v_tensor = self.critic_net(state_tensor)
120             df['v'] = v_tensor.detach().numpy()
121             df['next_v'] = df['v'].shift(-1).fillna(0.)
122             df['u'] = df['reward'] + self.gamma * df['next_v']
123             df['delta'] = df['u'] - df['v']
124             df['advantage'] = signal.lfilter([1.,], [1., -self.gamma],
125                     df['delta'][::-1])[::-1]
126
127             self.replayer.store(df)
128
129     def learn(self):
130         # replay expert experience
131         expert_states, expert_actions, _, _, _ = self.expert_replayer.sample()
132         expert_state_tensor = torch.as_tensor(expert_states, dtype=torch.float)
133         expert_action_tensor = torch.as_tensor(expert_actions,
134                 dtype=torch.float)
135
136         # replay novel experience
137         states, actions, ln_old_probs, advantages, returns = \
138                 self.replayer.sample()
139         state_tensor = torch.as_tensor(states, dtype=torch.float)
140         action_tensor = torch.as_tensor(actions, dtype=torch.float)
141         ln_old_prob_tensor = torch.as_tensor(ln_old_probs, dtype=torch.float)
142         advantage_tensor = torch.as_tensor(advantages, dtype=torch.float)
143         return_tensor = torch.as_tensor(returns,
144                 dtype=torch.float).unsqueeze(-1)
145
146         # standandize advantage
147         advantage_tensor = (advantage_tensor - advantage_tensor.mean()) / \
148                 advantage_tensor.std()
149
150         # update discriminator
151         expert_state_action_tensor = torch.cat(
152                 [expert_state_tensor, expert_action_tensor], dim=-1)
153         novel_state_action_tensor = torch.cat(
154                 [state_tensor, action_tensor], dim=-1)
155         expert_score_tensor = self.discriminator_net(
```

```
156              expert_state_action_tensor)
157         novel_score_tensor = self.discriminator_net(novel_state_action_tensor)
158         expert_loss_tensor = self.discriminator_loss(
159              expert_score_tensor, torch.zeros_like(expert_score_tensor))
160         novel_loss_tensor = self.discriminator_loss(
161              novel_score_tensor, torch.ones_like(novel_score_tensor))
162         discriminator_loss_tensor = expert_loss_tensor + novel_loss_tensor
163         self.discriminator_optimizer.zero_grad()
164         discriminator_loss_tensor.backward()
165         self.discriminator_optimizer.step()
166
167         # update actor
168         ln_pi_tensor = self.get_ln_prob_tensor(state_tensor, action_tensor)
169         surrogate_advantage_tensor = torch.exp(ln_pi_tensor -
170              ln_old_prob_tensor) * advantage_tensor
171         clip_times_advantage_tensor = 0.1 * surrogate_advantage_tensor
172         max_surrogate_advantage_tensor = advantage_tensor + \
173              torch.where(advantage_tensor > 0.,
174              clip_times_advantage_tensor, -clip_times_advantage_tensor)
175         clipped_surrogate_advantage_tensor = torch.min(
176              surrogate_advantage_tensor, max_surrogate_advantage_tensor)
177         actor_loss_tensor = -clipped_surrogate_advantage_tensor.mean()
178         self.actor_optimizer.zero_grad()
179         actor_loss_tensor.backward()
180         self.actor_optimizer.step()
181
182         # update critic
183         pred_tensor = self.critic_net(state_tensor)
184         critic_loss_tensor = self.critic_loss(pred_tensor, return_tensor)
185         self.critic_optimizer.zero_grad()
186         critic_loss_tensor.backward()
187         self.critic_optimizer.step()
188
189
190 agent = GAILPPOAgent(env, expert_agent)
```

Interaction between this agent and the environment once again uses Code 1.3.

16.5 Summary

- Inverse Reinforcement Learning (IRL) learns Reward Model (RM).
- Preference-based RL (PbRL) learns from preference data.
- Reinforcement Learning with Human Feedback (RLHF) learns from data provided by human.
- Imitation Learning (IL) does not have direct reward signals. IL learns from the transitions between expert policies and the environment.
- Behavior Cloning (BC) algorithm tries to maximize the likelihood of expert policy dataset. It essentially minimizes the KL divergence between the expert policy and the imitation policy.
- Adversarial imitation learning is more likely to reduce compounding error.
- Generative Adversarial Imitation Learning (GAIL) algorithm uses the discriminative network in Generative Adversarial Network (GAN). It essentially optimizes the JS divergence between the expert policy and imitation policy.
- Generative Pre-trained Transformer (GPT) is trained using BC, RLHF, PbRL, and IRL.

16.6 Exercises

16.6.1 Multiple Choices

16.1 On IL, choose the correct one: (   )

A. In an IL task, the agent can not obtain the reward directly from the environment.
B. In an IL task, the agent can not obtain the observation directly from the environment.
C. In an IL task, the agent can not obtain the state directly from the environment.

16.2 On IL, choose the correct one: (   )

A. BC algorithm tries to minimize the total variation distance between imitation policy and expert policy. Adversarial IL tries to minimize the JS divergence from imitation policy to expert policy.
B. BC algorithm tries to minimize the JS divergence from imitation policy to expert policy. Adversarial IL tries to minimize the KL divergence from imitation policy to expert policy.
C. BC algorithm tries to minimize the KL divergence from imitation policy to expert policy. Adversarial IL tries to minimize the JS divergence from imitation policy to expert policy.

16.3 On BC, choose the correct one: (   )

A. BC algorithm requires a mathematical model of the environment as the input.
B. BC algorithm requires a mathematical model of an expert policy as the input.
C. BC algorithm requires the ability to obtain the transition experience between an expert policy and the environment.

16.4 On GAIL, choose the correct one: (   )

A. GAIL introduces the generative network from GAN.
B. GAIL introduces the discriminative network from GAN.
C. GAIL introduces the generative network and the discriminative network from GAN.

16.5 On the training process of GPT, choose the correct one: (   )

A. It uses RLHF, where feedbacks are preference data.
B. It uses the reward signals provided by human to train a reward model, so it is an IRL algorithm.
C. It uses Behavior Cloning algorithm, which is an RL algorithm.

16.6 On the training process of GPT, choose the correct one: (   )

A. RLHF is used to learn values.
B. RLHF is used to learn policy.
C. RLHF is used to learn rewards.

16.6.2 Programming

16.7 Use the method in this chapter to solve PyBullet task `Walker2DBulletEnv-v0`. Since IL algorithms can not obtain the reward signal directly from the environment, your agent is not required to exceed the pre-defined reward threshold. An expert policy can be found on GitHub.

16.6.3 Mock Interview

16.8 What is RLHF? What forms of feedbacks can be used in RLHF?

16.9 Can you briefly introduce how GPT is trained?

16.10 What is Imitation Learning? What applications can IL be used? Can you give an IL use case example?

16.11 What is GAIL? What is the relationship between GAIL and GAN?

GPSR Compliance

The European Union's (EU) General Product Safety Regulation (GPSR) is a set of rules that requires consumer products to be safe and our obligations to ensure this.

If you have any concerns about our products, you can contact us on ProductSafety@springernature.com

In case Publisher is established outside the EU, the EU authorized representative is:

Springer Nature Customer Service Center GmbH
Europaplatz 3
69115 Heidelberg, Germany

The manufacturer's authorised representative in the EU is Springer
Nature Customer Service Centre GmbH, Europaplatz 3, 69115 Heidelberg,
Germany. If you have any concerns regarding our products, please
contact ProductSafety@springernature.com

Printed and bound by CPI Group (UK) Ltd, Croydon, CR0 4YY
29/04/2026
02099522-0007